Darwin's Racism, Sexism, and Idolization

Rui Diogo

Darwin's Racism, Sexism, and Idolization

Their Tragic Societal and Scientific Repercussions

Rui Diogo
College of Medicine
Howard University
Washington, DC, USA

ISBN 978-3-031-49054-5   ISBN 978-3-031-49055-2 (eBook)
https://doi.org/10.1007/978-3-031-49055-2

Familia selk´nam. Isla Grande, Tierra del Fuego. Fotografía de Martín Gusinde. 1920- 1923. En: "Fueguinos. Fotografías Siglos XIX y XX. Imágenes e Imaginarios del Fin del Mundo."Margarita Alvarado, Carolina Odone, Pedro Mege. Editorial Pehuén. S/F. (http://precolombino.cl/culturas-americanas/pueblos-originarios-de-chile/selk%C2%B4nam/)

This Springer imprint is published by the registered company Springer Nature Switzerland AG
The registered company address is: Gewerbestrasse 11, 6330 Cham, Switzerland

Paper in this product is recyclable.

When we identify where our privilege intersects with somebody else's oppression, we'll find our opportunities to make real change. (Ijeoma Oluo)
This book is dedicated to those that have suffered, or fought against, oppression, discrimination, racism, and misogyny in the past and present, including millions of people that went to the streets, across the globe, in the last years to protest for the rights of "others": women, minorities, and people that are different from the "norm." It is crucial to bring reality to history, science, and society, so the next generations will not be indoctrinated by factually inaccurate tales based on biases and stereotypes that have blinded, and continue to blind, so many scientists, politicians, and laypeople.

Preface

At all times there are people who do not think like others... that is, who do not think like those who do not think. (Marguerite Yourcenar)

One of the most fascinating aspects about Charles Darwin, more even than his astonishing life and travels, is that he provides an emblematic example of the human tendency to idealize, idolize, and create fictional heroes and deities. From scientists to historians, and ultimately to popular culture, there has been an idolization of both Darwin's evolutionary ideas and him as a person – often portrayed as a humble, kind, objective, bearded old naturalist that only cared about science and the natural world. The idealization, and sometimes even quasi-religious veneration, of Darwin indeed has almost no current parallel in science. Most scholars nowadays have no problems in recognizing that the theories of "scientific giants" such as Galileu or Newton were incomplete or even wrong and were not enough to understand the complexities of our physical cosmos. But numerous biologists continue to argue that Darwin, Darwinism, or Neo-Darwinism – which in a simplified way integrates Darwin's theories with subsequent genetic findings – are basically enough to understand nature and biological evolution. As if some of the hugely important biological findings made in the last decades, such as the discovery of DNA's structure, homeobox genes, countless human and non-human fossils, or ancient DNA, were not main game-changers.

So, this begs a question that is critical to understand science, beliefs, idolization, systemic racism and sexism, and our societies in general, which will be addressed in this book: why has Darwin been – and continues to be – idolized in a way that has almost no parallel, particularly in the West and specially among Western scholars? Many scholars continue to dramatize his life as if it were part of a Cervantes tale: as Quixote was accompanied by Sancho Panza in his fight against gigantic windmills, Darwin and his "bulldog" Huxley fought against the powerful God-fearing theists and Victorian status quo, often personalized by the Bishop Samuel Wilberforce and biologist Richard Owen.

Such dramatized fairytales ignore various critical points. Firstly, many of the "grand ideas" often attributed to Darwin were not merely the product of his "unique genius." Biological evolution was in the air much before Darwin was even born – being for instance part of his own grandfather's writings, Erasmus Darwin –, as was nature's "broom": natural selection. In fact, his contemporary Alfred Russel Wallace also developed an idea of evolution driven by natural selection, independently from Charles Darwin. Secondly, numerous evolutionary "facts" constructed by Darwin about the natural world and in particular about human evolution and humankind were factually wrong, in great part due to the crucial role played by Victorian biases and stereotypes in Darwin's life and works. These include the "facts" that women are mentally inferior, that there is a continuous suffocating war-like struggle-for-existence within nature and between "human" races, that Victorian society and culture were the pinnacle of evolution, and so on.

Thirdly, there was another major ideological "war" happening in England back then, which is often neglected in idealized accounts about Darwin. Namely, between those defending the revolutions and huge societal changes that were occurring on the other side of the English Channel and those fearing them, which included Darwin and his family. Darwin's ethnocentric, sexist, and hierarchical view of nature were music to the ears of his well-off family, friends, and colleagues and the cream of Victorian society in general, which were anxiously looking for a "scientific" confirmation of such a view in order to maintain their privileged position at the very top of Darwin's "chain-of-being."

Tragically, one of the major reasons that led to Darwin's immortality and that set him apart from scholars such as Wallace also led to some of the darkest societal repercussions of his works: Darwin's intellectual conservatism and *power of place*, and his successful use of simplistic, exaggerated, and sometimes factually inaccurate metaphors that were catchy and easily absorbed by both scholars and the general public. By doing so, he provided easy ammunition for simplistic populist political leaders, dictators, colonialists, and white supremacists to "scientifically" defend social hierarchies, sexism, discrimination, oppression, and segregation. Within the very few scholars that dared to recognize this fact so far, a typical argument used to then defend the idolized Darwin is that "back then" everybody was racist and sexist. However, this book will deconstruct this flawed argument by providing fascinating and enthralling case studies and travel descriptions undertaken by scholars such as Wallace and Alexander von Humboldt, who often praised the Indigenous peoples that so repulsed – and criticized the social hierarchies and Western imperialism that so marveled – Darwin.

Idolization always involves denial: in Darwin's case, that as a mere mortal, as all of us he had an ego, ambitions, and committed many logical incongruities. He had strengths and weaknesses, as we all do. The problem is that, contrary to other mere mortals, Darwin became the most famous and influential biologist of all times and is nowadays one of the most idolized scientists. Consequently, the ethnocentric and misogynistic "facts" constructed by him became deeply ingrained in academy and also in popular culture, leading to a plethora of tragic scientific and societal repercussions that continue to impact us today.

That is why I decided to write this book. In particular, because most books written about Darwin have been written either by historians of science or other scholars that are *not* evolutionary biologists – who tend to be those that most idolize Darwin – or were written by creationists – which often tend to demonize him. To my knowledge, this is the first book written by an evolutionary biologist that focus principally on Darwin's racism, sexism, and idolization and their scientific and societal repercussions, within a broad biological, anthropological, and historical context. The book's aim is therefore not at all to "cancel" Darwin or argue that he was always wrong. As we will see, clearly he was not: in general he was an extraordinary biologist– he was however a much less successful anthropologist due in great part to his Victorian biases. The aim of the book is therefore to discuss his writings, ideas, and their repercussions in a broader, encompassing way and considering what the factual evidence truly tell us about human evolution, "races," and genders, without taboos, omissions, idolization, or demonization. This is because if we do not acknowledge, emphasize, and correct the biases, prejudices, inaccuracies, idealizations, and abuses of our past, and merely continue to blindly idealize it, our kids will be condemned to undertake or suffer similar societal abuses in the future.

This book is accordingly written for the general public and in particular for the next generations. In order to reach people from various backgrounds, countries, and continents, I tried to reduce the use of jargon and to carefully choose what to include and, importantly, what *to not* include, in the book. Specifically, the book's aim is *not* to be an encyclopedic account repeating what has already been told again and again within the thousands of publications about Darwin, mostly in a very idealized way. Instead, I focused on what has often been omitted in such publications, on the "other side" of Darwin, putting those other aspects of Darwin's life and writings into a broader multidisciplinary context. This includes historical, biological, and anthropological data that show how many of the "facts" constructed by him about human evolution, "races," genders, and societies were deeply influenced by Victorian biases and prejudices and, critically, how they were often factually inaccurate and logically incongruent. In order to be fair and balanced, I opted to include, as much as possible, original excerpts from Darwin's own writings, from authors that idolized him, from the relatively few scholars that criticized him, and from politicians or others social actors that directly quoted his ideas to "scientifically" justify white supremacy, sexism, ethnocentrism, discrimination, or oppression. In a nutshell, my main preoccupation is to show exactly what was told by Darwin or by others that discussed, used, or abused his ideas, in a way that is as raw, transparent, and plain as possible, without exaggerating or dramatizing historical and scientific facts, without idolization or demonization. This is, plainly and simply, what this book is about.

Washington, DC, USA Rui Diogo

Acknowledgments

Common sense is the collection of prejudices acquired by age eighteen. (Albert Einstein)

I want to thank Alejandra Hurtado, the most amazing person I met in my life: she strengthened my empathy for "others" and drove me to fight even more the injustices created by the status quo. I am also very thankful to my parents, Valter Diogo and Fatima Boliqueime, for providing me the peace of mind to write this book during a pandemic. I also want to show gratitude to Howard University and its noble mission, against all types of discrimination. Being part of one of the most renowned historically Black Universities has allowed me to interact with people from all types of backgrounds, including underprivileged communities that are often underrepresented in sciences, and therefore to be more aware of and dedicated to address a plethora of societal problems. In particular, it has allowed me to at least partially escape to the academic bubble that entraps numerous scientists, who tend to forget that many of the grants and other funds that we use to do research are paid by taxpayers. Accordingly, we should make an effort to disseminate science, directly engage with the general public, and try to understand the longstanding and still prevailing fictional stories that have chiefly contributed to systemic racism, sexism, and dissemination in general, and to the underrepresentation of "others" in science, in particular.

I also would like to acknowledge the hundreds of colleagues, students, and friends with whom I had the privilege to collaborate with and discuss numerous topics covered in this book and, above all, to Michel Chardon, my former PhD advisor. He was a close friend, and the most humanist person I ever met, the "last true European humanist" as I often called him. Apart from everything I learned from him, and our endless philosophical discussions, long walks in forests and art museums, and short travels to taste delicious Belgian food, wine, and beer, the history of his life taught me a lot about how science really works. Contrary to the fairytales that we are indoctrinated to *believe* in, scientific recognition isn't based only, or sometimes even mostly, on scientific merit: in great part, it depends on the *power of place* and academic egos, networks, competition, gossip, and politics. Michel knew more about nature – about animals, plants, fungi, everything –, and in a less anthropocentric and biased way, than any biologist I met in my life. But compared to numerous scholars, he was a humble, non-egocentric, altruistic, non-political professor that taught and did research at the University of Liege, in Belgium, a small country. None of these things matches well with scientific fame, well on the contrary.

Accordingly, he died – coincidently, when I was writing part of this book, the 8th of April 2021 – in the same way in which he was born: essentially, unknown to the world, even in his own country. As in Orwell's book *1984*, it is as if he never existed: the remnant of an epoch that is no longer celebrated by many of the new supra-specialized technocrats of science. So, even in his death, the humanist professor did teach us a valuable lesson that is crucial within the present volume. This might seem to be a sad story, but what is truly more important – particularly for him, that is gone – is that he had a great life, did not have major regrets, and would have hated to be famous within academia anyway. Thanks for all the priceless lessons and everything else, Michel.

Michel Chardon

Contents

List of Figures

List of Boxes

Within the USA, this included the removal of public statues commemorating people that were "owners" of slaves, sold them, or contributed to oppress them, centuries ago. In European countries such as the UK, some of the targets were politicians that had promoted or participated directly in colonialism, imperialism, or slavery. An emblematic case that was, and continues to be, widely discussed in England concerns Cecil John Rhodes. He was a British mining magnate and politician that overtly and passionately supported British imperialism. Rhodes and his British South Africa Company founded the southern African territory of Rhodesia – named after him in 1895 – which included the countries now known as Zimbabwe and Zambia. More than 125 years later, there are still many statues commemorating Rhodes and what he symbolizes in numerous countries, including several cities of the UK. One of them is displayed at a place that is an emblematic symbol of both British power and Western scientific knowledge: Oxford University. After Floyd's tragic murder, many people claimed that it is outrageous that such an internationally renowned university still displays a statue of Rhodes, celebrating him as a "hero," well into the twenty-first century.

This example is one of the numerous powerful illustrations of how politics, society, indoctrination, and science are usually deeply interconnected. Strikingly, such interconnections are too often neglected within the TV shows, documentaries, museum exhibits, and even literature about, and in particular the idealized constructions of, "scientific giants" such as Charles Darwin, Freud, Newton, or Aristotle. Among the relatively few exceptions, one of the most pertinent is the short book *Biology As Ideology* published by one of the most brilliant and out-of-the-box biologists of the last decades, *Richard Lewontin*. As explained by him:

> Science is a social institution about which there is a great deal of misunderstanding, even among those who are part of it. We think that science is an institution, a set of methods, a set of people, a great body of knowledge that we call scientific, is somehow apart from the forces that rule our everyday lives and that govern the structure of our society… The problems that science deals with, the ideas that it uses in investigating those problems, even the so-called scientific results that come out of scientific investigation, are all deeply influenced by predispositions that derive from the society in which we live. Scientists do not begin life as scientists after all, but as social beings immersed in a family, a state, a productive structure, and they view nature through a lens that has been molded by their social experience.. More than that, those forces have the power to appropriate from science ideas that are particularly suited to the maintenance and continued prosperity of the social structures of which they are a part. So other social institutions have an input into science both in what is done and how it is thought about, and they take from science concepts and ideas that then support their institutions and make them seem legitimate and natural. It is this dual process – on the one hand, of the social influence and control of what scientists do and say, and, on the other hand, the use of what scientists do and say to further support the institutions of society--that is meant when we speak of science as ideology.

One of the central themes discussed in the present book is how this interconnectedness operates, both at an anthropological and sociological level, and how it helps to preserve the *status quo* and the discrimination and oppression of "others," using the example of Charles Darwin as the central case study (Fig. 1.1). This is because he is the most renowned and idealized biologist of all times. In fact, some scientists

Fig. 1.1 Charles Darwin is often depicted, both in popular culture and the scientific community, as an "old," "wise," "honest," "humble" "gentleman"

Fig. 1.2 Pope Darwin, a sketch on a letter from Huxley in 1868, concerning the visit of a German naturalist who wished to pay "his devotions at the shrine of Mr. Darwin"

have worshiped Darwin in such an intense way that he was literally depicted by some of his closest colleagues, anecdotally, as "Pope Darwin" (Fig. 1.2). Even historians of science that tend, in general, to be highly positive about Darwin and his scientific legacy recognize that, in a way, Darwin and Darwinism have become to be seen, and used, by countless scientists, poets, and other writers and artists, as a "secular" deity and a secular religion, respectively. A fascinating book highlighting this phenomenon is Michael Ruse's *Darwinism as Religion*. Due to the profound influence of Darwin's ideas within so many layers of society, particularly in Western countries, and the idealized and even quasi-religious way in which they are often portrayed, Darwin's writings had, and continue to have, a propound impact in the way in which both scholars and the broader public perceive the natural world, including the evolutionary history of our own species and its so-called "races," ethnicities, and genders. Ruse wrote:

> I argue that evolutionary thinking generally over the past 300 years of its existence, and Darwinian thinking in particular.. has taken on the form and role of a religion.. "Religion" is a somewhat elastic term, and I do not claim that evolutionists are committed to a god-hypothesis, or to a formal hierarchical system.. But I shall argue that in the way that evolution tries to speak to the nature of humans and their place in the scheme of things, we have a religion, or if you want to speak a little more cautiously a "secular religious perspective."

Ruse further elaborates this point in his 2019 book *A meaning to Life*, in which he explains how this "Darwinian religion" comprises both a creation story and a meaning to life, as most other religions do. That is, we originally evolved from very simple organisms – this part of the narrative is scientifically accurate – that were supposedly *inferior* to us and through biological evolution we *progressively* became the smarter, most dominant, most moral, and most special organisms – particularly those that are Europeans or European descendants. As in most religion stories, within this narrative, we are the chosen ones. The main difference is that we were not selected by a God but instead by Mother Nature and natural selection and that the notion of Providence is replaced by the notion of evolutionary progress, which does indeed play a crucial, central role in Darwin's writings, as we will see throughout this book. As put by Ruse:

> If we are to have a religion – secular or otherwise – we need an underlying metaphysic to hold it together. To make a picture. To confer meaning. This will be a kind of root metaphor. In the case of Christianity, although there are variations, we find our metaphysic, our root, in the idea of Providence. A Creator God, on whom we are totally dependent, who so loved us that for us He made the supreme sacrifice.. What does evolution have to offer in its stead? If not Providence, then what? Already, in looking at Erasmus Darwin, we have had strong intimations about what this might be. It is the idea that [Erasmus] Darwin expressed..: "Imperious man, who rules the bestial crowd.. Arose from rudiments of form and sense, *An embryon point, or microscopic ens*!". And then he tied it in with a more general philosophy of progress, telling us that the idea of organic progressive evolution "is analogous to the improving excellence observable in every part of the creation; such as the progressive increase of the wisdom and happiness of its inhabitants."
>
> Charles Darwin himself.. raised an Anglican.. would be looking for the big picture to make sense of all, and at first this clearly would be Providence. As his faith in the Christian Creator faded, his need for a big picture was no less, and progress slides readily in to take the place of Providence.. Thomas Henry Huxley [often known as "Darwin's Bulldog" for his advocacy of Charles Darwin's ideas].. was quite explicit that he was seeking a *new religion* to supplant the old, Christian religion.. Huxley's readers, friends and foes, saw what he was about, starting with the fact that he called his essays "Lay Sermons".. Huxley inspired others.. Huxley's grandson, the evolutionary biologist Julian Huxley.. was even

> keener to make a religion out of his science. He wrote a book called *Religion without Revelation*. In the pattern of the older man, grandfather Thomas Henry, Julian did not want to rid the world of religion. He wanted to change it for secular purposes. God must go, but what remains of religion is vital.

A key point of Ruse's book *A Meaning to Life* that is crucial for the present book is that such narratives of the Darwinian religion are not a thing of the past. They are still defended by many people from very different areas of knowledge, including some of the most renowned and influential scholars in the last decades, such as the psychologist and popular science author Steven Pinker and the Neo-Darwinist biologists Richard Dawkins and Edward Wilson. In particular, Ruse explains that these scholars reached such a *power of place* – that is such prominent role within the Western scientific community, media, and popular culture – not in spite of their belief in and dissemination of such narratives, but in great part *because* of that:

> Today, Dawkins stands in this tradition.. [stating that] "directionalist common sense surely wins on the very long time scale: once there was only blue-green slime and now there are sharp-eyed metazoa".. Having embraced computer technology early and enthusiastically, Dawkins slides easily into noting that, more and more, today's arms races rely on computer technology rather than brute power. In the animal world, Dawkins finds this translated into ever bigger and more efficient brains. Oh, what a surprise! We humans are the winners! .. Wilson is [also] open in his fervent belief in biological progress, [stating that] "the overall average across the history of life has moved from the simple and few to the more complex and numerous.. during the past billion years, animals as a whole evolved upward in body size, feeding and defensive techniques, brain and behavioral complexity, social organization, and precision of environmental control – in each case farther from the nonliving state than their simpler antecedents did.. progress, then, is a property of the evolution of life as a whole by almost any conceivable intuitive standard, including the acquisition of goals and intentions in the behavior of animals."
>
> [Within these narratives], meaning for the evolutionist is found in the upward rise of the history of life – monad to man. We humans are in some objective sense the winners, the top of the tree, of more value than other organisms. This is a function of many things, but our minds, our consciousness, our intelligence, are the all-important factors. We are in some sense more complex than other organisms. This complexity, in some way, plays itself out by making us thinking beings with an ability to understand our world and with our own powers of choice, of deciding between good and evil. This readily translates into prescriptions. In the biological world, we are to keep up the evolutionary process, at least not letting it decline and perhaps helping it ever upward. In the social realm, for remember that biological progress is a child of cultural progress, we are to make for a better society for one and for all. Thomas Henry Huxley did not sit on the London School Board by chance. Julian Huxley did not become director general of UNESCO by chance. Edward O. Wilson did not win numerous awards, like the Tyler Prize for Environmental Achievement, by chance. The endpoint is still in this world, but it has great value in itself and gives meaning to the lives of those who strive to realize it.

The present book further expands such discussions about the links between science, popular culture, and politics, focusing not only on the notion of biological progress and the key role it played in Charles Darwin's writings but also on how this and other similar inaccurate evolutionary narratives defended by him relate to his ethnocentric, racist, and sexist ideas about human evolution. Furthermore, the discussions on Darwin, the historical context of his life and writings, and the original links between his works and – as well as their subsequent impact on – politics, Western societies and indoctrination, and systemic racism and sexism will be complemented with analyses on other "scientific giants." This approach will set the stage for a broader understanding of these links, spanning several millennia, from Ancient Greece to the present time. Examples include Newton, Freud, Alfred Russel Wallace, Humboldt, Aristotle, Vesalius, and Galen, among many others. As we will see, some of these scholars constructed scientific "evidence" to support the idea that certain "races," or a certain gender, are "superior" or "favored" by nature, while some others tried to dismantle such ideas to at least some extent. We will see how the fact that Darwin is more idolized in Western countries and in particular by Western scholars than many of these and other "scientific giants" is in great part due to factors that are not necessarily related to scientific knowledge per se. These factors include the social position of Darwin's family, the networks established by his family and by him, politics, what he came to represent as a symbol within the context of the Victorian society, and so on. In this sense, it should be noted that in this book the terms Victorian era and Victorian society refer not only to the Victorian era proper – commonly considered to be from 1837 to 1901 – but also to the Victorian era *sensu lato*, including the so-called Pre-Victorian era and therefore spanning the whole life of Darwin, as he was born in 1809. This is an important point because Darwin's biases, prejudices, career, and fame indeed cannot be simply explained by the events that occurred during his adult life.

Indeed, as put by Janet Browne, who knows Darwin's works and their societal context more profoundly than almost any other scholar, in her book *Darwin Voyaging*: "[In Victorian society] scientific ideas and scientific fame did not come automatically to people who worked hard and collected insects.. a love of natural history could not, on its own, take a governess or a mill-worker to the top of the nineteenth-century intellectual tree.. nor can it, on its own, explain Darwin." As she further emphasized in her book *Power of Place*, Darwin's idolization did not occur despite his ethnocentric ideas and support for British colonialism, imperialism, and Victorian ideology, but in great part *because* of that. In other words, both Darwin's life in general and work in particular are profoundly related to contingency, including, to a great extent, the *power of place* occupied by him and his family. Similarly, the biases and prejudices that he had and that influenced his works were in great part deeply linked to the things he learned in school and from his family and close ones and the experiences he had when he was young, as well as to the privileged social position attained by his family much before he was even born. It is crucial to take this point into account to fully understand not only the life and writings of Darwin but also why there were and continue to be so often idolized in a way that has almost no parallel in science.

For instance, it would be almost unthinkable to hear a physicist, or for that matter a historian of science, argue that what Newton wrote about physics is basically *enough* to truly understand the complexity of the physical world and the cosmos. The information accumulated after the works of Newton – for instance, by Einstein – has shown that some of his theories were incomplete and sometimes even plain wrong. This obviously does not mean that when scientists and historians of science recognize this, they are saying that Newton was not bright or that he was always wrong, nor are they trying to "delete" or "cancel" Newton from history. They are merely stating a scientific fact and acknowledging the reality of the scientific process: that with new scientific studies and discoveries, it is normal that we acquire a more comprehensive understanding of the word and, often, that we disprove ideas that were previously accepted.

However, astonishingly, when it comes to Darwin – who died more than 140 years ago, in 1882 – numerous scholars continue to argue that his writings, or at least the way in which his ideas were subsequently combined with genetic concepts in the "Modern Synthesis" in the 1930s and 1940s, are essentially *enough* to have a pretty good understanding of biological evolution and the natural world. This topic was recently discussed in an interesting science article by Stephen Buranyi in the newspaper *Guardian*. He explained how the few scholars that dare to put into question some of the major evolutionary ideas defended by Darwin or within the *Modern Synthesis* are often quickly ostracized by other scientists or dismissed by them as "outcasts" or "misguided careerists." One example, among others given by Buranyi, concerns the reaction to a 2014 article published in the journal *Nature* that asked the question "Does evolutionary theory need a rethink?" As explained by Buranyi:

> Their answer was: "Yes, urgently." Each of the authors came from cutting-edge scientific subfields[of biology]. The authors called for a new understanding of evolution that could make room for such discoveries. The name they gave this new framework was rather bland – the Extended Evolutionary Synthesis (EES) – but their proposals were, to many fellow scientists, incendiary. In 2015, the Royal Society in London agreed to host *New Trends in Evolution*, a conference at which some of the article's authors would speak alongside a distinguished lineup of scientists. The aim was to discuss "new interpretations, new questions, a whole new causal structure for biology". But when the conference was announced, 23 fellows of the Royal Society, Britain's oldest and most prestigious scientific organisation, wrote a letter of protest to its then president, the Nobel laureate Sir Paul Nurse. "The fact that the society would hold a meeting that gave the public the idea that this stuff is mainstream is disgraceful," one of the signatories told me. Nurse was surprised by the reaction. "They thought I was giving it too much credibility," he told me. But, he said: "There's no harm in discussing things."
>
> Traditional evolutionary theorists were invited, but few showed up. Nick Barton, recipient of the 2008 Darwin-Wallace medal, evolutionary biology's highest honour, told me he "decided not to go because it would add more fuel to the strange enterprise." The influential biologists Brian and Deborah Charlesworth of the University of Edinburgh told me they didn't attend because they found the premise "irritating." The evolutionary theorist Jerry Coyne later wrote that the scientists behind the EES were playing "revolutionaries" to advance their own careers. One 2017 paper even suggested some of the theorists behind the EES were part of an "increasing post-truth tendency" within science. The personal attacks and insinuations against the scientists involved were "shocking" and "ugly," said one scientist.. What accounts for the ferocity of this backlash? For one thing, this is a battle of ideas over the fate of one of the grand theories that shaped the modern age. But it is also a struggle for professional recognition and status, about who gets to decide what is core and what is peripheral to the discipline.

As astutely emphasized by Buranyi, such ferocious backlashes and discussions are not at all merely about scientific facts. They are often strongly related to the *power of place*: professional status, social networks, who gets to "write" history and to decide who are the "triumphant" ones within such ideological and scientific disputes, and so on. After the death of Darwin and the subsequent development of the Modern Synthesis, thousands of biological and anthropological studies have been published, including critical discoveries about the structure of DNA, the genome of humans and several other species, ancient DNA, and of a huge number of human and nonhuman fossil taxa, as well as of thousands of living species. So, it does seem rather odd to argue that all those studies did not add anything truly relevant about, or made us rethink to at least some extent, the way in which Darwin or the Modern Synthesis portrayed biological evolution, human evolutionary history, or the natural world. As we will see in the next chapters, this is clearly not the case. For instance, multiple scientific studies have specifically contradicted many of Darwin's ideas about the natural world and in particular about human evolution, human "races," and supposed gender differences such as women being in general "intellectually inferior" to men.

Unfortunately, because of the type of ferocious backlash faced by those few scholars that dare to put into question Darwin's view of nature and human evolution, such scientists tend to prefer to mostly remain silent about these issues. Regarding the even fewer academics that do opt to overtly discuss these topics, they tend to be extremely cautious when they do so. Particularly when they discuss an issue that has been one of the major taboos concerning Darwin, since his death: whether his writings included, or not, factually inaccurate Western-centric, racist, or sexist ideas. An emblematic example of this concerns another recent *Guardian* article, entitled "How Should We Address Charles Darwin's Complicated Legacy?" The article, written in 2021 by Adam Rutherford, a half Guyanese Indian that was born and made his career as a geneticist in the UK, is fascinating, both because of its content and as an historical archive, due to its overall tone, what it says, and particularly what it *does not say*. On the one hand, the timing of the article is obviously related to the recent rise of the *Black Lives Matter* and *Me-Too* movements. In that sense, the article has the merit of bringing public attention to those facets of Darwin that are almost always omitted in the countless books, documentaries, and TV shows idolizing him as a "hero." However, on the other hand, precisely because Darwin is so idolized both as a national symbol of England and as a Western "scientific giant," it is palpable that even Rutherford, who is the author of the book *How to Argue with a Racist*, felt that he needed to be extremely careful in his approach to Darwin's legacy.

For instance, in the last part of the article Rutherford writes: "this is ultimately why *The Descent of Man* is my favourite Darwin book, because even the greatest of us are merely people – complex and flawed.. it is a deeply humanist book.. Darwin casts aside the idea that 'savage races' are distinct from the civilised, while using language that bears the indelible stamp of imperial dominance." It is deeply enlightening that, in 2021, after all the recent societal changes, the author of a book entitled

How to Argue with a Racist would still feel the need to end up an article about Darwin, racism, and sexism on such a "high" note, stating that his favorite Darwin's book is the *Descent of Man*. This is because, as we shall see throughout this book, the *Descent of Man* is undoubtedly the most scientifically flawed book of Darwin, including several inaccurate "facts" about how Westerns – and Victorians in particular – are superior mentally and morally to "savages," how the latter are "naturally" doomed to become either extinct or massacred by "civilized people," and how women are intellectually inferior to men.

Being a biologist and anthropologist and a teacher and researcher at Howard University – one of the most prominent historically Black universities of the USA – I have been particularly interested in, and exposed to, these topics for several years. Many of my papers, projects, as well as my 2021 book *Meaning of Life, Human Nature, and Delusions*, are focused on racism, sexism, the construction of factually inaccurate narratives, idolization, indoctrination, and scientific biases. That is why I was so interested in analyzing how scholars would react to the internationalization of the *Black Lives Matter* and *Me-Too* movements and to subsequent discussions about the societal legacy of "scientific giants," in particular of Darwin. Based on my previous research, I was not surprised to see that a significant portion of Western scholars would be rather careful, silent, or even defensive about such topics, "heroes," and "giants." However, I have to admit that I was somewhat surprised by the "ferocity" of the backlashes – using Buranyi's term – suffered by the very few scholars that did dare to overtly discuss such topics, particularly from their own academic colleagues.

I was also surprised by the fact that, within those backlashes, many argued that the discussions about such topics would lead to "canceling" culture or history or "heroes" or "scientific giants" such as Darwin, because this is the type of flawed argument that is more typically used by populist politicians and by laypeople that overtly defend the *status quo*. By definition, a person or group that reacts against, or tries to prevent, attempts to change the *status quo*, be it concerning a certain political or social system, or the dominant position of their group, community, or gender may be considered to be a *reactionary*. In stark contrast, science should, in theory, be precisely built on *new* information, ideas, and advances due to technological innovations that modify or disprove past knowledge, which should be reflected in new ways of seeing the cosmos, the natural world, and ourselves as a species or a group. This is the bulk of the scientific process, contrarily to, for instance, religious thinking, which often relies on assumptions that cannot be scientifically disproven, such as the postulation "Angels are real." Therefore, it does seem rather odd to see scientists defending the *status quo* so ferociously or adopting an overtly defensive reactionary type of stance about new discussions or evidence that put into question past knowledge and specific ideas or the societal legacy of past authors such as Darwin.

When Germany decided to remove the statues of Nazis such as Hitler from public spaces after the second world war, the idea was not to "cancel culture" or "delete Hitler" from history. Well on the contrary, Hitler continues to be intensively discussed in German textbooks, educational materials, documentaries, and history museums, precisely as a way to acknowledge the mistakes of the past and try to not repeat them. That is why the statues of Hitler are accordingly displayed together with appropriate educational materials in museums, where they are contextualized – not in public places, where they were originally put to glorify and commemorate him and his ideology. Within this framework, a tragic example recently reported in the media that plainly illustrates the fallacy of such "cancel" or "re-write" history arguments concerns the discoveries of dreadful details about what happened at the so-called Canadian and USA "boarding schools" for Native Americans. These discoveries have led the Canadian government to recognize that such schools were part of an attempt of "cultural genocide" against the local Native American populations. As summarized in a *BuzzFeed* news article entitled "Canada Is Mourning 215 Indigenous Children After Their Remains Were Found at a Residential School," there were over 130 residential schools operating across Canada for more than 100 years, many run by religious institutions. Thousands of indigenous children were taken from their parents and sent to those schools in an attempt to indoctrinate and "assimilate" them, until as recently as 1996. Many of these children were rife with abuse. As put in the article, "the Truth and Reconciliation Commission of Canada, which was formed to examine the legacy of the schools, has identified at least 4,100 children who died of disease or by accident in these schools.. it's estimated that the actual number of deaths could be more than 6,000." The same occurred in the USA, as well as in many other countries with "their own" indigenous people, or with colonies abroad.

Should the data obtained from these recent discoveries not be added to our knowledge about history, including in the textbooks that we use to teach the next generations? Is doing so truly "canceling" or "deleting" history, or is it instead actually telling a much more complete, less biased, less Western-centric version of it? History is not "written" in stone: if it was, why would there be countless historians and other researchers doing research about it, worldwide? Each new discovery should be used to "re-write" history, to render what we know about our history completer and more comprehensive. This should be especially the case concerning scientific research. Scientific texts written in Ancient Greece, or in the Victorian Era, did not refer to genes, or DNA, or human fossils such as Lucy – a female from the species *Australopithecus afarensis* that lived about 3.2 million years ago – because scholars did not know about those things then. But when scientists became aware of them, later, textbooks were accordingly "re-written" to add such crucial new information. Things change, we know more and more about our history, our biology, and evolutionary history, and *all* pieces of information need to be added in textbooks and other educational materials. Including those that might be inconvenient to – or comprise "darker" details about – our specific group, ethnicity, gender, country, political affiliation, societal "heroes," or "scientific giants."

In this sense, the idea for the present book came mostly from observing the contrast between the way in which a huge number of "others," as well as younger Westerners, are actively calling for a change of the prevailing *status quo*, and the much more passive, or even defensive, way in which many Western scholars are approaching current discussions on such "darker" details. Indeed, what is particularly interesting is that the huge contrast between the two types of reactions concerns not only a generational, ethnic, or gender gap but also a professional gap. Many of the most active supporters of the *Me Too* and *Black Lives Matter* movements tend to be from minority or underprivileged communities that are mostly underrepresented in academia, particularly within fields such as natural sciences or the history of sciences. Concerning Darwin in particular, the fact that Western scholars are among those that most idolize him and that, accordingly, tend to be more defensive about any public discussion about the "darker" side of his writings and societal legacy is profoundly remarkable at both an anthropological and sociological level. This is because it gives us a strong hint about why, and how, Darwin begun to be so idolized in the first place, as we shall see.

Moreover, this topic is also fascinating because it involves what seems to, at first sight, be a crucial paradox, which is one of the most neglected aspects regarding Darwin's idolization: the fact that this idolization involves narratives that are in many ways strikingly similar to those used by religious fundamentalists, as noted above. The irony of this apparent paradox is that many scientists and historians defend the idea that Darwin was a key actor within the historical decrease of the role played by religious fundamentalism and the belief in religious deities, within Western countries. But in many ways, the idealized way in which Darwin was, and continues to be, often portrayed within the academia, including in specialized scientific papers and textbooks, as well as in scientific TV shows and documentaries, does resemble that of a deity. For instance, the images that are often used within these scientific and media sources tend to portray Darwin as a humble, wise, kind, sage-bearded old naturalist (Fig. 1.1). Such a representation is, for example, similar to the typical images that are often used to display an omniscient, wise, kind, and sage-bearded God in Western religious buildings, such as the Sistine Chapel (Fig. 1.3). As noted earlier, Darwin's idolization has indeed been so prominent within scholars that he has been literally caricaturized as "Pope Darwin" by his own scientist peers, in the past (Fig. 1.2). As it will be shown throughout this book, the differences between religious and scientific imagery and narratives are indeed not

Fig. 1.3 The Creation of Adam, by Italian artist Michelangelo, a fresco painting that forms part of the Sistine Chapel's ceiling

always so marked as we tend to think. One of the most unfortunate outcomes of such idealized depictions of, and narratives about, Darwin is that many of those that idealize him, or blindly accept his ideas about human evolution, ethnicities, or genders or defend extreme versions of them, are actually doing a great disservice to him. As the saying goes, *they are more Papist than the Pope*. For what we know about Darwin, both from his public writings and private letters, it is unlikely that he would be happy to see the religious-like idealized way in which he, and his books, are frequently depicted in academia, scientific textbooks, TV shows, and documentaries several decades after his death.

Even more ironic and seemingly paradoxical is the fact that one of the main exceptions regarding the idolization of Darwin concerns, precisely, religious fundamentalists or creationists, who sometimes go to the other extreme within such religious-like narratives. That is, instead of idolization, they promote a demonization of Darwin because his evolutionary ideas go against the idea that species were created by God as they are today. Once again, this is just an apparent paradox because in reality it is a further example of how Darwin's idolization and demonization represent two faces of a same coin. This is a clear case in which the two extremes meet because both the idolized way in which Darwin is often portraited by Western scholars and the demonizing way in which many creationists depict him are powerful illustrations of the human tendency to build non-factual narratives. After all, we are the storyteller animal: we love fictional narratives, to hear, see, and read them - this is one of the defining characters of who we are, as a species. As put in Tallis' book *The Incurable Romantic*, idealization is a very common psychological phenomenon within humans: "it simplifies the word in order to reduce the anxiety caused by inconsistency and troublesome complexities.. it always incorporates a degree of *denial*, because in order to see someone as perfect we must deny the existence of their less favorable attributes." That is why many narratives that we build involve idealization and idolization of others – be them Gods, saints, or people such as loved ones, politicians, scientists, actors, or singers. Regarding "scientist giants" in particular, the idealized way in which they are depicted often involves the passive neglection, or even active omission, of the obvious fact that they were just humans, as anybody else, with their biases, prejudices, egos, ambitions, sense of self-esteem or self-importance, and personal intrigues and jealousies. These characteristics are, in turn, deeply linked to contingencies of life such as when and where those scientists were born, studied, lived, interacted with, and so on.

For instance, in the case of Darwin, he was deeply indoctrinated with and influenced by Victorian narratives and prejudices, which he absorbed in his youth and then carried with him to the boat – the famous *Beagle* – that took him to various continents in his fantastic voyage around the globe. Such narratives and prejudices played a huge role in the way in which he perceived and described the non-Western cultures that he encountered in that voyage, or the supposed intellectual differences between people from different groups, or between women and men. Subsequently, the fact that Darwin's writings supposedly provided scientific support for the Victorian narratives and prejudices that profoundly influenced his own writings played a crucial role within his subsequent idealization within the Victorian society, elite and leadership, exemplifying the complex, interactive role between indoctrination, science, politics, and society. Accordingly, as Darwin begun to be more and more idealized, a significant portion of Western scientists, politicians, and laypeople became deeply influenced by his writings or used them to justify the biased way in which they perceived women and non-Western societies. Within various societal strata, such as male Western scholars, the biased depictions provided in Darwin's works begun to be accepted as dogmas and this phenomenon played a critical role in the further spreading and perpetuation of Western-centric and sexist-biased narratives and prejudices.

Another interesting parallel between the idolization of scientists such as Darwin and religious thinking that is often neglected in the literature is that, as many religious fundamentalists actually often do not have a deep theological knowledge, a great part of the scholars – and of those within the broader public and the media – that idolize Darwin in reality actually never read his books, commonly not even one of them. Within the broader public, and even among academics, what percentage of people have truly read the books of Darwin, and in particular the two books in which he focused on human evolution, *The Descent of Man* and *The Expression of the Emotions in Man and Animals*? The percentage is very likely extremely low because within the very few that did read any of his books, this mostly applies to the most famous one, *On the Origin of Species*, which does not discuss human evolution. An even smaller percentage of people read the books published by those few scholars that do have a deeper knowledge about Darwin's life, his works, letters, and the societal context in which they were written and why they became as influential as they did. Such scholars include historians of science such as Desmond, Moore, and Browne, as we shall see. As brilliantly put by Joyce Cary, "it is a tragedy of the world that no one knows what he doesn't know – the less a man knows, the more sure it is that he knows everything." It has always been like that.

In fact, most people, including the vast majority of academics, indeed do not know that those scholars that have studied Darwin's life and works in greatest detail do often recognize that Darwin's writings include numerous logical inconsistencies, problematic inaccurate societal statements, and biased narratives and prejudices that had enduring societal repercussions. Within the authors mentioned just above, some emblematic examples are Desmond and Moore's outstanding book *Darwin – The Life of a Tormented Evolutionist*, and Browne's exceptional books *Charles Darwin Voyaging* and *Power of Place*. The latter author, Janet Browne, was an associate editor of the early volumes of *The Correspondence of Charles Darwin*, and her books were awarded several prizes, including the James Tait Black award for non-fiction in 2004, the W. H. Heinemann Prize from the Royal Literary Society, and the Pfizer Prize from the History of Science Society. As an example of the very detailed type of work these scholars do regarding Darwin's life, writings, and societal legacy, she is now exploring the history of Darwin's impact on popular culture and societal repercussions, from the time of his death to the present time.

The books of these and other scholars therefore provide crucial hints to address a central question: why is Darwin's idolization and the ferocious backlashes against any criticism about the ethnocentric and sexist ideas defended in his *Descent* and *Expression* books, usually undertaken by Westerners, and in particular by male scholars that, in general, did *not* read those two books? The dramatized type of narrative used in the very first page of a book published by one of the most renowned Western historians of science, Richards' 1987 *Darwin and the Emergence of Evolutionary Theories of Mind and Behavior*, gives us a clue to answer this question:

> The scene is familiar and emblematic of the Darwinian revolution. Thomas Henry Huxley confronts Bishop Samuel Wilberforce at the Oxford meeting of the British Association in 1860, a few months after publication of [Darwin's] the *Origin of Species*. Wilberforce, armored in righteousness and crammed in biology by Richard Owen, represents orthodoxy both in religion and science. Huxley, Darwin's partisan and an intellect of dazzling agility, stands for the new scientific order. Wilberforce spoke first. An auditor recalled much later, in 1898, that the Bishop intoned "for full half an hour with inimitable spirit, emptiness, and unfairness".. Wilberforce strode into disaster. Huxley slapped his knee and whispered to his companion, "The Lord hath delivered him into mine hands". "On this", the correspondent continued, "*Huxley slowly and deliberately arose.. a slight tall figure, stern and pale, very quiet and very grave, he stood before us and spoke those tremendous words – words which no one seems sure of now, nor, I think, could remember just after they were spoken, for their meaning took away our breath, though it left us in no doubt as to what it was.. he was not ashamed to have a monkey for his ancestor; but he would be ashamed to be connected with a man who used great gifts to obscure the truth.. no one doubted his meaning, and the effect was tremendous.. one lady fainted and had to be carried out; I, for one, jumped out of my seat*". Huxley conquered, there is no doubt, and Darwinism remains triumphant in our own time.

A dramatic start, a happy ending, and a powerful narrative and symbolism: Darwinism, symbolizing the "noble" rational evolutionist movement, won against the "dark" forces, the powerful irrationality "armored in righteousness" and representing "orthodoxy." Not only it won: It "conquered," and "remains triumphant" until today. One very informative aspect regarding this narrative is that the tone used by the correspondent, and by Richards himself, shows that they perceive themselves as part of the "triumphant" side. They feel identified with it: *they are not seeing, or documenting, history as outsiders, as neutral viewers – they see themselves as the winners of this historical moment.* Within this still prevailing narrative, Darwin and his "bulldog" Huxley are accordingly often portrayed, in countless other books and media sources, somewhat as Don Quixote and Sancho Panza, within an ideological "war" between the "noble" rational atheist evolutionists *versus* the powerful God-fearing irrationals. As a way to personalize and dramatize this tale even more, Wilberforce and Owen are often portrayed as the symbol of the Victorian *status quo*: the system, the authority. At the beginning of another chapter of the same book, Richards reinforces this narrative: "Darwinian theory, as everyone knows, crushed nineteenth-century belief in a spiritually dominated universe and purged nature of intelligent design and moral purpose."

The problem is that this powerful narrative has several flaws, as we will see throughout this book. For instance, as noted in Desmond and Moore's *Darwin – The Life of a Tormented Evolutionist*, when Darwin died in 1882, he had a state funeral in Westminster Abbey. This Abbey is an emblematic symbol of British power and privilege, where numerous coronations and royal weddings and funerals of the most powerful and privileged members of British society have taken place. One emblematic example of this was extensively reported in the news just a few weeks before I wrote these lines: the funeral of the longest-reigning monarch in British history, Queen Elizabeth II, was undertaken at this Abbey. Even more revealingly, during Darwin's funeral at this famous Abbey, he was specifically commemorated by the cream of the Victorian society – including many religious leaders – as the symbol of "English success in conquering nature and civilizing the globe during Victoria's long reign." This is not what one expects to see in the funeral of someone that is portrayed as a Don Quixote fighting against the powerful windmills of the Victorian society, orthodoxy, and *status quo*. In fact, what is especially interesting regarding Richards's dramatization of the "ideological war" between Darwin and Huxley *versus* Wilberforce and Owen is that Richards actually used it to call attention to a central thesis of his book that is, itself, often ignored within the literature about Darwin. Namely, contrarily to the oversimplified accounts about Darwin that are frequently provided in the literature, Darwin's books actually did *not* argue that nature was "morally meaningless." As noted by Richards, such simplistic "characterizations of Darwin's accomplishment control our perception of the late nineteenth and early twentieth centuries.. but.. they grievously distort historical reality." This is because, as put by him, Darwin's theories instead "scientifically reconstructed nature with a moral spine," and Darwin's most prominent followers accordingly often had deep religious or spiritual feelings or believed in a mindful universe or in a progressive divine "world-plan." As astutely pointed out by Richards, in face of the evidence provided in his book, "it might be thought anomalous that any scholars would still cultivate the vintage belief that Darwinism and religious conviction were fundamentally opposed."

Interestingly, despite recognizing this crucial fact, Richards himself fails to recognize another major flaw of the dramatized narrative about the "triumph" of Huxley and Darwin *versus* religious leaders such as Bishop Wilberforce: most people today, including in Richard's country, the USA, continue to define themselves as believers or spiritual – only about 18% of the world's population say there are not (Fig. 1.4). Therefore, as was the case during the Victorian Era before Darwin published his books, it is actually Wilberforce that continues to be "triumphant" today, in the sense that most people in the globe continue to agree more with his creationist ideas than with the evolutionary ideas of Darwin and Huxley. In fact, as we shall see, empirical data indicate that the percentage of religious people across the globe, overall, is actually increasing, not the other way round (Fig. 5.1). Richard's statements on this topic therefore highlight, once again, how biased humans – including excellent scholars as himself – can be, due to our tendency to see our own group as the "norm" or the "triumphant" one. This tendency often leads us to ignore "others" and their points of view. In other words, when Western scholars such as Richards enthusiastically state that Darwin and his evolutionary ideas "conquered" and "remains triumphant in our own time" they are seeing history mostly from their own perspective – that of Western scholars (see also Box 1.1).

A recent example illustrates how scholars frequently continue to unconsciously embark on such biased narratives about "ideological wars" between what they perceive to be their group – the "triumphant" Western scientists, as symbolized by the idealized version of Darwin that they often construct – *versus* the other group, the

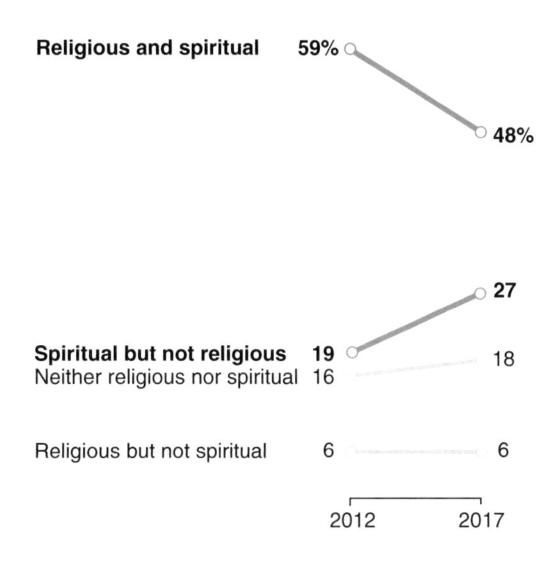

Fig. 1.4 Pew Research Center survey of US adults done in 2017, reporting an increase of people who see themselves as spiritual but not religious

Box 1.1 Plato, Aristotle, Paley, "Purpose" and Biases in Western Thought

Lovejoy's 1936 book *The Great Chain of Being* provides a superb introduction to the omnipresence of teleology, including the notion of a chain of being – also called ladder of life – in Western thought (Fig. 1.7). "Chain of being" refers to the Latin notion of "*scala naturae*": that there is a "progress" from "lower forms," such as plants, to nonhuman animals, and then to humans, which are the culmination point of such a "progression" toward perfection. This notion is profoundly related to our quest to understand humankind – including its "subgroups," or imaginary "races" – and its place in nature. Turner's 2007 book *The Tinkerer's Accomplice* provides a succinct historical background on the idea of intentionality in nature, referring, for example, to platonic teleology, in which crabs were seen as made by God to offer particular moral examples to teach humans.

Aristotle saw purpose in a different, more dynamic way, focusing on physiology: for instance, an animal is supposed to do something and, when taken out of that context, say a fish taken out of water, it will intentionally try to go back to it or to adapt to a new niche – for example, to life on land. Achtner, in a 2009 book chapter, summarizes this topic about the teleological origins of Western thought and science, and about Charles Darwin as well as Paley, who

(continued)

influenced Darwin so much. However, Achtner commits again some of the fallacies that Western scholars often commit when discussing this issue. For instance, he uses a type of "winner" narrative that, as noted above, is typical of Western writers and omits the fact that many passages of Darwin's writings are also teleological. In that sense, Achtner's summary is a powerful example that what some scholars write is highly influenced by how they *like to perceive the truth and themselves* – in this case, as nonteleological, non-biased scholars that are not affected by beliefs or prejudices, and so on:

> Aristotle.. coined the notion of teleology as a result of his observations in nature in general and of organic life in biology in particular. Interestingly, he had a dispute on the teleological development in nature with Empedokles, who claimed that all development in nature is driven by pure chance, thus foreshadowing the debate in the nineteenth century between Darwin and the representatives of natural theology based on teleology. In antiquity Aristotle won the battle, in the nineteenth century the revenant of Empedokles, Charles Darwin, was the winner. Due to the Aristotelian reception by theologians and philosophers in the Middle Ages teleology becomes part of the scientific canon.. In the seventeenth century, teleology was completely replaced by the concept of law of nature within the Scientific Revolution.
>
> It was only in biology that teleology survived as a scientific concept as late as the nineteenth century. Nineteenth century famous biologists like Gerhard Oncken (1800–1884) or Johannes Müller (1801–1858) took it still seriously as scientific. As is generally known, British natural theology was based on the teleological interpretation of nature. The long-standing tradition of natural theology found its climax in the work of William Paley (1743–1805) and his book *Natural Theology*. The studied theologian Charles Darwin owes his decisive inspiration to coin "adaptation" or "fitness" as a scientific term to him. In natural theology adaptation and fitness was interpreted as the result of divine creation. Darwin replaced this theological interpretation of adaptation as a result of the divine creator by a scientific explanation, in which the evolutionary process is governed solely by pure chance. Thus teleology was extinguished from biology and theology suffered because the argument of divine design could no longer be sustained. The problem arose whether or not the evolutionary process of creation based on chance could be reconciled with the idea of divine creation and teleological providence.

creationists. It concerns the rather bellicose reactions to a 2021 editorial published in the journal *Science*. In that editorial, entitled "The Descent of Man – 150 Years On," the anthropologist Agustin Fuentes stated that Darwin's works, in particular the *Descent*, included several factually inaccurate ethnocentric, racist, and sexist assertions. As put by Fuentes, this statement can be easily confirmed by both scholars and laypeople because the *Descent* is freely available online for everyone to read it. Within the numerous harsh reactions by Western scientists to the editorial, Fuentes was often portrayed as a "radical," an outcast within the scientific community, or even somehow as a "traitor" or an "heretic" that is playing into the hands of the "enemy," the creationists. For instance, in an e-letter later published in the same journal, 12 scientists, including various renowned ones, reacted to Fuentes' editorial using statements such as: "*we fear that Fuentes' vituperative exposition will encourage a spectrum of anti-evolution voices.*" That is, according to this type of war-like reasoning, in order to not let the opponent group – the creationists – defend their omniscient and morally flawless God, or to not "lose" people to that group, we scientists should create our own idealized omniscient, flawless, and uncriticizable deity: Pope Darwin. This is not how the scientific process and scholarly discussions should be: scientists should not be afraid of facts, or be so closed to any type of criticism of their scientific heroes. The fact that Darwin's books included ethnocentric and sexist ideas has nothing to do with – and should not be censored because of fear of – creationists and their religious ideas.

Within such a defensive stance adopted by many Western scholars – particularly by evolutionary biologists – when they feel that "their" group, symbolized by Darwin, is being attacked, a typical line of defense used by them is to state that the "darker" aspects of Darwin's writings are merely the "product of his epoch." As we shall see, in reality things are much more complex and nuanced than this. For instance, paradoxically those same scholars often defend that the "brighter" ideas of Darwin, such as the one about evolution being driven by natural selection, are the product of his "unique genius." We see this type of mental gymnastics, circular reasoning, and logical incongruities over and over within several publications, educational materials, and documentaries idolizing Darwin, further evidencing the parallel with religious thinking. Everything good has to be *God's work*, everything bad has to come from the *Devil*: only "good" things can come from Pope Darwin's own mind, so all the "darker" ones have to be exclusively blamed on other people, such as those of his Victorian society.

This book will deconstruct these flawed arguments in several ways. For instance, it will show that while on the one hand most European men at that time did defend racist and sexist ideas that greatly influenced Darwin, as noted above, on the other hand some eminent scholars that lived at the same time and even within the same society as Darwin were able to deconstruct those ideas. As brilliantly put by Marguerite Yourcenar, there are obviously always such exceptions, if not there would never be societal changes: "*at all times there are people who do not think like others.. that is, who do not think like those who do not think.*" Indeed, two of the most famous naturalists that have also traveled around the planet and personally met with Darwin – Alfred Russel Wallace and Alexander von Humboldt – often praised indigenous peoples and overtly condemned colonialist hierarchies or Western imperialism, in sharp contrast to Darwin. The case of Wallace is particularly relevant because he was also British, also lived in the Victorian era, and also developed a theory of evolution driven by natural selection, independently from Darwin. Still, contrary to Darwin, Wallace was able to deconstruct, to a certain extent, many of the Victorian ideas that were prevalent at the time: for example, apart from openly criticizing British imperialism, he also proclaimed to be a socialist and a feminist. As we shall see, Wallace's public and very vocal stance on these topics, *contra* the *status quo* defended by most within the cream of Victorian society and scientific community, is precisely one of the major reasons – together with many others – why Wallace did not became as idealized as Darwin by the Victorian society, and why there is no *Wallacism* today. That is, Wallace's stance against imperialism and his more positive views about non-Westerns and their lifeways, as well as about women in general, were not as appealing to the Victorian elite as were Darwin's more conservative, ethnocentric, hierarchical, and sexist view of nature.

These introductory notes allow us to start having a deeper understanding on why the "darker" aspects of Darwin's life and works that are recognized by historians of science such as Desmond, Moore, Browne, and others continue to be so often ignored, neglected, or discarded by most scholars and by the media and broader public in general. One major reason, which led me to write this book, is that the very extensive and meticulous historical accounts provided by such authors are mostly read by other historians of science, or by scholars who work on fields such as biology or the relatively very few laypeople particularly interested on those fields. Moreover, the "darker" facts about Darwin's writings discussed in those authors' books are often diluted within a huge amount of information about other topics, such as personal details about Darwin's family, colleagues, his wife, children, and so on. Another major reason has to do, as we have seen, with the fact that idolization – be it of Darwin or anyone else – involves confirmation biases and the self-reinforcement of narratives that are often dogmatically accepted by those that idolize, which tend to ignore or even deny the facts that do not fit within those narratives. That is, even among the relatively few people that actually read such specialized history of science books about Darwin, there is often an unconscious tendency to dismiss, or give less importance, to the "darker" aspects about his writings that do not match the idolized narratives constructed about him. A third main reason, which is deeply related to the second one and was also mentioned above, is that many Western scientists feel that any criticism to Darwin is a criticism to themselves, to "their" group: Darwin's "triumph" is their "triumph"; his "flaws" are their "flaws."

That is why I decided to write this book, which is focused primarily on Darwin's ethnocentrism and sexism and their historical context and societal legacy, and targeted for both academics and the broader public in general. In a nutshell, to my knowledge this is the first book written by an evolutionary biologist for the broader public that focus principally on Darwin's writings about humans and human evolution, including his ideas about different human groups and genders, the societal repercussions of these ideas, and whether they are or not supported by factual scientific data. Specifically, the book analyzes *what Darwin explicitly wrote in both his private and public writings* within an historical, anthropological, sociological, and biological multidisciplinary context, in order to contribute to broader public discussions on systemic racism and sexism and to a less biased understanding of human evolution, history, science, and society. The aim of this book is therefore not at all to criticize Darwin, or to argue that all his scientific ideas were wrong, or to "cancel" him. Well on the contrary, it is instead to *complement* the information that is often provided in countless books, educational materials, and media sources that idealize Darwin, which are the ones that have *often neglected or literally omitted* key aspects of his writings, as we shall see. We, scholars, should no longer continue to bury our heads in the sand and ignore or be defensive about any criticisms made about Darwin and other "scientific giants," enclosed in our own academic bubble. In particular concerning Darwin, by reacting so defensively or even harshly against any attempt to discuss and fully understand his ethnocentric and sexist ideas and their societal repercussions within the public arena, we are doing a disservice to science and to the whole society. This is because if we are not able to openly discuss and face the past, including the "darkest" aspects of such "scientific giants" and their writings, we are condemned to continue to blindly accept and repeat its prevailing inaccurate narratives, biases, and stereotypes, affecting the next generations. By doing so, those generations will be indoctrinated with such narratives and prejudices and therefore will likely commit or suffer the same type of abuses or atrocities that were undertaken because, or justified by, such biased ethnocentric, racist, or sexist tales in the past.

In this sense, ultimately this book is mainly written for the *next* generations, both directly by being read by them and indirectly, by reaching those teachers that will teach them, scholars that will write other books that they will read, or media agents or film makers that will produce TV shows and documentaries that they will see. That is why the book is written in an easy-to-read way, avoiding jargon, so it can be read and used by the broader public, instructors, disseminators of science, people working on the media, professors at high schools and universities, filmmakers, historians, and anyone interested in such topics and, crucially, in bringing and discussing them into the public sphere. That is also why Darwin is, in a sense, just a central case study that will be discussed together with many other topics in this book, in order to address questions with broader societal implications, such as the vicious circle of systemic racism and sexism, and the links between indoctrination, science, politics, and society.

Early Life and Societal Contingencies, Bubbles, and Biases

> *It is remarkable how Darwin rediscovers among beasts and plants the society of England, with its division of labour, competition, opening up of new markets, inventions, and the Malthusian struggle for existence.* (Karl Marx)

In order to provide a broader context to understand the topics briefly introduced above, it is now time to travel to Darwin's epoch. And we will do that by starting at the *end* of his life. Why? Because that is when he wrote his short autobiography, in chronological order as his life has unfolded. That autobiography is particularly interesting because it shows what he *chose* to mention about his life and works, and, importantly, what he *omitted* about them. He wrote it mainly as a private memoir for his family but, being aware of how huge his reputation was, particularly within the Victorian society and the scientific community, he very likely knew that it would likely be read by many people, as was indeed the case. In fact, 5 years after he died, in 1887, his son Francis Darwin edited and published it, together with a collection of letters also written by his father, in a book entitled *The Complete Life and Letters of Charles Darwin*.

In a somewhat similar pattern, in this section we will combine part of Darwin's autobiography with excerpts of his personal notebooks, letters, and other materials, together with texts of his scientific books. By doing this, we can have a better idea of not only the chronological sequence of key events within Darwin's life but also of how they, and his personal opinions and views about politics and the British society, deeply influenced his scientific writings. Moreover, this format also allows us to briefly introduce and discuss some of the most relevant evolutionary ideas proposed by Darwin, which will then be further discussed in subsequent chapters. As we will see, by complementing what Darwin published in his books with what he wrote privately, either to himself in the form of notebooks or to individuals he personally knew in the form of letters, one learns fascinating details about his thought process, societal bubbles, biases, preoccupations, beliefs, and prejudices.

Let us thus now go back to the end of the nineteenth century, to the very words that Darwin used, then, to describe certain key events and achievements of his fascinating life. In this part of his autobiography, he is taking us even further back on a journey to the beginning of the nineteenth century, in England, when he was born:

> I was born at Shrewsbury on February 12th, 1809.. My mother died in July 1817, when I was a little over eight years old, and it is odd that I can remember hardly anything about her except her death-bed, her black velvet gown, and her curiously constructed work-table. In the spring of this same year I was sent to a day-school in Shrewsbury, where I stayed a year. I have been told that I was much slower in learning than my younger sister Catherine.. Nothing could have been worse for the development of my mind than Dr. Butler's school, as it was strictly classical, nothing else being taught, except a little ancient geography and history.
>
> The school as a means of education to me was simply a blank. During my whole life I have been singularly incapable of mastering any language. Especial attention was paid to verse-making, and this I could never do well. I had many friends, and got together a good collection of old verses, which by patching together, sometimes aided by other boys, I could work into any subject. Much attention was paid to learning by heart the lessons of the previous day; this I could effect with great facility, learning forty or fifty lines of Virgil or Homer, whilst I was in morning chapel; but this exercise was utterly useless, for every verse was forgotten in forty-eight hours. I was not idle, and with the exception of versification, generally worked conscientiously at my classics, not using cribs. The sole pleasure I ever received from such studies, was from some of the odes of Horace, which I admired greatly.
>
> When I left the school I was for my age neither high nor low in it; and I believe that I was considered by all my masters and by my father as a very ordinary boy, rather below the common standard in intellect. To my deep mortification my father once said to me, "You care for nothing but shooting, dogs, and rat-catching, and you will be a disgrace to yourself and all your family." But my father, who was the kindest man I ever knew and whose memory I love with all my heart, must have been angry and somewhat unjust when he used such words. Looking back as well as I can at my character during my school life, the only qualities which at this period promised well for the future, were, that I had strong and diver-

> sified tastes, much zeal for whatever interested me, and a keen pleasure in understanding any complex subject or thing.
>
> I was taught Euclid by a private tutor, and I distinctly remember the intense satisfaction which the clear geometrical proofs gave me. I remember, with equal distinctness, the delight which my uncle gave me (the father of Francis Galton) by explaining the principle of the vernier of a barometer with respect to diversified tastes, independently of science, I was fond of reading various books, and I used to sit for hours reading the historical plays of Shakespeare, generally in an old window in the thick walls of the school. I read also other poetry, such as Thomson's "*Seasons*," and the recently published poems of Byron and Scott. I mention this because later in life I wholly lost, to my great regret, all pleasure from poetry of any kind, including Shakespeare.

Considering what we now know about Darwin's career as well as what has been told about him by many authors since then, this excerpt is remarkable for several reasons. One fascinating topic, which is recurrent in Darwin's autobiography, concerns a crucial issue discussed in Gladwell's 2013 book *David and Goliath*. That is, the fact that a significant portion of the "greatest" of all times – artists, writers, scientists, athletes, and so on – started as pretty "average" or even "mediocre" competitors or students – or, paraphrasing Darwin, even refer to themselves as "slower in learning." The case of Darwin does not seem as dramatic as many described by Gladwell but still somewhat fits into the general pattern described by Gladwell in the sense that, as recognized by Darwin himself, very likely nobody would bet that a kid as he was, in his childhood, would become such a "scientific giant." On the one hand, the case of Darwin is, in a certain way, similar to that of other "greats," such as Einstein, in the sense that, for such creative minds, school – as a means of education – can often be perceived as "simply a blank," as put by Darwin. But on the other hand, the case of Darwin is rather different from that of many other scientific "giants" because it was not only school that was perceived by him as "dull," particularly when he was young, but also to a certain extent when he was older as we shall see: for him a plethora of things – including many fields of knowledge – were seen as "dull" or uninteresting. As recognized in his autobiography:

> As I was doing no good at school, my father wisely took me away at a rather earlier age than usual, and sent me (Oct. 1825) to Edinburgh University with my brother, where I stayed for two years or sessions. My brother was completing his medical studies, though I do not believe he ever really intended to practise, and I was sent there to commence them. But soon after this period I became convinced from various small circumstances that my father would leave me property enough to subsist on with some comfort, though I never imagined that I should be so rich a man as I am; but my belief was sufficient to check any strenuous efforts to learn medicine.
>
> The instruction at Edinburgh was altogether by lectures, and these were intolerably dull, with the exception of those on chemistry by Hope; but to my mind there are no advantages and many disadvantages in lectures compared with reading. Dr. Duncan's lectures on *Materia Medica* at 8 o'clock on a winter's morning are something fearful to remember. Dr.—— made his lectures on human anatomy as dull as he was himself, and the subject disgusted me… I also attended regularly the clinical wards in the hospital. Some of the cases distressed me a good deal, and I still have vivid pictures before me of some of them; but I was not so foolish as to allow this to lessen my attendance. I cannot understand why this part of my medical course did not interest me in a greater degree..

> I also attended on two occasions the operating theatre in the hospital at Edinburgh, and saw two very bad operations, one on a child, but I rushed away before they were completed. Nor did I ever attend again, for hardly any inducement would have been strong enough to make me do so; this being long before the blessed days of chloroform. The two cases fairly haunted me for many a long year.. I was also a member of the Royal Medical Society, and attended pretty regularly; but as the subjects were exclusively medical, I did not much care about them. Much rubbish was talked there, but there were some good speakers, of whom the best was the present Sir J. Kay-Shuttleworth. Dr. Grant took me occasionally to the meetings of the Wernerian Society, where various papers on natural history were read, discussed, and afterwards published in the "*Transactions*."
>
> I heard Audubon deliver there some interesting discourses on the habits of N. American birds, sneering somewhat unjustly at Waterton.. During my second year at Edinburgh I attended ——'s lectures on Geology and Zoology, but they were incredibly dull. The sole effect they produced on me was the determination never as long as I lived to read a book on Geology, or in any way to study the science.. One of my autumnal visits to Maer in 1827 was memorable from meeting there Sir J. Mackintosh, who was the best converser I ever listened to. I heard afterwards with a glow of pride that he had said, "There is something in that young man that interests me." This must have been chiefly due to his perceiving that I listened with much interest to everything which he said, for I was as ignorant as a pig about his subjects of history, politics, and moral philosophy.

"Dull," "dull," "dull," and "dull": so many things, topics, and whole areas of knowledge were seen as "dull" by him. As noted in Desmond and Moore's 1994 book *Darwin – the life of a tormented evolutionist*, when Darwin was a student at Edinburgh's university, he "did not even bother to join the library," for a full year, in 1826. That is, regarding the *specific topics* that did interest him, such as the specific traits of plants and animals, he could analyze them in an astonishing detailed way, as he did regarding the fauna and flora that he encountered during his Beagle voyage. However, as recognized by him and attested by the many books he wrote, most of the topics that he was passionate about and that he discussed extensively in his works concerned mostly three fields of knowledge: biology, geology and, in a much less successful way as we shall see, anthropology (as defined today). On some occasions, he did refer to topics from other fields, such as psychology, in his writings, but he never wrote a book that was more philosophical, or historical, or about arts, or about the cosmos, for instance. In this sense, Darwin's case contrasts that of many other "scientific giants," such as Aristotle, or Da Vinci, for instance. What was Da Vinci? An artist? A biologist? An engineer? A cartographer? It is difficult to answer because he was all of those things, and much more: he had a profound interest in not only several different fields of science but also in numerous other, nonscientific, areas.

This also applies to "scientific giants" that were alive at the same time as Darwin was, for instance, Johann Wolfgang Goethe (1749–1832), who was a scientist and naturalist as Darwin was but also a poet, philosopher, diplomat, and civil servant, or Alexander von Humboldt (1769–1859: Fig. 1.5). Humboldt was also a scientist and naturalist but, contrary to Darwin, he wrote books about a plethora of nonbiological and nongeological subjects, being consensually considered a polymath, as discussed in Wulf's 2015 *The Invention of Nature*. An even more recent example that further shows that Darwin's overspecialization was not an obligatory product of the increased scientific knowledge gained in the last centuries is Stephen

Fig. 1.5 Alexander von Humboldt is often considered to be one of the greatest polymaths of the German Romantic age

Jay Gould (1941–2002). Apart from being a biologist, Gould wrote books about a plethora of other topics, including religion – *Rocks of Ages: Science and Religion in the Fullness of Life*; racism – *The Mismeasure of Man*; history – *Time's Arrow, Time's Cycle*; art – *Crossing Over Where Art and Science Meet*; the end of the last millennium – *Questioning the Millennium*; and even baseball – *Triumph and Tragedy in Mudville: A Lifelong Passion for Baseball.*

The above discussion and comparisons emphasize two important points that might surprise those that have not read Darwin's books and autobiography and that have read or heard about him from books, documentaries, educational materials, or media articles that tend to idealize him. The first point is that Darwin was *not* at all a polymath, very far from that, as recognized in his own autobiography. The second point, related to the first, is that Darwin's case contrasts with that of several other "scientific giants," such as the ones mentioned just above, in the sense that, contrary to him, many of them did not like school because it involved memorizing specific topics instead of "learning everything about everything." As stated by Darwin, he actually liked to, and was very good at, memorizing very specific information "with great facility," such as "learning 40 or 50 lines of Virgil or Homer." This was indeed one of the main weapons that he used to beat "Goliath" later in life: his amazing memory, the capacity to recollect a huge amount of specific information he gathered in his trips and his readings, primarily about biology and geology. He was particularly good at this, and he did read with much "zeal" a huge number of books and papers about very specific details within these two fields of science. In addition, he was very good at putting together such details. His most brilliant ideas are precisely a product of all these key qualities. However, as recognized in Richards' 1987 book, within the minority of historians and scientists that do not idealize Darwin, a criticism that is often made about him is precisely that he was a "great assembler of facts" but "a poor joiner of ideas." However, as Richards, I do not fully agree with this criticism, in the sense that such a statement is not really the best way to define the major weakness of Darwin. As Darwin wrote in his autobiography – precisely as a defense to this criticism about him, which he had already heard from various scholars during his life – his *Origin* shows that he *was* able to join specific facts to build his broader, and brilliant, ideas about biological evolution driven by natural selection.

In this regard, a detailed analysis of Darwin's life, notebooks, letters, and books points out that one of the major problems of Darwin's writings is instead related to his *lack of depth* concerning a wide range of non-biological and non-geological topics, including the complex historical, anthropological and sociological aspects that led the origin, and subsequent perpetuation, of systemic racism and sexism, or of narratives about progress such as the ones that he often used in his books. Indeed, as we will see, this lack of depth was one of the many factors that did not allow him to have the needed wider multidisciplinary historical, societal, and philosophical framework to deconstruct his Victorian biases and prejudices and to separate them from his scientific ideas. In this sense, it is revealing that even authors that tend to idealize Darwin often concede, as Richards did, that Darwin "was not possessed of the genius, say, of Huxley, whose swiftness of insight often made the older man [Darwin] uncomfortable." "But," as argued by Richards, "genius has its varieties.. Darwin's own definition, which he offered in the *Descent of Man*, suggests another kind: [Darwin wrote that] 'genius has been declared by a great authority to be patience; and patience, in this sense, means unflinching, undaunted perseverance'." As put by Richards, "this describes, not accidentally, Darwin's own mental character."

Richards claims that patience and intellectual conservatism were the major weapons of Darwin: "the salient feature of his mental style that, I believe, contributed to his adoption of the evolutionary hypothesis was, paradoxically, his intellectual conservatism." He added: "unless this trait of his psychology be understood, other aspects of his developmental history will remain opaque.. his conservative style, his persistence in retaining and modifying ideas rather than simply dropping and replacing them" were critical in Darwin's work. This is true, no doubt about it: however, unfortunately, these traits were not only related to the "best" sides of Darwin's work and legacy but also to some of the darkest ones, as we shall see. On the one hand, Darwin's persistence in retaining and modifying ideas rather than simply dropping and replacing them was for sure critical for his most outstanding and original book, the *Origin*. But on the other hand, Darwin's persistence in retaining, and not being able to deconstruct, his Victorian bubble biases even when they were clearly contradicted by what he actually saw during his Beagle voyage, linked to his intellectual conservatism and difficulty – or lack of curiosity – to step out of that bubble, led to some of his most problematic scientific flaws. For instance, these were among the major reasons that led him to extrapolate the whole natural world from what he saw in his Victorian society.

Regarding this topic, Michael Ruse, who has also a profound admiration for Darwin, addressed the following questions in 1993: "Charles Darwin is a mystery man.. was he a great scientist, really great I mean, of the caliber of Albert Einstein, that everyone accepts as having been a genius? Or was he perhaps like some of the prominent figures of molecular biology – smart and ambitious, but lucky in having been the person around when important conceptual moves and empirical discoveries were there to be made?" He added: "Was he even a bit thick, a man who hit on his theory but really had no idea of what he had grasped? 'Yes' answers to all of these questions can be found in the literature." And Janet Browne, who as explained above has studied Darwin's life and letters in more detail than most other scholars, and that accordingly has a much more nuanced view of Darwin's works, plainly recognized in her book *Voyaging* that "no other thinker shook Victorian England as deeply as Charles Darwin with his theory of evolution by natural selection.. but Darwin was the most unspectacular person of all time.. his personality did not seem to match the incisive brilliance other people saw in his writings."

This recurrent topic has also been the focus of numerous specialized papers written by some of the most prominent biologists of our time, such as Adam Wilkins, who wrote a 2009 paper entitled: "Charles Darwin: Genius or Plodder?" Wilkins started by recognizing that "there is no doubt about the magnitude of Charles Darwin's contributions to science.. there has, however, been a long-running debate about how brilliant he was.. his kind of intelligence was clearly different from that of the great physicists who are deemed geniuses." "His most apparent qualities were," he added, "thoroughness and doggedness, qualities that seem the antithesis of brilliance.. this leads one to wonder whether Darwin, to use Francis Crick's description.. was actually a 'plodder,' albeit an exceptionally productive and lucky one." However, Wilkins then applies the type of mental gymnastics that is so often applied within works about Darwin. That is, as summarized in the abstract of his paper, he concludes "*that the apparent discrepancy between Darwin's achievements and his seemingly pedestrian way of thinking reveals nothing to Darwin's discredit but rather a too narrow and inappropriate set of criteria for 'genius'*." In other words, if the idealized Darwin does not fit in the typical definition of "genius," the problem has to be that definition because we all *know – or better say, were dogmatically told* – that Darwin was a genius, right? If the mountain will not come to Muhammad, then Muhammad must go to the mountain.

To be clear, nobody is arguing here that it is a problem, per se, if a scientist is not a genius, or focus his/her attention almost exclusively on one or two specific topics, contrary to the case of many other "scientific giants" such as Galileu, Da Vinci, Aristotle, Copernicus, Avicenna, or Hypatia (Fig. 1.6). The vast majority of scientists do not define themselves as geniuses and have no problem in recognizing – and many are even proud of saying – that they have a narrow focus of research, and still many of them are able to publish excellent work within their fields of interest, including some that became Nobel Prize awardees. However, in the specific case of Darwin, this was a major problem because he wrote not only about worms, or finches, but about the evolution of life in general and, importantly, about our own evolutionary history, the evolution of our societies, cultures, ethnic groups, and genders. That is, about topics

Fig. 1.6 "Death of Hypatia in Alexandria." Hypatia was a brilliant public speaker and scholar and wrote on mathematics and astronomy, inventing the astrolabe for ship navigation and devices for measuring the density of fluids

that are extremely complex and that are so close to our "heart" that they tend to be plagued by our own biases, prejudices, and stereotypes, and accordingly have moreover far wider societal repercussions than a book about, let us say, the beaks of finches would have. Perhaps an easier way to explain what I mean about Darwin's lack of depth about such societal issues is to refer to an interview by Ori Givati that I read today in the Portuguese newspaper *O Publico* (May 24, 2021). He is an Israeli ex-soldier that did service in the West-Bank – where most people are Palestinian – and that is now part of the organization *Breaking The Silence*. He explains how both Israeli Jews and Palestinians are indoctrinated to see the "others" as the enemy: most people believe these narratives because they "are educated in a system – the school, the media, families, everything around us prepares us" that makes them think this way. Therefore, most people see such narratives and the acts to which they lead "as legitimate." As he recognizes, he also "saw this as legitimate," when he was a soldier, and only later in life he was able to deconstruct these indoctrinated ideas because "it takes a lot of *active effort* to start *thinking critically* about this.. because what we hear all our life" are the "us *versus* other" narratives.

As defined today, *critical reasoning* is the effort to actively and astutely conceptualize, analyze, question, and appraise beliefs and ideas. In contrast, *dogmatic thinking* or *dogmatism* refers to the acceptance of unquestioned information *without the intervention of active thought or criticism.* To some extent, this is what Darwin somewhat failed to do: to do an *active effort* to read the type of books that would give him the needed broader sociological, historical, and philosophical background to be able to *critically think* about the Victorian indoctrination that he undertook during his early life. When Darwin stated, in his scientific writings, that women are intellectually inferior to men, or that savages are repugnant and are "inferior" to and have a lower morality than Europeans, or that hierarchical socioeconomic systems are better, he was indeed falling into the trap of dogmatic thinking. That is, he was taken as a given many of the biased narratives and beliefs that he heard or read about early in life at home, in school, or from people around him, without truly questioning them, even when they were contradicted by the very people he encountered and occurrences he faced during his Beagle travel. That is why he wrote things like the ones below as if they were "scientific observations" and "evolutionary facts" about the "others," the "savages," such as the Fuegians that he met at Tierra del Fuego in the Southern region of South America (Figs. 1.13 and 1.14):

> The Captain sent a boat with a large party of officers to communicate with the Fuegians.. When we landed the party looked rather alarmed, but continued talking & making gestures with great rapidity. It was without exception the most curious & interesting spectacle I ever beheld. I would not have believed how entire the difference between savage & civilized man is. It is greater than between a wild & domesticated animal, in as much as in man there is greater power of improvement. The chief spokesman was old & appeared to be head of the family; the three others were young powerful men & about 6 feet high. From their dress &c &c they resembled the representations of *Devils on the Stage*, for instance, in *Der Freischutz*. The old man had a white feather cap; from under which, black long hair hung round his face. The skin is dirty copper colour.. the only garment was a large guanaco skin, with the hair outside. This was merely thrown over their shoulders, one arm & leg being bare; for any exercise they must be absolutely naked..
>
> Their very attitudes were abject, & the expression distrustful, surprised & startled: Having given them some red cloth, which they immediately placed round their necks, we became good friends. This was shown by the old man patting our breasts & making something like the same noise which people do when feeding chickens. I walked with the old man & this demonstration was repeated between us several times: at last he gave me three hard slaps on the breast & back at the same time, & making most curious noises. He then bared his bosom for me to return the compliment, which being done, he seemed highly pleased. Their language does not deserve to be called articulate: Capt. Cook says it is like a man clearing his throat; to which may be added another very hoarse man trying to shout & a third encouraging a horse with that peculiar noise which is made in one side of the mouth. Imagine these sounds & a few gutterals mingled with them, & there will be as near an approximation to their language as any European may expect to obtain..
>
> They are excellent mimics, if you cough or yawn or make any odd motion they immediately imitate you.— Some of the officers began to squint & make monkey like faces; but one of the young men, whose face was painted black with white band over his eyes was most successful in making still more hideous grimaces. When a song was struck up, I thought they would have fallen down with astonishment; & with equal delight they viewed our dancing and immediately began themselves to waltz with one of the officers.. If their dress & appearance is miserable, their manner of living is still more so.. I believe if the world was searched, no lower grade of man could be found.

That is also why the part about human evolution of Darwin's 1871 *Descent of Man* – and not, for instance, his 1881 *Worms* book – is markedly biased, including several racist, sexist, and ethnocentric factually inaccurate assertions. Similarly, his accounts on human evolution in his 1872 book *The Expression of the Emotions in Man and Animals* contain more biased factual inaccuracies than, for instance, his works about barnacles. The fact that the *Descent* and the *Expression* were precisely the ones – together with his 1859 *Origin* book – that became so revered in the academy as well as within the media, broader public and political leaders, indicates that it was in great part *because – not despite –* of the fact that they contain such ethnocentric and sexist biased "facts." It is not a coincidence that, apart from the *Origin*, among all the other books of Darwin, it is precisely the *Descent* that is commemorated by Western scholars, as evidenced by the huge recent commemorations of the 150th anniversary of the first edition of that book. There were no such commemorations in the 150th anniversary of his monograph about *The structure and distribution of coral reefs*, published in 1842, nor about his monograph about *Volcanic Islands*, published in 1844. The fact that his two most biased and scientifically inaccurate books, the *Descent* and the *Expression*, are much more revered and commemorated than his much more brilliant and scientifically accurate books about barnacles, volcanoes, coral reefs, and worms indeed tells us a great deal about the power of human, societal, and scientific biases.

This is an important point because one needs to recognize that the darker sides of the legacy of Darwin's writings are not only related to his societal biases. They are also linked to the people that subsequently idolized him and his works, including not only Western scholars but also Western media, politicians, and other social actors that used his ideas *and* idolization as ideological weapons to justify their own biases and prejudices. In fact, in those cases in which such biases and prejudices led politicians to justify, or directly undertake, societal atrocities, clearly they are the ones that should be mostly blamed. Having said that, scientists and historians of science should also be criticized because they provided something that was critically needed for those politicians: the idealization of Darwin. If Darwin and his ideas – including his most flawed societal ones – had not been idealized and literally sometimes even worshiped by so many scholars, such political leaders and other social actors would not have been able to use Pope Darwin to "scientifically" justify their factually inaccurate racist and sexist ideas.

In this sense, it is important to note that while Darwin's *Origin* is less biased and overall more scientifically accurate than the *Descent* and *The expression*, it does include some problematic general ideas that have also been used to promote and justify social inequalities and atrocities. This is because, despite almost not referring to humans directly, that book was anyway highly influenced by anthropocentric concepts and Victorian ideas. This includes exaggerated parallels between artificial selection – mostly purportedly made by humans – and natural selection, as well as the repeated extrapolation to biological evolution as a whole from Victorian biases and narratives. For instance, from capitalistic ideas about competition and selfishness within a brutal, soulless "struggle-for-existence." This point was even recognized by Karl Marx, who, as we will see, noted that: "it is remarkable how Darwin rediscovers among beasts and plants the society of England, with its division of labour, competition, opening up of new markets, inventions, and the Malthusian struggle for existence."

Marx is referring to a concept of Thomas Robert Malthus that was particularly influential in Victorian society and therefore in Darwin's works: a population will increase exponentially if unchecked, while the resources tend to only increase arithmetically, so the checks that will exist on population growth will result in a brutal struggle-for-existence. Or, as put by Nordenskiold in *The history of biology*: "from the beginning Darwin's theory was an obvious ally to liberalism; it was at once a means of elevating the doctrine of free competition, which had been one of the most vital cornerstones of the movement of progress, to the rank of a natural law, and similarly the leading principle of liberalism, progress, was confirmed by the new theory." That is, "the deeper down the origin of human culture was placed, the higher were the hopes that could be entertained for its future possibilities.. it was no wonder, then, that the liberal-minded were enthusiastic; Darwinism must be true, nothing else was possible."

As explained in James Suzman's 2021 book *Work – A Deep History from the Stone Age to the Age of Robots*, Malthus' models were themselves influenced by preconceived biased ideas, such as the principle that people working in the fields would tend to reproduce as much as they could because of their uncontrollable lust. It is therefore not a surprise that many of such ideas and models proved to be wrong. For example, as noted by Suzman, Malthus "radically underestimated the extent to which food production in the fossil-fuel era would keep pace with a surging global population" and "failed to anticipate the trend in industrialised societies towards steadily declining birth rates that began almost as soon as his essay was published." Another emblematic example in which Darwin's preconceived ideas were based on false premises concerns the evolutionary concepts and "facts" that he included in the *Descent* about women. He repeatedly stated, in that book as well as in his private writings, that women had lower intellectual powers, imagination, and ingenuity than men and that women "have never advanced the science." While in the latter statement he was talking about a case that referred to the field of physiology in particular, his writings clearly indicated that women's supposed inferior mental skills and "ignorance," as well as their "nature" in general, did not help them to become top scientists.

These "facts" are historically and scientifically factually wrong. Women are not mentally "inferior" to men, and at the time in which Darwin wrote such sexist statements, many women had already achieved prominence in science, written books with powerful imagination, and been successful political leaders. A clear example of such a prominent scientist was Hypatia, a brilliant Hellenistic philosopher, mathematician, and astronomer (Fig. 1.7). She was born in 360AD and lived in Egypt – then part of the Eastern Roman Empire – where she taught astronomy and

Fig. 1.7 "Ascent of Life," by F. Besnier, 1886, illustrating the notion of a ladder of life

philosophy. Darwin did not refer to her, or other female prominent scholars and writers that will be discussed in Chap. 5, nor did he even consider the numerous true historical reasons that prevented countless other bright women to become prominent scientists, writers, and politicians. For instance, he did not discuss the fact that during about two-thirds of his life (1809–1882), women simply could not go to university in England. For instance, only in 1868 the University of London's Senate voted to admit women to sit the General Examination, becoming the first English university to accept women. While Darwin had the privilege to go to university – actually to two very prestigious universities, Edinburgh and Cambridge – despite considering a lot of the classes there as "dull", *all English women* were forbidden to do so, even if they were extremely motivated to do so. This is what male privileges look like: but Darwin's biases, prejudices, and lack of curiosity about, and interest to examine in detail, such broader societal questions "outside-of-the-box" did not allow him to *see* those male privileges and take them into account within his writings, ideas, and evolutionary theories.

The same applies to the statements, in Darwin's scientific books, about the Victorian society as a whole. For instance, in the *Descent* when he discusses sexual selection and evolution, he defended the idea that the marked difference of gender roles typical of the Victorian society was probably the most "progressed" type of social organization of the sexes. He also wrote that Victorian society was the more "progressed" type of society within the "highest" species of all, *Homo sapiens*. What a coincidence, of all the millions of species, and of all the numerous human groups, that his own species and, within it, its own societal group was the pinnacle of biological evolution. This is a typical flaw made by most people, including prominent thinkers or leaders: they tend to "conclude" that their own group is the "superior" one. We rarely see an African American leader proclaiming white supremacy, or a Hindu leader proclaiming Muslim supremacy, for example.

Darwin's biases and prejudices were aggravated by the contingencies related to the fact that he lived in a particularly privileged male Victorian bubble. With a few exceptions, he was almost completely unexposed to non-"white" people until his Beagle voyage and was also not so exposed to intellectual conversations with women while at Edinburgh and Cambridge, in great part because women could not attend those universities. One of the major exceptions concerns a "black taxidermist" referred in his autobiography, who apparently taught him some taxidermy techniques. Apart from that, he was basically "well acquainted with several young men" – "white" men, mostly "well-off" within the hierarchical Victorian society – as put in his autobiography. However, the fact that he was privileged also means, on the other hand, that he *could* have been able to deconstruct the Victorian biases and prejudices if he had made a greater effort to go out of that bubble and "think-outside-of-the-box." This is because that privileged position allowed him to have very easy access to one of the most effective tools to deconstruct factually inaccurate narratives, which the poorest people could hardly get at that time: books, cheap or expensive, national or international. Despite the fact that he perceived countless topics and fields of knowledge as "dull" or uninteresting, he *did* have access and did read a considerable number of books, including about nonbiological and nongeological topics, in particular in his adulthood.

As recognized in his autobiography, he read a lot of Victorian novels and "non-intellectual" stuff when he was an adult, but we do know that he did read sometimes about subjects such as philosophy and religion and about the quest for the "truth," particularly after the death of loved ones. As explained in Quammen's 2006 *The reluctant Mr. Darwin*, he read "philosophical and scriptural topics, ranging from Hume, Locke, and Adam Smith to James Martineau's *Rationale of Religious Enquiry*, and from Paley, Herschel, and Ray to John Abercrombie's *Inquiries Concerning the Intellectual Powers and the Investigation of Truth.*" He also "took some interest in the works of Francis Newman, a Latin professor whose elder brother, John Henry, had turned Catholic and would eventually become Cardinal Newman.. he read Newman's *History of the Hebrew Monarchy*, which critiqued the Old Testament for its dubious historicity." Accordingly, as noted in Gruber's 1981 *Darwin on Man*, Darwin did write "on man, mind, and materialism.. [in his] M and N notebooks and associated documents.. [in which it] can be seen Darwin's struggle towards the materialistic philosophy of biology." However, he never published those M and N notebooks precisely because "he himself had seemingly dismissed [them] as 'full of metaphysics on morals and speculations on expression'." Why was Darwin so uncomfortable with and reticent about "metaphysics" – which is today defined as the branch of philosophy that examines the fundamental nature of reality – after all? In order to try to understand this point and further analyze how his youth influenced his evolutionary ideas and scientific works in general, let us thus go back in time again: now, to the time he went to Cambridge, as per his autobiography.

Cambridge, Purpose, and Design

One must be very naïve or dishonest to imagine that men choose their beliefs independently of their situation. (Claude Lévi-Strauss)

After having spent two sessions in Edinburgh, my father perceived, or he heard from my sisters, that I did not like the thought of being a physician, so he proposed that I should become a clergyman.. Considering how fiercely I have been attacked by the orthodox, it seems ludicrous that I once intended to be a clergyman.. As it was decided that I should be a clergyman, it was necessary that I should go to one of the English universities and take a degree; but as I had never opened a classical book since leaving school, I found to my dismay, that in the two intervening years I had actually forgotten, incredible as it may appear, almost everything which I had learnt, even to some few of the Greek letters. I did not therefore proceed to Cambridge at the usual time in October, but worked with a private tutor in Shrewsbury, and went to Cambridge after the Christmas vacation, early in 1828.

During the three years which I spent at Cambridge my time was wasted, as far as the academical studies were concerned, as completely as at Edinburgh and at school. I attempted mathematics, and even went during the summer of 1828 with a private tutor (a very dull man) to Barmouth, but I got on very slowly. The work was repugnant to me, chiefly from my not being able to see any meaning in the early steps in algebra. This impatience was very foolish, and in after years I have deeply regretted that I did not proceed far enough at

> least to understand something of the great leading principles of mathematics, for men thus endowed seem to have an extra sense. But I do not believe that I should ever have succeeded beyond a very low grade. With respect to Classics I did nothing except attend a few compulsory college lectures, and the attendance was almost nominal.
>
> In my second year I had to work for a month or two to pass the Little-Go, which I did easily.. In order to pass the B.A. examination, it was also necessary to get up Paley's "*Evidences of Christianity*" and his "*Moral Philosophy*." This was done in a thorough manner, and I am convinced that I could have written out the whole of the "*Evidences*" with perfect correctness, but not of course in the clear language of Paley. The logic of this book and, as I may add, of his "Natural Theology," gave me as much delight as did Euclid. The careful study of these works, without attempting to learn any part by rote, was the only part of the academical course which, as I then felt and as I still believe, was of the least use to me in the education of my mind. I did not at that time trouble myself about Paley's premises; and taking these on trust, I was charmed and convinced by the long line of argumentation.
>
> By answering well the examination questions in Paley, by doing Euclid well, and by not failing miserably in Classics, I gained a good place among the *oi polloi* or crowd of men who do not go in for honours.. But no pursuit at Cambridge was followed with nearly so much eagerness or gave me so much pleasure as collecting beetles. It was the mere passion for collecting, for I did not dissect them, and rarely compared their external characters with published descriptions, but got them named anyhow.. Public lectures on several branches were given in the University, attendance being quite voluntary; but I was so sickened with lectures at Edinburgh that I did not even attend Sedgwick's eloquent and interesting lectures. Had I done so I should probably have become a geologist earlier than I did.

When we read what Darwin wrote about himself, we can see that the idea that he was humble is at least partly corroborated by what he wrote, but that the notion that he was an omniscient "sage," as he is so often portrayed in popular culture, was clearly not constructed by him. Instead, it was mainly constructed – *contra* what he wrote about himself – by subsequent scholars that started to be more Papist than the Pope and to *believe*, or to lead others to believe, that he was such a quasi-religious omniscient deity. Indeed, according to the description given by him in his autobiography, Darwin was everything but a polymath sage. For instance, he explains that he was far more interested in collecting beetles than learning "dull" things at the University. Such passion for beetles and the natural world is repeated countless times in books idolizing Darwin, and no doubt about it, he was a passionate and very successful, brilliant naturalist: he was one of the top naturalists of all times. But what those books almost always fail to mention is that such a passion was combined – or was perhaps also part of the cause for? – his lack of interest about many other fields of knowledge, in particular until his Beagle travels. And this is a crucial point because this was in part what prevented him to acquire the needed broader multidisciplinary background and depth to fully grasp the complex and diverse societies he encountered in his Beagle voyage as well as the vaster implications and societal repercussions of the scientific ideas that he wrote during that voyage.

What Darwin read – and did *not* read – before boarding the Beagle was critical for the way he perceived the natural world and other humans. For instance, the fact that one of the authors that most fascinated him at Cambridge was William Paley is key. This is because Paley, and in particular, his "watchmaker analogy," is one of the most famous examples of the type of teleological way of thinking that tends to characterize humans. Teleological narratives are related to the notion that things have a "special" or "cosmic" purpose or use. The word teleology builds on the Greek *telos* – "purpose," or "end" – and *logia* – "a branch of learning" (Box 1.1). The "watchmaker" analogy argues that a *design* implies a *designer*. That is, when you find a watch on the ground, you will assume that it was made by an intelligent designer – such as a human – so the only way to explain the origin of humanity is to assume that it was made by a creator deity. Paley's "watchmaker analogy" played a crucial role in natural theology at Darwin's epoch. In a simplified way, natural theology can be defined as a type of theology – that is, of study of nature and the divine – that provided arguments for God's existence based on reason and observation of the natural world. In other words, this is basically the same type of reasoning that is still used today to support arguments for the existence of God or of an intelligent/perfect design of the cosmos, including by countless current creationists that define themselves as "intelligent designers." To criticize Darwin and his ideas, they first construct the "fact" that there is a "fine-tuning" of nature and the cosmos and then use that "fact" to assert that only God could have made things that way.

Humans tend to think about final purposes: everything happens for a reason, as recurrently repeated in Hollywood movies or bestseller books. Within science and philosophy, scholars often defend this idea by stating that "nature does nothing in vain" – a notion already defended by Aristotle thousands of years ago (Box 1.1). The example of Aristotle's ideas, which in a way influenced so much Darwin's own teleological ideas, is particularly interesting because the influence – and idolization – of Aristotle's ideas, particularly until the Renaissance, is indeed similar to the way Darwin's ideas have been idolized in the last centuries. Aristotle's idea that "nature does nothing in vain" continues to be used in documentaries about plants, animals, fungi, bacteria, and even nonliving organisms such as viruses, as attested by many narratives put forward during the recent Covid-19 pandemic. Even within top documentaries about the natural world, such as those narrated by David Attenborough, we frequently hear things such as "the wings of birds were made to fly," which are teleological, and scientifically inaccurate, assertions.

Firstly, wings were not "made," they just evolved: nobody "made" the wings of birds, they are not the result of a "masterplan" of a purposeful God or *Mother Nature*. They are just the result of the evolution of several different types of tissues and cells, and obviously none of them had any idea that they would eventually all combine to form bird wings, millions of years after they started to evolve. There is never a final goal, or purpose, within the evolution of anatomical structures. Secondly, and related to this point, the main original function that the combination of those tissues and cells that ultimately formed wings conferred to the ancestors of modern birds was, very likely, *not* flying. The wings of birds include bones and muscles that were already present in their non-winged ancestors, and the very first anatomical complexes that are often called "bird wings" probably were related to functions other than flying, such as helping in thermoregulation. That is, the ability to keep body temperature within certain boundaries, even when the surrounding temperature was very different. This is what biologists nowadays call an "exaptation": something that was originally related to an adaptation to a function A and that later became adapted to a function B. Such a type of tinkering is very common in evolution, further showing the fallacy of stating that a certain structure X was "made" for one of its current functions. That is not at all how the natural world is: it is instead a very biased, teleological, and factually inaccurate way that most humans – including, to some extent, Darwin – use to describe nature.

This is a crucial point, because, contrary to what is often said about Darwin (see Box 1.1), he *did not* remove teleology from the natural world, despite showing that no supernatural creator was needed to explain the origin of organisms. That is, one of the most emblematic paradoxes – or better said, logical inconsistencies – of Darwin's works is that while they did remove the religious teleological "watchmaker" analogy from the scientific field of biology, Darwin was unable to remove teleology from the way he – and thus his followers – saw the natural world. For example, Darwin *did use* countless teleological factually inaccurate terms not only about nature as a whole but also to scientifically legitimize sexist and ethnocentric ideas. Regarding human evolution and human "races," he often used terms such as the "progress" from "lower" to "higher" groups or societies or morality, which were directly related to the teleological notion of chain of being that Ancient Greek authors such as Aristotle had put forward millennia before (Fig. 1.7, Box 1.1). Darwin also applied such terms to nonhuman organisms. As an example, apart from using terms such as "higher and lower forms of life," he explicitly wrote in his notebook B that "there must be progressive development; for instance none of the vertebrata [animals with vertebrae] could exist without plants & insects had been created; but on the other hand creations of small animals must have gone on since from parasitical nature of insects & worms." Such ideas are factually wrong in the sense that there is nothing such as "progressive development" in the natural world. No group – or for that matter, gender – is "higher" or "better," nor "lower" or "worse" than other groups; these are just teleological and moral constructions created by our human brains.

At first, it could seem a paradox that Charles Darwin was so fascinated, and influenced, by Paley's teleological "argument from design," because some of Paley's writings were done as a response to the evolutionary ideas of Charles' grandfather, Erasmus Darwin (Fig. 1.8), particularly those put forward in Erasmus' book *Zoonomia*. However, if we dig a bit deeper, we can understand that this is not a paradox. This is because, as noted in Gruber's 1981 *Darwin on Man*, Erasmus was also fascinated by the "argument from design," the key difference being that contrary to Paley he mainly ascribed such "design" to Mother Nature, precisely as his grandson Charles did decades later:

> Both Darwins, but especially the grandson, were deeply influenced in their view of change and struggle by the argument from design. [Within] this teleological view.. the entire course of evolution was seen as a series of small readjustments on the part of a self-regulating system, nature as a whole.. The Darwins were not Natural Theologians, but they did try to study nature as a whole. And they both had an attitude toward nature that might be called *worshipful* – reverential, enthusiastic, and poetic.. In a very general sense, both Erasmus and Charles accepted the utilitarian ethic.. For Erasmus.. "the greatest good for the greatest number" was translated into a "greatest happiness principle," about which he wrote on more than one occasion.

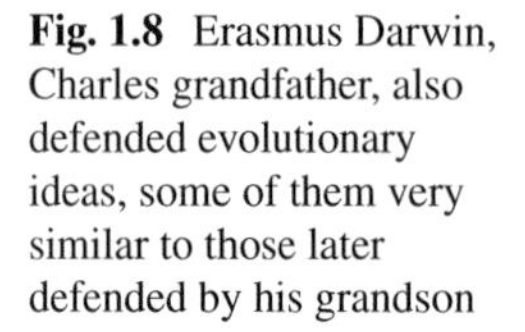
Fig. 1.8 Erasmus Darwin, Charles grandfather, also defended evolutionary ideas, some of them very similar to those later defended by his grandson

> In *Zoonomia* he expressed his *belief* "in the progressive increase of the wisdom and happiness" of the inhabitants of the earth.. In the *Temple of Nature*, he wrote that "the sum total of the happiness of organized nature is probably increased.. when one large old animal dies, and is converted into many young ones.. [His grandson] Charles discussed the sources of happiness in the M notebook. His argument there is similar to his grandfather's. Higher thought processes yield greater happiness.. perhaps too hopefully, he writes of aggressive impulses that 'with lesser intellect they might be necessary and no doubt were preservative and are now, like all other structures slowly vanishing.' He sounded the same note of evolutionary optimism in *Descent*: in discussing the conflict in each between social and anti-social tendencies, he concluded that through natural evolutionary processes".. the struggle between our *higher and lower impulses* will be less severe, and *virtue will be triumphant*.

In other words, Charles wrote notes to remind himself that he should not fall into the trap of teleology, but ultimately he could also not resist it, repeatedly using sentences in scientific writings about biological "progress," "higher and lower impulses," how "virtue will be triumphant," and so on. Virtue of whom? Of humans, the "higher" animal? Or of bacteria? Or of viruses, such as Covid-19? And why would virtue be ultimately triumphant within nature? How can there be a direction toward "virtue," in general, if his idea of natural selection in theory relates instead to local adaptations that depend on very specific local physiological, ecological, geographic, and environmental conditions, which in turn are at least partially related to chance and contingency? The crucial point that needs to be emphasized here, which is almost always omitted in idealized accounts about Darwin, is that these are examples of "scientific facts" that Darwin constructed about biological evolution that are not based on any kind of empirical data or objective observation. When he was in, let us say, the Galapagos, he did not observe tortoises getting "happier and happier," or becoming "more virtuous." Instead, what we see in Darwin's writings is a theme that is recurrent in science, and a central topic within the present book: the profound links between teleology, biases, prejudices, and scientific ideas. Darwin himself, later in life – after publishing the *Origin* and the *Descent* – admitted that he was very "pleased" to be recognized as a teleological writer. For instance, in a 1874 letter to Asa Gray – who had published in the same year an article for the scientific journal *Nature* that was essentially a tribute to Darwin – he stated: "What you say about Teleology ['let us recognise Darwin's great service to Natural Science in bringing back to it Teleology: so that, instead of Morphology versus Teleology, we shall have Morphology wedded to Teleology'] pleases me especially, and I do not think any one else has ever noticed the point.. I have always said you were the man to hit the nail on the head."

This theological adaptationist tradition continues to be followed by many anatomists nowadays. In a simplified way, adaptationism is a Darwinian view that many anatomical or psychological traits of organisms are evolutionary adaptations. As explained in my 2017a book *Nature driven by organismal behavior*, in the twentieth century many of those that were more Papist than Pope Darwin began to defend more extreme adaptationist views of evolution than those defended by Darwin, for instance, arguing that the *vast majority* of traits are adaptations. Although in the last decades things have changed a bit, such a position continues to be defended by many scholars, particularly in fields such as evolutionary medicine, evolutionary psychology, or behavioral ecology. As we will see in the next chapters, the main problem of such extreme adaptationist views is that they often involve a non-falsifiable – and thus in that sense a quasi-religious – type of circular reasoning in which traits of organisms are seen as "useful adaptation" a priori, as pointed out by Stephen Jay Gould.

Coming back to the teleological way in which both Erasmus and Charles Darwin constructed an "Omniscient Mother Nature" – including *worshipfulness* of and *belief* in Her and related terms such as the *Temple of Nature* – also lead to two other major problems. One is the exaggerated importance of natural selection and, related to this, the exaggerated parallel that Charles made between artificial selection and natural selection, as noted above. This parallel does not truly hold up, scientifically, precisely because artificial selection by humans is often made within a plan and with a defined final goal – for instance, humans want to make tomatoes bigger, or more colorful – while biological evolution in general does not follow such "master-plans" nor such final goals. Even books that tend to idealize Charles Darwin, such as Richards' 1987 book discussed above, do recognize that he often exaggerated his teleological theories and metaphors. For instance, Richards stated: "I think more potent forces led him [Darwin] to 'exaggerate' the importance of natural selection." Actually even Charles recognized that he often fell into the teleological trap of personifying Mother Nature. In the *variorum* edition of the *Origin* – he produced six editions of that book during his lifetime – he wrote: "so again it is difficult to avoid personifying the word Nature; but I mean by Nature, only the aggregate action and product of many natural laws, and by laws the sequence of events as ascertained by us.. with a little familiarity such superficial objections will be forgotten."

The problem is that the personalized and teleological way in which Charles referred to Mother Nature was not forgotten, nor did it become a minor historical point. Instead, as a *main* repercussion of his works, and the way they often became to be idolized, such a teleological way of seeing Mother Nature has been largely exaggerated. Importantly, in Slotten's book *The Heretic in Darwin's Court – The*

Fig. 1.9 Alfred Russel Wallace was a British naturalist, explorer, anthropologist, biologist, illustrator, and a famous geographer that independently conceived the theory of evolution through natural selection, at about the same time that Charles Darwin did

Life of Alfred Russel Wallace, he noted that Wallace (Fig. 1.9) did sometimes criticize the teleological circular reasoning of adaptationists. As an example, in one of the articles that Wallace wrote during his travel to Asia about his observations of orangutans, he stated: "do you mean to assert, then, some of my readers will indignantly ask, that this animal, or any animal, is provided with organs and appendages which serve no material or physical purpose?" Answering to his own question, Wallace noted that "naturalists are too apt to imagine, when they cannot discover, a use for everything in nature." More than this: the other major problem of seeing "design" or "fine-tuning" in nature and assigning it to the action of an omniscient Mother Nature within a suffocating, continuous, oppressive "struggle-for-life," as Darwin sometimes did and many scholars continue to do, is that one is portraying evolution in an erroneous way. This is because data from many fields have unequivocally shown that neutral, or even slightly detrimental, features are very common in nature. As we will see, biological evolution can never truly lead to optimal "designs" because all organisms living today are derived from other organisms, so there are developmental and evolutionary constraints that never allow them to fully adapt to their new environments. That is, they can never become an "optimal design" to the current conditions: at the maximum, they can only turn into "good enough" sketches.

By insistently repeating such factually wrong notions of fine-tuning and optimality in nature, scientists are creating a construction of the natural world that is much more consistent with the factually inaccurate ideas of creationists. If let us say humans are optimal bipeds or whales are optimal aquatic animals, it is much more logical to argue that this is because they were made by God to be so, than saying that they optimally do these things *despite* the fact that their ancestors were doing completely different things for billions of years. However, we do continue to read and hear, for instance in textbooks and documentaries about the human body, and in particular within popular culture, that our human bodies are "perfect machines." As I always say to my medical students when they are learning human anatomy, both anatomically and physiologically our bodies are actually full of features that make *no sense at all*, from an engineering point of view, precisely because our deep evolutionary past. Otherwise, if we were indeed "perfect machines", they would very likely not have a job as physicians when they complete their medical education.

We have been evolving for billions of years, and there are several evolutionary limitations – constraints – that do not allow certain structures or organisms to suddenly adapt in a perfect way to a completely new function or type of life. That is how things are. A book published by my colleagues and me in 2016b, *Learning and Understanding Human Anatomy and Pathology*, provided several examples of such "shortcomings" within our bodies. It is precisely because of such constraints and shortcomings that whales, for instance, have to go up to the surface of the water to take a breath, each and every time they need to breathe, despite the fact they that live in aquatic environments since millions of years ago. Humans and whales are not all optimally "designed" to live in the way they do nowadays: they are simply *good enough* to live in that way. Whales are most of the time under water, and just come up to take a breath from time to time, so it is OK, or *good enough*, for them to do so in the sense that they did not become extinct despite doing so, particularly because they mainly have no predators, except our own species, at times. They are OK because contrary to what Darwin and in particular subsequent Darwinian ultra-adaptationists argued, in the natural world frequently there is no implacable, suffocating "struggle-for-existence" making natural selection to finish off animals that are not perfectly adapted to their environments. This planet has a lot of natural resources – particularly the seas, which fortunately we have destroyed less than the continents, so far – for whales to be able to eat, reproduce, and survive *enough* to not become extinct despite their need to come up to the surface every time they breathe. Contrary to such a "struggle-for-existence" view of nature, this also applies to similar shortcomings – designated as morpho-ecological, morpho-behavioral, or eco-behavioral mismatches in my 2017a book *Evolution Driven by Organismal Behavior* – of a huge number of other organisms.

One fascinating point about Darwin, which is almost never discussed in the literature, is that concerning this topic he also constructed some factually inaccurate theories and oversimplified some evolutionary ideas because he mixed *certain* teleological and religious-like ideas with science *while* trying at the same time to distance himself from *other types* of religious or quasi-religious ideas. As explained in Gruber's *Darwin on Man*:

> The idea of adaptation as such is something which flows all through the thought of Paley, whom Darwin admired so much, and permeated the thinking of everyone studying natural history during Darwin's Cambridge days. [This idea] was the central theme of the argument from Design. The idea of continuity in nature occurs in many places in the history of human thought. *Natura non facit saltum* – nature makes no jumps – was a guiding motto for generations of evolutionists and protoevolutionists. But Darwin encountered it in a sharp and interesting form, posed as an alternative of terrible import: nature makes no jumps, but God does. Therefore, if we want to know whether something that interests us is of natural origin or supernatural, we must ask: did it arise gradually out of that which came before, or suddenly without any evident natural cause?

That is, Darwin insisted that biological evolution had to be mainly gradual – minor changes accumulating slowly in time – in part because in a way he *needed* to defend such a concept in order to show that what he perceived as mainly a "fine-tuning" occurred naturally within the laws of Mother Nature, and not due to God's supernatural powers and sudden creations. This was a brilliant idea of Darwin, and he was very successful at making, repeating, and widely disseminating it. Regarding the part about biological evolution being often a gradual process, he was partially right, although we now know that sometimes evolution can indeed have a "jumpy" pulse, as, for instance, in cases of evolutionary saltation related to sudden and large mutational changes, as we will see below. However, the part about a suffocating "struggle-for-existence" in general leading to a "fine-tuning" is more problematic, as we have seen. In fact, a more factually accurate, as well as more efficient, way to explain to the general public that evolution has nothing to do with supernatural powers was that astutely used by *Stephen Jay Gould* in the last decades of the twentieth century. He emphasized, using a plethora of case studies of evolutionary constraints and mismatches from several different biological organisms, that most animals actually have several evolutionary shortcomings that would make no sense at all if they were made by an omniscient God for their current life ways. If God created whales as they are now, to live in water, why would He make them with lungs similar to those of terrestrial mammals and oblige them to go to the surface to take a gulp of air, while creating most other aquatic organisms in a way in which they can breathe in water, as fish do?

Moreover, by focusing instead on how the natural world is actually often *not* finely tuned, Gould did not *need* to defend the idea that nature never or almost never "makes no jumps," in order to refute the intervention of supernatural powers, as Darwin defended. In fact, Gould and his colleagues argued that a detailed analysis of the fossil record shows that major evolutionary changes sometimes occur slowly, while in other occasions they occur faster within geological time. In particular, they argued that when certain new forms of life appear in evolution, their population will tend to become stable and thus show little evolutionary change for most of their geological history. However, in some of these cases, evolutionary and developmental constraints can be broken, leading to new, relatively fast evolutionary changes – a concept named punctuated equilibrium. Related to this concept, we now know, from data obtained from fossils and from genetic studies, that for instance small changes in certain regions of DNA – such as a homeobox genes – can indeed lead to relatively rapid and drastic saltational evolutionary changes, as noted above. In most cases, such drastic new forms of life would be selected against and thus would not be viable evolutionarily, as wisely pointed out by Darwin. But on some occasions *they do succeed*, contrary to Darwin's motto *Natura non facit saltum*. That is, such cases – designated as "*hopeful monsters*" within the biological literature – are indeed *good enough* to initially survive within their environments, and then eventually even to thrive, as it happened with many fascinating hopeful monsters living with us right now, such as chameleons (Fig. 2.8). I discussed this topic and provided several other examples of such thriving "hopeful monsters," in the book *Evolution Driven by Organismal Behavior*. We will come back to this issue in Chap. 2.

Fascinatingly, when I was doing research for the present book, I found interesting – historically and scientifically – links and differences between this concept of "hopeful monsters" – which in a way contradicts some of Darwin's general ideas about the natural world – and certain specific ideas of Darwin regarding natural selection. In a sense, we can designate the latter as related to a concept that could be somewhat defined as "hopeful variants." The discussion of these links, which to my knowledge has been mostly neglected in the literature, takes us back to a critical point: contrary to what is often suggested in many accounts idolizing Darwin, the teleological notion of "Mother Nature," the concept of biological evolution – back then often called transmutation – and even the notion of natural selection were all in the air well before Darwin was born. As noted by Gruber:

> As a conservative force in nature, the idea of natural selection had a long history before Darwin, both as "*nature's broom*" and as "*nature's policeman*." It was long known that deviants from a species norm – sports, or monsters, or mutants, depending on the historical context – tend to be less fit than their more typical brothers and sisters. In the struggle for existence, natural selection, "nature's broom," operates to remove these weaklings and ugly ducklings. It was also recognized that there exists a balance of nature, and that various natural mechanisms of population control, "nature's police," operate to maintain it. Predating Malthus by some forty years, Linnaeus had emphasized these points and had seen that man was part of the balance of nature.
>
> He wrote, in his essay on the *polity of nature*: "I know not by what intervention of nature or by what law man's numbers are kept within fitting bounds.. it is, however, true that the most contagious diseases usually rage to a greater degree in thickly populated regions, and I am inclined to think that war occurs where there is the greatest superfluity of people.. at least it would seem that, where the population increases too much, concord and the necessities of life decrease, and envy and malignancy toward neighbors abound.. *thus it is a war of all against all*". All that Darwin had to do was to see that, although most mutants are at a disadvantage, some few are better adapted to their environment, and that over the long haul of geological time the perpetual struggle for existence permits these few to leave their marks on the course of evolution. That was "all" the change Darwin had to make. As we inspect the numerous appearances and disappearances of the idea of natural selection in Darwin's notebooks, we can see that his problem was not so much to discover the idea as to discover its significance for evolutionary theory – to rescue it from its concealment in a variety of hiding places.

There are three crucial points in this short excerpt. First, the highly teleological characterization of nature – "the polity of nature" – and of natural selection – nature's "broom" and "policeman" predated – was very similar to the teleological portrayals subsequently often used by Erasmus and Charles Darwin. Second, this passage highlights Charles' astuteness: while he was deeply influenced by such ideas that were already in the air and had been discussed before his birth by his predecessors, he could connect them in a way in which almost nobody else – except Wallace – could, back then. That is, based on his painstaking, brilliant observations of the natural world and in particular of biological variations, he saw a "positive," "creative" role for natural selection, instead of seeing it merely as a "negative" "broom" that put new deviants from a species norm in the garbage, discarding them from evolution. In other words, in some cases, if those new variants happened to be more "fit" within certain specific local conditions, the "broom" would select them and slowly throw the remaining, non-deviant ones to the garbage. In this sense, Gruber's characterization of Darwin's remarkable ideas was a bit misleading because Darwin's brilliant idea did not concern "hopeful monsters/mutants" – as we have seen. Darwin in general did not like the idea of evolutionary "jumps." Darwin's astuteness rather concerned the fact that he gave a huge importance to what I defined above as "hopeful variants," within his view of evolution. By selecting some such "hopeful variants," natural selection ultimately would gradually and mostly slowly – according to his ideas – lead to new forms of life. In this way, he also distinguished himself from previous authors that would exclusively recur to Lamarck's "use-disuse idea" – for instance, that giraffes could, over a long time of straining to reach high branches, develop an elongated neck – or to vitalistic ideas – that is, that cells, or tissues, had a mysterious "vital force" – to explain biological evolution. Darwin did partially accept Lamarck's "use-disuse" idea, but he had the wisdom to create an additional and *very different way* – that was non-vitalistic and did not invoke supernatural Gods – to explain the creation of new life forms: *the natural selection of such "hopeful variants."*

The third critical point about the above passage is that it shows us that the "warlike" notion of "struggle-for-existence" of Malthus that so deeply influenced Darwin's ideas was in the air much before Malthus' wrote about it, being defended by authors such as the biologist and physician Carl Linnaeus (1707–1778), who defined this "war" as a "war of all against all." It is however important to point out that, for reasons that will be discussed below, Darwin used this "struggle-for-existence" notion in a more exaggerated way than did Linnaeus and many other biologists, to a point in which this notion became one of the central concepts of Darwin's books, as recognized by Darwin. As an aside, in his autobiography Darwin stated: "I may here also confess that as a little boy I was much given to inventing deliberate falsehoods, and this was always done for the sake of causing excitement." This is just an anecdote, obviously, but it is interesting that, as we shall see later, in certain cases Darwin did seem to play such a "exaggeration card" to render his evolutionary theories more plausible, or dramatic, or sometimes to just make them more appealing or readable to the greater public. Something that many other scholars have done, and continue to do, when they want to reach the broader public. Indeed, one of the few points in which both the countless scholars that idolize Darwin and the creationists that demonize him tend to agree is that Darwin's books were particularly effective at causing excitement, including within the general public, through the use of appealing metaphors and, in certain cases, of such hyperboles.

Being an extremely effective science communicator is surely one of the strengths of Darwin. However, when such type of metaphors and hyperboles concerned sentences such as the "Preservation of Favoured Races" – which was the original subtitle of his most famous *On the Origin of Species* book – or ethnocentric sentences about savages being morally inferior or catchy sexist terms to characterize women such as their supposed sexual passivity or coyness, these can be easily used by racist or sexist leaders to support and disseminate their ideas and agendas. That is, to convince the laypeople, leaders could now easily say something like: "look, Darwin, that scientist that you all know, considered the greatest biologist ever, provided incontestable scientific facts in his books supporting the (sexist and racist) things I am telling you." That women are inferior, or that non-Europeans are primitive or immoral. Or that there is a suffocating, eternal "struggle-for-existence" that will ultimately lead to the natural extinction or massacre of some "races" and to the domination of the more "fit" and "smart" – the Westerns, or course. They could say those things, and many of them did say them, and used them to justify and promote their agendas, as we will see throughout this book. This combination of being an excellent disseminator and at the same time defending and propagating ethnocentric and sexist Victorian ideas as if they were scientific facts is one of the most problematic aspects of Darwin's writings about human evolution precisely because of their broader societal implications. That is, many other scientists did defend such ideas back then, but they did not have the same societal impact because they were not so good as science disseminators or because they did not have the privileged power of place that Darwin did. On the other hand, the minority of scientists that did not share such ideas did not have the same skills as science disseminators or, either because they lacked such as power of place or precisely because the things they were saying were not convenient to the Victorian' elites, their message was not further promoted by those elites anyway, so they did not have the same societal impact. As noted above, while *Darwinism* and *Social Darwinism* are terms known by a huge number of people, including nonbiologists such as economists, sociologists, psychologists, and even a significant portion of the broader public, the same cannot be said about *Wallacism or Social Wallacism.* In fact, the very existence and frequent use, until the present day, of the term *Social Darwinism* – which we will discuss in detail below – in politics and fields such as economics and sociology clearly emphasizes that Darwin's ideas did have and continue to have a huge impact in our societies, at numerous levels.

As we will see, one of the darkest uses of some of Darwin's ideas and metaphors concerns their employment in the development of eugenic ideas, which in a simplified way were related to aims to increase the "genetic quality" of a human group or of our whole species. Many of the scholars and politicians defending these ideas did not necessarily defend that one should protect "superior" people or "races" or exclude those judged to be inferior. But many did, including in England where eugenics began to be particularly in vogue in the second half of the twentieth century, and globally, with the use of some of the most extreme eugenic ideas within the Nazi ideology that lead to the second world war. The Nazi plans to genocide groups such as Jews and gypsies are one of the darkest illustrations of the danger of using, promoting, and disseminating factually inaccurate ethnocentric, racist, and eugenic ideas (Fig. 3.1).

For what we know about Darwin, it is almost sure that he would not support, well on the contrary, such distorted extreme Nazi ideas and that he would be extremely sad and outraged to know that various Nazi leaders used his name, works, or ideas to justify or promote them. So, when certain groups that try to demonize Darwin, such as creationists, say things like "Darwin led to Nazism" or even "Darwin was a

Nazi," they are being not only extremely biased but literally doing historically factually inaccurate statements. At the maximum, what one can say, factually, is that some of Darwin's ideas, metaphors, and hyperboles unfortunately provided easy ammunition to be subsequently used by groups such as white supremacists, including some leaders of the USA based *Ku Klux Klan* or of the Nazi party in Germany, as we shall see. One example is the fact that Darwin did defend, as if it was a scientific fact, the idea that most "savage" non-European groups, such as Australian aborigines, were doomed to extinction or extermination as this was the "natural" outcome of how Mother Nature and its natural selection "broom" operated. So, if Mother Nature does it, there is surely nothing wrong in doing so, argued numerous eugenicists and racist politicians and colonialists. Many of them directly quoted Darwin's works to justify the sterilization or mass killings or even genocides of "others" by arguing that they were just "helping" to expedite Mother Nature's task, who as stated by Darwin in the *Origin*, is always "rejecting that which is bad, preserving and adding up all that is good."

Among eugenicists, Francis Galton, one of the most prominent ones in England during the last decades of the second half of the nineteenth century, frequently cited Darwin and his evolutionary ideas. Galton's life is particularly revealing not only because he was both a cousin of Charles Darwin and a prominent eugenicist but also because it had a lot to do with the David *versus* Goliath topic discussed above. Moreover, his life is an additional powerful example of how such topics can be strongly linked with societal contingencies, privileges, and biases and how these links can, in turn, further influence science and society, in a circular, self-reinforcing way. Some of these points were astutely noted in Richards' 1987 book about *Darwin*:

> Galton had his own familial debt to acknowledge. He avowed that scales fell from his eyes when he read the *Origin of Species*. At once, he related in his autobiography, he brushed aside the occluding film of religious belief, and began to pursue exact science. His studies of heredity, which date from the mid-1860s, undoubtedly had the stimulus of Darwin's work. But they likely had a more personal motive as well. Like Darwin, Galton had been chosen by his father for a medical career. After a hospital apprenticeship at age sixteen, Francis's desire to study mathematics persuaded Samuel Tertius Galton to send his son to Cambridge. The boy had shown promise, even if his abilities had been somewhat magnified by his eight older brothers and sisters.
>
> At university, Galton enjoyed the company of other well-financed young men, and stood a bit in awe of those who tested well in mathematics. At the beginning of his third year, he started preparing his father for the possibility that he might not take his exam for honors.. [he wrote letters that] meant to cushion the blow to his father and to his own.. As Galton's heart palpitations and giddiness grew worse, his father took him out of university for a long vacation. He finally received an A.B. (finishing forty-fourth in the medical list and third in the mathematical) and continued on in medicine. But when his father died leaving him a nice inheritance, Galton, again in the footsteps of his cousin, gave up medicine for travel. For one who had failed to achieve eminence at university, the doctrine that genius had biological roots and could not be earned in schoolboy labor must have had an appeal.

In other words, by being a cousin of "Pope Darwin" and therefore being pressured by the comparisons with his cousin and the fact that he did not "achieve eminence at university," Galton had particularly strong motivations to follow such an "appealing" scientific idea and agenda. By doing so, he could prove that he was still a genius – the "David" Galton wanted to become an immortal giant, and so he did. But was this ascendence to immortality solely related to merit, or to a true pursuit of "truth," or to his genius? Or was it related at least partially also to the *power of place*, in the sense that he, as Charles, were privileged Victorian males from well-off and well-connected families within a country and epoch in which class played an enormous role within the achievement of success? Perhaps some of the well-offs that start as "Davids" and then became "immortal" giants do so not necessarily as a *reaction* to the difficulties or problems they faced during youth, as the story goes. Perhaps in several cases they become giants instead *despite* of them simply because these difficulties or problems were not big enough to jeopardize the huge societal privileges and advantages they had from the very beginning of their lives, which gave them a huge head start even if they would not "achieve eminence at university." Be that as it may, within the context of the present book what is critical is to see how the mental process that led Galton to construct inaccurate "facts" to "scientifically" support his agenda and the notion that "genius had biological roots" had so much to do instead with societal biases and contingences related to *where* and *how* one is born, to the *power of place*. As put by Richards:

> As his wife's diary records, Galton began a statistical analysis of the hereditary transmission of human mental ability in 1864. The aim of his essay "*Hereditary Talent and Character*" published the following year, was to show that conspicuous mental talent – for science, mathematics, literature, painting, and law – ran in families, that mind and character were biologically transmitted. To demonstrate this, Galton searched biographical dictionaries of distinguished people, the roster of lord chancellors, the list of senior classics at Cambridge, and the roll of past presidents of the British Association. He discovered that the men recorded there tended, beyond the average in the population, to have close relatives who were also of noted intellect. His figures implied "that when a parent has achieved great eminence, his son will be placed in a more favorable position for advancement, than if he had been the son of an ordinary person". Galton pursued this discovery in subsequent works.. [which] as he admitted, all merely substantiated and elaborated his original essay..
>
> But even with the laws of heredity still obscure, Galton's analysis suggested that hope for continued progress of human reason and morals was well founded. "What an extraordinary effect might be produced on our race," he mused in Platonic reverie, "if its object was to unite in marriage those who possessed the finest and most suitable nature, mental, moral, and physical!" But such a consummation, Galton feared, would not come about unless men took evolution into their own hands. For the melancholy fact stood plain: not only did men propagate their virtues, but they transmitted their vices as well – "craving for drink, or for gambling, strong sexual passion, a proclivity to pauperism, to crimes of violence, and to crimes of fraud". As Galton surveyed his own time, he perceived that civilization actually had a retarding effect on the natural selection of the best. The poverty of men of good character is, he declared, "more adverse to early marriages than is natural bad temper, or inferiority of intellect". The best hope for the unabated advance of human mind was enlightened social policy that encouraged early marriage of men and women of talent and set obstacles to the egregious propagation of intellectual and moral paupers..

It is indeed truly fascinating – and deeply desolating, at the same time – to see how societal and scientific biases are so profoundly connected. Galton reported that "when a parent has achieved great eminence, his son will be placed in a more favorable position for advancement, than if he had been the son of an ordinary person."

So, how did he interpret this, "scientifically"? By stating that this meant that "mental talent ran in families." What a coincidence that a well-off Victorian white male with such an agenda got to such a conclusion that both supported his agenda and justified the privileges given to him by his power of place. Would it not be more likely that this meant that within the highly hierarchical Victorian society it was much more difficult for poor people to achieve societal and scientific prominence even if they had "mental talent," than it was for those that already had such a head start with some many advantages? After all, at the time he was writing his works there was a particularly appropriate case study in England that could give him a good hint to contradict his statements: the royal family. One becomes part of the royal family not because of any specific "mental talent" or merit, but simply because one is a descendant of, or is married with, people that just happen to be part of the royal family.

Did Galton cite, within his mental exercise, this case study and thousands of other similar ones that were occurring in front of his eyes, at that time in his own country? No: he merely tries to see what he *wanted to see*, even if that was factually wrong as shown by all those case studies. For him, there was never any doubt, even before he gathered or constructed his "facts": the rich should be praised, and the poor should be blamed, so the fact that the latter often do not achieve societal prominence has to be their fault, not of a highly hierarchical society. Accordingly, this has to mean that the poor just lack the mental powers of the rich and are moreover inclined to "craving for drink, or for gambling, strong sexual passion, a proclivity to pauperism, to crimes of violence, and to crimes of fraud." So, the million-dollar question is: by mainly writing such ideas plagued by biases and factual inaccuracies, how did Galton reach "immortality"? Well, mainly because of the combination between his power of place and the fact that the majority of the prominent scientists and leaders in the England back then were also privileged "white" males that had a huge motivation to blindly accept, promote, and use such ideas.

What is particularly disturbing about all this is that such factually inaccurate ideas promoted by privileged scientists or well-off families with an agenda became to be not only defended by the privileged but also interiorized by the underprivileged due to indoctrination and the cult of power and inequalities, to this very day. Take the *American Dream* as an example. Many truly *believe and publicly defend the idea* that everybody can become a CEO, and that if you do not, it is exclusively your fault, even if you were born in absolute poverty. Such an erroneous narrative omits the fact that no country would be able to be composed only by CEOs. Others have to work for them, and the vast majority of those that will work for the CEOs are precisely those that come from underprivileged families or communities. The system was built in that way, and despite some very noble changes that have been occurring after the Victorian era in this regard, numerous empirical studies show that social mobility – which can be defined as the movement in time of families, individuals, or other social units between positions of varying advantage in the system of social stratification of a certain society – is still quite limited in England and the USA.

In fact, narratives such as those of the *American Dream* play a crucial role within indoctrination, the safeguarding of the *status quo*, and the legitimization of privilege and of the privileged. Hollywood movies love to show stories about the *American Dream*, particularly those that are "based on true facts." But they often omit the fact that while one out of a million or more of those born in absolute poverty within the USA can become a multibillionaire, all the others do not. That is, this does not represent the rule, but rather the *exception that confirms the rule*, as shown over and over by empirical social mobility studies. These societal fairytales that are repeated over and over in Hollywood movies and TV shows are however highly appealing not only to the privileged – who can *believe* that their privileges are completely related to merit and deserved – but also to those from the middle class – as they can argue that they are better off and thus have more merits than the poor –, and even to the deprived, who can *dream* that they can become CEOs one day, that this just depends on them.

Moreover, such a narrative about the *American Dream* is also particularly appealing to those that want to come to the USA from other countries, as well as to those that are already here and use *American Dream* or *America First* narratives to exclude "others," "aliens," and "foreigners." This is exemplified by the huge controls done, or walls built, in the southern borders of the USA – or by the very tough immigration laws of Europe. If we let "them" come, "they" will invade us and destroy *our* way of living with their "vices," exactly as Galton suggested, more than a century ago. Darwin himself wrote similar ideas, in the *Descent*, sometimes in a particularly harsh way, as we shall see. This was recognized by authors such as Richards, who wrote that "even Darwin thought that those favored by nature – the intellectually superior, the morally upright (and generally those of the appropriate social class) – were those fit to survive.. if they did not, it was an anomaly, which men should correct, or nature herself eventually would." As he further explained, "none of the scientists who considered the problem of human evolution attempted to separate nature from the ghost of the since-departed Deity – and Darwin no more than the rest.. the traits they valued most – intellect and moral rectitude – had been ascribed by hallowed tradition to a beneficent God." Hence, "if nature (or nature's God) granted these favors, they could not fail to be valuable properties.. the unfit gained an advantage, it was supposed, only because highly 'artificial' schemes (e.g., the *Poor Laws* [made to help the underprivileged]) had fouled the well-designed machinery of nature." As brilliantly put by Richards, "by our contemporary lights, of course, if the lower-class vandals of the population were trampling the tender buds of the Victorian best and brightest, it might be because they were actually superior in what mattered most in classic 'Darwinian' terms – survival."

The logical incongruities of such eugenicist, racist, or exclusionary fairytales always fascinate me: within such "war-like" narratives such as Darwin's evolutionary "struggle-for-existence," the ones that deem themselves to be "biologically superior," such as Galton, are often panicking because they say they would lose the "fight" against the "others" unless something is done about it. Hitler got inspiration from such confusions, incongruities, and factual inaccuracies, arguing, for instance, that if Arians let Jews, or gypsies, or people with incapacities thrive, they would be outcompeted by them, so Arians needed to do something in order to prevent this to happen. But were Arians not supposed to be "biologically superior," after all? So why would they not prevail, naturally, against the "others," without needing to do something to "prevent" the opposite to happen? The reality is that such narratives are so unscientific and illogical that scholars such as Galton and political leaders such as white supremacists often did not even cared about these incongruities, or did absurd mental gymnastics to justify them because the point was never about logics at all, but simply about the preservation of the *status quo* and of their privileges.

In this sense, one of the darkest legacies of Darwin's works, particularly of the exaggerated war-like "struggle-for-existence" notion that he defended in his *Origin* and then applied to his ideas about human evolution in the *Descent*, is that many eugenicists were indeed strongly influenced by these notions and ideas. In fact, one of the major taboos about Darwin, which makes most historians and scholars feel particularly uncomfortable – including myself, no doubt about that – is that apart from writing works that influenced many eugenicists, Darwin also explicitly *agreed*, and directly cited, many of the ideas of his cousin Galton in the *Descent*. That is, this was clearly not a case in which some "good" ideas of Darwin were "badly" used subsequently by scholars: instead, it was an interactive process in which Darwin's works both influenced and were influenced by, and overtly supported, eugenicists such as Galton. Darwin's *Descent* includes some passages that are disturbingly similar to not only many passages of Galton's works, but also to ideas subsequently defended by leaders such as Hitler, for instance, in his *Mein Kampf*. For instance, Darwin wrote:

> In regard to the moral qualities, some elimination of the worst dispositions is always in progress even in the most civilised nations. Malefactors are executed, or imprisoned for long periods, so that they cannot freely transmit their bad qualities. Melancholic and insane persons are confined, or commit suicide. Violent and quarrelsome men often come to a bloody end. Restless men who will not follow any steady occupation—and this relic of barbarism is a great check to civilisation—emigrate to newly-settled countries, where they prove useful pioneers. Intemperance is so highly destructive, that the expectation of life of the intemperate, at the age, for instance, of thirty, is only 13·8 years; whilst for the rural labourers of England at the same age it is 40·59 years.
>
> Profligate women bear few children, and profligate men rarely marry; both suffer from disease. In the breeding of domestic animals, the elimination of those individuals, though few in number, which are in any marked manner inferior, is by no means an unimportant element towards success. This especially holds good with injurious characters which tend to reappear through reversion, such as blackness in sheep; and with mankind some of the worst dispositions, which occasionally without any assignable cause make their appearance in families, may perhaps be reversions to a savage state, from which we are not removed by very many generations. This view seems indeed recognised in the common expression that such men are the black sheep of the family.
>
> A most important obstacle in civilised countries to an increase in the number of men of a superior class has been strongly urged by Mr. Greg and Mr. Galton, namely, the fact that the very poor and reckless, who are often degraded by vice, almost invariably marry early, whilst the careful and frugal, who are generally otherwise virtuous, marry late in life, so that they may be able to support themselves and their children in comfort. Those who marry early produce within a given period not only a greater number of generations, but, as shewn by Dr. Duncan, they produce many more children. The children, moreover, that are born by mothers during the prime of life are heavier and larger, and therefore probably more vigorous, than those born at other periods.

> Thus the reckless, degraded, and often vicious members of society, tend to increase at a quicker rate than the provident and generally virtuous members. Or as Mr. Greg puts the case: "The careless, squalid, unaspiring Irishman multiplies like rabbits: the frugal, foreseeing, self-respecting, ambitious Scot, stern in his morality, spiritual in his faith, sagacious and disciplined in his intelligence, passes his best years in struggle and in celibacy, marries late, and leaves few behind him. Given a land originally peopled by a thousand Saxons and a thousand Celts—and in a dozen generations five-sixths of the population would be Celts, but five-sixths of the property, of the power, of the intellect, would belong to the one-sixth of Saxons that remained. In the eternal 'struggle for existence,' it would be the inferior and less favoured race that had prevailed—and prevailed by virtue not of its good qualities but of its faults."

Although it is difficult to choose which sentence is most unfortunate, factually inaccurate, or tragic, within this passage, many of them – such as "the careless, squalid, uninspiring Irishman multiplies like rabbits," which Darwin directly quotes *and* supports – do disconcertingly resemble the type of tone often used by white supremacists or fascists. We will discuss this specific topic in detail in Chap. 3. For now, as this topic and such notions about "savage states" are directly related to the way in which Darwin often described the indigenous populations that he encountered in his Beagle travels, let us go back in time again and join Darwin in one of the most famous boats and travels ever done, using his autobiography as our guide.

The Beagle, Slavery, Progress, and Racism

> *Out beyond ideas of wrongdoing and rightdoing there is a field, I'll meet you there.* (Rumi)

> Early in my school days a boy had a copy of the "*Wonders of the World*," which I often read, and disputed with other boys about the veracity of some of the statements; and I believe that this book first gave me a wish to travel in remote countries, which was ultimately fulfilled by the voyage of the "Beagle".. [Many years later] on returning home from my short geological tour in North Wales, I found a letter from Henslow, informing me that Captain Fitz-Roy was willing to give up part of his own cabin to any young man who would volunteer to go with him without pay as naturalist to the Voyage of the "Beagle".. I was instantly eager to accept the offer, but my father strongly objected, adding the words, fortunate for me, "If you can find any man of common sense who advises you to go I will give my consent." So I wrote that evening and refused the offer. On the next morning I went to Maer to be ready for September 1st, and, whilst out shooting, my uncle (Josiah Wedgwood.) sent for me, offering to drive me over to Shrewsbury and talk with my father, as my uncle thought it would be wise in me to accept the offer.
>
> My father always maintained that he was one of the most sensible men in the world, and he at once consented in the kindest manner. I had been rather extravagant at Cambridge, and to console my father, said, "that I should be deuced clever to spend more than my allowance whilst on board the 'Beagle';" but he answered with a smile, "But they tell me you are very clever".. Fitz-Roy's character was a singular one, with very many noble features: he was devoted to his duty, generous to a fault, bold, determined, and indomitably energetic, and an ardent friend to all under his sway. He would undertake any sort of trouble to assist those whom he thought deserved assistance. He was a handsome man, strikingly like a gentleman, with highly courteous manners.

The latter sentence is interesting, within the context of Darwin's autobiography and some of the sexist ideas of his books. This is because, as astutely pointed out by many feminists, in his writings Darwin used adjectives such as "handsome," "strikingly," or "highly" much more frequently to refer to males than to females, both for humans and for other species. This topic was discussed in detail in books such as Evans' *Darwin and Women* and Hamlin's *From Eve to Evolution*, which we will discuss later. Terms such as "strikingly like a gentleman," cavemen, and so on are part of a discourse created by men to basically take out agency from women. For instance, women were – and continue to be – almost never included in portraits about human evolution, such as those concerning the invention of fire (Fig. 1.10), production of stone tools (Fig. 1.11), or cave paintings (Fig. 1.12). After all, there were only *cavemen* back then, right? Contrarily to his brilliance as a biologist, the fact that in societal terms Darwin tended to be conservative and sexist lead him to neglect otherwise evident facts that would contradict the Victorian *status quo*.

The major exception to this rule concerns slavery. Not surprisingly, this is precisely the major societal issue that most accounts idealizing Darwin tend to focus on, often in a highly romanticized and individualized way, not considering its contextual framework. As noted above, regarding Darwin's public *and* scientific support of several sexist and ethnocentric ideas and of colonialism, British imperialism, and social hierarchies and inequities, scholars often blame the Victorian society – he

Fig. 1.10 "*Homo erectus* and fire"

Fig. 1.11 "The earliest Manufacture and Polishing of Flints," by Emile Bayard for Louis Figuier's *L'homme primitif*, 1870

Fig. 1.12 "Cro-Magnon artists," by Charles Knight, 1924

was just a product of his time, case closed. But regarding the "good" part, his opposition to slavery, then many accounts attribute this exclusively to his merit and sagacity, a product of his mind alone. Within such an idolized narrative, such accounts often portray this as a personal fight of Darwin, that he *alone*, in his Beagle travels, suddenly realized how wrong slavery was, and since then began to fight against it passionately. However, as almost always, reality lies in between such extreme narratives – or, as famously stated by Rumi, "Out beyond ideas of wrongdoing and rightdoing there is a field, I'll meet you there."

That is, his ideas about colonialism, gender differences, imperialism, and so on are of course highly influenced by his society, but he could have deconstructed them, at least partially, as others did at that time; regarding his stance against slavery, it was highly influenced by the antislavery ideas of his family, but he does have the merit to have *chosen* to attack slavery publicly and vocally as he did. He was not a lone fighter against slavery: apart from his own family, a huge number of people were already against slavery in England at that time. In fact, the antislavery movement was so prominent at that time that it led to the abolition of slavery in England in 1833, through the *Slavery Abolition Act* – an act of Parliament that abolished slavery in most British colonies. That is, this happened when Darwin was only 25 years old, traveling in the Beagle – and 6 years before he publish the first book in which he publicly wrote against slavery, the *Voyage of the Beagle*. Many decades earlier, before he was even born, his well-off family was already very vocal against slavery. So, chronologically, one can say that his antislavery ideas basically followed the tradition of his family and were in agreement with a position that was becoming very prominent within the cream of Victorian society, to which his family belonged. Still, he clearly had the merit of not only defending those ideas publicly but also of being extremely empathetic toward the slaves that he met in his travels, that is something very noble that should be highly praised about Darwin.

If one analyzes these topics in a broader perspective, one can say that Darwin carried both his antislavery ideas as well as his sexist, ethnocentric and imperialist Victorian ideas to the Beagle, and he *chose* to actively support all of them "scientifically," for better or worse. The fact that Darwin mainly confirmed and "scientifically" justified all these preconceived societal ideas that he carried to the Beagle further contributed to some of the major logical incongruities that plagued his works. For instance, concerning his noble stance against slavery, he justified this scientifically by asserting that no animal or human is evolved to be a "natural slave." However, his opposition to slavery and forced labor clearly conflicted with his support – both in his private writings and his scientific books – for colonialism, British imperialism, and social hierarchies because at least some stages of colonialism and imperialism almost always involve some type of forced work by the "others" that are colonized. In order to further examine this issue, let us go back to Darwin's autobiography and see what he wrote about his Beagle voyage and his opposition to slavery:

> Fitz-Roy's temper was a most unfortunate one. It was usually worst in the early morning, and with his eagle eye he could generally detect something amiss about the ship, and was then unsparing in his blame. He was very kind to me, but was a man very difficult to live with on the intimate terms which necessarily followed from our messing by ourselves in the same cabin. We had several quarrels; for instance, early in the voyage at Bahia, in Brazil, he defended and praised slavery, which I abominated, and told me that he had just visited a great slave-owner, who had called up many of his slaves and asked them whether they were happy, and whether they wished to be free, and all answered "No." I then asked him, perhaps with a sneer, whether he thought that the answer of slaves in the presence of their master was worth anything? This made him excessively angry, and he said that as I doubted his word we could not live any longer together.
>
> I thought that I should have been compelled to leave the ship; but as soon as the news spread, which it did quickly, as the captain sent for the first lieutenant to assuage his anger by abusing me, I was deeply gratified by receiving an invitation from all the gun-room officers to mess with them. But after a few hours Fitz-Roy showed his usual magnanimity by sending an officer to me with an apology and a request that I would continue to live with him.. The voyage of the "Beagle" has been by far the most important event in my life, and has determined my whole career; yet it depended on so small a circumstance as my uncle offering to drive me thirty miles to Shrewsbury, which few uncles would have done, and on such a trifle as the shape of my nose. I have always felt that I owe to the voyage the first real training or education of my mind; I was led to attend closely to several branches of natural history, and thus my powers of observation were improved, though they were always fairly developed.

This excerpt is related to two very interesting aspects about Darwin's life and works. One concerns the contingencies and randomness of life in general, as well as of science. Contrary to what is often portrayed in scientific and historical accounts that idealize scientists, including Darwin, scientific success and recognition are not merely – and sometimes even not mainly – related to scientific merit and the "search for the truth." Darwin not only was on the right place at the right time but was also part of a privileged, well-off, well-connected Victorian family – *the power of place*, as attested by his statement that his Beagle voyage "depended on so small a circumstance as my uncle offering to drive me 30 miles to Shrewsbury." Indeed, the HMS Beagle was not a boat for poor curious adventurers: it was a Cherokee-class 10-gun brig-sloop of the Royal Navy that is often said to have taken part in celebrations of the coronation of King George IV. The boat took part in three major survey expeditions led by a Vice-Admiral – Robert FitzRoy. Darwin's voyages were part of the second of those second expeditions.

The second interesting aspect concerns the logical incongruities of Darwin's ideas and writings and how they were related to Darwin's power of place, societal conservatism and to contingencies, as astutely pointed out in Browne's *Charles Darwin voyaging*. As noted above, the fact that he defended, on the one hand, British imperialism and colonialism and Victorian ethnocentrism and, on the other hand, the abolition of slavery are in a way, at least apparently, logical incongruities. But Bowne argued that they were also, in a sense, actually the two sides of a same coin (see also Box 1.2). That is, she explains that at that time within the cream of Victorian society – including Darwin's family – such anti-slavery ideas were by themselves a defense of English and Victorian superiority because they were mainly a criticism of the slavery done by "other," non-English "white" males. For instance, as seen in Darwin's excerpt above, he harshly criticized slavery in Brazil, a region that was colonized by the Portuguese until it got its independence in 1822. As we will see, when he later traveled to lands colonized by the English, his writings have a completely different tone, often idealizing the English empire and its type of colonialism and the "amazing" towns and farmhouses the English built in those "savage," "dangerous," and "brutal" far-away lands. Such accounts therefore were in line with Victorian narratives that were prominent at that time. For instance, that the "cosmic purpose" of the "superior" Victorian society was to bring "progress" to both the "savages" *and* the continental European countries, by "liberating" them from primitive and barbarian traits, such as slavery. As elegantly put by Browne:

> Landing at Bahia (Salvador).. Darwin was horrified to find himself in a full-fledged slave society. His hackles automatically rose. Portugal was still legally transporting Africans to Brazil, and no steps towards emancipation would be made until 1858. The scene was a "scandal to Christian nations," Darwin declared to Henslow. Slavery inflamed all his most passionately held beliefs about human nature. It was the one social issue guaranteed to upset and annoy him throughout his long life. He had no problems with the idea of self-interested commercial expansion through plantation crops, for example, or the grinding poverty, factory children, and indentured servants to be found at home. Like the majority of people he knew, he bowed to the capitalist ethos of the British ruling classes without a qualm.
>
> But the actual state of slavery agitated powerfully humanitarian instincts in him. He could not think about it, let alone see it for himself, without boiling over in righteous anger. Moreover, his Darwin-Wedgwood family background entitled and prepared him to express strong condemnation. Each of his grandfathers had taken a pronounced abolitionist stand in the 1780s and 1790s, as did his father, brother, and sisters in relation to the nineteenth-century emancipation movement, and he now felt similar crusading emotions coursing in his veins. In fact, the first Josiah Wedgwood and the first Erasmus Darwin were central figures in rousing public opinion in Britain during the crucial anti-slavery years at the end of the eighteenth century, along with family friends.. and other influential Whig campaigners. Wedgwood's design, in blue-and-white china, of a Negro in chains below the famous phrase "*Am I not a man and a brother?*" (which he coined) affirmed all the outraged feelings of the era and was adopted by the Anti-Slavery Society soon after 1787. (It must also be said that the mass production of these cameos was an astute commercial move capitalising on the philanthropic feelings surging through upper- and middle-class Britain).
>
> These stirring words and images were an integral part of Darwin's up bringing. He was more than familiar with the history and achievements of the anti-slavery campaign, and had met some of the main protagonists as a boy at Uncle Jos's house. The milestones that marked the progress of emancipation were to him family events, telescoped and romanticised in Wedgwood and Darwin household legend to make it seem as if the poet and the potter had forced abolition just the day before yesterday. It was in truth a long-drawn-out process, still unfinished in Darwin's time, first introduced to Parliament as a motion to inquire into the transport and sale of slaves as early as 1788. Britain banned trading in its dominions only in 1807.
>
> The effects were not seen until 1811, when Henry Brougham carried a bill that made human cargo a capital offence. Passing the laws did not resolve the problems either. Abolishing British trade merely aggravated the traffic run by other nations, especially since France, Spain, and Portugal delayed taking legal action for years. Nor did abolition make life any easier for existing slaves in British dominions. By 1820 or so, it was plain to stalwarts of the anti-trading movement like William Wilberforce that actual slave-holding should be abolished in British territories overseas. Under his aegis, and whipped up by an

> avalanche of printed material constituting Britain's first experience of mass public propaganda, the reforming zeal which began life as a minority principle among a small band of liberals and evangelicals became a national, pervasive obsession that captured the imagination and energies of millions. Britain's civilising mission was in large part thereafter portrayed as a drive to purge the "uncivilised" world of its pagan, slaving habits.

It is also important to note that while Darwin, as many of his family members before him, should be highly praised for his antislavery stance, in his published books such a noble antislavery stance is often combined with a patronizing way of talking about slaves. For example, in some cases he suggested that slaves had "happy and contented lives" (see also Box 1.2), as he did in his *Voyage of the Beagle*:

> April 13th.—After three days' travelling we arrived at Socego [Brazil], the estate of Senhor Manuel Figuireda, a relation of one of our party. The house was simple, and, though like a barn in form, was well suited to the climate.. During the meals, it was the employment of a man to drive out of the room sundry old hounds, and dozens of little black children, which crawled in together, at every opportunity. As long as the idea of slavery could be banished, there was something exceedingly fascinating in this simple and patriarchal style of living: it was such a perfect retirement and independence from the rest of the world. As soon as any stranger is seen arriving, a large bell is set tolling, and generally some small cannon are fired. The event is thus announced to the rocks and woods, but to nothing else. One morning I walked out an hour before daylight to admire the solemn stillness of the scene; at last, the silence was broken by the morning hymn, raised on high by the whole body of the blacks; and in this manner their daily work is generally begun. On such fazendas as these, I have no doubt the slaves pass happy and contented lives. On Saturday and Sunday they work for themselves, and in this fertile climate the labour of two days is sufficient to support a man and his family for the whole week.

Box 1.2 Darwin, Slavery, Colonialism, and Logical Inconsistencies

Browne's *Charles Darwin Voyaging* provides an interesting discussion on these subjects, which is broader and more profound than the oversimplifications that are often included in most accounts about Darwin:

> Darwin had moral inconsistencies of his own about slavery that were just as glib as the ones FitzRoy let slip in Bahia, and in any case, few nineteenth-century figures escaped the quagmire of stolid British hypocrisy about slavery. Anti-slavery movements had become an irresistible form of cultural imperialism in the early years of the century, full of images of national superiority. Not only that, but the continued persistence of slavery in the United States, South America, Africa, and elsewhere provided the British with an undeniable sense of self-righteousness. Anti-slavery turned into a national talisman which distinguished the good (the British) from the bad (the rest). In the light of Darwin's real abhorrence of slavery, these cultural dissimulations have often been passed over without remark. Yet Darwin saw no incongruity in the wide-scale employment of servants, for example, and was as patronising as the next man about the lowly nature of the uneducated masses in Britain or the indigenous people he met in various parts of the globe.
>
> His father's fortune was built on the backs of entrepreneurial companies that exploited cheap labour – a family business sense which continued unabated in his own later endorsement of joint-stock railway companies. Both grandfathers also belittled the role of human labour in advancing British prosperity. Though

(continued)

Wedgwood's employees were in principle free to come and go, they were in practice tied to his cottages, to his insurance societies, and to his wages. In truth, with more than fifteen thousand people living at the Etruria works at the time of Wedgwood's death in 1795, the site resembled nothing so much as a displaced plantation town with its big house and separate workers' quarters. And Darwin never questioned the legal ties binding FitzRoy's seamen – officers and boys as well as sailors – to naval service. The money-making classes of Britain perpetuated forms of human bondage which seemed to many critics merely a variant of slavery – the new system of slavery that Lord John Russell decried in 1840.

However, everything of an ethical nature taught to Darwin at home was designed to inspire horror at the idea of buying and selling human beings as chattels. His father, his Wedgwood aunts and uncles, and particularly his sisters Caroline, Catherine, and Susan, who were active in charitable organisations throughout their lives, gave him a firm belief in the essential unity of mankind. He hated slavery with all the outraged zeal that a freeborn Englishman could muster. Like countless others, his commitment was buoyed up within a general cultural framework of British heroism and Christian ardour in confronting and suppressing this barbaric custom in hostile regions of the world.. The ephemeral nature of civilisation struck him most forcefully when he compared the three Anglicised Fuegians [transported in the Beagle after spending a few years in England] with members of their original tribes. These people had been transformed into virtual Europeans in less than four years. Jemmy Button polished his shoes and made jokes, Fuegia Basket wore jewellery and an English bonnet.

Their language, faith, and aspirations had altered, as he believed, to reflect an English education, short and basic as it was; and they themselves seemed to feel markedly different from their former compatriots.. When Darwin likened that process to the difference between wild and domesticated animals he consequently employed far more than a simple farmyard metaphor: he saw what he thought was an authentic transformation of personality from aboriginal brutishness to the softer, tamer, more civilised nature of Western humanity, "domestic" in all its senses. Reflections like these were underpinned by his strong commitment to the *idea of progress* -progress, the theme of the age. In all areas of English life, a large proportion of people believed in a universal and inbuilt tendency towards improvement: improvement in manufacturing, in commercial life, in science and art, in technology, and in social relations.

Even those individuals who brought the absence of progress to public notice – as in brutal working conditions, the exploitation of women and children, and the lack of general enfranchisement – were committed to the idea that improvement was possible; and philanthropic groups felt responsible for providing the means of improvement to groups that could not effect it for themselves. This could be seen most obviously in missionary work overseas.. For FitzRoy, and then for Darwin, such a concept of cultural progress became real during the Beagle voyage in the most dramatic fashion imaginable. Jemmy, York, and Fuegia were virtually English. "What a scale of improvement is comprehended between the faculties of a Fuegian savage & a Sir Isaac Newton," wrote Darwin in his diary in amazement.

"To those who have never seen man in his savage state," echoed FitzRoy, it was "one of the most painfully interesting sights to his civilised brother." A voyage such as the Beagle's which took them to remote countries inhabited by races scarcely known in Europe brought the notion of progress home with a vengeance. The ability to change like this, to advance from a state of "savagery" to the world of English manners or, alternatively, to degenerate into a hardened slave-master or berry-

(continued)

picking refugee, was to FitzRoy and Darwin, as to many others in pre-Victorian society, a natural feature of the human condition. Though a vast gulf lay between civilised and uncivilised races, it always remained a possibility that the gulf could be bridged. The possibility of change linked all mankind, according to circumstance, along the arc of culture.

Another critical point that needs to be made is that while Darwin's pro-colonialist, imperialist, and ethnocentric views were similar to those made by many Victorian scholars back then, there is a major difference between Darwin's writings and those of most other Victorian scientists about those topics. This is because Darwin was *one of the very few* Victorian – and European, for that matter – scholars that, mainly due to his *power of place*, was fortunate to travel, see by himself, and directly interact with various non-European indigenous groups around the globe. One thing is to live in the nineteenth century in England in a Victorian bubble and to read everywhere that "savages" were brutish cannibalistic human beings and that Europeans were kind, civilized, and "morally superior" to them, and therefore to *accept* such racist fairytales as dogmas. Another very different thing is that someone such as Darwin, who travelled and could directly observe how such ideas were factually inaccurate, and moreover saw by himself the atrocious atrocities and massacres that Europeans were doing in regions such as South America and Australia, continued to not only *believe* in such fairytales but also to even construct further "facts" to support them.

As put by Desmond and Moore in their 1994 *Darwin* book, Darwin "knew that they [the 'savages'] were being destroyed by the white man's scourges, measles and liquor." But despite knowing about all these atrocities, massacres, and hideous use and abuse of indigenous peoples by Europeans, he did continue to repeat, not only in private letters and notebooks but also in his scientific books, that "white civilization" was vastly "superior" to "savage" brutish "uncivilized habits." This bring us back to another critical aspect that is also often omitted within the countless publications that idolize Darwin: various scholars that lived and did science during Darwin's epoch, particularly some of the very few that were also fortunate to directly interact with non-Europeans in their travels, were able to escape such white male ethnocentric European fairytales and the intellectual chains created by them. As noted above, one of the most emblematic examples was the "heretic" Alfred Russel Wallace (Fig. 1.9) – as explained in Slotten's book *The Heretic*:

Alfred Russel Wallace was born on January 8, 1823, in Usk, Wales, the eighth child of Thomas Vere and Mary Anne Wallace. By a quirk in the registry, his middle name was misspelled "Russel" and was never corrected. The Wallaces, devout Anglicans, were of the middle class, but Thomas Wallace, a lawyer who never practiced his profession, had squandered his inheritance of £500 a year as a result of a series of poor business decisions. To minimize expenses, in 1818 or 1819 he moved his family from Hertford, a town north of London, to Usk, a remote and picturesque market town in southeastern Wales, where rents were low and prices of goods half those of London.. In 1828 Mary Anne Wallace inherited money from her stepmother and the Wallaces returned to Hertford, where Alfred received his only formal education. At the local grammar school, he was taught elementary French but not enough Latin to make sense of Virgil, Cicero, or the other great Roman writers..

In *My Life*, his autobiography, he writes that he acquired more knowledge from his father and older brothers than from his schoolmasters. From his father, who read the plays of Shakespeare and other classics to the family in the evenings, he developed a love of all types of literature. By the age of thirteen, he had read *Tom Jones*, *Don Quixote*, *Paradise Lost*, and the *Inferno* – demonstrating not only a precocious intellect but also a high degree of self-motivation.. But prosperity and tranquility were to be short-lived. In 1832 Alfred's twenty-two-year-old sister, Eliza, died from tuberculosis, a devastating blow to his parents, who already had endured the deaths of three other daughters. Four years later, Thomas Wallace lost the last of his personal savings as a result of bad real-estate investments.

Shortly thereafter, Alfred's maternal uncle, the executor of the Greenell family's estate, declared bankruptcy, having (without anyone's knowledge or consent) borrowed against the Wallace children's small legacies and Mary Anne's modest inheritance to settle his debts. Once again driven to the brink of destitution, Thomas and Mary Anne Wallace and their youngest child, Herbert, moved from their comfortable house in Hertford to a small red-brick cottage in the village of Hoddesdon, a few miles south, to be near Fanny, their only remaining daughter, who at the age of twenty-four had taken a job as a governess to help support the family. For thirteen-year-old Alfred, the reversal in the family fortunes altered the course of his life. His father could no longer afford to pay for his education.

Moreover, there was no room for him in the new cottage. Around Christmas 1836, Thomas removed him from school and packed him off to London to board with John, an apprentice carpenter at a builder's yard, where he was expected to make his own way in the world. The abrupt relocation to London seemed not to be too traumatic, however, and Alfred quickly adjusted. He shared a room and bed with John on Robert Street, off Hampstead Road, a five-minute walk from the workshop of a Mr. Webster, John's employer (and future father-in-law). At first he was not expected to work; he was merely an observer of working-class life. But it was an eye-opening experience. Here in London, the rudiments of his social conscience were awakened. At John's instigation, for the next six months he was exposed to the radical ideas of working-class men at the London Mechanics' Institute, one of several such institutions of higher learning scattered throughout the British Isles and established by forwardthinking entrepreneurs who needed skilled, educated men to manage their factories. Fired by the egalitarian teachings of the Welsh socialist and philanthropist Robert Owen, Alfred rejected the artificial constraints of the English class system, which pigeonholed every citizen and blocked the lower classes from sharing power and wealth with the ruling elite.

Despite the fact that – or perhaps precisely also *because* – Wallace rejected the artificial constraints of the English class system, even when he later became a prominent Victorian scientist, he never truly reached the *power of place* and adulation by the Victorian society that Darwin enjoyed. Further confirming the huge role played by contingencies and biases in science, one can say that in a sense even the very different way in which Darwin and Wallace reacted to, and wrote about, indigenous people was in great part related to the markedly different social position occupied by their families within Victorian society, before they were even born. For example, as extensively documented in Brantlinger's 2003 *Dark Vanishings: Discourse on the Extinction of Primitive Races*, Darwin was a particularly active and prominent defender of the view commonly held in the nineteenth century that "savage" races/groups were doomed to extinction or extermination:

Darwin's notion of the savage condition is distinctly Hobbesian. He agrees with his friend and ally, John Lubbock, whom he quotes in *Descent*: "It is not too much to say that the horrible dread of unknown evil hangs like a thick cloud over savage life, and embitters every pleasure". In the famous conclusion to *Descent*, Darwin suggests that the gulf between savages and civilized humans is almost unbridgeable. "He who has seen a savage in his native land will not feel much shame," he writes, "if forced to acknowledge that the blood of some more humble creature flows in his veins". Darwin's reaction to the Fuegians [indigenous habitants of Patagonia] was one of repulsion and disbelief that such primitive humans could be his "fellow creatures". He puzzled: "One's mind hurries back over past centuries, and then asks, could our progenitors have been men like these?"

And his instinctive response was no: "men, whose very signs and expressions are less intelligible to us than those of domesticated animals; men who do not possess the instinct of those animals, nor yet appear to boast of human reason, or at least of arts consequent upon that reason.. I do not believe it is possible to describe or paint the difference between savage and civilized man.. It is the difference between a wild and tame animal," only greater, "inasmuch as in man there is a greater power of improvement." Influenced by both Malthus and Charles Lyell, Darwin, in *Voyage*, stresses the extinction of species and of primitive races more than he puzzles about their origin. "Certainly," he writes, "no fact in the long history of the world is so startling as the wide and repeated exterminations of its inhabitants". In his *Journal of Researches*, Darwin writes: "All the [Tasmanian] aborigines have been removed to an island in Bass's Straits [sic], so that Van Diemen's Land enjoys the great advantage of being free from a native population" – a sentence he omitted from *Voyage*. But he adds, in language that reappears in *Voyage*: "This most cruel step [of removal of the Tasmanians] seems to have been quite unavoidable, as the only means of stopping a fearful succession of robberies, burnings, and murders, committed by the blacks; and which sooner or later would have ended in their utter destruction."

"Van Diemen's Land enjoys the great advantage of being free from a native population": this is one among many examples of quotes of Darwin that continues to be completely omitted in the innumerable books that idealize him, sometimes to the point of saying that he "cared" a lot and even that he was highly "empathetic" toward non-European indigenous populations. For instance, Strager's 2016 book *A Modest Genius* about Darwin – which, by the way, has a foreword by Sarah Darwin, Charles' great-great-granddaughter – omits such quotes about indigenous people and references about the many factually inaccurate ethnocentric and sexist evolutionary ideas defended in Darwin's books. A main paradox that results from such omissions is that by doing so the authors of such books about Darwin are actually forced to exclude a huge number of passages from various manuscripts written by the very scientist they are idealizing, as noted above. That is why, apart from Darwin's writings, and the publications of many creationists trying to demonize Darwin, one can often only read about such ethnocentric and sexist quotes or ideas of Darwin in books published by historians or scientists that are *not* mainly focused on Darwin, as is precisely the case of Brantlinger's *Dark Vanishings*. For instance, in that book, one can read passages of Darwin's writings that, as put by Brantlinger', "could easily have come from Robert Knox," one of the pioneers of scientific racism in Britain:

Both in *Voyage* and in his later writings, Darwin at times looked forward to the complete triumph of civilized over primitive or "lower" races. "How long will the wretched inhabitants of N.W. Australia go on blinking their eyes without extermination?" Darwin wondered in 1839; and in 1860 he wrote that "the white man is 'improving off the face of the earth' even races nearly his equals". Further, the progress of the world seemed to dictate not just the peaceful transformation of "savagery" into its opposite but, for better or worse, its *violent liquidation*: [Darwin wrote] "looking to the world at no very distant date, what an endless number of the lower races will have been eliminated by the higher civilized races throughout the world". In short, Darwin accepts his predecessors' basic views about the extinction of primitive races without much question. Thus, minus his evolutionary speculations about the fate of "the anthropomorphous apes," this statement from *Descent* could as easily have come from Robert Knox or James Hunt as from Darwin: "at some future period, not very distant as measured by centuries, the civilized races of man will almost certainly exterminate, and replace, the savage races throughout the world.. at the same time the anthropomorphous apes.. will no doubt [also] be exterminated.. the break between man and his nearest allies will then be wider, for it will intervene then be wider, for it will intervene between man in a more civilized state, as we may hope, even than the Caucasian, and some ape as low as a baboon, instead of as now between the negro or Australian and the gorilla"..

In any event, whether consciously or not, Darwin often applies the *active, violent rhetoric of the discourse about racial extinction* to processes that are long-term and nonviolent. Further, as Peter Bowler notes, "the mechanism of the survival of the fittest could be used to justify a more ruthless approach toward conquered peoples, in which extinction was both a symbol and a consequence of inferiority". After all, the [original] subtitle of *The Origin of Species by Means of Natural Selection* reads: *The Preservation of Favored Races in the Struggle for Life*. Although "races" here refers to varieties and species of nonhuman organisms, in Darwinism and Politics (1889) D. G. Ritchie not unfairly claims that Darwin "looks forward to the elimination of the lower races by the higher civilized races throughout the world" and also that social Darwinism is a "scheme of salvation for the elect by the damnation of the vast majority" whether abroad or at home.

Now, compare this excerpt with the title and synopsis – which is, importantly, written for the broader public – of the 2009 book *Darwin's Sacred Cause: Race, Slavery and the Quest for Human Origins* by Desmond and Moore. As noted above, these authors are more nuanced about, and tend to idealize less, Darwin than most other authors of books focuses on Darwin. But they still propagate the idea that he was particularly kind and positive about "others," to the point of stating that he even had a "sacred cause" to preserve the dignity of all humankind against the idea that whites were superior? Did he really do so in his published books? Did he do that when he stated that "evil hangs like a thick cloud over savage life," or when he suggested in the *Descent* that the gulf between savages and civilized humans is almost unbridgeable (see also Box 3.3)? Or was it when he referred to the Fuegians and asked if such "savages" could be his "fellow creatures" (Fig. 1.13)? Perhaps it was when he answered that question with a resounding *no* because "men, whose very signs and expressions are less intelligible to us than those of domesticated animals.. who do not.. yet appear to boast of human reason, or at least of arts consequent upon that reason.. I do not believe it is possible to describe or paint the difference between savage and civilized man.. it is the difference between a wild and tame animal," only greater.

If someone had the quasi-religious "sacred cause" to finish once for all with the "arrogance" of white men, as the synopsis of that book suggests, the things that Darwin publicly wrote, as "scientific" facts, in his own books do not seem the best way to promote such cause. Or did Darwin endorse this cause when he wrote that Fuegians were "the most abject and miserable creatures I anywhere beheld.. the difference between a domesticated and a wild animal is far more strikingly marked in man: in the naked barbarian, with his body coated with paint, whose very gestures, whether they may be peaceable or hostile are unintelligible, with difficulty we see a

Fig. 1.13 The Yaghan people are one of the indigenous groups of the Southern Cone, who are regarded as the southernmost peoples in the world. In the nineteenth century, they were known as Fuegians by the English-speaking world because their traditional territory includes the islands south of Isla Grande de Tierra del Fuego, extending their presence into Cape Horn

fellow-creature." Or when he asserted that he felt "quite a disgust at the very sound of the voices of these miserable savages." Or wrote, after Jemmy Button (Fig. 1.14) – a Fuegian, actually named O'rundel'lico – refused to go back to England, that "they have far too much sense not to see the vast superiority of civilized over uncivilized habits; and yet I am afraid to the latter they must return." There is clearly a mismatch when one compares such idealization narratives about Darwin to what he actually wrote in his books. Another example concerns how such narratives tend to portray Darwin as an "objective" scientist that was above all passionate about observing the facts of nature, while many of the ethnocentric things he wrote about the brutish "habits of savages" have nothing to do with what he truly observed in his Beagle voyage, as noted above. Did Darwin truly see the "savages" doing atrocities that were more horrible than the enslavement, massacres, and other brutalities that the European colonizers were doing to non-Westerns? If not, why would he write the things he wrote, what is the explanation? Could it be that he *perceived* the "savages" – let us say the Fuegians, which were among the groups that most repulsed him – as more "brutish" and "barbaric" than they truly were due to the biases and prejudices that he carried to the Beagle even before he actually interacted with them? Desmond and Moore, in their 1994 *Darwin* book, give us some hints to answer to this question:

Fig. 1.14 Jemmy Button in 1831 and 1834

The scenery was spectacular. They proceeded along the Beagle Channel, tracking a granite ridge of mountains, the backbone of Tierra del Fuego.. While dining on shore half a mile away, they heard a "thundering crash" as a huge mass of ice fell from its face. The impact sent "great rolling waves" racing towards their flotilla. Darwin was quick to act. He and others seized the boats, hauling them to safety just as the first breaker crashed down.. Darwin had acted less out of bravado than fear. Without the boats, he reflected, "how dangerous would our lot have been, surrounded.. by *hostile Savages* & deprived of.. provisions". Some Fuegians were indeed hostile. With courage "like that of a wild beast," they menaced the party's overnight camp, and armed guards were posted. Charles, keeping watch, shivered at his vulnerability in this land. "The quiet of the night is only interrupted by the heavy breathing of the men & the cry of the night birds – the occasional distant bark of a dog reminds one that the Fuegians may be prowling, close to the tents, ready for a fatal rush". Everyone *believed* that they were among cannibals.

Jemmy [a Fuegian], when asked about the practice, had given the expected answer, stating tongue-in-cheek that "in winter they sometimes eat the women". No one questioned his authority, and the party dreaded what they might find back at "the Settlement". Returning along the Channel after nine days, they found that Matthews and the rest had suffered from *little more than bullying and systematic looting*. All their possessions had been divided indiscriminately among the Fuegians. Here was further evidence of *co-operative savagery*, Darwin surmised. "The perfect equality of all the inhabitants will for many years prevent their civilization". Until "some chief" rises who "by his power" can heap up possessions for himself, "there must be an end to all hopes of bettering their condition".. But "they have far too much sense not to see the vast superiority of civilized over uncivilized habits".. [therefore] eventually the example of their prosperity [of the three Fuegians that previously came from England in the Beagle] might alter the habits of many "savage inhabitants", and the whole Fuegian coast would welcome English sailors.

Indeed, a very relevant hint: "everyone *believed* that they were among cannibals.. [but] after nine days, they found that Matthews and the rest had *suffered from little more than bullying and systematic looting*: all their possessions had been divided indiscriminately among the Fuegians." Let us analyze this point in some detail. A group of people A, the Fuegians, do "little more than bullying" and looting people that invaded their own lands and moreover, after doing so, distribute the looted possessions in a remarkably egalitarian way, among themselves. In contrast, a group of people B, the Europeans, traveled to other continents to colonize, oppress, and enslave – and often massacre – hundreds of millions of people that were living in those continents, including the Fuegians. So, how did Darwin, who is often constructed as one of the greatest and most objective observers of the natural world that moreover supposedly had a quasi-religious "sacred cause" to defend "others," compared those groups A and B with regard to their biology, society, and morals, based on his own observations? He "scientifically" concluded that group A, the Fuegians, were barbaric brutish "savages" and that group B, the Europeans – his own group – were "civilized" and vastly "superior" both biologically and morally to the Fuegians, who had "far too much sense not to see the vast superiority of civilized over uncivilized habits." Even worse, he asserted that group A, the Fuegians, should have been smart enough to "alter" their "habits" by following the example of the "educated" Fuegians that came back from England and therefore to gladly "welcome" the kind and altruistic visits of those that are the pinnacle of biological evolution: that is, to "welcome English sailors" (see also Boxes 1.4 and 3.3).

That is, contrary to the brilliant and objective way in which he often observed, as a biologist, many biological aspects and behaviors of non-human organisms, he frequently failed to do so as an anthropologist about its own species, constructing inaccurate "facts" to support Victorian biases and fears about "savages." Paraphrasing George Orwell, "*to see what is in front of one's nose needs a constant struggle*." Another part of the above excerpt stresses this point: Darwin's statement that "the perfect equality of all the inhabitants will for many years prevent their civilization".. unless "some chief" rises who "by his power" can heap up possessions for himself. That is, something that would in theory be seen by many people as "morally noble" – being egalitarian – was deemed by Darwin to be "bad" because it would prevent something that he, a "civilized" Victorian man, accepted a priori to be good: hierarchical "civilizations." Accordingly, within this Victorian narrative, "some chief" rising and "by his power" heaping up the possessions of those of his own group for himself was deemed to be something "good," leading to "progress" and "civilization."

If an anthropologist would have observed in detail and scientifically studied, in a truly objective way, the behavior of the Fuegians back then it would become clear that that within their culture what Darwin called "looting" had nothing to do with what Westerns often define as conscious looting or robbery. There can be no conscious looting if there is no concept of private property, which is a concept that is often meaningless for many small hunter-gatherer nomadic groups, as explained in Kelly's 2013 *The Lifeways of Hunter-Gatherers*. Although there are obviously exceptions, nomadic human groups that are mainly egalitarian, such as the Fuegians were, tend to share most food and other items. This is not because they are "noble savages" that just obey to Mother Nature, which is both a romantic and racist narrative mainly created by Westerns and other agricultural societies – in the sense that it removes agency from hunter-gatherer "savages", depicting them as if they are just passive and impotent against "nature." Contrary to such erroneous narratives, such societies often *actively enforce egalitarianism* by promoting a type of indoctrination that emphasizes the *cult of sharing and of social equities*, instead of the cult of power and social hierarchies emphasized by the indoctrination of most agricultural "civilizations," including Darwin's Victorian society. For instance, such small nomadic groups often do this by educating children to share since an early age, by asking for gifts that will later be reciprocated, or by humiliating or ostracizing people that do not follow such norms or that are abusive and put at risk the common good of the group.

In this sense, as the English saw the Fuegians as "immoral," it is likely that the Fuegians also saw the English as "immoral." This is because what the English sailors were doing – not sharing things with the Fuegians – is something that is often considered by small nomadic groups to be among the most depraved things one can do. Apart from the ethnocentrism of Darwin's writings, the excerpt above also highlights some of his misogynistic ideas about human evolution. For instance, when Darwin asserts that the "savages" can never "progress" toward "civilization" unless they have a *male chief*. In this and many other passages of his writings about human evolution, he never even raised the possibility that a woman could become the chief or leader of such groups – something that, as an aside, is ironic because when he wrote his books, he himself was being ruled by a female: Queen Victoria. This, in turn, leads us to another crucial point that is overemphasized by the countless publications idealizing Darwin, and even in books of more nuanced authors such as Desmond and Moore, including their 2009 book *Darwin's Sacred Cause*: the "noble" and even "sacred" way in which he defended the unity of humanity. As we have seen, in some of his writings, Darwin explicitly stated that it was difficult to accept that disgusting "savages" such as the Fuegians could be his "fellow creatures." Therefore, his defense of humanity's unity had clearly less to do with such a "*sacred cause*" or any type of particular empathy toward the indigenous people that so often repulsed him than with the fact that such a unity was a mandatory prerequisite for his general theory of biological evolution. As we shall see, even evolutionists that are almost consensually described as racist, such as Ernst Haeckel – a German biologist that Darwin profoundly admired – also defended humanity's unity: they had to. There was no other way for their evolutionary ideas to make any sense. In other words, what authors such as Haeckel did was, in a way, to try to have a cake and eat it too: they could easily defend humankind's unity without putting in question their racist ideas because they could simply use the old notion of ladder of life to do so (Fig. 1.7).

In this sense, what they did was similar – but in a reverse order – to what Christian religious European leaders did back then. For the latter, humanity's unity was also a prerequisite for their religious narratives, as they *believed* that God created only one Adam and one Eve. So, while evolutionists such as Darwin and Haeckel constructed "facts" to support the idea that there was an evolutionary ladder-of-life leading to the "higher" and most "progressed" group of all – Europeans – Christian theists constructed the idea that the direction of the ladder-of-life was the opposite. That is, from the "higher whites," such as Adam and Eve, made directly by God, there was then a "degeneration" leading to the "savages." Instead of the inaccurate "facts" created within the umbrella of evolution and natural selection by Darwin and Haeckel, such theists created religious "facts" to explain such degeneration, such as the punishment for committing a sin, or for losing their faith in the Christian God, and so on. In this sense, these narratives were moreover also strikingly similar in the sense that Darwin assigned the evolutionary rise within the ladder of life to the "broom" of a teleologized Mother Nature's – that is, to natural selection – while the Christian theists assigned the fall within that same ladder to the will, or the failure to follow the Word, of a teleologized deity, God.

Such types of ethnocentric and racist narratives, both scientific and religious ones, were used in the nineteenth and first half of the twentieth centuries by Europeans to justify and undertake atrocious massacres of non-European indigenous peoples such as Fuegians and Australian aborigines (Fig. 1.15). Fortunately, those atrocities were not enough to fulfill the predictions of scholars such as Darwin's that such indigenous peoples were in general naturally doomed to become extinct or be exterminated by Europeans. Some did become extinct mainly due to huge massacres, such as the Fuegians. But multiple others – including the Australian aborigines to which Darwin explicitly referred to when he discussed such ideas – were not, despite also suffering huge atrocities and massacres at the hands of Europeans (Fig. 1.15). In fact, while I am writing these lines, in 2021, exactly 150 years after the publication of his *Descent of Man*, Australian aborigines are still here, among us, *blinking their eyes as they have done for dozens of thousands of years*. Many of them were forced or socially pressured – most of them – or have voluntarily chosen to adopt the lifeways of the European colonizers and their descendants. But many of them were able to continue to follow and respect their ancestral traditions, and some of them are prominent thinkers, artists, and singers, often using those traditions in their performances and works.

Contrary to such "scientific" predictions about their demise, and despite continuous attempts of cultural "assimilation" or even cultural genocide done by Europeans during centuries, there is nowadays an increasing fascination about Australian aboriginal traditions not only within the aborigine communities but also within the European descendants living in Australia and in other parts of the globe. This is attested, for instance, by the growing number of books, documentaries, movies, and concerts that have been done in the last years about those traditions, many of them by Australian aborigines – many of them receiving several awards and being commercially quite successful (Fig. 1.16). In summary, one can say that regarding many non-European indigenous groups – unfortunately, not all – the test of time answered

Fig. 1.15 Australian aborigine prisoners in chains at Wyndham Prison, after a massacre that took place at Forrest River (Oombulgurri massacre) in 1926

such predictions of "natural extinction" with a resounding *no*: groups such as the Australian aborigines, and their ancestral cultural traditions and ideas, were *not* an evolutionary dead end just waiting to be *naturally vanished* by Mother Nature's broom.

This topic brings us back to the discussion about whether or not this type of biased ethnocentric predictions postulated by Darwin were truly merely a "product of his time" shared by all Victorian intellectuals and scholars. As noted above, one of the most relevant case studies contradicting such a "product of his time" type of argument is Alfred Russel Wallace, the Victorian naturalist that also travelled around the world and independently developed a theory of evolution by natural selection similar to that postulated by Darwin (Fig. 1.9). As explained in Brantlinger's *Dark*

Fig. 1.16 Didgeridoo player Ŋalkan Munuŋgurr

Vanishings, with respect to the way Wallace perceived, and emphasized with, non-European indigenous groups, there was indeed a stark contrast with Darwin:

> For his part the co-discoverer of natural selection, Alfred Russel Wallace, was something of a romantic – a Rousseauistic Darwinian, perhaps. In contrast to Darwin's.. Hobbesian view of savagery, Wallace developed an appreciation of the virtues of "savage life". Thus, while in Borneo, he [Wallace] could write: "The more I see of uncivilised people, the better I think of human nature on the whole, and the essential differences between so-called civilised and savage man seem to disappear". So, too, Wallace's ethnographic accounts of the likely futures of many of the "natives" and "primitive races" he observed during his global travels are often explicitly elegiac.
>
> In *Narrative of Travels on the Amazon and Rio Negro* (1853), Wallace writes that "the Indians" are peaceful, affectionate, ingenious, and civilizable: "they seem capable of being formed, by education and good government, into a peaceable and civilised community." This positive result, however, seems unlikely, because "they are exposed to the influence of the refuse of Brazilian society, and will probably, before many years, be reduced to the condition of the other half-civilised Indians of the country, who seem to have lost the good qualities of savage life, and gained only the vices of civilisation".. Wallace may have been something of a "crank" or an intellectual faddist, but he was also less willing than Darwin, Huxley, Spencer.. to biologize – that is, naturalize – capitalism, war, imperialism, and the extinction of the noncivilized races through the supposed progress of white civilization.

This last powerful sentence astutely summarizes the main difference between Darwin and Wallace concerning their attitudes and writings regarding "others." With respect to Brantlinger's characterization of Wallace as a Rousseauist and Darwin as a Hobbesian, this might be an appropriate simplification if one considers that the so-called Rousseau *versus* Hobbes debate, which is often linked to the "primitivist" notion of "noble savages" *versus* the anti-primitivist view that "savages" are brutish, is typically oversimplified (see Box 1.3).

Box 1.3 Rousseau-Hobbes, Adam-Eve, and Primitivism Versus Antiprimitivism

Moser's 1998 book *Ancestral Images – The Iconography of Human Origins* includes numerous beautiful illustrations as well as a text that is highly informative and interesting regarding the so-called Rousseau *versus* Hobbes debate. The book highlights that Ancient Greeks and Romans already had these two major types of views about human nature, which in a way still remain until today within some areas of knowledge. For instance, the Greek poet Hesiod defended a primitivist framework that shared some similarities with Rousseau's "noble savage," while the Greek poet Lucretius defended a more antiprimitivist framework that resembles in some ways the ideas of Hobbes about "brutish savages." Moreover, the book brilliantly connects such views with religious stories about, and iconography displaying, Adam and Eve, in a further example of the deep links between scholarly narratives and discussions and religious thinking. Such an example is, in turn, also very interesting because it highlights how such narratives are often incongruent and paradoxical. That is, although Christian theologists in general started to accept the *scala naturae* view of life and "progress" defended by most Ancient Greeks (Fig. 1.7), they tended to see the lives of Adam and Eve before eating the apple as a paradisiacal example of existence in the Garden of Eden. In other words, within this religious narrative, before eating the apple Adam and Eve were mainly "noble savages," a view that is in a way more in agreement with the primitivist ideas of writers such as Hesiod and Rousseau, than with the notion of *scala naturae* and its related notion of "progress" defended by Lucretius and Hobbes. Moser explains:

> The poet Hesiod was one of the first writers of the western tradition to provide an explanation of our human beginnings. This was outlined in *Works and days*, written around the eighth century BC, which was an account of Greek cultural history. Here Hesiod presented the legend of the *Five Races* or generations of the world, in which the successive races of humankind are described. First was the Golden race, who were peaceful and happy, living in a natural state without hardship. They were succeeded by the Silver race, who represented a deterioration in that they were inferior physically and morally and they began to fight with one another. Third was the Bronze race, who represented a further decline in that they became violent and began to eat flesh. Fourth was the race of Heroes, who did not represent a further

(continued)

deterioration in culture, being noble and righteous. They were followed by the Iron race, who were savage, libidinous and cruel. With this final race of humanity there was a complete physical and moral decline. The Five Races scheme, with its metallic ages, incorporated elements from other creation myths, especially Babylonian, Persian and Asiatic ones. The symbolism of the metals is important in that it embodies a sense of cultural evolution, later to become a fundamental feature of scientific schemes of prehistory. However, cultural evolution was certainly not the underlying theme of Hesiod's scheme. His scheme was one of degeneration, in which the most distant past was conceived of in the most positive light and the most recent past was considered to represent a decayed state. Thus, in the Five Races legend, history was characterized not by progress, but rather by decline.

An alternative to Hesiod's scheme was outlined by those who argued that human evolution was characterized by progress rather than decline. The major proponent of this view was the Greek poet and philosopher Lucretius, who, in his *De rerum natura* of the first century BC, developed the idea that history was characterized by gradual social and technological advance. Emphasizing the brutal and primitive nature of our beginnings, Lucretius informs us that the first humans were little different from beasts.. thus we learn that the first ancestors were born of the earth, and that they were large, muscular and strong. They did not live as couples, but rather lived like animals in a pack. The skills of working the land were unknown to them, as was the knowledge of metals and the ability to grow and maintain crops. Individuals thought only of themselves; they did not share or give to others. Male and female were united not by love but by lust.

These two different perspectives have been described as primitivism and anti-primitivism. While the primitivist view was nostalgic and glorified a simple idyllic existence, the anti-primitivist view appreciated the advances technology had brought. It is the latter that is most closely aligned with the development of scientific views of prehistory. In many ways the life of Adam and Eve in the Garden of Eden resembled the descriptions of the life of the Golden race in the Five Ages of Hesiod. Like the people of the Golden race, Adam and Eve did not need clothes, shelter or weapons because there were no threats to their life. While the Creation and the Golden Age were both very popular as visual themes, the biblical story was more prolifically illustrated throughout the early Christian and medieval periods. Similarities between the classical and biblical schemes can also be seen in the story of the expulsion from Eden, which led to a degradation in the life of the first couple, and the model of Hesiod, which embodied the concept of regression from an early idyllic state. The scenes of Adam and Eve labouring after their expulsion are also important. It is here that Adam is represented as a digger toiling on the land and Eve as a nurturer raising infants. The other point to note in the labouring imagery is the standardization of gender roles. The representation of the division of labour in this way was to become a dominant theme in the imagery of human ancestry.

In Slotten's book *The Heretic*, it is further shown how Wallace often enjoyed to be among – and even "crashed parties" of – non-European indigenous people, particularly Native Americans, frequently describing them or their cultures as "far superior" to Westerns in some ways, in contrast to Darwin:

> [When Wallace was in Javita, an Amazonian locality] occasionally he would attend an Indian festa [party], where everyone drank freely and danced. Yet he was not unhappy. He noted that he was the only white man among two hundred peaceful, "half-wild" inhabitants who, when they were not reveling, were piously occupied in cleaning their churches, streets, and homes. He admired the young girls – "far superior in their graceful forms" to English village girls of the same age, whom they resembled "save in their dusky skin" – and envied the nakedness of the boys, their heads, bodies, and limbs unrestricted by the burdensome clothes of the modern world. The villagers of Javita lived simply, their only luxury being salt. He reflected on the miseries he had left behind in England, where many who were worse off "in physical and moral health" than the Brazilian Indians longed for nothing but material wealth. While in a state of "excited indignation against civilised life in general", he put his thoughts into verse: "*Rather than live a man like one of these, I'd be an Indian here, and live content to fish, and hunt, and paddle my canoe, and see my children grow, like young wild fawns, in health of body and in peace of mind, rich without wealth, and happy without gold!*"

He wrote to Spruce from Javíta that he was enjoying himself "amazingly" in a romantic and unexplored country far beyond the region of the ague.. He was fascinated by the Indian diet. Along with the usual yams, sweet potatoes, peppers, cassava bread, and various fruits, they regularly ate insects. Six orders were deemed edible: hymenoptera, neuroptera, homoptera, coleoptera, aptera, and annelida (or segmented worms—including earthworms and leeches—now no longer classified as an insect order). One delicacy was the great-headed red ant, which at certain seasons swarmed in the thousands from enormous mounds and created a stir of excitement in the villages; everyone rushed out and gathered as many as they could into baskets and calabashes. The ants were eaten alive: they were grasped by the head, as one might hold a strawberry, and the egg-rich abdomen was bitten off and eaten along with farinha. Wallace found that the great-headed red ant was more to his liking when roasted or smoked and sprinkled with salt. Termites, which the natives chased out of their nests with long blades of grass, were also considered a delicacy.
Slotten also explained that Wallace was so fascinated by traveling to such remote areas and in particular by being among "others" that, just 4 days after arriving to England from his long trip to South America, he decided to travel again:

> Although Wallace had sworn never to travel again, he changed his mind after only four days ashore. In a letter to Richard Spruce, he spoke of undertaking another expedition. "How I begin to envy you in that glorious country where 'the sun shines for ever unchangeably bright'," he wrote. "Fifty times since I left Pará have I vowed.. never to trust myself more on the ocean. But good resolutions soon fade, and I am already doubtful whether the Andes or the Philippines are to be the scene of my next wanderings." Humboldt once remarked that the joy of finding oneself back in civilization would be short-lived if one had learned "to feel deeply the marvels of tropical nature." The memory of what one had endured quickly faded, he observed, and soon the weary traveler began to plan another journey. Humboldt was right.

Importantly, Wallace was just one of the many Victorian naturalists that traveled around the world back then and that had a way to see and interact with "others" that was markedly different from that of Darwin. For instance, as shown in McCalman's

2009 book *Darwin's Armada*, this applied to some extent to Joseph Hooker, who "unlike Darwin.. loved the landscape and was respectfully impressed by Maori culture" when he was in New Zealand during his Antarctic voyage. And even to the young Thomas Huxley, who later became to be known as Darwin's closest ally and "bulldog" and also included several ethnocentric passages in his writings about human evolution. Contrary to those later writings – which are often described by many scholars as those of a proud Hobbesian – earlier in life Huxley was markedly more open-minded to and empathetic toward "others" than Darwin was, for instance during his travel in the HMS Rattlesnake. As an illustration, regarding the Papuans Huxley stated that they "seem happy, the means of subsistence are abundant, the air warm and balmy, they are untroubled by 'the malady of thought', and, so far as I can see it, civilization as we call it, would be rather a curse than a blessing to them" (see also Box 1.4). In fact, McCalman explicitly points out that, at that time, Huxley's "attitude to missionaries differed sharply from Darwin's.. Huxley thought it better for the Papuans to 'walk familiarly with the devils they have, than to take to themselves the seven worse, which during a long period of transition, will infallibly follow in the train of the white man, his commerce and his missionaries'."

Box 1.4 Young Huxley's and Young Darwin's Sharply Different Attitude to "Others"

In *Darwin's Armada*, McCalman's description of Huxley's encounters with Papuans illustrates how, early in their lives, the way Huxley and Darwin reacted to non-Europeans was significantly different. Crucially for the context of the present book, McCalman also stresses how the horror stories that Europeans constructed – and often used as a justification to colonize "cannibalistic headhunting" brutish "savages" – were often disconnected from reality:

> When *HMS Rattlesnake* sailed out of Port Jackson on 8 May 1849 to survey the south-east coast of New Guinea and its neighbouring Louisiade Archipelago.. Thomas Huxley couldn't wait. Having been denied a place on the Kennedy expedition, he was thrilled to have another opportunity to test his mettle as a scientific explorer. The reputation of Papuan peoples for headhunting and cannibalism was one of the stock fabulations of South Seas travel. Huxley had read the warnings of the seventeenth-century English travel writer Samuel Purchas: "Heere be those blacke people, called Os Papuas.. man-eaters and sorcerers among whom devils walk familiarly as companions". Captain Stanley was also familiar with such cannibal tales and had enjoyed arousing his cousin Louisa's anxieties by half joking about the risks he'd face. "From all I can learn", he wrote in May 1849, "they are rather partial to human flesh [and] look upon the white man as a dainty morsel and prize his head above all things as an ornament for the doors of their houses"..
>
> On 16 June, Stanley at last permitted them [Huxley and others] to land on Piron Island, "to open a communication with the natives".. Huxley was keen to befriend the Papuans-smallish, copper-coloured men and women with pleasing expressions and "large fuzzy heads of hair with combs one foot long.. stuck in their remarkable

(continued)

coiffure". They were sociable and willing to barter ornaments, tortoiseshell, yams and coconuts for iron and red cloth.. On 5 July, Stanley.. accompanied two heavily armed boats ashore at what they believed was Chaumont Island. The sailors could see dusky shapes flitting among the coconut plantations, but rather than engage with them, the captain remained sitting in the boat at the water's edge, "looking as stupid as a stockfish".

Eventually Huxley could stand the inaction no longer: he waded ashore, broke off some green branches as a peace sign, and began to dance up and down on the foreshore, an to dance up and down on the foreshore, an action that produced instant laugher and reciprocation from the Papuans. Soon the supposed cannibals were making balletic springs and jumps along the beach. "Tamoo", as Huxley called himself, consolidated his pacific intentions by squatting on the sand to sketch a couple of the cheerful warriors. As all parties began to barter, Huxley quietly wandered down a path to a small village, where he sketched one of the ornamented timber houses.. Huxley was not particularly surprised by this ruckus; he was no sentimentalist and regarded the Papuans as children who could be expected to use "foul" as well as fair means to get what they wanted. He drew a half-comical sketch of the warriors being tumbled in their canoes by the grapeshot. If threatened personally, Huxley said, he wouldn't hesitate to shoot the transgressors dead with his pistol.

On the other hand, he always insisted that interactions with native peoples be governed by the same rational criteria that one would use with Europeans. The *Rattlesnake*'s sailors needed to be careful he thought but not needed to be careful, he thought, but not paranoid. Huxley had already noted, for example, how "the blackies behaved very honestly with us, not attempting to take anything without giving a proper equivalent". He was furious when his colleagues behaved unjustly, such as the marine sergeant who casually fired his gun close to some trading canoes. Above all, Huxley believed that a particular fracas, like the one at Joannet Island, was no excuse for engaging in a generalised overreaction against Papuan traders. But overreaction was what they got. Stanley named the place Treacherous Bay, and cited the incident in his log as "a most convincing proof" that native peoples would attack "the moment they saw their numerical force was greater than ours".

From this point he insisted on dispatching armed cutters to break up trader flotillas on any pretext, even when the locals showed such obvious signs of friendliness as bringing their women and children along in the canoes. It never occurred to Stanley, Huxley fumed, to consider that the natives might be "in doubt of our intentions". A Papuan had only to wave a spear in the air or spit on the ground and the captain acted as if they'd been threatened with a bloody battle. "What a Sir Joshua Windbag the little man is!" Trading canoes became steadily scarcer, especially after Captain Yule-infected by Stanley's paranoia-opened fire on another harmless group of traders and killed two of them. At beautiful Brumer Island, where they spent a fortnight surveying, Huxley and his colleagues were permitted to touch shore only twice, and then only for a couple of hours each time, despite the conspicuous friendliness of the inhabitants..

Most of all, Huxley could not forgive this "brute", coward and "little fiend" for stopping him and his colleagues from investigating the culture of "the human inhabitants of Papua", and so from contributing to "the young but rapidly growing science of ethnography". This discipline had been named only a decade earlier, and it encompassed the scientific description of races or nations of men, including their customs, values and differences values and differences. Like Darwin before him, Huxley believed that the naturalist's domain should extend to studying mankind as

(continued)

much as birds, rocks, plants, fish and animals. From the small contact he'd managed to make with Papuan peoples, Huxley had been impressed by "their invariable gentleness towards each other; the kind treatment of their women; the cleanliness of their persons and of their dwellings; their progress in the useful arts, as exhibited in the pottery, cloth, cordage, pets, sails and weapons of all sorts.. in the ingeniously built houses and canoes.. the perseverance and design displayed in many of their carved works".

Off Brumer Island, he summed up his overall impressions: "[the Islanders] seem happy, the means of subsistence are abundant, the air warm and balmy, they are untroubled by 'the malady of thought', and, so far as I can see it, civilization as we call it, would be rather a curse than a blessing to them". As a political radical and religious sceptic, *his attitude to missionaries differed sharply from Darwin's*. Huxley thought it better for the Papuans to "walk familiarly with the devils they have, than to take to themselves the seven worse, which during a long period of transition, will infallibly follow in the train of the white man, his commerce and his missionaries".

Apart from the English scholars mentioned in the paragraphs above, there were obviously many non-English European scholars, including prominent naturalists, that also did "*not think like others*," to paraphrase again Marguerite Yourcenar. Alexander Humboldt – who died in 1859, the year Darwin's *Origin* was published, and who was one of Darwin's heroes as noted above – is one of the most emblematic examples, as we will see in Chap. 3. I am of course not arguing that Wallace and Humboldt – and Huxley, in his youth – were not racist or ethnocentric. There are historical records that indicate that some of the things they wrote, said, and did were racist, as we shall see below. I am just calling attention to the important fact that *in general* the tone of their descriptions about their encounters with "others" were far less biased, prejudiced, and ethnocentric than that of Darwin's *Voyage of the Beagle* and *Descent of Man*. This is further attested by the historical fact that Humboldt was – and continues to be, in a way – celebrated in many parts of South America for his highly favorable ideas about the lifeways of Native Americans and support for the end of their colonialization and oppression at the hands of European states. That is why South American revolutionaries such as Simon Bolívar – who continues to be seen by many South Americans as a hero for his involvement in the nineteenth-century Hispanic independence movements – often publicly praised Humboldt and used his name to promote their anti-colonialist stance.

In contrast, while some European revolutionary theorists such as Marx did use – in a biased way – some of Darwin's ideas, such as those about the struggle-for-existence, to promote the fight against the oppression of workers, for the vast majority of non-European revolutionary leaders such as Bolívar using Darwin's name and writings would be a political suicide. This is particularly so because Bolívar would need the support of South American indigenous groups, and most of what Darwin wrote about those groups was negative, as we have seen. In other words, it would be unthinkable for people such as Bolívar to use Darwin's name to promote the fight for independence from Europeans because many non-Westerners precisely saw – and continue to see, correctly so in that sense – Darwin as an icon of Western imperialism and scientific racism. To give just an example, among many, when I went to Chile and Argentina to give a series of scientific talks, I learned that many local university professors mention Darwin's racism, sexism, ethnocentrism, and defense of colonialism and imperialism in their courses, without taboos, in stark contrast to what often happens in most European universities. In that sense, what such professors teach – one of the most prominent ones being Virginia Abdala, a particularly renowned Argentinean evolutionary biologist – is closer to the historical truth than what most Western professors teach, concerning Darwin ideas about human societies and evolution. We will further discuss these topics below. Before that, let us travel again to the past, to the years after Darwin returned to England, using again his autobiography as a guide.

The *Origin*, Scientific Egos, Rivalries, Networks, and Immortality

> *I would love to believe that when I die I will live again, that some thinking, feeling, remembering part of me will continue.. but as much as I want to believe that, and despite the ancient and worldwide cultural traditions that assert an afterlife, I know of nothing to suggest that it is more than wishful thinking.* (Carl Sagan)

After Darwin returned to England from the Beagle voyage, in October 1836, he began his "systematic enquiry" on the diversity, and ultimately the origin, of species. In his autobiography, he provides some interesting notes about this investigation. A critical idea that Darwin internalized through indoctrination and continued to accept as a dogma after his trip around the world and that became one of the most central ideas within this enquiry was the war-like Malthusian notion of struggle-for-existence. He recognized this point, in his autobiography:

> In October 1838, that is, fifteen months after I had begun my systematic enquiry, I happened to read for amusement "*Malthus on Population*," and being well prepared to appreciate the struggle for existence which everywhere goes on from long-continued observation of the habits of animals and plants, it at once struck me that under these circumstances favourable variations would tend to be preserved, and unfavourable ones to be destroyed. The result of this would be the formation of new species. Here then I had at last got a theory by which to work; but I was so anxious to avoid prejudice, that I determined not for some time to write even the briefest sketch of it. In June 1842 I first allowed myself the satisfaction of writing a very brief abstract of my theory in pencil in 35 pages; and this was enlarged during the summer of 1844 into one of 230 pages, which I had fairly copied out and still possess. Early in 1856 Lyell advised me to write out my views pretty fully, and I began at once to do so on a scale three or four times as extensive as that which was afterwards followed in my "*Origin of Species*"; yet it was only an abstract of the materials which I had collected, and I got through about half the work on this scale. But my plans were overthrown, for early in the summer of 1858 Mr. Wallace, who was then in the Malay archipelago, sent me an essay "*On the Tendency of Varieties to depart indefinitely from the Original Type*"; and this essay contained exactly the same theory as mine. Mr. Wallace expressed the wish that if I thought well of his essay, I should sent it to Lyell for perusal.
>
> The circumstances under which I consented at the request of Lyell and Hooker to allow of an abstract from my MS., together with a letter to Asa Gray, dated September 5, 1857, to

> be published at the same time with Wallace's Essay, are given in the "*Journal of the Proceedings of the Linnean Society*," 1858, page 45. I was at first very unwilling to consent, as I thought Mr. Wallace might consider my doing so unjustifiable, for I did not then know how generous and noble was his disposition. The extract from my MS. and the letter to Asa Gray had neither been intended for publication, and were badly written. Mr. Wallace's essay, on the other hand, was admirably expressed and quite clear.
>
> Nevertheless, our joint productions excited very little attention, and the only published notice of them which I can remember was by Professor Haughton of Dublin, whose verdict was that all that was new in them was false, and what was true was old. This shows how necessary it is that any new view should be explained at considerable length in order to arouse public attention.. I gained much by my delay in publishing from about 1839, when the theory was clearly conceived, to 1859; and I lost nothing by it, for I cared very little whether men attributed most originality to me or Wallace; and his essay no doubt aided in the reception of the theory.

In a way, this last sentence is misleading and provides some support to the assertion made in Browne's *Voyaging* that it seems as if Darwin used his autobiography to either omit certain specific facts or to "clean" others in order to somehow influence the way he would be judged by his family and loved ones, and ultimately by history. After all, when he was writing his autobiography, he *knew* that he and his scientific ideas could achieve *immortality*. Therefore, as suggested by Browne, it is quite possible that during those last years before his death, he knew that what he would write, particularly in his autobiography, could be important for the way in which his life and work would be seen after he was gone. As we shall see, various scholars have argued that in a sense Darwin wanted to be seen as the "Newton of Biology" – he was, indeed, buried near to Newton. In fact, Darwin explicitly referred to Newton's gravity at the very end of his opus magnum, the *Origin*, making a direct parallel between his theory of evolution by natural selection and Newton's idea of gravity, as we shall discuss below. In this regard, the cases of Darwin and Newton are also similar, and are different to those of many other "scientific giants," because they already knew, at the end of their lives, that they would very likely be remembered as "giants." This leads us to discuss a crucial topic that is often neglected in accounts about such "giants," due to the tendency to idealize them as if they were purely moved by their love of science or the pursuit of truth, and not as well by their own egos and ambitions, contrary to the almost totality of other human beings.

In the case of Darwin, as of other "giants," such a narrative is clearly *not* the full story, as demonstrated by several examples and by a detailed analysis of his own private writings and letters. For example, such writings contradict the statement he made in the above excerpt of his autobiography – "I cared very little whether men attributed most originality [of the theory of evolution trough natural selection] to me or Wallace." Quammen's *The reluctant Mr. Darwin* is among the few accounts about Darwin that discuss this topic in a detailed and balanced way, so it is worthy to pay special attention to that book. In particular, Quammen's book reminds us that science is very far from being an objective, altruistic quest for reality in which scientists happily stand on the shoulders of other scholars without any kind of jealousy or disputes of ego or about who was the first to have a certain idea or do a certain invention. Instead, many times – but not always, fortunately, this has to be emphasized – science can be a dirty quest for fame, involving the use of tricky strategies, social networks, privileges, and the power of place.

The quest for "originality" between Darwin's and Wallace's theory of evolution by natural selection was not as dirty and heated as many others have been within academia. One of the reasons for this was because Wallace was likely aware that he could not, or at least should not, embark in such a public fight, because when he developed his theory he was mainly an outsider that was both physically – he was literally in a faraway jungle – and intellectually disconnected from the societal and scientific Victorian networks in which Darwin thrived. As we will see, it seems very likely that Wallace was aware of some of the tricks done by some among the cream of the Victorian scientific community regarding this issue of "originality" but that he decided to not be hostile – and actually to almost never refer to them – due to the weak *power of place* that he had when he first wrote his theory. In that sense, Darwin also did not have a reason to be publicly hostile toward Wallace, well on the contrary, as he recognized that Wallace was always very easy-going concerning that issue. For a significant portion of their lives, they therefore got along very well and even became friends for several years until they began to distance themselves because of *other* scientific issues, as we will see. Notably, even after that, Wallace always continued to state that the idea of evolution by natural selection was mainly Darwin's, despite the fact that although Darwin did start to develop this idea before him, he was actually the first to finish a scientific manuscript exposing it in a clear way.

In order to try to understand Wallace's stance about this "originality" issue, let us see what Quammen wrote about it, including some minor – but significant, and historically crucial – tricks and dishonesties done by Darwin's scientific network in England while Wallace was traveling:

> The abundance of naturally occurring variation within species was a crucial clue to the transmutation mystery, unnoticed by most naturalists of the day. Darwin needed eight years with barnacles, following five years of travel and ten years of study, to awaken him about variation in the wild. Wallace saw it sooner because, besides being an alert observer, he was a commercial collector, hungry and broke.. Wallace's paper from Sarawak, about the "law" regulating the "introduction" of new species, was published in September 1855. It created no sensation, but it did generate some murmurs. Wallace's agent, Samuel Stevens, told Wallace about several London naturalists who had groused that he should stop theorizing and stick to collecting facts. Charles Lyell, on the other hand, found the paper intriguing..
>
> Darwin's opinion was different. He read the paper around that time and made some notes for his own memory, as he routinely did with his eclectic research reading. That was with his eclectic research reading. That was Darwin's way, methodical and thorough; he chewed through huge amounts of material, swallowed the good bits, spit out the rotten stuff and the husks. Wallace's paper tasted like husk. It discussed geographical distribution, Darwin recorded, but offered "nothing very new". It used the simile of a branching tree ("my simile", in Darwin's jealous view) to represent affinities and diversity in nature. It mentioned rudimentary organs, though to what point?
>
> And the Galápagos comment – about how those peculiar creatures and curious patterns had never received "even a conjectural explanation" – didn't pass unnoticed. Darwin may

> even have winced, knowing it was true. He hadn't risked any explanation in the *Journal*, but.. give a man time. Well, all right, he'd had time. Still, not enough. And what did Mr. Wallace know of the complex considerations? Rather than arguing the point in his mind, or rising to this small provocation as a challenge, Darwin dismissed Wallace's whole effort. He saw no real explanatory value to the "law" of juxtapositions and he heard nothing in the vague language except a rehash of old-fashioned natural theology.
>
> Now if Wallace had scratched the word "creation" and spoken instead about "generation" of new species, Darwin told himself, he could agree with the paper. So far as it went. But Wallace hadn't used any such word. "It seems all creation with him", Darwin judged, and went back to his pigeons. He sent off his letters to Thwaites, Layard, and those others on the list, including "R. Wallace". I would be most grateful, he told them, for any skins of chickens, pigeons, rabbits, or ducks.. [Later] from Lombok.. [Wallace did send] a local variant of the barnyard duck. Wallace's note to the agent explained: "The domestic duck var. is for Mr. Darwin". Please forward. It's hard to say whether that duck ever reached Darwin. If so, he was presumably grateful but not surprised. He had come to expect a high degree of generous cooperation from the people (especially those below him in social status) he called on for research assistance.

Here, we see a critical point that was also made in Browne's books *Charles Darwin Voyaging* and *Power of Place*: Darwin was not always as humble as most books idealizing him tend to portray him. For example, when it concerned people that he perceived to be below his social and scientific status, such as those from which he was used to *receive* things such as "skins of chickens, pigeons, rabbits, or ducks." Until he received Wallace's letter containing a manuscript explaining Wallace's theory of evolution by natural selection, in 1858, Darwin seemingly mainly saw Wallace in a somewhat patronizing way, as pointed out by Browne and Quammen. This likely aggravated the way Darwin felt, when he receive that letter in 1858: how could someone such as Wallace, who was so far – in terms of scientific recognition and physical presence – from the cream of the Victorian scientific society, jeopardize his scientific ambition of being the first to tell the world about "his doctrines"? As explained by Quammen, prior to receiving that letter, Darwin had recognized that he was rather divided between this ambition and his scientific duty: "he told Lyell '*I rather hate the idea of writing for priority.. yet I certainly shd. be vexed if any one were to publish my doctrines before me*'.. that sentence captures it: he hated the idea of writing for priority, but dammit he did want priority." Quammen further noted that a letter – dated as 1 May 1857, written by Darwin – demonstrates that Darwin was, in a way, likely *actively* trying to make sure that such "priority" was not taken from him:

> There's another odd comment in that letter, showing Darwin's sensitivity about how long he had delayed. After noting the similarity of their views, and the rarity of such concord between two theorizing naturalists, he stroked a dash on the page, as though clearing his throat. Then he wrote: "This summer will make the 20th year (!) since I opened my first notebook, on the question how & in what way do species & question how & in what way do species & varieties differ from each other". At last, Darwin intimated, he had found the answer.. He couldn't possibly explain this idea in a mere letter, he told Wallace, too complicated. "I am now preparing my work for publication, but I find the subject so very large, that though I have written many chapters, I do not suppose I shall go to press for two years". He was wheedling for time and consideration. Although he still didn't take Wallace quite seriously – not seriously enough – he felt mildly wary. With its histrionic exclamation point,

> Darwin's remark was an assertion of his own interests, precedence, and claims. A male dog makes the same sort of assertion, raising his leg to mark a tree. Wallace's nose must have been off, because he didn't get the hint.
>
> On or about June 18, 1858, another mailing from Alfred Wallace arrived at Darwin's front door. It came, like the others, from somewhere in the Malay Archipelago. It had been four months in transit on a series of boats. This envelope was bulkier than usual, containing a manuscript as well as a letter. Darwin opened it. Scanning the letter, reading the enclosure, he felt a nauseating surge of emotions that began with surprise and swelled quickly toward despair. His big book at this point was still a work in progress, two-thirds written and growing more unwieldy every day. Meanwhile his young pen pal, Wallace, had independently conceived the idea of evolution by natural selection. Wallace's manuscript was titled "*On the Tendency of Varieties to Depart Indefinitely from the Original Type.*"
>
> It comprised about twenty pages of lucid, easy prose, written out in the author's hand. Its cardinal point, signaled in the title, was that the difference between species (as a category) and variety (as a category) is merely a difference of degree. That is, the amount of variation seen between varieties within a species is not inherently limited; rather, within a species is not inherently limited; rather, those increments can accumulate boundlessly until a variety splits away, becoming a distinct species unto itself. The manuscript posited "a general principle in nature" causing many varieties to do exactly that. And they don't just split from the parent species, Wallace asserted; they compete against it, sometimes outlive it, and eventually give rise to still other varieties differing more and more from the original type. Wallace, unlike Darwin, had coined no name for this "general principle." But his manuscript built a case for it with logic very similar to Darwin's own.

"Darwin's remark was an assertion of his own interests, precedence, and claims: a male dog makes the same sort of assertion, raising his leg to mark a tree": unfortunately, the example of Darwin is just one among multiple ones, within science. If science was really all about a "pure" quest to discover reality, so others know about and benefit from it, as it is often portrayed within popular culture and the media, why would such "priority" issues matter so much to scientists? Why would Darwin care if the world would learn about evolution driven by natural selection from Wallace's manuscript or from his book, if his sole "pure" goal would be for the world to know about the reality of biological evolution? Yes, science can be a *candle in the dark* and has been very useful for a plethora of things – leading, for instance, to the creation of vaccines and medication that save the lives of dozens of millions of people every year, as brilliantly argued by *Carl Sagan*. But scientists should not be idolized: they are just humans, with their own biases, egos, and ambitions, and many of them did do mistakes that had dark societal consequences as well. They should also not be demonized – a trend that is unfortunately gaining ground recently, with the rise of the notion of "fake-news" and of anti-vaccination and other similar types of so-called 'anti-intellectual' movements, even at a time in which scientists and vaccines are literally saving the life of millions of people during the Covid-19 pandemic.

This reminds me of a funny, but revealing, story told by Kerry Daynes at the very end of her 2019 book *The Dark Side of the Mind: True Stories from My Life as a Forensic Psychologist*. She says that within a group of psychologists and patients addressing the control of rage, two of her colleagues engaged in a harsh discussion, and fought, in front of the patients. The patients begun to laugh, and later one of them told her that this episode was important because it led him to the "firm conclusion that psychologists are also humans." Yes, they are. One of the most interesting points of Kerry's book is that it precisely shows how the decisions of psychologists, and scholars in general, are indeed profoundly influenced – and affected – by the contingencies of life, including their personal biases, prejudices, egos, and ambitions. Perhaps, the most powerful example to attest this point is that every year there are *individual* Nobel prizes and countless other *personal awards* being given to scientists. For many scientists, a Noble price is the biggest recognition they can achieve: often, it is a sign that they have reached the Holy Grail of intellectual immortality. As we have seen, such a *personal ambition* was clearly recognized in Darwin's personal letters: "*I rather hate the idea of writing for priority.. yet I certainly shd. be vexed if any one were to publish my doctrines before me*." Now, compare this sentence with what Darwin wrote in his autobiography: "*I cared very little whether men attributed most originality to me or Wallace*." Two very different statements written by the very same person. One was written in a personal, private letter *when* the "priority" issues were truly happening. The other was written decades later, in Darwin's autobiography, probably reflecting the fact that he knew that this is how *in theory* things should have ideally happened and how he likely *wanted* future readers to think it had happened, as suggested by Browne.

So, let us analyze in some more detail what *truly, factually happened*, in 1858, after Darwin received the letter from Wallace including Wallace's manuscript about the theory of evolution by natural selection. As explained by Quammen:

> Darwin felt crushed. He had only himself to blame. His dilatoriness, his perfectionism, his big mouth. Suddenly he was trapped, flattened, between the demands of honor and the claims of self-interest. He howled with pain. "Your words have come true with a vengeance", he wrote Lyell, "that I shd. be forestalled". Enclosed is a manuscript that Wallace asks me to send you, said Darwin. It's well worth reading. It's also, he added glumly, the closest thing to a précis of my own theory. (In the panic of the moment, he was overlooking a significant difference: Wallace focused on competition between varieties, not between individuals – that is, selection of one group versus another, not selection. of individuals within a group.) "I never saw a more striking coincidence", Darwin moaned. Even some of the phrases Wallace used, such as the "struggle for existence", echoed what Darwin had already written into the draft of his big book. Wallace hadn't asked him to help get the manuscript published, Darwin noted, only to share it with Lyell; but of course Darwin would write Wallace immediately and offer to send it to any journal.
>
> "So all my originality, whatever it may amount to", he whined, "will be smashed". In the meantime Darwin had heard back from Lyell with some thoughts about how the Wallace dilemma could be handled. What did Darwin have on paper, Lyell wondered, that might testify to his priority of discovery? Well, there was the manuscript essay of 1844, which Hooker had read; also a six-paragraph summary of the theory, which he'd sent last year to the botanist Asa Gray, his trusted correspondent at Harvard. These unpublished but witnessed writings were proof that he'd conceived the whole idea long ago, solitarily, and stolen nothing from Wallace. "I shd. be extremely glad now to publish a sketch of my general views in about a dozen pages or so", he told Lyell. "But I cannot persuade myself that I can do so honourably".
>
> He worried that receiving the Wallace manuscript – which he hadn't asked for, after all – put him in a bind. He would rather burn his own book-in-progress, he said, than be seen

> as behaving shabbily. But was it too late to publish a summary of his views and say he was doing so on the advice (two years earlier) of Lyell? He repeated: "If I could honourably publish.." No, he couldn't persuade himself that it was okay; but implicitly he begged Lyell and Hooker to do the persuading. Altogether, he was fuddled with anguish. He hated himself for thinking about such stuff while his children were battling for their lives. "This is a trumpery letter", he ended, "influenced by trumpery feelings". But the feelings wouldn't go away. Lyell and Hooker took their cue. Within days, serving him faithfully as friends, serving science by their lights, serving justice more dubiously, they cooked up an arrangement that rescued the situation – or at least, it rescued Darwin's interests. They certainly couldn't ignore Wallace's paper entirely and connive to see Darwin given credit alone; that would have been dishonorable, unprofessional, and scandalous when the truth came out. Instead they devised and sponsored a joint presentation of Wallace's manuscript and Darwin's unpublished work. This peculiar duet would occur at the next meeting of the Linnean Society, one of London's better scientific associations, of which Hooker, Lyell, and Darwin were all governing members. Darwin consented to the arrangement, sending Hooker his 1844 essay and the six-paragraph summary he'd written for Gray.

An arrangement so that Darwin could have his "priority," after all. As so it was: that is why almost everybody from the general public nowadays associates the theory of evolution by natural selection with Darwin, and almost everybody has heard about Darwinism and almost nobody talks about *Wallacism*. Regarding this issue, an important point needs to be made. Within the absurdity and vanity of the quest for – and fights about – scientific "priority" concerning a new idea or the discovery of a new fossil, or species, or gene, within academia, the *critical* aspect that is often used to determine scientific originality is *who was the first to publish that idea or discovery*, rather than who was the first to start thinking about it. Everybody can say that they were thinking about an idea that was awarded with a scientific Nobel prize decades before the winner or winners of that prize started doing so. But obviously that will not bring you immortality because that assertion cannot be proven – and, even if it was, that would not count because the idea was not published before the recipients of the prize published their idea, anyway. If we therefore apply this common scientific practice to the case of Darwin and Wallace, when Darwin received, in 1858, Wallace's letter with Wallace's manuscript about evolution by natural selection, the reality is that Darwin had *not* finished his manuscript about the same idea. He was going to publish that idea in his *Origin* book, which was neither complete nor ready to be published.

Darwin published that book only a year later, in 1859, and that was in great part precisely because of the rush created by his desire to have "priority" over this issue, after he received Wallace's manuscript, as recognized in Darwin's letters. So, purely in terms of academic procedure and fairness, officially the priority for that idea should have been attributed to Wallace's 1958 manuscript, if Lyell would have sent it for publication instead of getting involved in the weird "arrangement" to suddenly publish a Darwin–Wallace manuscript. If, for instance, Wallace had sent his 1858 manuscript to let us say a French or German scientist or scientific academy or journal without the knowledge of Darwin or his close Victorian colleagues, and the manuscript had been published some weeks or months later, today the precedence of the theory of evolution by natural selection would very likely be attributed to Wallace's manuscript. The fact that Darwin had started to develop his idea before

Wallace did, as indicated by the historical records available, would not be a decisive factor for that precedence attribution. Of course, we do not know whether Wallace's manuscript would have become or not as influential as Darwin's subsequent *Origin* book became, worldwide. Probably not, because of both the type of detail and prose of Darwin's book and also the *Power of Place* of Darwin, among other factors. But that has nothing to do with scientific "priority," as explained above: nowadays, if we discover that a certain species described by an author A in a 1900 article published in a renowned scientific journal or book was actually previously described in 1888 by an obscure author B in an obscure scientific journal, author B would still be considered the one that "originally" described that species.

As explained by Slotten in *The Heretic*, various authors have argued that, apart from being involved in such a weird arrangement concerning this topic of priority, Darwin might have actually changed a few details of his own theory after he received the manuscript of Wallace:

> Not all historians of science are satisfied with this account of a famous historical conjuncture [the arrangement between Darwin, Hooker and Lyell]. John Langdon Brooks has cast a less flattering light on the events of 1858. He believes that Darwin behaved in an underhanded manner to secure his place in history, outmaneuvering Wallace, who should be recognized as the first person to announce a complete theory of evolution. Brooks's investigation focuses on a sliver of time that Darwin specialists have elected to pass over. What really happened to Wallace's paper, he asks, after it was mailed from Ternate in March 1858? Brooks assiduously works out the elaborate journey of Wallace's essay to Down House and concludes that it could have arrived in England as early as May 18. (H. L. McKinney provides evidence that a letter to Henry Bates's brother, Frederick, dated March 2, 1858, but presumably mailed on the same day as the essay to Darwin, arrived in Leicester on June 3.) Thus Darwin possessed Wallace's essay for at least two weeks – and perhaps for as long as a month – prior to his notation of its receipt in his private journal and the letter to Lyell. During this time period, Brooks boldly asserts, Darwin engaged in a bit of intellectual piracy, revising his notions about natural selection and divergence after reexamining Wallace's Sarawak Law paper, "*On the Law Which Has Regulated the Introduction of New Species*", and studying the new essay.
>
> These revisions, eventually embodied in the later chapters of *The Origin of Species*, resemble Wallace's conceptions more than his own previous sketchy ideas, which did not make as striking a connection among extinction, intermediate forms, and the natural system of classification. That resemblance, he claims, is greater than Darwin scholars have acknowledged, and he spends a good portion of his book comparing sections of the *Origin* and Wallace's essays to substantiate his claim of intellectual theft. Without Wallace's two papers, according to Brooks, Darwin could not have completed the *Origin*. Darwin not only plagiarized Wallace but never gave him proper credit for his critical contributions in any edition of the *Origin* or in his autobiography. Arnold Brackman, who shares Brooks's conviction, is even less sanguine about the outcome of the July 1, 1858, meeting of the Linnean Society. Brackman accuses Darwin, Hooker, and Lyell of conspiring to rob Wallace of his rightful claim to priority.
>
> It is a classic tale of class power, he says, one more example of an elite group of men trampling a lower-class rival. Lost letters, a missing manuscript, pirated doctrines, behind-the-scenes maneuvering, and *mea culpa* letters to the offended party all lend themselves to assertions of foul play – Wallace and Darwin à la Mozart and Salieri. But such views oversimplify the events of 1858 and presume that the participants possessed a greater knowledge of the future than they actually had. Despite their eloquent advocacy, Brooks and Brackman have failed to convince leading scholars that Darwin plagiarized Wallace, though many have criticized Darwin (in retrospect) for inadequately referencing Wallace's work in future editions of the *Origin*.

About this topic, one should say that a more detailed analysis is clearly needed to test the idea that Darwin might have, or not, adapted some of his ideas to match what was written in – and/or to try to distinguish some of his ideas from those of – the manuscript that Wallace sent him in 1858. Be that as it may, this would not mean that Darwin committed plagiarism because Darwin was undoubtedly developing his ideas well before 1958, as extensively documented in his notebooks and letters. It would just mean that he could have been slightly influenced by Wallace's manuscript, that is all. After all, the only historical fact that we are sure of about this issue on whether Darwin changed or not some parts of his manuscript after he received Wallace's is that Darwin had an advantage that Wallace did not have. That is, that he was able to read Wallace's complete manuscript in 1858 before he finished his 1959 book, while Wallace obviously did not have the possibility to read Darwin's manuscript before he wrote his. Be that as it may, until such much needed comparative works between the ideas of Wallace's 1958 manuscript and the chronological order in which Darwin developed the specific ideas of his 1959 book, I would tend to agree with Slotten. That is, that as far as *we know* what was historically taken from Wallace's manuscript was its priority as the first manuscript that was completed about the idea of evolution by natural selection, with the issue about Darwin changing some of his ideas after that not being as relevant within that big-picture historical context.

So, having this in mind, let us now see the way in which Wallace reacted to the priority "arrangement" made by Darwin, Hooker, and Lyell's about *his 1858 manuscript*, in *its absence* and *without his knowledge*. As put by Slotten, Wallace's reaction might be at first sight surprising, for most people, because he mainly repeated, until the end of his life, that regarding the theory of evolution driven by natural selection, "all credit [was] to Darwin; none for himself":

> Wallace.. summarized his findings from Menado in an article written in October 1859 and published the following April in *Ibis*. One remarkable statement he made had nothing to do with his own observations. For the first time, he made it clear that the theory of natural selection was Darwin's theory, not his own. Regarding the maleo's unusual behavior, he said, "For a perfect solution of the problem we must, however, have recourse to Mr. Darwin's principle of 'natural selection', and need not then despair of arriving at a complete and true 'theory of instinct'." It was at this point that he seems to have closed his species notebook. If he had any lingering hopes of completing his own major work on the subject, they came to an end with Darwin's announcement. Whatever internal conflict he may have had about the priority issue had been resolved. The position that he took on that occasion was the position he stuck to for the rest of his life: all credit to Darwin; none for himself.

Why did Wallace react in such a way? About this topic, Quammen emphasized a crucial point that is often not recognized in accounts about Darwin: that apart from not being at all involved in the "arrangement" between Lyell, Hooker, and Darwin, Wallace actually received, subsequently, letters from Darwin that included some "untruthful" statements about it:

> Wallace, on the other hand, didn't consent to the joint reading (at least, not in advance); he couldn't, because no one consulted him. He was still doing fieldwork in the eastern islands,

> unreachable on short notice, far out of the loop. Nobody seems to have asked Lyell and Hooker: Gentlemen, what's the allfired hurry? Nobody suggested that Darwin, having waited twenty years to publish, might reasonably wait another six months for Wallace's acquiescence. It was a done deal before anyone thought to quibble. The reason for hurry, I think, was that Lyell, Hooker, and Darwin all felt some embarrassment about this high-handed bestowal of shared credit, and they knew that delay might bring complications. So there was no delay. The insiders moved deftly and fast. The details were settled in a flurry of overnight letters between London and Downe. Hooker chose an excerpt from Darwin's 1844 essay and inserted that, along with the Gray summary and Wallace's manuscript, into an already full agenda for the Linnean Society meeting.
>
> These three statements were ordered alphabetically by author – Darwin's two, followed by Wallace's. On the evening of July 1, 1858, the Darwin-Wallace material and five other papers were read to an audience of about thirty people. Hooker and Lyell attended. By coincidence, so did Samuel Stevens, who may have wondered how this Wallace paper got to London without passing through his hands. The two authors were absent. By hindsight you might view them as "conspicuously absent", although Wallace's non-presence wasn't notable at the time. He didn't belong to the Linnean Society. His voice was admitted like the crawk of an exotic parrot, interesting and indelicate.. he was unaware of the event in London.
>
> Darwin, acutely aware, missed the Linnean meeting, too. He was home in Downe with a dead child and a bad case of ambivalence. The most remarkable thing about Darwin-Wallace night at the Linnean Society is how little immediately came of it. No general discussion followed the reading of papers. No one stood up in response to what Darwin and Wallace proposed and said, That's brilliant! or That's outrageous! Tea was served, probably. There was some private chat. And then the Linnean fellows went home. The foundations of science had shifted beneath their feet but they didn't notice. Why not? This is hard to know. Possibly it was because the excerpts from Darwin and the paper from Wallace focused on the circumstances and details of the mechanism, natural selection, not on its larger significance. The word "transmutation" wasn't mentioned by either author, let alone the word "evolution"..
>
> In the ears of a careless listener, on a hot July night, during an overlong meeting, the Darwin-Wallace readings with their roundabout logic may have seemed to involve merely varieties and variation. Another reason that the audience missed the point may have been that those Linnean fellows generally weren't asking themselves the question – How do species change, one into another? – that Darwin and Wallace were answering.. Wallace got news of the arrangement, by letter, when he returned to his base in Ternate. There wasn't one letter but two – from Darwin and from Hooker. Darwin's contained Hooker's as an enclosure, leaving Hooker to do the main explaining. Darwin was understandably abashed and tried to portray himself as a passive party swept along by events. (Later he would assure Wallace that "I had absolutely nothing whatever to do in leading Lyell & Hooker to what they thought a fair course of action", a claim that was weaselly at best and arguably *untrue*, given his strong hints and lamentations to both men.. he would also misstate the dating of his own excerpts in the Darwin-Wallace package, telling Wallace that they'd been "written in 1839 now just 20 years ago!" In fact, they'd been written in 1844 and 1857.) Both letters to Wallace have been lost, but Darwin mentioned elsewhere that he considered Hooker's to be "perfect, quite clear & most courteous" in presenting the *fait accompli*.

Fait accompli is a French term that is often used to describe "an action which is completed before those affected by it are in a position to query or reverse it": Quammen's choice of words thus does reflect what happened, in a way, back then. As the saying goes, the rest is history.

The *Descent*, Sexual Selection, and Human "Races"

> *The beauty of anti-racism is that you don't have to pretend to be free of racism to be an anti-racist.. anti-racism is the commitment to fight racism wherever you find it, including in yourself.. and it's the only way forward.* (Ijeoma Oluo)

About the time after the publication of his 1959 *Origin* book, Darwin wrote, in his autobiography:

> My "*Descent of Man*" was published in February, 1871. As soon as I had become, in the year 1837 or 1838, convinced that species were mutable productions, I could not avoid the belief that man must come under the same law. Accordingly I collected notes on the subject for my own satisfaction, and not for a long time with any intention of publishing. Although in the "*Origin of Species*" the derivation of any particular species is never discussed, yet I thought it best, in order that no honourable man should accuse me of concealing my views, to add that by the work "light would be thrown on the origin of man and his history". It would have been useless and injurious to the success of the book to have paraded, without giving any evidence, my conviction with respect to his origin.
>
> But when I found that many naturalists fully accepted the doctrine of the evolution of species, it seemed to me advisable to work up such notes as I possessed, and to publish a special treatise on the origin of man. I was the more glad to do so, as it gave me an opportunity of fully discussing sexual selection – a subject which had always greatly interested me. This subject, and that of the variation of our domestic productions, together with the causes and laws of variation, inheritance, and the intercrossing of plants, are the sole subjects which I have been able to write about in full, so as to use all the materials which I have collected. The "*Descent of Man*" took me three years to write, but then as usual some of this time was lost by ill health, and some was consumed by preparing new editions and other minor works. A second and largely corrected edition of the "*Descent*" appeared in 1874.

The *Descent of Man* is Darwin's most controversial book. Controversial not only because it was the book in which he extensively developed the idea that humans derived from other animals and compared different living human groups but also because it included several factually inaccurate ethnocentric and sexist statements. Darwin was amazingly efficient and brilliant at observing and putting together, as a biologist and naturalist, very specific and often previously unnoted details about nonhuman organisms and geological formations. However, as explained above, as an anthropologist, his personal biases and prejudices, amplified by a certain lack of depth regarding other fields of knowledge and about some particularly complex societal issues, led him to commit such factual inaccuracies about human evolution. In a simplified way, and using modern terms, he was a hugely successful biologist, but clearly not as thriving as an anthropologist – or, for that matter, as a psychologist, as stated by Gruber:

> In reconstructing the private path Darwin took to his theory of evolution, we have used as our guideposts the rich manuscripts he left, and as our clearly visible beacon the theory at which he indubitably arrived. In the case of his psychological thought, we must be either prudent or more speculative. The manuscripts available are fewer and the ultimate extended public statement is almost entirely lacking. In short, Darwin, by choosing to remain a biologist, "failed" to become a systematic psychologist.

This idea, that Darwin was much more successful studying other organisms than he was at observing his own species, is somehow consensual within the relatively few historical and scientific books that do not idolize him. For instance, Quammen, in his *The reluctant Mr. Darwin*, also stated that "the *Descent of Man* is really two books smooshed into one, as its full title admits: *The Descent of Man, and Selection in Relation to Sex*.. the smooshing is far from seamless; there's a big, lumpy transition after the first seven chapters, right where Descent gives way to Sex." He further stated that, "humanity's descent from an animal lineage was one of Darwin's boldest ideas, true, but that book on the subject isn't one of his best efforts.. published in 1871, intended as a complement to *The Origin of Species*, it doesn't have the same sharp focus, inexorable momentum, and magisterial power." Having said that, one should give a huge credit to Darwin for being brave to write a whole book about two topics that were so tabooed back then: human evolution and sex. To give just an example of how this was so, Gregor Johann Mendel, born in 1822, began his studies on heredity using mice at St. Thomas's Abbey but his bishop did not like the idea of having one of his friars studying sex, and particularly animal sex. This was one of the various reasons that led Mendel to subsequently mainly use plants for his studies. Mendel is reputed to have said, "I turned from animal breeding to plant breeding.. you see, the bishop did not understand that plants also have sex." Ironically, the fact that almost nobody knew about Mendel's sex genetic studies back then, including Darwin, led to the lack of something crucial in Wallace's and Darwin's theories about evolution by natural selection. Namely, the knowledge about genes and how they are passed from parents to offspring, and therefore about one of the major sources of a critical, central theme in Darwin's works, biological variation.

The marked brilliant biologist versus less accomplished anthropologist dichotomy is particularly noticeable in Darwin's *Descent* because the part about sexual selection of nonhuman animals is scientifically excellent, while the parts about sexual selection in humans and about human evolution are much weaker, scientifically. Some of the factually inaccurate "facts" and ideas included in those weaker parts, such as that women's have "lower" mental capacities than men, were also included in Darwin's private notebooks. As noted by Gruber, "in the M and N notebooks, when he compared himself with his sisters and in other passages, he seemed to be saying that he had a highly developed imagination [which he considered to be the product of 'higher mental powers'], good powers of attention, and enough of the other faculties to form a useful repertoire." As emphasized by Gruber, Darwin's writings, including his published books, often also included rather inaccurate and logically inconsistent assertions about innate versus acquired mental capacities and behaviors in different human groups. Furthermore, Darwin's Victorian bubble made him confuse "mental powers" and the mere acceptation of socially constructed Victorian morals, conducts, and notions of "progress." This is clear, for instance, when he stated in the *Descent* "that man has *risen*, trough by slow and interrupted steps, from a *lowly* condition to the *highest* standard as yet attained by him in knowledge, morals, and religion." Not surprisingly, for him the highest standard was that displayed by the Victorian morals and protestant beliefs of the English white "gentlemen."

Apart from the reasons mentioned above, another major factor that contributed to the inclusion of many inaccurate "facts" and ideas within Darwin's accounts on human evolution concerned the exaggerated – and often incorrect – parallels that he made between artificial selection made by humans and the races derived from it – for instance, those of dogs – and the evolution of so-called human "races." It is often argued that the English noun "race," used from the mid-1500s on, probably came from the Italian word *razza* – meaning a biological species or kind – which in turn might be related to the Old French *haraz*. This latter term referred to horses and mares kept for breeding and might have been related, in turn, to the Arabic word *faras*, or horse. Although there is still some controversy about some of these etymological connections, what is important to emphasize here is that using the biological term race – as applied biologically to subgroups of domestic animals such as dogs or horses – to designate groups such as "blacks," "whites," "Europeans," "Africans," or "Asians" is scientifically wrong. This has been pointed out by many biologists and anthropologists since decades ago, being, for instance, officially recognized by the *American Anthropological Association*. However, the media and general public often continue to wrongly use terms such as "black race" or "white race," in great part precisely because numerous prominent scientists have incorrectly used such terms, for centuries – strikingly, some still use, including Nobel laureates, as we shall see below.

In the case of Darwin, it is true that back then scientists were not aware about many of the specificities known today regarding the evolutionary tree of our species. However, Darwin should have been aware of the fact that using domestic animals such as dogs, horses, or cattle as models for the evolution of human groups would be far stretched. For instance, "fullblood" cattle are considered to be "fully pedigreed" because we *know* that every ancestor is in theory registered in a "herdbook" and that he/she tends to show the typical characteristics of that specific "breed." But obviously nothing like this applies to "black" or "white" people, for many reasons. Firstly, Europeans came from Africa, so this means that some ancestors of the so-called "whites" were "blacks" and, accordingly, that the group defined as "blacks" would not include all their descendants. Secondly, in a country like the USA today, a huge number of "whites" do not have exclusively European ancestors and an enormous number of African Americans do not have exclusively African ancestors. So, the problem is not only that Darwin used the Victorian society as *the* model for the pinnacle of human evolution and Victorian ideas as a model for biological evolution in general, as when he applied Malthusian and capitalistic ideas to create *general* evolutionary ideas. The problem is also that, although he had the brilliance to understand that what he designated as "artificial selection" could provide some clues about "natural selection," he overemphasized such links. One of the reasons for that is related to the personalized and teleologized way in which he often depicted "Mother Nature," which led him to often confuse Her "broom" – natural selection – to the artificial selection broom used by our species. The latter often has a *final, conscious goal*, planned by humans, such as selecting bigger tomatoes or a type of dogs that are particularly good for hunting – such a final, conscious goal is obviously lacking in natural selection. Apart from this crucial difference, and related to it, another major reason why changes provoked by artificial selection often occur faster and in a more marked way than those occurring during natural selection is that within the former, humans can make sure that organisms A and B do not reproduce, even if they want to do so. For instance, the domestic "fullblood" cattle concept mentioned above would very rarely apply to wild animals of a same species that coexist in a same geographic area and that *can and want* to reproduce with each other.

These points can be illustrated by a comparison between dogs and wolfs. Dogs came from wolves, and thus their origin is more recent than that of wolves. But, compared to wolves, dogs are far more diverse in size, shape, and mental attributes: there is a vast number of physically very different dog breeds, contrary to wolves. This is because dog owners or breeders have precisely used techniques – for instance, castration – that guarantee that even if different dog breeds interact with each other physically, let us say in dog parks, they would not reproduce with each other. Of course, nothing similar to this occurs without human intervention in nature. That is, separation of subgroups of species, either by physical barriers leading to allopatric speciation or by anatomical, physiological, or behavioral barriers leading to sympatric speciation, do occur without the intervention of our species. But these phenomena occur due to mechanisms that are completely different to those planned and undertaken by humans, such as castration and other techniques.

What about the so-called "racial" or "ethnic" human groups that exist nowadays? Firstly, as noted above, although some of them may purposely *plan or try* to only reproduce between them, almost all of them ultimately include at least some members that reproduce with members of other groups, as consistently shown by genetic studies. Actually, such studies have also shown that members of our species even reproduced in the past with those of other human groups that are sometimes considered to be from other species, such as Neanderthals. Each of us has at least some genes that derived from Neanderthals, due to such sexual intercourses. Therefore, not even *Homo sapiens* can be truly considered biologically to be "fullblood." So, talking about "pure or fullblood human races" such as "blacks" or "yellows" or "whites" really makes no biologically sense at all. Such groupings are nothing more than just social – often racist – human constructions, which were often created to unify a certain group of people by excluding "others" using factually inaccurate narratives such as the teleological notions of "progress," "purpose," and the ladder of life.

In many private letters, Darwin explicitly recognized that he did apply his concepts of artificial selection to his ideas about the evolution of humans in particular, because humans are "said to be domesticated." The human self-domestication hypothesis is nowadays increasingly accepted among biologists and anthropologists, so this idea that humans somehow domesticated themselves is not a problem per se, well on the contrary: Darwin's observations are, in this respect, brilliant. The problem is rather that he somewhat exaggerated the links between natural and artificial selection as well as the importance of sexual selection for the evolution of human "races." The theoretical link between artificial selection and sexual selection *sensu* Darwin is evident: contrary to what he defined as natural selection, in both artificial and sexual selection, the selector is an *organismal agent* – humans in the former, females or males of the group being selected in the latter. A reason why Darwin often overstated the importance of sexual selection to explain the evolution of different human "races" was because he accepted, a priori, the dogmatic view accepted back then that such "races" were real. So, Darwin was going in circles to try to explain, unsuccessfully, something that he assumed to be right from the start, but that was factually inaccurate. This circular reasoning becomes evident when we analyze and compare his published books – in particular the *Descent* – with his private letters, such as the one he wrote to Alfred Russel Wallace in 1864, in which he stated: "I suspect that a sort of sexual selection has been the most powerful means of changing the races of man." As explained in Desmond and Moore's 1994 book, Wallace did not agree – correctly so – with this idea of Darwin, and this was one of the various scientific reasons that started to drive the two men apart, together with other reasons, including their very different societal and political views:

> Wallace's naturally selected group morality was leading society in a very *unDarwinian direction*. The old socialist peered optimistically at the millennium under this moral regime: everybody will "work out his own happiness"; policing will be unnecessary, freedom will be the order of the day, "since the well balanced moral faculties will never permit any one to transgress on the equal freedom of others"; coercive governments will wither ("every man will know how to govern himself"), the lot to be "replaced by voluntary associations for all beneficial public purposes". On that upbeat, oddly anarchist note – intelligent selection leading to an egalitarian society – Wallace closed. Darwin was presumably nonplussed to find the path to utopia paved by his science.
>
> He told Wallace that the brain/body dichotomy was "grand and most eloquently done", but he demurred on the abating of selection and played dumb on the politics. As he reasoned, the Australian savage is subject to selection, given his "constant battles". And English society will stay vital and progressive only through unimpeded competition. The sickly and degenerate deserve to be scythed down, he believed, even as he sent subscriptions to the Downe charities to maintain his own paternal order and worried about his sons' in-bred ailments. He decried "primogeniture for destroying Natural Selection" even as he had Lubbock set up his eldest William in the banking business. Wallace shook his head. Wars did not pick the fit, for the "strongest and bravest" die first. Nor could he see much to "sexual selection", with each race choosing mates according to its own standards of beauty..
>
> Wallace and Darwin were differing more and more. Darwin had sexual selection replacing God the artist, just as natural selection ousted God the architect. Brutes were their own breeders, forming fancy varieties; humans sculpted themselves by selecting mates. But for every brilliant bird that Darwin ascribed to sexual selection, Wallace described another that was due to natural selection. Dull-coloured female birds are camouflaged to survive sitting in an open nest. Gaudy moths, distasteful to their predators, wear warning colours; still others mimic them. Wallace rejected sexual selection as the "main agent" forming the human races. Natural selection was quite equal to the task. To Darwin this came as "the heaviest blow possible". No amount of evidence on birds and insects would turn Wallace. Darwin now played the apostate devil, arguing against the very theory that he had spent his life supporting. Wallace for his part was turning out the more single-minded adherent of natural selection, more Darwinian than Darwin.. [Accordingly, Wallace also criticized] Darwin's claim that the European aristocracy is handsomer than the middle classes. Mere "manner" and refinement among the leisured classes were being "confused" with beauty. Politics was coming between the two men.

Wallace's statement that Darwin exaggerated the Malthusian war-like notion of struggle-for-existence within his evolutionary theories, as well as the importance of sexual selection within human evolution, is exceptionally interesting and revealing about how science and scientific biases operate. This is because Wallace's statement was in general scientifically accurate, as we now know on the basis of a plethora of available empirical biological and anthropological evidence. For instance, Wallace pointed out that Darwin's sexual selection theory for the origin of "races" had serious flaws because tribalism led people from a certain "race" to often choose to reproduce with members of the same "race," so that should mainly prevent new "races" from forming. Accordingly, Wallace proposed that natural selection was the main factor leading to the formation of new "races." And we now *know* that traits that were often historically used – wrongly so – to distinguish "races," such as the color of the skin, are indeed mainly correlated with environmental and geographic items such as the incidence of ultraviolet rays. That is, I am not arguing that *some of those traits* were not *partially* determined by sexual selection – that might well have been the case – I am merely saying that according to the evidence available that does not seem to have been the main factor in at many cases.

For instance, we know that people that happen to be born in locations near the Equator with a lighter tone of skin, for example, due to genetic conditions such as those occurring in African albinos, tend to be more at risk of skin cancer. So, clearly there was a huge external – environmental – selective pressure leading to a more common occurrence of a darker color of skin in those regions. It actually seems somewhat puzzling that when Darwin was writing the *Descent,* he somewhat minimized the importance of such a correlation between environmental factors and the development of so-called "racial" traits, particularly because this correlation provided a hugely strong support for the theory of evolution by natural selection that he developed earlier in the *Origin*. In a way, it is as if the Darwin that wrote the *Origin* was not the Darwin that wrote the *Descent*, further evidencing the logical incongruities characterizing the "*brilliant biologist vs less accomplished anthropologist*" dichotomy mentioned above.

Why would Darwin minimize the crucial role played by external environment in human evolution? Was it because if he recognized that natural selection leading to *local adaptations* was a crucial component within human evolution, then his Victorian ideas about human evolution being mainly a history of "progress" or about 'whites' being 'superior' to 'blacks' would not make sense? Or was it because within the capitalistic notions that were so in vogue in Victorian England by then – for instance, that one needs to be a *productive* member of society, an *active agent*, and so on – it would be somewhat problematic to accept that human evolution was in great part the mere result of a combination of contingent and random events? In this sense, it is interesting to see that this biologist–anthropologist dichotomy, which characterized Darwin during his whole career, also affected Wallace, but mainly only after a certain point of his life. Namely, while in his first decades as a naturalist Wallace's ideas about human evolution were in general consistent with his ideas about how other organisms evolved, and thus more accurate in that sense than some of Darwin's ideas about human evolution, Wallace later began to stray off. For example, after having criticized – rightly so – Darwin for stating that natural selection was very likely much less important than sexual selection within human evolution, Wallace began to argue that human evolution could actually not be explained at all by natural selection. He started to claim that the evolutionary history of the human brain could only be explained by "something else" and to be interested in areas such as spiritualism, as we shall see. In that regard, one can say that the two creators of the theory of evolution by natural selection could not avoid falling into the trap of thinking that there was something unique about humans, to the point that both of them later in life argued that natural selection could not be the main driver of human evolution. Such are the paradoxes created by anthropocentrism.

Within the biased assertions that Darwin made about humans in the *Descent*, one that became particularly influential – and problematic, with respect to its societal repercussions, to this very day – is related to the application of his exaggerated notion of a war-like struggle-for-existence to human "races." For instance, he wrote that "*when two races of men meet they act precisely like two species of animals – they fight, eat each other*." Apart from the fact that this assertion is factually inaccurate because there are no different human "races," phenomena such as globalization clearly show that humans from different countries, ethnic, and even religious groups do not always fight with each other. I am writing these lines in DC, a city that is a true melting pot, and I do not see African Americans, or European or Asian descendants, or Christians or Muslims or Hindus or Jews, fighting against, and "eating," each other all the time, here.

Actually, the fact that some members of some of these groups, or others, have engaged in wars, mass killings, or even genocides in the last century was, in many cases, precisely mainly due to ethnocentric constructions about "us *versus* them," rather than to any "natural" drive condemning humans to "naturally" behave like brutal killers (see also Box 3.5). Significantly, some of these social biased constructions, such as those related to Nazi ideologies, were profoundly influenced by such war-like "us *versus* them" struggle-for-existence narratives, as we shall see. Browne astutely discussed, in her *Power of Place*, how the recurrent application of such biases and circular reasoning in Darwin's works often lead him to come "full circle" and end up supporting the very societal and Malthusian ideas that provided the framework for his evolutionary theories in the first place:

> More obviously, social economists seized on parallels between the organic kingdom and political economy. Competition, struggle, adaptation, success, and extinction – all these concepts moved freely between both domains. They were the Malthusian parallels on which Darwin had first drawn when composing his theory. While many commentators of the period remained divided on Malthus's meaning for human society – to those with working-class sympathies Malthus's principle merely blamed the poor for being poor, a marked contrast to those who applauded it for encouraging responsibility and self improvement – there could be no denying the concept's status. In one sense it could be said that Malthus's images were turning *full circle*, for Darwin applied political economy to biology, and now

these biological ideas were being reintegrated back into political economy, seemingly providing a "natural" account of the way human populations and social economies were thought to work.

Malthus's principles were biologised and then reabsorbed into economic thought. In another sense, the social and the biological were scarcely separable. Malthus's remarks did not so much travel back and forth as exist already embedded in the same cultural context. Either way, Malthus's doctrines looked like incontrovertible laws of nature to a nation steeped in competitive economic activity, buoyed up with Samuel Smiles's anthems of self-help, adaptation, struggle, and survival, and as a political body fully engaged in territorial and commercial expansion. "*It is remarkable how Darwin rediscovers among beasts and plants the society of England, with its division of labour, competition, opening up of new markets, inventions, and the Malthusian struggle for existence*", remarked Karl Marx in a letter to Engels in 1862.

It is indeed striking that, as noted above, Marx – who was not a biologist – was able to easily detect something that is obvious but that many scientists, including many evolutionary biologists, are not able – or do not want – to recognize today, about 160 years later. Marx was against Darwin's defense of the "well-off," the Victorian highly hierarchical system, and the supposed supremacy of English imperialism. But he quickly realized that, by cherry-picking, defending, and using certain biased evolutionary ideas of Darwin, he could promote his own societal concepts and agenda. As noted by Browne, Marx literally proclaimed that "Darwin's work" suited his "purpose": "*although developed in the crude English fashion, this is the book which in the field of natural history, provides the basis for our views*," he said to Engels. He repeated the same comment to Ferdinand Lassalle, stating that "*Darwin's work is most important and suits my purpose in that it provides a basis in natural science for the historical class struggle*." To a certain extent, this is very common within the interactive links between biased scientists, politicians, and other societal actors: they influence themselves, often in a biased and self-reinforcing way. This was, for instance, the case when European scientists defended that Africans were "inferior": politicians used this idea to defend colonialism, and many within the broader public of their societies mainly accepted such ideas and therefore supported colonialism.

Importantly, some of Wallace's criticisms of Darwin for applying such Victorian biases and Malthusian war-like struggle-for-existence notions to his evolutionary ideas, as well as for overemphasizing the role played by sexual selection in human evolutionary history in particular, are very similar to some of the points made nowadays by more and more biologists defending the need for an *Extended Evolutionary Synthesis*. For instance, in contrast to Darwin's ideas, Wallace's theories were mainly based on group selection – something that is increasingly being recognized as critical within biological evolution and as crucial to understand the evolutionary history of cooperation and altruism. Obviously, Darwin recognized that cooperation and altruism exist in, and are important within, nature, including human evolution. However, the more selfish, individualistic, struggle-for-existence view of biological evolution that he proposed based on Victorian sociopolitical ideas and ideologies, and the further exaggeration of this idea by subsequent scholars, played a huge role within the elaboration of the Evolutionary synthesis. In *Darwin and the emergence of evolutionary theories of mind and behavior*, Richards recognized some of these key differences between the ideas of Darwin and Wallace:

> Though Darwin believed that competitive struggle brought civilized reason up from a primitive state, his ideas did not completely converge with those of Wallace, despite his own observation to the contrary. The differences may appear slight, but they grew in significance during the latter part of the 1800s. The first difference concerns the unit of selection in human mental evolution. Wallace argued in his 1804 paper: "*Capacity for acting in concert, for protection and for the acquisition of food and shelter; sympathy, which leads all in turn to assist each other; the sense of right, which checks depredations upon our fellows [etc.]... are all qualities that from their earliest appearance must have been for the benefit of each community, and would, therefore, have become the subjects of 'natural selection'.. Tribes in which such mental and moral qualities were dominant, would therefore have an advantage in the struggle for existence over other tribes in which they were less developed, would live and maintain their numbers, while the others would decrease and finally succumb*".
>
> [Therefore] Wallace assumed that selection would operate on whole communities and tribes, since the traits selected would confer benefit primarily on the group rather than the individual. He thus appears to have endorsed group selection, though undoubtedly without detecting the miasma of difficulties infecting that concept. Darwin, by contrast, hinted at nothing beyond individual selection in his letters to Wallace and Lyell. Darwin's initial failure to recognize this difference may have been due to his own acceptance of community selection in the *Origin*. In Darwin's version of community selection, however, the unit was a group of *related* individuals.

While at a societal level, Darwin almost always sided with the powerful, the factory bosses, the privileged, and the colonizers, Wallace did indeed tend to have more empathy toward the underprivileged, the colonized, and the oppressed, although there are of course many exemptions to this oversimplification, as explained above. Wallace's views contributed to the fact that he did not became as respected as Darwin by the Victorian elite because these were not at all the type of views that the "well-off" were looking for to justify the continuation of the *status quo* and of their privileged lives. But many other factors were also at play, including the numerous ways in which Wallace "shot himself on the foot" later in his life, as for example when he began to put his own ideas about natural selection in question, regarding human evolution – Slotten's book provides a fascinating account about how Wallace began to "go astray":

> Traditional religious dogma was no longer acceptable, yet the new scientific materialism, with its faith in nothing but blind natural forces, was profoundly unsettling. Spiritualism's peculiar blend of science, philosophy, and concern with the supernatural offered a middle ground for agnostics, and it was to spiritualism that Wallace turned for answers to life's greatest mysteries. Natural selection could explain the origin of species, including the origin of humankind, but Wallace felt it could not explain the origin of our moral and intellectual nature, consciousness, life, or the origin of the universe. Wallace believed that he had found evidence of Mind or Intelligence behind natural laws, and of an Intelligent Designer manipulating those laws for a higher purpose: humanity's spiritual evolution after death..
>
> This was more than Darwin could bear. Wallace's beliefs opened a chasm between the two men that has survived them. Although Darwin did not dispute Wallace's co-origination of his theory, he began to doubt Wallace's scientific judgment. At times, he tried to undermine Wallace's credibility, most famously in rebuttals against him in *The Descent of Man*, but he did not have to try very hard: Wallace had undermined his own scientific credibility

more effectively than any of his enemies – or friends ever did. His fervent socialism and baffling stance against the efficacy of the smallpox vaccine did not enhance his reputation. By making scientific meetings and popular journals his bully pulpit for views that only a minority shared, Wallace alienated many of his fellow scientists. The result was that much of his later work was dismissed as the work of a crank, a man of science who had mysteriously been led astray, a view still held today.

In Browne's *The Power of Place*, she relates these topics with Darwin's fascination with the importance of sexual selection in human evolution. Namely, the discussions and facts provided in her book lead us to raise the following crucial question: apart from his Victorian biases and anthropocentrism, were there also *specific personal* reasons that led Darwin to exaggerate the importance of sexual selection in human evolution? That is, did he do this in the *Descent* – published 12 years after the *Origin* – in order to further stress the fact that his theories were markedly different from those of previous authors, in particular from those of Wallace? Browne wrote:

At the centre lay Darwin's idea of sexual selection. This was his special contribution to the evolutionary story of mankind, his answer to Wallace, Lyell, and others, and to all the reviewers and critics of the previous twelve years. "I do not intend to assert that sexual selection will account for all the differences between the races", he wrote in his book. Nonetheless, he felt certain that it was "the main agent in forming the races of man". Sexual selection was "the most powerful means of changing the races of man that I know". In brief, Darwin claimed that human beings were like animals in that they possess many trifling features that are preserved and developed solely because they contribute to reproductive success. Just as peacocks had developed tail feathers to enhance their chances in the mating game, so humans had developed characteristic traits that promoted individual reproductive success.

These traits were fluid, changeable, and not directly related to adaptation and survival. But Darwin pushed this claim far beyond the mere acquisition of secondary sexual characteristics. By these means he thought he could also explain the divergent geographical and behavioural attributes of human beings, such as skin colour, hair texture, maternal feelings, bravery, social cohesion, and so forth. Preference for certain skin colours was a good example. Men would chose wives according to localised ideas of beauty, he suggested. The skin colour of a population would gradually shift as a consequence. Similarly, sexual selection among humans could enhance mental traits such as maternal love, bravery, altruism, obedience, hard work, and the "ingenuity" of any given population.. In effect, humanity made itself by producing and preserving differences, a process that broadly mirrored his understanding of artificial selection in which farmers chose traits for "use or ornament", impressing their own taste or judgement on organisms. He ventured onto thorny ground when he analysed human societies in this way. His naturalism explicitly cast the notion of race into evolutionary and biological terms, reinforcing contemporary ideas of a racial hierarchy that replicated the ranking of animals.

And he had no scruple in using the cultural inequalities between populations to substantiate his evolutionary hypothesis. Darwin certainly believed that the moral and cultural principles of his own people, and of his own day, were by far the highest that had emerged in evolutionary history. He believed that biology supported the marriage bond. He believed in innate male intellectual superiority, honed by the selective pressures of eons of hunting and fighting. The possibility of female choice among humans hardly ruffled the surface of his argument, although he repeatedly claimed that female choice was the primary motor for sexual selection in animals. Primitive societies, he conceded, may be matriarchal or polygamous.

However, he regarded this as an unsophisticated state of affairs, barely one step removed from animals. Advanced human society, to Darwin's mind, was patriarchal, based on what was then assumed about primate behaviour and the so called "natural" structure of civilised societies. For Darwin, it was self-evident that in civilised regimes men did the choosing. A limited number of women might sometimes be in a position to choose their mate (he was perhaps thinking of heiresses, or royalty, or beautiful heroines in novels). But his vision of mating behaviour was an explicit expression of his class and gender. His personality was evident too. His description of courting practices in *The Descent of Man* gave a romanticised picture of "rustics" at a country fair, "courting and quarrelling over a pretty girl, like birds at one of their places of assemblage". For him, Victorian males set the evolutionary compass.

The *Descent* and the Malthusian "War-Like" Struggle-for-Existence

The histories of mankind are histories only of the higher classes. (Thomas Malthus)

Gruber's *Darwin on Man* highlights how Victorian sexist biases influenced not only Darwin's books about human evolution – particularly the *Descent* – but also his theories about biological evolution in general. For instance, Gruber points out that within the "ensemble of metaphors" used by Darwin, three "come easily to mind": the "free market of Adam Smith," "war" – related to the Malthusian notion of struggle-for-existence – and "artificial selection" – complementing natural selection, both linked to the notion of "survival-of-the-fittest." These three concepts were particularly prominent in the Victorian era. The powerful "survival-of-the-fittest" concept, which later became – and still is – so often used by white supremacists, racists, and social Darwinists, was originally popularized by Herbert Spencer. As explained in Black's *Against the Weak*:

> In the 1850s, agnostic English philosopher Herbert Spencer published *Social Statics*, asserting that man and society, in truth, followed the laws of cold science, not the will of a caring, almighty God. Spencer popularized a powerful new term: "survival of the fittest". He declared that man and society were evolving according to their inherited nature. Through evolution, the "fittest" would naturally continue to perfect society. And the "unfit" would naturally become more impoverished, less educated and ultimately die off, as well they should. Indeed, Spencer saw the misery and starvation of the pauper classes as an inevitable decree of a "far-seeing benevolence", that is, the laws of nature. He unambiguously insisted, "the whole effort of nature is to get rid of such, and to make room for better.. if they are not sufficiently complete to live, they die, and it is best they should die". Spencer left no room for doubt, declaring, "all imperfection must disappear". As such, he completely denounced charity and instead extolled the purifying elimination of the "unfit". The unfit, he argued, were predestined by their nature to an existence of downwardly spiraling degradation. As social and economic gulfs created greater generation-to-generation disease and dreariness among the increasing poor, and as new philosophies suggested society would only improve when the unwashed classes faded away, a third voice entered the debate.
>
> That new voice was the voice of hereditary science. In 1859, some years after Spencer began to use the term "survival of the fittest", the naturalist Charles Darwin summed up years of observation in a lengthy abstract entitled *The Origin of Species*. Darwin espoused "natural selection" as the survival process governing most living things in a world of limited

resources and changing environments. He confirmed that his theory "is the doctrine of Malthus applied with manifold force to the whole animal and vegetable kingdoms; for in this case, there can be no artificial increase of food, and no prudential restraint from marriage". Darwin was writing about a "natural world" distinct from man. But it wasn't long before leading thinkers were distilling the ideas of Malthus, Spencer and Darwin into a new concept, bearing a name never used by Darwin himself: social Darwinism. Now social planners were rallying around the notion that in the struggle to survive in a harsh world, many humans were not only less worthy, many were actually destined to wither away as a rite of progress. To preserve the weak and the needy was, in essence, an unnatural act.

Social Darwinism is, in a way, a normalization and naturalization of poverty, starvation, social hierarchies, inequities, subjugation, and exploitation: these items are seen as "natural," as the norm, a *fait accompli*. However, as noted above, anthropological, archeological, and historical data show that this was not at all the norm, but rather an exception, during 99.8% of our 7-million-year evolutionary history, until the rise of major sedentary agglomerations and agriculture in various regions of the globe, since about 15,000 years ago. Scott's 2017 *Against the Grain* is a superb book that summarizes a huge amount of data that show the fallacies of the narratives we often heard in school about brutish, non-civilized "savages" or barbarians that had a horrible *all-or-nothing* type of life, always at the brink of starvation. Compared to the earlier agricultural societies, hunter-gatherer "savages" or "barbarians" actually often had a more relaxed lifestyle, in general, because if a certain food item would eventually not be available, they would have a plethora of other options available. In contrast, many earlier agricultural societies depended on a few crops, and sometimes even a single crop, which could become scarce for a series of reasons, such as grasshopper plagues, or a higher dryness of the environment. Even less than two centuries ago, millions of people of a so-called "civilized" country suffered enormously because of problems related to a single crop. The Great Famine, or Irish Potato Famine, led to mass starvation and disease in Ireland from 1845 to 1852, with about 1 million deaths and more than a million people leaving the country. Similar stories, and often even much more horrible ones, about plagues, starvations, massive death, huge migrations, and so on, are recurrent in texts written in agricultural regions thousands of years ago – the Bible is an example, among countless ones.

These topics are related to Social Darwinism in other manners as well. For example, as we will see, the countries in which Social Darwinism has permeated the most their socio-economic system, tend to have problems such as a very high number of average work hours per week, or chronic stress, sleep deprivation, suicides, anxiety attacks, and so on. One of the reasons is precisely because many people in such countries, particularly the underprivileged ones, do feel that they are often facing *all-or-nothing* situations in which everything they have built in their lives, or dreamed about, may collapse suddenly. And unfortunately, it often does, indeed. Such as when a professor, or nurse, in the USA can eventually become homeless in just a few weeks because they do not have enough money to pay both the huge medical bills and the mortgage when they discover they have a cancer or very harsh chronic diseases. I personally know many of such cases. In Washington DC, where

Fig. 1.17 Scientific studies using blood markers and other methods have shown that people from so-called "traditional" groups such as the Tsimane of Bolivia often have, on average, less heart attacks, stress, anxiety, burnouts, and depression than people living in so-called "developed" societies and particularly in big cities such as New York or Tokyo

I live, many people like to go to Dupont Circle to play chess with homeless people, and this type of stories is recurrent, when they talk about their lives. I have written extensively about these topics in my last book, *Meaning of Life, Human Nature, and Delusions*, and we will refer below to some data related to them, so I will not further discuss these topics in this section. It suffices to say that numerous studies of physiological markers and body organs have empirically shown that, in comparison with more "traditional" societies such as the *Tsimane* of Bolivia (Fig. 1.17), people living in "developed" countries such as the USA or Japan tend to have, on average, more stress, anxiety, heart attacks, depression, or burnout.

Fascinatingly, even authors idealizing Darwin recognize not only this fallacy but also various other problems related to Social Darwinism, as explained by Richards. However, as noted by him, "Darwin's enthusiasts.. [then] often exculpate him by settling the human evolutionary debt on Spencer – urging that what goes under the rubric of social Darwinism ought really to be called 'social Spencerianism'." Another example of the type of mental gymnastics that is often used in publications idolizing Darwin, related to the narrative that only "good" things can come from

Pope Darwin, so all "bad" ones have to come from other people. From *it's not God's fault, it's the Devil's* to *it's not Darwin's fault, it's Spencer's*, so *it's not Social Darwinism, it's Social Spencerianism*: Darwin's idolization in a nutshell. Unfortunately, such inaccurate idealized narratives slowly become seen as reality, as a dogma, entrenched in popular culture. For instance, this narrative was followed in Black's outstanding book *Against the Weak*, which is an otherwise very well-documented book written by an excellent author that is not at all a typical "Darwin's enthusiast."

This narrative includes the omission of many historical facts, including what Darwin truly wrote and the books he published. For instance, while Spencer's ideas – including his 1850 book *Social Statics* and notion of "survival-of-the-fittest" – were published years before Darwin's 1859 *Origin*, one cannot neglect the crucial fact that later in life Spencer – and his works – were themselves highly influenced by Darwin's ideas. Spencer himself made this very clear. Moreover, Darwin *actively decided* to coined Spencer's term "survival-of-the-fittest," in his writings because this concept was very important in the way Darwin constructed his notion of biological evolution. That is, although Darwin and Spencer were surely not, at a personal level, the "best pals," their ideas and writings were nerveless feeding each other, in an interacting, circular, self-reinforcing process. Darwin's 1871 *Descent* plainly highlights the tragic culmination of these biased and, in this sense, unscientific process: Darwin not only repeatedly cited and supported Spencer ideas but also constructed *factually inaccurate "scientific facts" and "theories"* to further support and expand such ideas to the natural world as a whole.

The end result is that some of the passages of the *Descent*, a book that was and still is often seen as a scientific "masterpiece," are indeed plagued by social Darwinist narratives, some of them being very close to eugenicist ideas. Indeed, tragically, regrettably, and shamefully for evolutionary biologists such as me, some of those passages indeed resemble many of those used in hard-core eugenicist publications and even in Hitler's 1925 *Mein Kampf*, published just 54 years after the *Descent*, as noted above. Moreover, regarding this topic, such a simplistic idealization of Darwin and demonization of Spencer often omits the "other side" of Spencer, as explained in Richards's 1987 book. As Darwin had different sides that often were logically incongruent – supporting racist and pro-colonialist views and, at the same time, being anti-slavery – Spencer did have a very different side, which is sometimes referred to as "utopian socialism" because in some ways it was almost the opposite of the dark "Social Spencerianism" side that so many scholars are so inclined to emphasize in order to idealize Darwin. Idealization of something often involves demonization of something else. As Richard noted:

> Spencer's evolutionary ideas, to which I will specifically turn in a moment, began to smolder during the 1840s, initially ignited by his reading of phrenology and Lyell's account of Lamarck's views. The evolutionary theory adumbrated in *Social Statics* served as a vehicle to give some scientific substance to his Utopian socialism. Spencer believed that the progressive development of man toward a perfect society was a logical necessity. He argued that individuals were ever adapting to different social circumstances through inherited use and disuse. Increasing adaptation, on its side, improved social relations and circumstances,

> thus creating a new environment against which individuals would, per force, continue to adapt. This developmental dynamic would, in Spencer's estimation, have two consequences, one social and one moral. Progressive adaptation would, first, produce greater specialization within society, each part of the social organism becoming more articulated and adapted to particular functions.. Adaptation to the social state had, in Spencer's estimation, a further consequence, this a moral one.. [it] produced an evanescence of evil and a social organization with the most elaborate "subdivision of labor" – that is, an extreme mutual dependence, because of adaptation, but with each individual exercising his faculties in complete freedom, since all pernicious desires will have been purged from society..
>
> In a passage that might be mistaken for an excerpt from the *Communist Manifesto*, Spencer declared: "*All arrangements, however, which disguise the evils entailed by the present inequitable relationship of mankind to the soil postpone the day of rectification.. a 'generous Poor Law' is openly advocated as the best means of pacifying an irritated people.. workhouses are used to mitigate the more acute symptoms of social unhealthiness.. parish pay is hush money.. whoever, then, desires the radical cure of national maladies, but especially of this atrophy of one class and hypertrophy of another, consequent upon unjust land tenure, cannot consistently advocate any kind of compromise*".. Is this socialism or even communism? The critic of the *North British Review* thought so. He declared that.. "Mr. Spencer but repeats the.. notion also it is that history is an evolution of the doctrine of equal rights, and that the goal to which the human race tends is that of anarchy, or the absence of all forms of government". But Spencer even went further.. since he included both women and children in this social evolution, arguing that they should achieve equal rights with men. For the reviewer this latter proposal conjured up nightmares of domestic anarchy.

It should however be emphasized that Spencer's utopian socialism used some of the same type of biased notions that Darwin's works used in the *Descent*. For example, the idea that the marked division of labor so typical of the industrial revolution and the Victorian Era was the most "elaborate" or "progressed" type of social organization. On the other end, this utopian socialism side of Spencer is clearly different from the "social Darwinism" side of Darwin, or from any other side of Darwin, in the sense that Darwin's ideas – both expressed in private and public writings – tended to be markedly conservative regarding societal matters, as noted above. Darwin defended social hierarchies and the Victorian *status quo* and, contrary to what happened to Spencer, nobody would describe Darwin as someone who proclaimed "communist" or "anarchist" ideas.

This topic about social Darwinism and Darwin's legacy leads us to discuss another powerful example of how Darwin's idealization is often done in the literature and the type of mental gymnastics that is involved in doing so. While Darwin's enthusiasts have no problem to proclaim that Darwinism "conquered" and remains "triumphant" and "highly influential" not only in Western countries but also in the globe as a whole, they then argue that people like David Duke, Hitler, or other prominent white supremacists did not know, cared about, or were influenced by Darwin's ideas. Once again, they want to have the cake and eat it too. If one proclaims that Darwinism "conquered" and remains "triumphant" and "highly influential" to this day, how can one then argue that magically no prominent political or societal leader – particularly among white supremacists – was influenced by his ethnocentric, racist, and sexist ideas or at least used them as easy ammunition to "scientifically" justify theirs? As viruses such as Covid-19 do not infect only people of a certain ethnicity or with certain political views, widespread scientific theories, such as geocentrism and heliocentrism – that is, that earth or the sun are the center of the solar system respectively – or Darwinism, cannot be known by or influence exclusively the "good" guys.

There is ample historical evidence concerning the critical links between scientific racism and sexism, politics, and societal repercussions that unambiguously shows that this is not how things truly are, except within flawed idealized narratives. In this specific case, there *are* numerous writings and recorded speeches of people like Hitler and David Duke and they do refer to evolutionary ideas and metaphors that were repeated over and over by Darwin – such as the notion of a "struggle-for-existence" in the natural world – in order to "scientifically" legitimize their white supremacist and racist ideas within the general public, as we shall see in detail in Chap. 3. A very different thing is to engage in a never-ending discussion about non-testable topics such as: would Darwin be happy to know that people like Hitler and David Duke used such ideas and metaphors, and directly cited passages from his own books as Duke did, to justify the horrible things they said and did, if he would come back from his grave? I personally think he would not, based on what we *know* about Darwin. For instance, he was never a major fan of violence, although he did publicly support, in his books – as a further example of his typical logical incongruities – things that necessarily involved a lot of physical and mental violence, such as British colonialism and imperialism. But this is just an educated guess, which obviously will never be possible to be tested because Darwin will never come back from his grave. Such unanswerable and untestable questions are in a way unproductive and, importantly, somewhat distracting because they divert us from the historical, testable, and extremely well-documented fact that racists, white supremacists, and misogynists *did* use the easy ammunition provided by Darwin, often directly citing his works.

This is indeed another major problem characterizing countless accounts idealizing Darwin: they often engage in such distracting and untestable assertions and end up by diverting the public's attention from such unquestionable historical facts by simply stating things such as "I am sure Darwin would never approve this." Firstly, this is not true – nobody can be truly *sure* of what Darwin would do, or which political ideas he would support if he was alive today. Secondly, even if Darwin would not support the white supremacist ideas of let us say David Duke, the *fact* is that Duke did directly cited passages of Darwin's *Descent* to "scientifically" support his racist and white supremacist ideas, which were then accepted by millions of people in the U.S. An emblematic example of how scholars often engage in such distracting mental gymnastics instead of referring to, or despite recognizing, these historical *facts* is precisely Gruber's *Darwin to Man*. For instance, in the excerpt below, Gruber provides a fascinating, and in a way somewhat amusing, example of this:

> [About] struggle, Charles Darwin is often, and I believe incorrectly, characterized as the biological theorist of struggle in the sense of hostile warfare among living beings. One even encounters the argument that the biological doctrine of survival of the fittest justifies war among nations. Nothing could be further from either Charles' or Erasmus' [Charles's grandfather] views. It is true that they both adopted the language of warfare from time to

> time, and in critical passages describing the struggle for survival. Nevertheless, it is clear that their usage was metaphoric and that the metaphor must be qualified in several important ways..
>
> Any industrious reader can entertain himself by finding passages in Charles' work that will seem to contradict the above remarks. He used many metaphors, among them the metaphor of war among men. On balance, the metaphor of war was foreign to his mature view of nature, and insofar as it crept into his thought it may even have hindered him in the development of his more central theme of inventive variation and selection. Darwin's great love of Milton's poetry at one period of his life seems to me to argue against the view of Darwin as reluctant to accept the notion of titanic struggle in nature. Darwin tells us that during the Beagle voyage Milton's *Paradise Lost* accompanied him everywhere, the only book so cherished. The core of *Paradise Lost* is the struggle between good and evil, and the fall of man. Assuming that Darwin took all this to heart (we ought at least to allow the possibility that he read Milton in a more recreational spirit), we can only guess what it meant in his life: perhaps the Weltschmerz of early manhood. The years of the voyage were enough to cure him of any predilection he may have had for this polarized view of reality.

The mental gymnastics employed in this excerpt offers a mesmerizing case study to analyze the mechanisms of idealization. First, Gruber recognizes that Darwin used countless times the evolutionary "struggle-for-existence" metaphor and often associated this figure of speech with the "metaphor of war." But then, in a surprising twist worthy of a thriller movie, he tells us that Darwin was surely cured "of any predilection he may have had for this polarized view of reality" trough his "great love of Milton's poetry." That is, Gruber minimizes the fact that "any industrious reader" can find many passages in Darwin's works that use a war-like concept of evolutionary struggle-for-existence by simply noting that Darwin.. read Milton's poetry. Case closed. Apparently, everybody knows that those that read Milton's poetry are always kind, loving, caring human beings.

This argument makes no sense at all, obviously, because many people that did huge atrocities, including those that cited Darwin's works to justify them, did read poetry, or liked art in general. The more extreme, and painstaking, example of this is Hitler: he liked art and even painted. Moreover, such flawed narratives just emphasize the absurdity of the other side of idealization: demonization. As if someone that did horrible atrocities could not be, at same time, able to like art, following a narrative of absolute "evil" counterbalancing absolute "good."

Furthermore, Gruber's argument makes no chronological sense. The *Origin* and *Descent* passages in which Darwin used countless times the evolutionary war-like "struggle-for-existence" metaphor were written *after* his Beagle voyage, so clearly those voyages did not "cure him" of this predilection. This predilection started from the very moment Darwin read Malthus and other authors proclaiming Malthusian ideas and lasted until the very end of his life. As put by Browne in *Darwin Voyaging*, Malthus had given Darwin a "fierce and destructive sword." As we have seen above, Darwin stated that "*when two races of men meet they act precisely like two species of animals – they fight, eat each other, bring diseases to each other &c., but then comes the more deadly struggle, namely which have the best fitted organizations, or instincts (ie intellect in man) to gain the day*." The fact that Malthus' societal ideas and then Darwin's evolutionary ideas about a suffocating, omnipresent, implacable struggle-for-existence was – and continues to be, in a way – so widely and blindly accepted for such a long time within Western scholars and popular culture is one of the most powerful examples of the scientific and societal repercussions of Darwin's idolization.

In Richards' 1987 book about *Darwin*, he explains that internalist historical explanations tend to focus more on, let us say, psychological traits of the scientists being discussed, while externalist ones tend to be more centered on external social and economic substructures. As we have seen above, a fascinating point about Darwin's idolization is that it often involves a logically incongruent combination of both. In a simplified way, what is perceived as "positive" aspects of Darwin's ideas, such as his theory of natural selection, is mainly attributed to his "genius," individually. In contrast, in the much fewer common cases in which scholars recognize "problems" or "negative" aspects of Darwin's ideas, these are typically attributed to his society, or to others, such as Spencer. As usual, reality lies in between. On the one hand, Darwin's theory of evolution by natural selection is a product of both his peculiar mind and way of doing things, *as well as* of his cultural and societal background. The proof of this is that while such a theory did not exist for millions of years of human evolution, it was developed at the same time by two men from exactly the same country at almost at the same time: by him and by Wallace. On the other hand, Darwin's "scientific" support for racist and ethnocentric ideas, as well as for colonialism, social hierarchies, and British imperialism, was obviously also related to the time and society in which he lived *and* of his own way of being, such as his marked social conservatism and inability to escape from such Victorian narratives and biases and the disgust he felt when he saw slaves being mistreated, during his travels.

Even top historians such as Richards – who is otherwise often so sober, astute, and incisive in his books about scientists other than Darwin – fall into that trap when Darwin is the historical figure being discussed. For instance, Richards recognized that "the Darwinian image of man is that of a competitively isolated individual.. [but] some historians and philosophers, while acknowledging this as the image of man forged by nineteenth-century evolutionary theory, are uneasy.. like Eiseley, [who wrote] that 'man was not Darwin's best subject'." Richards further noted that "Himmelfarb, with less reserve, simply maintains that Darwin's efforts in biopsychology display his 'failures of logic and crudities of imagination'." However, after saying this, Richards then criticizes such a view, and states: "*man, I hope to show, was indeed Darwin's best subject – and Spencer's as well.. their scientific and philosophic considerations were penetrating and sophisticated*."

One wonders if Darwin's ideas about non-Western peoples are really his "best," "penetrating and sophisticated" ideas. When Darwin wrote "I do not believe it is possible to describe or paint the difference between savage and civilized man.. it is the difference between a wild and tame animal," only greater, is this really one of the "best" ideas that Darwin produced? Is his statement that "man is more courageous, pugnacious, and energetic than woman, and has a more inventive genius" truly representative of his most "sophisticated" ideas? Fortunately, Richards was wrong about this, as we have seen. Darwin's ideas about human evolution, diversity, and genders are not at all his best, well on the contrary. His observations and ideas about specific traits of nonhuman organisms, such as the petals of plants or the beaks of birds, and his theory of evolution by natural selection, those are the things that truly show us Darwin's genius: they are among the most brilliant ideas ever formulated within the field of biology. By trying so hard to idealize Darwin by stating that his anthropological observations are "Darwin's best," authors such as Richards are ultimately doing instead a major disfavor to Darwin and his otherwise mainly brilliant work about non-human organisms.

In several occasions, such as in a letter from Darwin to Lyell – 18 June 1958 – Darwin makes it crystal clear that the war-like Malthusian "struggle-for-existence" notion was not just a crucial part of his theory of evolution by natural selection but that the *whole theory depended* on it. Darwin wrote: "I explained to you here very briefly my views of 'natural selection' depending on the Struggle for existence." Among the numerous examples showing this, one that I find particularly curious, and emblematic, concerns Darwin's reference to the "constant struggle" between sharks and fishermen. As noted by Gruber, "halfway between the Cape Verde Islands and the northeastern coast of Brazil.. in his diary he merely wrote 'they [seamen] caught a greater number of fine large fish and would have succeeded much better had not the sharks broken so many of their hooks and lines'". Seven years later, in Darwin's *Journal*, this became "the sharks and the seamen in the boats maintained a constant struggle which should secure the greater share of the prey." Indeed, a very different, and exaggerated, tone. Yes, sometimes sharks break fishermen's hooks and lines, in some regions of the globe. But saying that there is a "constant struggle" between them to "secure the greater share of the prey" is in actual fact a highly inflated statement, to say the least.

Apart from his Victorian biases, and strategic use of such metaphors to strengthen specific critical concepts defended in his theories, why did Darwin so actively and repeatedly use such simplistic, and sometimes even factually inaccurate, metaphors in his works? One of the reasons for him to do so was probably linked to both his eloquence as a science disseminator and his desire for immortality, which are ultimately deeply related. He was a brilliant writer and *knew* that a keyway to bring attention to his works was to use such catchy metaphors. By doing so, his peers and the general public would more easily remember his work and general ideas, as they clearly did, and continue to do. The use of catchy – and, unfortunately, many times erroneous – metaphors is a strategy commonly used by many scholars and writers, particularly by those that are more prominent, often in part precisely because they are so good at creating and using them. Being a prominent scientist is not only about having brilliant ideas but also about the power of place and the ability to disseminate them to the greatest number of people. Current examples of this are prominent scholars such as Richard Dawkins or Steven Pinker, which are almost omnipresent within the scientific media. The "butterfly" effect, "struggle-for-existence," "survival-of-the-fittest," DNA "blueprint," "mitochondrial Eve," or talking about the immune system as a police force, or blood vessels as highways, or natural selection as Mother's Nature "broom," or computers as brains, or the nervous system as trees, are examples of powerful metaphors to convey ideas, no doubt about this. But unfortunately, many of these metaphors are oversimplified or exaggerated, or scientifically inappropriate, and some are even factually inaccurate. In a way, the evolutionary notion of "survival-of-the-fittest" is a mix of the former three, while the evolutionary war-like concept of "struggle-for-existence" or of a purposeful Mother Nature's "broom" is literally partially inaccurate scientifically, as seen above.

There are some indications that Darwin's repeated emphasis of the "struggle-for-existence" evolutionary metaphor was also related to the very specific contingencies concerning his personal life and health. Quammen explains that when Darwin was writing the *Origin* during the 1840s, he was often feeling sick and thought he could even die before the book's publication. Darwin "told Emma in the testamentary letter, she should please raise the offer to £500 [for an appropriate person to finish, improve, and edit the work].. and if that didn't suffice, he wrote, just publish the thing as it is." For "all he knew, the gut-heaving, head-blurring symptoms of his chronic illness might turn acute at any time, and he could be dead of some unknown ailment within a year." While he was fearing for his life and plaguing his book with the struggle-for-existence concept, we would write letters – for example to Leonard Jenyns – specifically talking about this concept. At a time, he was thinking about his health struggle and possible early death, and in such letters he would ask things such as: how severely do struggle and early death limit population increase for any given species? In addition, we know that Darwin's evolutionary "struggle-for-existence" metaphor was also related to other, broader contingencies of his life because the very same notion had been employed, also in an evolutionary way, by his grandfather Erasmus Darwin. As seen above, the similarity of Charles and Erasmus Darwin's ideas does not concern only this metaphor, but many other strikingly similar evolutionary concepts. As pointed out by Gruber:

> Erasmus Darwin is known as the grandfather of another famous evolutionist, as the author of an evolutionary theory quite similar to and fifteen years earlier than Lamarck's, as a founder of the Lunar Society, as a famous physician, and as the composer of thousands of lines of verse, mostly rhyming couplets popularizing quite accurately the scientific knowledge of his day.. How can we understand the extraordinary fact that grandfather and grandson generated influential theories of organic evolution? The basic answer is that the Darwins' evolutionary thought was not simply a special scientific hypothesis that happened to clash with accepted religious ideas on certain matters of fact. It was, rather, the product of, and an essential part of, a *Weltanschauung* closely linked to the making of the industrial revolution and the political revolutions, notably the French, those great historic currents spanning the years 1776–1848..

> To clarify the frame of reference that Charles Darwin brought with him, first to his university education and later to his scientific work, would be conceptually simpler if this framework were complete prior to the intellectual developments it is intended to explain. The task is complicated by the fact that the framework evolves in the life history, and is changed by the very ideas we are trying to explain. Charles did not begin his university education or his scientific career as a *tabula rasa* [blank state] he came with some family tradition, some conception of himself, and some ideas.
>
> [We therefore need] to turn our attention to a sketch of some ideas shared by grandfather and grandson.. [For instance] there are two aspects of [Erasmus' ideas].. which are really transformed in Charles' treatment of them, because of his explicit and thorough development of the theory of natural selection. The first of these is the mechanism of sexual selection. Erasmus touched on it, and expressed the basic idea quite clearly. Having described the struggle among males for sexual access to the female in various species, he concluded: "The final cause of this contest amongst the males seems to be, that the strongest and most active animal should propagate the species, which should thence become improved". In Charles' thought, sexual selection was one specialized aspect of the much more general process of natural selection.
>
> Systematically placed in this more general context, the concept gains much greater power and clarity. Charles Darwin dealt with it briefly in the *Origin* and in great and masterful detail in Part II of his work on man, whose full title is *The Descent of Man*, and *Selection in Relation to Sex*. Finally, both men gave some attention to the theme of fertility. Erasmus, in a highly Malthusian passage, wrote of the explosive potential for population growth inherent in the reproductive mechanisms of all species, and concluded: "*All these, increasing by successive birth, Would each o'erpeople ocean, air, and earth.. So human progenies, if unrestrain'd, By climate friended, and by food sustain'd.. would spread Erelong, and deluge their terraqueous bed.. But war, and pestilence, disease, and dearth, Sweep the superfluous myriads from the earth.. The births and deaths contend with equal strife, And every pore of Nature teems with Life*".

"Charles did not begin his university education or his scientific career as a *tabula rasa*; he came with some family tradition, some conception of himself, and some ideas," including those from his own grandfather. Interestingly, Charles almost never refers to this obvious fact, and specifically to the evolutionary ideas of his grandfather, in his autobiography. He also almost never does this point even in the books that he published during his life. That is, Darwin minimizes, or literally omits, the crucial importance of the contingencies of life – and particularly of his *power of place, and the fact that many of his evolutionary ideas were very similar to those of his grandfather* – in his writings. Most scientists do so, not only to maximize their own intellectual merit but also because science is supposed to be objective and unbiased, and thus to not be influenced by such contingencies of life, a specific culture, being born at a certain geographical place and time, having a certain grandfather, and so on. But, of course, this is almost never truly so. Take, for instance, the fact that Darwin defended, in the *Descent*, that women are sexually selected by men, something that was inconsistent with his – correct – idea that in mammals, including apes, in general the females are the ones that select males. Why did he say that females of our evolutionary lineage were different from those of most other mammalian groups? A major reason for this was that within Victorian manly made narratives, women were typically seen as *passive* players of society – including as sexually passive – so surely they could not possibly be the *active players*, choosing their partners: it had to be the other way round.

This was indeed a topic that was used – correctly so – by many feminists to criticize the misogynistic ideas defended in Darwin's books, particularly in the *Descent*. As noted in Kimberly Hamlin's 2014 *From Eve to Evolution*: "the two main tenets of sexual selection theory then were male battle and female choice of sexual mates; however, Darwin asserted that among humans, men, not women, selected mates, an observation that puzzled many nineteenth-century reformers because it seemed to contradict Darwin's otherwise firm belief in the animal-human continuum." The way Darwin talked about his idea of human male selection clearly shows how he saw women as merely passive sexual players, exactly as they were commonly seen in the Victorian Era – incorrectly so, as we will see in Chap. 4. Darwin wrote: "It could never have been anticipated.. that the power to charm the female has sometimes been more important than the power to conquer other males in battle." Basically, human evolution was chiefly a manly thing: men conquered sexually passive women *and* conquered other men in battles. Interestingly, on some occasions, such biases, prejudices, and/or logical inconsistencies were recognized by Darwin himself. For instance, in his notebooks, he called "absurd" the very notion of "biological progress" that he used so often in his books, and particularly in the *Descent* concerning human evolution, as noted by Gruber:

> As he came to accept modern geological views of a constantly changing order in the physical world, a contradiction within his point of view developed as follows: each species was adapted to its milieu; the milieu was undergoing constant change; and yet the species were changeless. Darwin probably began to feel this contradiction during the final months of the voyage, as he was going over his notes and organizing his materials. It was not until July 1837, ten months after returning to England, that he began his first notebook on "*Transmutation of Species*." It was over a year after that, in September 1838, that the role of natural selection in evolution began to be clear to him. But Darwin did not await this clarification before applying his still chaotic and murky ideas to man. In 1837 Darwin believed that the chief agent of evolutionary change was the direct influence of the changing physical world. Accordingly, in the very first passage of his transmutation notebooks Darwin writes, "even mind and instinct becomes influenced" by changes in the physical environment. A few pages further on, his feeling that man is part of the evolving natural order is reflected in an interesting argument by analogy from man to the rest of that order: "Each species changes. Does it *progress*. Man gains ideas. The simplest cannot help becoming more complicated; and if we look to first origin, there must be *progress*." A few months later Darwin makes a different comment on the inevitability of progress, showing that he has come to see some danger in this anthropocentrism: "It is absurd to talk of one animal being higher than another. We consider those, where the cerebral structure, intellectual faculties, most developed, as highest. A bee doubtless would where the instincts were".

Brain sizes and domestication provide another example of the logical inconsistencies of the *Descent* and of how the facts that Darwin the meticulous biologist recognized in nonhuman organisms were often contradicted by the biased ideas of Darwin the anthropologist. About rabbits, he wrote: "I have shewn that the brains of domestic rabbits are considerably reduced in bulk, in comparison with those of the wild rabbit or hare; and this may be attributed to them having been closely confined during many generations, so that they have exerted their intellect, instincts, senses and voluntary movements but little." Although there are exceptions, Darwin's assertion is in general right, as in various species of domestic animals, the brain is proportionally smaller than those of wild animals of the same species or of closely related wild species. However, in the very same book, when it came to talk about human "progress," "civilization," and wildness, he completely reverted the story. Within the idea of human self-domestication and using his rabbit example, the next logical step would be for him to assert that those that he called "wild savages," such as the Fuegians, "exerted their intellect, instincts, senses and voluntary movements" more than did Victorian people that were "closely confined" during a substantial portion of their lives to their homes or were doing very repetitive tasks confined into their workspaces.

Did Darwin say that? Of course not. Europeans have to be superior, women have to be inferior, so he changed the narratives accordingly, once again, when it came to human evolution. Contrary to most nonhuman animals, a sedentary and "closely confined" lifestyle leads humans to have higher intellects, instincts, and senses, not the other way round, he argued.

Obviously, this does not mean that those humans that scholars designated as "non-civilized" have necessarily bigger brains, and moreover having bigger brains does not lead unavoidably to being "smarter" or having a higher "intellect." On the one hand, we do know that the "non-civilized" Neanderthals often had, in proportion, a bigger brain that many of us "civilized" people have. On the other hand, empirical studies based on autopsies show that people that spend more years in schools – for instance, that completed university versus those that only completed high school – tend to have more developed or complex connections in areas implied with the understanding of language. And this is the major point, which brings us again to the Darwin biologist versus anthropologist mismatch. Within his general natural selection idea, selection refers only to a specific trait or group of traits at a specific moment in time and geographical area. So, it is not surprising that people that go more to school, a place in which language, reading, writing, and so on plays a huge role, may have developed more certain *specific* traits within certain *specific* areas of the brain related to language. But this does not mean that such so-called "highly-educated civilized" people are overall "better" or "smarter" or "superior" than people living in non-agricultural groups, as wild rabbits or "wild" Neanderthals having bigger brains in proportion than domestic rabbits and "civilized" humans does not make the former "better" or "superior" or "smarter" than the latter. There is no "better" or "worse" or "superior" or "inferior" within the natural world, contrary to the terminology recurrently used by Darwin the anthropologist to refer to humans in the *Descent*. There are just local adaptations, local environments, specific traits, and so on: as Darwin the biologist often – but not always, it should be stressed – postulated when he discussed the evolution of nonhuman organisms in the *Origin*.

The *Expression*, Altruism, Morals, Inequities, and Indoctrination

No person, I think, ever saw a herd of buffalo, of which a few were fat and the great majority lean. No person ever saw a flock of birds, of which two or three were swimming in grease, and the others all skin and bone. (Henry George)

In 1872, just 1 year after publishing the *Descent*, Darwin published another book that mainly focused on humans and that he mainly ignored when he wrote his autobiography, just a few years after that. The only thing he said about it was: "My book on the *Expression of the Emotions in Men and Animals* was published in the autumn of 1872.. I had intended to give only a chapter on the subject in the *Descent of Man*, but as soon as I began to put my notes together, I saw that it would require a separate treatise." Why did Darwin mainly ignore this book in his autobiography? We will never be able to know for sure, but it might have been, at least in part, because, as put by historians of science such as Richards, this was by far Darwin's most "disconcerting" book. In his epoch, it was widely criticized by numerous Victorians, even by some of Darwin's closer scientific allies. While the *Descen*t is by far Darwin's most unfortunate book because of its numerous erroneous racist, ethnocentric, sexist, and war-like struggle-for-life ideas and because of their subsequent societal and scientific repercussions, the *Expression* is, scientifically speaking, his weakest book. This is because whereas the *Descent* part about humans was plagued by factually inaccurate ideas, the parts about evolutionary vestiges and sexual selection in other animals were quite strong. But the *Expression* was mainly plagued by "speculations" – something that Darwin so harshly criticized when it was done by other authors – biased assertions, and inaccurate passages, as recognized by many of his Victorian peers and subsequent scholars.

The *Expression* focused on very complex societal, sociological, and psychological topics concerning, for instance, emotions, morals, and ethics. But for some reason Darwin – who was often very cautious when it came to publishing in general, and took a huge amount of time to publish his *Origin* – took much less time to publish the *Expression*, just 1 year after the *Descent*. According to various historians of science – such as Richards in his 1987 book – Darwin just spent a bit more than 20 months putting his notes together and writing the whole book. Whether this was truly so, or not, the fact is that compared to many of his other books, the *Expression* does read a bit like a somewhat rushed book. There are many theories about why Darwin published his two books about humans in just 2 years. These include the scientific pressure he was getting to do so, and, related to this, the attacks that Wallace did later in his life, arguing that the theories that Darwin put forward in the *Origin* could not truly explain the evolution of humans and their mind, emotions, and morals. Because Richards discussed these topics in detail in his 1987 book, which is precisely about the emergence of evolutionary theories of mind and behavior and mainly focuses on Darwin, a substantial portion of this short section is based on that 1987 book.

As seen above, one major problem that Darwin had when he was trying to provide a solid, well-grounded discussion on the evolution of morality and traits such as altruism – as compared to Wallace, before he started to "go astray" – concerned the different ways in which the two saw natural selection. Darwin's theories about natural selection tended to focus more on individuals, or on groups of *related* individuals – "community selection," which somewhat resembles the current term "kin selection." In contrast, Wallace's ideas focused more on what we now define as group selection, that is, to whole populations or even major evolutionary groups formed by them. In this sense, Wallace's ideas are closer to those defended in the last decades by scholars such as *Stephen Jay Gould* and, more recently, by many evolutionary biologists stressing the need for a more comprehensive, and inclusive, Extended Evolutionary Synthesis. According to such scholars, it is crucial to recognize that selection can operate at many levels, including genes, cells, individual organisms, populations, species, and even larger groups. For them, group selection is particularly critical to explain the evolution of features such as morality, altruism, religion, and so on.

For instance, why do catholic priests take celibacy vows that ultimately lead them to not have kids and thus to *not* propagate their own genes? Many authors that define themselves as "Darwinian or Neo-Darwinian" defend that a major reason might be kin selection. That is, that the families of those catholic priests benefit from this, so ultimately the genes of their brothers, or sisters, or other family members are more propagated. However, there are no major, conclusive studies showing that, in different geographical areas, epochs, or contexts, this is truly often the case. Instead, what anthropological, historical, psychological, and sociological studies seem to indicate is that this might rather benefit a larger, non-kin-related, group: the Catholic Church and its followers as a whole. For instance, by enforcing and following laws, norms, or practices that are particularly difficult to follow and that are evolutionarily detrimental to certain individuals within the group and even to their families or loved ones, the size or power of the religious group or sect can actually grow. As explained by Hood and colleagues in the fascinating 2009 book *The Psychology of Religion*, this is because the members of that group are sending a strong signal to the other members of that group. Basically, they are saying: look, you can trust me, I am not a cheater, I am not having sex, or eating pork, my whole life, you can see how serious I am about this, believe me. After all, a major characteristic of religions is precisely that they tend to transcend families, and even ethnic groups: you can be one of *us*, either if you are Bantu, or a Native American, or a Scandinavian, you just need to trust *us* and *believe* in *our God*, or *Gods*.

There are some passages of Darwin's *Descent* in which he does seem to be hinting at something similar to such a notion of group selection, probably at least in part because of Wallace's attacks and also because he had carefully read Wallace's 1864 paper "The Origin of Human Races" before he published the *Descent*. Indeed, it is ironic that while in popular culture Darwin is wrongly portrayed as the first to write about natural selection and how it applied to human evolution, others finished manuscripts about these topics before he did. As noted above, Darwin had the privilege to read Wallace's 1858 manuscript about evolution by natural selection before he published the *Origin* in 1859, and to read both Wallace's 1864 paper and Huxley's 1863 book *Evidence as to Man's Place in Nature* before he published the *Descent* in 1871 and the *Expression* in 1872. In that 1864 paper, Wallace proposed that some moral qualities could confer an advantage to a group in their competition with other "tribes." About these topics, Darwin wrote, in the *Descent*:

> It must not be forgotten that although a high standard of morality gives but a slight or no advantage to each individual man and his children over the other men of the same tribe, yet an enhancement in the standard of morality and increase in the number of well – endowed men will certainly give an immense advantage to one tribe over another. There can be no doubt that a tribe including many members who from possessing in a high degree the spirit of patriotism, fidelity, obedience, courage, and sympathy, were always ready to give aid to each other and to sacrifice themselves for the common good, would be victorious over most other tribes, and this would be natural selection.

There are some major problems with this passage. One concerns the combination of Darwin's Victorian biases and the incongruities between them and some of his broader evolutionary theories. This is clearly patent in the last sentence, which is in great part the result of an indoctrination to the cult of power and social hierarchies that neglects the reality of human history. When people in Egypt fought wars and built pyramids thousands of years ago, they were surely not doing this for the *common good*, but mainly for the "good" of pharaohs, religious leaders, and others of the cream of the Egyptian society. Many of those that built the pyramids were slaves that were forced to do so and that lived particularly harsh lives. They did not build the pyramids altruistically, but because the pharaohs and other leaders obliged them to do so. This also applied, for a long time, to the British empire, which had slaves and forced millions of people to work in its colonies supposedly for the "good" of Britain, but truly particularly for the good of its monarchy, the well-off, and the cream of the British society. The same applies today, although obviously in a different way, to millions of African kids that are obliged – either by their parents or by others, or because of their extreme poverty – to work for 10, 15, or even more hours a day, for instance in mines, to extract components needed for *our* Western cellphones, computers, and so on. Such kids are not doing that for the common good of their ethnic groups, nor of the globe as a whole, but in great part for the "good" of the Western societies that colonized and oppressed their ancestors centuries ago. This is part of the vicious cycle of subjugation and poverty created by colonization, oppression, and exploitation. That is why it was naive – and logically inconsistent, in a way – for Darwin to say that he was against "slavery," while at the same time he was publicly so enthusiastic – including in his scientific books, as we have seen – about British imperialism and colonialism. Indeed, history tells us that since the rise of so-called "early civilizations" and thus of extreme inequities and social hierarchies, a group fighting wars or engaging in imperialism or colonialism and "being victorious over most other tribes" often had nothing to do with the pursuit of attainting overall "common good," but rather the "good" of that group and in particular of the cream of its society. All the major early civilizations engaged in forced labor, subjugation, oppression, and so on. The naive notion that groups such as the Ancient Egyptians, Assyrians, or Romans fought together and "sacrificed themselves" for the *common good* neglects the fact that not only most people that fought were obliged to do so but also that most of the leaders that obliged them to do so never truly "sacrificed themselves." As they often do not do today, when they sit in the Kremlin and send their "people" to fight and die for their wars and interests. As astutely put by Paul Valery, contrary to such naive and factually wrong assertions, war between "civilizations" is usually "*a massacre of people who don't know each other for the profit of people who know each other but don't massacre each other*."

A further example of such logical incongruencies is that while in the *Descent* Darwin postulated that in general those societies that failed to focus on the *common good* would perish, in the *Voyage* he had defended the opposite, when he stated: "the perfect equality of all the inhabitants will for many years prevent their civilization.. until some chief rises, who by his power might be able to keep to himself such presents as animals &c &c, there must be an end to all hopes of bettering their condition." According to the ideas he defended in the *Descent*, societies such as the Fuegians would be the pinnacle of evolution, as within the evolutionary history of our species nobody truly regards the *common good* as highly as small nomadic hunter-gatherer egalitarian societies tend to do. One of the many reasons for that is precisely *because* they often prevent the rise of chiefs and of the "1%" that keep "presents" to themselves and that oblige others to mainly work or fight for their "good" even if almost nobody wants to do so. Not because there are "noble savages" but because that would often be highly detrimental to small nomadic groups.

One of the most detailed – and brilliant – books about this topic is Scheidel's *The Great Leveler: Violence and the History of Inequality from the Stone Age to the Twenty-First Century*. Based on an incredible amount of historical and economic empirical data, Scheidel states that within those groups in which there was a rise of agriculture and "civilizations," subsequently there was *almost always* an increase of inequities. The major exceptions have been due to what he calls the "four horsemen of leveling": mass-mobilization warfare, transformative revolutions, state collapse, and catastrophic plagues. In fact, in contrast to the narratives that we are often indoctrinated to *believe* in school – which have been repeatedly defended by so many scholars, such as Darwin – stable Western societies usually tend to follow the same trend, contributing to the rise of inequalities. This also applies to more recent stable Western democracies, which are often designated – mostly by the people that live in them – as the "best of all types of socio-political government." As shown by Scheibel, even during the last decades of stability and absence of war in Western European and North American countries, inequality has risen significantly. To give just one among countless examples, within the USA, the richer 1% had 8.2% of "wealth" in 1980, 13% in 1990, and 17.5% in 2010, with the lowest value recorded within the last half century being 7.7%, in 1973. As noted by Scheibel, although these numbers may vary within different studies and economic markers, there is a clear general trend for Western countries, and the whole globe, within the last 50 years.

Scheibel's book is very likely the kind of book that, if published during the Victorian Era, would likely have been mainly disregarded or even highly criticized by well-off Victorians, such as Darwin. The type of facts given in that book does not match well with the type of stories that the cream of Victorian society liked to read and *believe*. For instance, about cultural and moral "progress" and how the British Empire was the pinnacle of that "progress" and thus had a greater "common good" than did "brutish savages" such as the Fuegians. Well into the twenty-first century, we are still often indoctrinated to *believe* in some of these, and many others, types of erroneous Eurocentric and ethnocentric narratives, from a young age, in school, TV, at home, by the media, and so on. Tragically, in face of such indoctrination, it is often amazingly difficult, for many people, to distinguish reality from fiction – in particular, racist, sexist, Eurocentric, or ethnocentric fiction. One of the most efficient ways to try to escape from such indoctrination is to read things written by "others," from their own perspective, or by someone from our in-group that does "*not think like those who do not think*," as put by Marguerite Yourcenar. For instance, concerning topics such as the Fuegians, we Westerners were indoctrinated to believe that – and we do still often hear things like – Southern Europeans "discovered" America. This is obviously historically wrong, as the first people that "discovered" America were of course the Native Americans, many thousands of years before Europeans went to America. Actually, technically, Native Americans went to lands that are now part of the European territory much before any European went to America. This is because about 4500 or 4000 years ago – approximately 2500–2000 BCE – Native Americans went from what is today Canada to Greenland, which is nowadays officially an autonomous territory within the Kingdom of Denmark. Not only that, studies indicate that about five centuries before Southern Europeans went to America, the Scandinavians – namely the Norse Vikings – had already likely done so: historical data indicate that the first Europeans to be born in America were Vikings, not Southern Europeans. Why is this not so often emphasized in popular culture, and in school, in many Western countries? Well, because in our countries the Vikings have been often portrayed as brutish "barbarians" and "pagans," contrary to the "civilized" Christians of Portugal, Spain, and Italy.

Also, contrary to how narratives in Western countries tend to focus on the "technological" superiority of Europeans, as if history was static, when Greenland was inhabited by both Native Americans – the Thule, a group that arrived in Greenland after the Vikings did so in 980, and that were the ancestors of the Inuit (Fig. 1.18) – and Scandinavians, it was the Thule that "outcompeted" the Vikings, not the other way round. They did so because they had a technology that was more efficient to thrive in the extremely difficult local conditions they faced, for example to hunt aquatic animals such as seals, narwhals, and walruses. While the Vikings left southern Greenland in the fifteenth century and Scandinavians only consistently resettled it in the eighteenth century, the Inuit continued to live there until today. Qaanaaq, today a major city of northwestern Greenland, was formerly known as Thule or New

Fig. 1.18 The Inuit: present in North America and Greenland, some North-American Indian groups called them "Eskimos" – as many people still do today – meaning "people who eat raw meat," but they often call themselves, and prefer to be called, Inuit, which means "people"

Thule. This example powerfully shows the fallacy of ethnocentric and racist narratives about a group A, or its culture, being "superior" to group B, or it's culture. Not only there is no "superior" or "inferior" in the natural world, but things are not static, they change over time. In contrast to what happened before the fifteenth century, in the nineteenth century the technology of the Scandinavians allowed them to dominate the Inuits, to the point that in 1814 the Treaty of Kiel awarded Greenland to Denmark. Times change, things change: organisms, groups, cultures, and societies change. That should have been precisely one of the major societal lessons that the knowledge about biological evolution should have taught us.

A similar example, related to these ones, concerns the fact that, technically, Polynesians also went to lands that are now officially part of America – namely, Rapa Nui, or Easter Island, which is part of Chile, around 1380 – more than a century before Christopher Columbus did so, in 1492. Furthermore, as summarized some months ago by Carl Zimmer in a *New York Times* article entitled "Some Polynesians carry DNA of ancient Native Americans,," there is now growing evidence, from various areas of science including genetics, indicating that Polynesians might well have gone all the way to continental South America as early as 1150. There were already strong indications that this might have been so, from anthropological and archeological studies, well before the publication of more recent genetic studies. But, once again, for many it was difficult to even envisage the idea that "brutish savage Polynesians" were able to reach continental America before "civilized Christians" did. More empirical data is needed to test the hypothesis that

Polynesians did go all the way or not to continental America, but the important point is that technically, Columbus only went to lands that are now officially part of American countries *after* three groups had previously done so – Native Americans, Vikings, and Polynesians – and *much after* – possibly up to 4000 years after – Native Americans went to what is now part of Denmark.

Coming back to the type of narratives defended by Darwin in the *Descent* and the *Expression* about the evolutionary history of morality and traits such as altruism, and about which type of selection would favor the evolution of such features, Richards noted that:

> There is one significant problem that Darwin's contemporaries missed, though modern biologists have not. This has to do with an apparent disparity between the model for community selection and its application to human groups. Community selection, as Darwin defined it in the *Origin of Species*, works because the community members are *related*: neuter workers in a hive of bees arc siblings, so any communally beneficial traits can be passed on to future generations through the queen. But this is not usually the case in human communities. Darwin seems to have been aware of the difficulty, since he obliquely offered several suggestions that mitigated the problem, if they did not quite eliminate it. Darwin believed that the social instincts themselves were extensions of the "parental and filial affections" which could be explained by natural selection. So altruistic behavior might initially gain a foothold in a group whose members were indeed closely related, fie knew also, of course, that small tribes would consist of only a few clans. He alluded to this fact when he considered that the inventiveness and rational acumen of some members of a tribe could benefit the whole group in competition with other tribes. In so fortunate a tribe, the new inventions and clever plans would spread by means of imitation. This would then give the tribe a selective advantage..
>
> He seems not to have realized that in a tribe without sufficient relatedness, the next generation would not have, in the case of his example, any greater representation of intelligent members (though on average no less). Traits a tribe copies from its few geniuses might give it a selective advantage, but not necessarily an effectively heritable one. Only under certain conditions might group selection prove a force in evolution – for example, small groups that remain stable, large numbers of them in competition, little migration between groups, sufficient biological relatedness of members, small disadvantage to members initiating the group-enhancing trait, and so forth. Darwin seems not to have been completely oblivious to these problematic requirements, though not focally aware cither – hardly surprising, except to hagiographers. Once, however, tribes had been established with highly developed social instincts and an enlarged mental capacity, then several non-selection mechanisms, Darwin believed, would continue to hone the moral sense. And as the evolution of mind reached a pitch where individuals of different societies and nations would begin to recognize each other as brothers, members of the same global tribe, then quite automatically social sympathy would also be mutually extended. So in this way, what began in our ancestors as primitive instincts, might evolve, Darwin supposed, into a respect for the moral nature of all men.

We see here, once again, how Darwin use erroneous concepts such as "better" or "universal" morals. Usually, what each human group calls "morals" are just social constructions within that specific group, so in this sense "morals" can't be "better" or "worse," nor they are in "universal" to all human groups. The morals of Catholic Church are not necessarily the same as those of Theravāda Buddhism, and surely not the same as those of the Fuegians that Darwin met in Tierra del Fuego. This applies not only to humans but also to other organisms. For instance, cuckoos have been evolutionarily successful by undertaking a type of niche construction that led them to evolve parasitic-like features such as laying their eggs in the nests of other bird species, so they are incubated by the foster parents, who rear the young cuckoos. If something similar was done by a certain group of humans, people from another group could say that the individuals that trick the foster parents are doing something "morally wrong." But it does not make sense blaming the cuckoos, or any other organisms with parasitic-like features, for being "immoral," or telling them that they should change their behavior, accordingly. Cuckoos are just living and evolving, as all organisms are, and are actually an example of evolutionary success: within the reality of the natural world, *things are just what they are*.

Similarly, let us suppose that, at a certain time – for instance, without refrigeration – and geographical region – for instance, in a very hot area – in which a religion emerged, eating a certain type of food, such as pork meat, would be riskier, in terms of health. As a crisis management system, in order to reduce such medical risks, the leaders of that religion could create taboos, social norms, or "commandments" so its followers would not eat pork. But, centuries later, with the invention of refrigeration and so on, even if eating pork meat would not be so risky anymore, medically speaking, the followers of that religion could continue to follow and reinforce that rule, in order to preserve a strong "trust-me" religious signal and an "in-group" marker to distinguish themselves from the "others," as explained above. By then, the members of that religion could actually *believe*, or force others from that in-group to *believe* or behave as if they believed, that eating pork meat is truly bad, and that their morals and "commandments" are "better" than those of "others." However, as in the case of cuckoos, within the reality of the natural world, it makes no sense to say that the members of that religion in particular – or, for that matter, of any other religion, or any non-religious group – are "better" or "worse" or have "higher" or "more noble" beliefs or morals. In human evolution, many beliefs and taboos have likely started from specific adaptations to local environments or challenges faced within those environments, or to distinguish a religion form another one, so talking about "universal" or "universally" better morals is just a fallacy.

But, instead of recognizing this fact, after postulating his ideas on local adaptations in the *Origin* and after having meeting in person the Fuegians and many other non-Victorian groups of humans during his Beagle travel, Darwin stated that the Fuegians had fewer noble morals or were inferior because they did not have Victorian morals. For instance, he did this when he complained that the Fuegians were "looters," while he could instead have praised them for being egalitarian. That is, the biases he carried from England made him affirm that Victorian morals were the more evolved, progressed on the planet: essentially, they were the right ones, the *model* that should be used as a guide to know what or who is "bad" or "good." In other words, he did not only see human "races" and anatomies as part of a "ladder-of-life": morals were also part of it, with "bad" ones at the bottom, and the "best" ones – the Victorian ones, of course – at the top.

Darwin's *biases* and *beliefs* are manifested over and over in the *Descent* and the *Expression.* He referred to the beliefs of certain non-Western groups as "strange superstitions and customs" but then used the terms "religion" and "high morality" in a positive way when he referred to European Christian beliefs: our beliefs are better than yours. Darwin's *beliefs* also lead to another major problem characterizing his theories concerning morals and ethics: the notion that morals can become innate. In an oversimplified way, having an innate morality would be as if suddenly, a newborn baby would already have a tendency to accept what Jesus Christ taught his disciples. Interestingly, such inaccurate ideas are somewhat similar, in a way, to current – and relatively popular – ones defended by authors such as Steven Pinker, who in *Better Angels* and other books repeats the idea that there is a global "*expanding moral circle*." Of course, for Pinker, as for Darwin or Spencer more than a century ago, this "expanding" movement is ultimately leading to our own, Western, morality, the pinnacle of evolution. In his *First Principles*, Spencer wrote that "evolution can end only in the establishment of the greatest perfection and the most complete happiness": this sentence could well be a summary of Pinker's idea of an expanding moral circle ultimately leading to our best – Western – "angels." Basically, nothing new under the sun, just repetitions of very similar ethnocentric factually inaccurate ideas based on the fallacious notion of a linear ladder-of-life involving not only natural but also societal and moral progress.

Fascinatingly, despite these fallacies, within the idolization of Darwin many scholars continue to use terms such as "Darwin's morality" or "Darwin's ethics," or to cite Darwin's ideas about these topics, to highlight how much Darwin was a polymath sage that combined his knowledge about biological evolution in general with his deep knowledge about other topics such as these ones concerning cultural evolution. This is happening right now, as I write these lines, at many talks and meetings about cultural evolution and similar issues, to celebrate the *Descent*'s 150th anniversary. Darwin was not the first to try to provide an explanation for the evolution of morality or altruism, contrary to what is often said in such publications and talks. As noted above, others such as Wallace had tried to do so, for instance, in his 1864 paper "The Origin of Human Races." In some ways, they did this more successfully than Darwin, and even before Wallace others had discussed such topics under an evolutionary or a quasi-evolutionary framework, such as Spencer. It is precisely because many English scholars were already discussing such topics for years that they were intellectually well prepared to criticize Darwin's take on such topics in his *Descent* and *Expression* books.

> The *Athenaeum* reacted more predictably. Its critic judged the chief merit of the book to be the accumulation of a vast array of facts, but found Darwin at his "feeblest" in attempting to give an evolutionary account of the moral and intellectual faculties. The *Times* also complimented Darwin for the drudge's work of gathering observations, and likewise regarded him as "quite out of his clement" in treating mind and morals.. The account in the *Pall Mall Gazette*, while critical, yet struck a more conciliatory note. It peaked Darwin's interest in discovering the identity of its anonymous author. The writer had several insightful objections to Darwin's theory of conscience. First, he attempted to correct a misconception Darwin had about John Stuart Mill's theory of morals..

> Even for an evolutionist, the reviewer pointed out, "the foundations of morality, the distinctions of right and wrong, arc deeply laid in the very conditions of social existence," which had to be understood and reflectively negotiated. Darwin.. admitted his blunder in ascribing to Mill the notion that greatest happiness acted as foundation of the moral sense; yet he held fast to his criticism of Mill, that the philosopher did regard moral feeling as acquired, not innate.. The reviewer for the *Quarterly*, however, descended from that high critical plane, not only to bury the *Descent*, but also to shovel dirt in Darwin's face.. the reviewer confessed acute chagrin in pointing out the "grave defects and serious short comings" of Darwin's arguments, but honesty compelled him. Certainly he could not overlook Darwin's "singular dogmatism", which led the eminent naturalist to assert what required proof, to beg all the important questions, and to fawn over his own supporters. The reviewer allowed the *Descent* bulged with examples, but had to report that "Mr. Darwin's power of reasoning seems to be in an inverse ratio to his power of observation."

These criticisms about Darwin's take on these topics are revealing. Firstly, they show us that although Darwin was already idolized back then – after all, the Pope Darwin illustration shown in Fig. 1.2 was done when he was alive – his idolization was not as marked among scientists as it has been in the last decades. Actually, people from the general public, or the media, or even scientists other than those within the natural sciences, often do not know that for many decades after Darwin's death many were predicting, or even proclaiming, that Darwinism would probably fade away. Back then, many scientists were more prone to accept Lamarck's ideas or vitalistic ideas, for instance. In a simplified way, one can say that it was mainly in the 1920s and 1930s, before the second world war, with the rise of genetics, the Modern Synthesis, and eugenics, that Darwinism began to gain a huge momentum and that Darwin's idolization begun to take full force.

Secondly, related to this first point, the criticisms about Darwin seen in the citation above contradict a typical line of defense used by Darwin's idolaters concerning Darwin's less "noble" or "bright" ideas: that those ideas have nothing to do with Darwin himself, but to something else, for instance, that they were just an inevitable product of the epoch in which he lived. Such criticisms emphasize that back then many scholars did *not* think like Darwin. For example, numerous scholars *knew* that morals could *not* be innate. In reality, as recognized by Richards, the criticisms became particularly harsher and more widespread after Darwin published *The Expression*, in which he also defended the idea that emotional responses in general did not have any major evolutionary use. Because emotions were one of the central topics of that book, many scholars that had been thinking about such topics for a long time harshly criticized the fact that Darwin did not see any major evolutionary role for them and therefore concluded that he did not even have to invoke natural selection to explain their occurrence in nature. Within other factors, Darwin's intellectual conservatism and Victorian biases seem to have contributed at least in part to these erroneous assertions about the evolutionary origins and role of emotions. This is because, within the cream of Victorian society the public display of emotions was often seen in a rather negative way, as if displaying them was something "bad," unnecessary, or futile.

Another rather weird concept about these issues defended by Darwin in the *Expression* was the "*principle of antithesis*." As put by Richards, this concept "held that when certain actions were linked with a particular state of mind, the appearance of an opposite state would tend to elicit behavior of an opposite kind.. for example, a dog, which stands rigid with hair erect and tail stiff when hostile, will crouch low with back bent and tail curled when affectionately disposed." Darwin assumed that the link between the opposite disposition and the corresponding behavior would gradually "become hereditary through long practice." About this, Richards commented: "why he felt it necessary to add this last provision, since he thought contrary emotional states naturally evoked contrary behavior, is hard to say.. it seems but another attempt to secure the evolutionary foundations of all patterned behavior." Another critical principle that Darwin proposed in the *Expression* was that of *community selection*, which we briefly discussed above. As put by Richards, "community selection.. fixed moral obligation most strongly toward one's family, with the moral bonds becoming weaker as they stretched to remote kin, neighbors, other community members, and men in general." Darwin "conceived this hierarchy as established not by wise men but by a wiser nature, which, as he put it in the *Origin*, '*is daily and hourly scrutinizing, throughout the world, every variation, even the slightest; rejecting that which is bad, preserving and adding up all that is good*'."

This last sentence highlights many of the major problems concerning Darwin's view of not only human evolution, but also of life in this planet in general, including the teleological and personalized way in which he portrayed Mother Nature and "her" commitment to an exaggerated Malthusian "war-like" struggle-for-existence, "scrutinizing" everything, every time. In this sense, the Mother Nature portrayed by Darwin was in a way actually more omniscient, omnipresent, or micromanager than even some of the Gods portrayed by various religions. And also more kind than those Gods that are depicted as momentarily in rage or violent or revengeful - Darwin's Mother Nature always rejects "that which is bad," "preserving and adding up all that is good." There is however one particular aspect in which the personified Mother Nature constructed by Darwin is very similar to most of those Gods: she is intellectually conservative, as they tend to be, and exactly as Darwin was. As pertinently put by Richard Herley, humans had created their Gods – or their Mother Nature – in their own image, not the other way around. For example, as pointed out by Richards, Darwin *believed* that humans "did act without intending their own pleasure: when they acted instinctually or from habit, for instance, or when they disinterestedly sought the welfare of another.. according to Darwin's evolutionary ethics, acting from pleasure never indicated that an action was moral, but it often marked immoral behavior." *Acting from pleasure is often immoral*: the God of the Old Testament would very likely be pleased with such a concept.

Apart from the point that, contrary to religious or quasi-religious tales, in nature that is no "good" or "bad" or "moral" or "immoral," in a strictly scientific point of view makes no sense to assume that pleasure would necessarily be correlated with mostly "bad" evolutionary behaviors. Everything obviously depends on the specific moment, place, and context in which the pleasureful event would occur. Sex is often deeply related to pleasure, and sex leads to something that is surely evolutionarily "good" according to Darwin's theory of evolution by natural selection: reproduction. Without sex, sexual organisms would just become extinct. Similarly, many of the things that we and other organisms do are obviously driven by pleasure: the pleasure of eating the food we like, and so on. So, saying that seeking for pleasure, or having "primal" or "savage" bodily desires, is "bad" or "immoral" would lead to an evolutionary paradox, as recognized by Spencer in a letter to Huxley. He wrote: "[I insisted that there is] a *non-moral character of Nature*.. pointing out that for 99 hundredths of the time life has existed on the earth (or one might say 999 thousandths) the success has been confined to those beings which, from a human point of view would be called criminal." Indeed, if one would take the Victorian society as a model of "morality," as Darwin often did, then 99.999% of the organisms that have existed and still exist in this planet, as well as 99.9% of human evolutionary history, would be highly immoral. In a nutshell, as noted above and astutely put by Michael Shermer in his book *The Science of Good and Evil*: "*if there were no humans there would be no evil*."

Concerning the way in which the ideas of Darwin's *Descent* and *Expression* about humanity, morality, and behavior were seen since his death until the end of the twentieth century, Richards noted:

> Contemporary Darwinism focuses a stark image of man. Through its lens we have come to perceive man as a completely material being, whose reason traces the narrow paths of fixed brain circuits, whose religious sentiments bespeak the need for conformity rather than a passion after transcendence, and whose moral feeling, driven from below by selfish genes, quickens to secret pleasure for self rather than to the welfare of others.. The received view of Darwinian man, though a potent icon in modern culture, does not resemble that image shaped by Darwin, Spencer, and the Darwinians writing in the last part of the nineteenth century.. Rather that original image became refracted and transformed into the specter that hovers over contemporary debates about sociobiology. The transformation came as the result of powerful disciplinary and social forces acting during the early part of this century, forces which initially inhibited the further development of evolutionary theories of mind and behavior, but which in the end served to recast the image of Darwinian man. The strongest of these forces erupted in the social sciences. James, Baldwin, and Morgan – three principal contributors to the Anglo-American development in evolutionary biopsychology – had entered the twentieth century with philosophy on their minds..
>
> As evolutionary theories of mind became elaborated into grand philosophy instead of being stitched into the fabric of advancing science, they lost their attraction for those new scientists of the *laboratory craft*. John Watson, the progenitor of modern behaviorism in psychology, was the very model of the laboratory man. He originally came to the University of Chicago to study with John Dewey. Dewey's philosophical naturalism rippled with evolutionary ideas and seems initially to have stimulated Watson to follow a particular research path. But the gauzy character of Dewey's speculations could not hold the intellectual enthusiasms of this empirically directed researcher. "I never knew what he was talking about then", Watson recalled in his autobiographical sketch, "and unfortunately for me, I still don't know". Watson retained the democratic and pragmatic part of the Deweyan vision, but the decidedly evolutionary pan faded. In 1924 when he had become the recognized spokesman for the new behavioral technology, Watson exclaimed: "give me a dozen healthy infants, well-formed, and my own specified world to bring them up in and I'll guarantee to take any one at random and train him to become any type of specialist I might select – doctor, lawyer, artist, merchant-chief and, yes, even beggarman and thief, regardless of his talents, penchants, tendencies, abilities, vocations, and race of his ancestors."

> The democratic potential implied by Watson's behaviorism struck a resonant cord in the hearts of most of his American readers. Much of the academic world, which perhaps allowed for Madison Avenue hyperbole, and large portions of the lay public believed his boast. At least they thought every good American boy might grow up to be president, if he worked hard and received the proper training. To set children along the right path, Watson offered advice to the mothers of America in a number of popular articles and in a Spock-like baby book, *Psychological care of infant and child*. The message of these publications was clear: the mind of the child had no inherited groves that determined its station in life; it could be put in the hands of the behavioral technologist, or the well-informed mother. The triumph of behaviorism in psychology was one of the signal causes of the decline in theorizing – at least in English-speaking countries – about the evolution of mind and behavior.

There are two interesting points in this excerpt. Firstly, it is fascinating to see that, within this context, some of the worse traditions of "ultra-Darwinism" and "ultra-adaptationism" are still very much in vogue within many social scientists, as well as numerous biologists that deal with the study of human and nonhuman behavior. Evolutionary psychology, in particular, but also fields such as behavioral ecology are particularly powerful examples of this.

Evolutionary psychology narratives have become so influential within scientific narratives of human evolution, particularly within the media and popular press, that various other fields, such as evolutionary medicine – including evolutionary psychiatry – have been partially contaminated by such ultra-adaptationist tales, as we shall see in Chap. 2. The second point concerns the rise of the "*new science of the laboratory craft*," as well as the related scientific obsession with "big" numbers or "big data." If instead of merely talking about psychological ideas and concepts, one talks about "experimental psychology" and in particular shows a lot of numbers, statistics, and attractive graphs, then this often becomes to be seen under a panache of an "exact" science, a "serious," grantable field. Something somewhat similar happened with Darwin: apart from his *power place*, one of the critical features that sets him apart from other early evolutionary biologists, according to various historians of science, is precisely the fact that he was both an observer of nature and an experimenter. He did several studies and experiments with pigeons, such as crossing those with dissimilar characteristics to generate different offspring.

A particularly revealing statement made by Richards is the following: "evolutionary approaches to human mind and behavior received another check during the early decades of.. [the 20th] century.. the blow came from anthropology.. the newly burgeoning science of human culture turned its back on evolution because of the ethnocentrism and racism that evolutionary theory seemed to imply." Darwin's ideas on human evolution, put forward in his *Descent*, did not *seem* to imply ethnocentrism and racism: *they implied them, all the way, because they were in great part based on them*. The million-dollar-question is, therefore: why so many Western scientists and historians fail to recognize, or try so much to minimize, this obvious fact?

I face this important question almost on a daily basis, as a professor and researcher at Howard University, one of the most eminent so-called "black" universities within the U.S. For instance, when I try to include evolutionary concepts in my lectures for the students at Howard's College of Medicine, I often face a major problem. The vast majority of those students are African Americans, and a significant part of them does not agree – and many do not even like to hear – that humans came from other animals, particularly from apes and, therefore, from monkeys. As I write a lot about these topics, I am very interested in hearing why this is so. And, year after year, the new students tell me that their discomfort about evolutionary ideas is not only – and for many not even mostly – related to their religious beliefs, or to any special animosity toward apes or monkeys. Instead, many of them tell me that they associate evolutionism, and Darwinism in particular, with racism. They have nothing against apes per se, but instead against the way numerous scientists compared in the past apes to African humans and their own descendants, when they suggested, for instance, that they were more similar to those apes than to people from Europe. Due to indoctrination – and further showing the importance, and influence, of such erroneous biased narratives created by scientists – such racist views indeed became widespread among the general public and even the media – in particular within "whites," obviously, being extremely damaging to non-"whites." One just needs to remember how just a few years ago, in 2009, there was a widespread dissemination and discussion of a graphic, violent cartoon – among many others, it needs to be said – suggesting that Barack Obama was like a chimpanzee – actually, a dead, assassinated one (Fig. 1.19).

So, can we say that the Howard University students, or many non-"white" Westerners as well as a huge number of non-Westerners, are wrong when they associate Darwinism with racism or with scientific racism? As will be explained in Chap. 3, scientific racism has a very long story, which started much before Darwin was alive, at the beginning of the eighteenth century with biologists such as Linnaeus. However, as we are seeing throughout this book, Darwin's publications, particularly the *Voyage* and the *Descent*, indeed played a huge role within scientific racism, legitimizing many of its key ideas. Let us just think about this: Darwin literally wrote, as if this was a "scientific" fact, in a "scientific" book, that numerous groups of Africans or African descendants – that is, of people such as the majority of Howard University's students – were not only morally "inferior" to "whites" but also likely "naturally" doomed to become *extinct or massacred*. So, how can one think that people from such groups would now, all of the sudden, find the "light" and embrace and revere Darwin and Darwinism, as if this tragic side of science, of evolutionary biology, and of Darwinism, had never truly happened and did not have any societal repercussions that ultimately affected their ancestors? Pretending that this was so, that none of these things happened – as many of those that now use "cancel culture" arguments to defend the *status quo* do – is what would truly be a cancellation of a crucial part of our Western culture, history, and science.

Sadly, as it often happens, many within the cream of our societies, particularly those that never had to see or feel life under the prism of those that are disfavored, underprivileged, or discriminated, do not even try to be empathetic and understand

Fig. 1.19 This 2009 *New York Post* cartoon became widely seen and discussed in the USA, as well as other countries. Depicting two white male police officers shooting of a chimpanzee, the cartoon connected two stories that were being discussed in the press back then: Barack Obama – African-American, then President of the USA – was signing of a stimulus bill and there had been a shooting of a chimpanzee after it attacked a Connecticut woman

the point of view of "others" and their skepticism about evolutionary biology, Darwin or Darwinism. They instead are quick to conclude that such students might be "backward" or "unscientific," that they just do not want to accept – or, worse, cannot even understand – scientific "facts." But the reality is that many of such students are truly mostly uncomfortable – and rightly so – with *the factually inaccurate racist* statements that scholars like Darwin did, and therefore also with the countless evolutionary biologists that have idolized, and continue to idealize, Darwin. Just think about James Watson, who as noted above is a Nobel laureate biologist that has a remarkably long record of defending appalling inaccurate sexist and racist evolutionary ideas, and you get an idea of why such students might feel uncomfortable about both those ideas and the fields of the people proclaiming them (see also Box 3.7). So, is the "solution" simply to aim for the "inclusion" of such minority or underprivileged or discriminated communities in "science," that is, to "integrate" them in fields in which markedly racist and sexist people such as Watson continue to still be highly influential? Or should it also involve having those fields of science/the researchers working on them, recognizing the racist and sexist past of those fields, and to speak up against such biased inaccurate evolutionary ideas? Instead of simply criticizing "others" for not being able to "see" the Holy Grail of "our" "noble science," it is also up to us, scientists, to recognize such biases, prejudices, inaccuracies, idealizations, and abuses of our fields, and to not react defensively to those "others" every time that they refer to such biases and idealizations.

Ashes to Ashes, Dust to Dust, Immortality, and the Newton of Biology

What spurred.. (Darwin's) move to natural selection was the strongly felt need to be the Newton of biology – to find a cause for the change. (Michael Ruse)

There is a small passage in Darwin's autobiography that is often not discussed in the literature or is just usually mentioned as an anecdotal aside, but that is very informative within the context of a theme that is recurring in Darwin's autobiography, as we have seen above: the use of the word "dull" or "uninteresting" to refer to many books, professors, and scientists. This is because in that passage he does this even when he refers to the very moment when he finally met, on 29 January 1842, Alexander von Humboldt, who was one of his youth heroes, who inspired him to travel around the planet and observe its nature (Fig. 1.5). As previously explained, Humboldt was a polymath that was fascinated by numerous scientific and nonscientific topics. So, at least in theory, Humboldt is probably one of the last names that would come to one's mind when one thinks about a scholar that would be uninteresting to meet and talk with. As there are almost no detailed historical records about how that meeting went, we cannot really know what Humboldt and Darwin talked – or did *not* talk – about. The only thing we know is that Darwin just wrote this about the meeting: "I once met at breakfast at Sir R. Murchison's house, the illustrious Humboldt, who honored me by expressing a wish to see me.. I was a little disappointed with the great man, but my anticipations probably were too high. I can remember nothing distinctly about our interview, except that Humboldt was very cheerful and talked much."

Apart from this, what we also do *know* is that, 3 years after this meeting, in October 1845, Darwin wrote to Lyell a negative comment about the English translation of Humboldt's *Cosmos*: "Have you read '*Cosmos*' yet? The English Translation is wretched, & the semi-metaphisico-poetico-descriptions in the first part are barely intelligible; but I think the volcanic discussion well worth your attention; it has astonished me by its vigour & information." That passage of Darwin's autobiography and the content of this letter therefore take us back, full circle, on how we started this chapter. That is, although this is often unnoticed within the idealized literature about Darwin, there is indeed a recurring pattern within Darwin's private writings – his autobiography, letters, diaries, and notebooks. On the one hand, he had an almost obsessive fascination for biological and geological details – such as, let us say a discussion about a volcano. Scientifically, this is obviously something that should be praised, science cannot be done without the observation and gathering of detailed, specific natural facts. Darwin was amazingly good at doing both: at observing details and at reading an endless number of publications made by others about other details observed by them. Darwin refers to this strength of his in the autobiography:

> Having said thus much about my manner of writing, I will add that with my large books I spend a good deal of time over the general arrangement of the matter. I first make the rudest outline in two or three pages, and then a larger one in several pages, a few words or one word standing for a whole discussion or series of facts. Each one of these headings is again enlarged and often transferred before I begin to write in extenso. As in several of my books facts observed by others have been very extensively used, and as I have always had several quite distinct subjects in hand at the same time, I may mention that I keep from thirty to forty large portfolios, in cabinets with labelled shelves, into which I can at once put a detached reference or memorandum. I have bought many books, and at their ends I make an index of all the facts that concern my work; or, if the book is not my own, write out a separate abstract, and of such abstracts I have a large drawer full. Before beginning on any subject I look to all the short indexes and make a general and classified index, and by taking the one or more proper portfolios I have all the information collected during my life ready for use.

However, on the other hand, Darwin did not have the same interest about broader, more theoretical, multidisciplinary discussions such as the ones that he designated as "metaphysical speculations" or concerning societal topics such as the history of racism and sexism and how "others" perceive the world. This would not be so problematic if, for instance, Darwin had only written about volcanos or plants. But this became indeed a major problem, with huge societal repercussions, for a scholar that wrote two books about human evolution, "human races," morality, emotions, the women's role in society, and so on, and that became highly influential within various fields of science and layers of society, since then. This is because this lack of depth led him, as we have seen and will discuss in further detail in the chapters below, to often not being able to distinguish between what he had been indoctrinated to *believe* to be "the best" or "most noble" or "more moral" people and way of living – the Victorian "normality" – and the reality of the natural world and of our species in particular, in all its complexity. Even when he had the unique opportunity to see the facts just before his eyes, as when he encountered "others" in his voyages, such as the Fuegians, he often "saw" their evolutionary "inferiority" and their "immorality" under a Victorian perspective. He used the Victorian model not only as the "standard" but, even more problematic, as the pinnacle of evolution. As we have seen, this markedly contrasts with the way other scientists that interacted with Darwin, such as Wallace and Humboldt – and the young Huxley – reacted to, and described, "others," which tended to be less ethnocentric and more empathetic, although there were exceptions, of course. In this sense, it is interesting that Darwin recognized, in his autobiography, that he "could never have succeeded in metaphysics" and "abstract thoughts," and that he was aware that he was, already at his time, often criticized for being a "good observer, but.. [having] no power of reasoning."

Obviously, the criticism that he had "no power of reasoning" – which is sometimes still made nowadays, by some creationists – is unfair and nonsensical. As he pointed out about that criticism, he was actually amazingly successful at connecting the "things which easily escape attention" to other scientists. That is why only Wallace and him were able to fully connect two ideas that were already in the air – transmutation and natural selection – in the way they did, contrary to all the other scholars that were alive back then, including those that attacked his "reasoning" skills. Already aware, in the last years of his life, that he would likely become "immortal" – but probably not expecting that he would be so idolized by historians and scientists – Darwin actually spent the very last pages of the autobiography recognizing his "abstract-thinking" weaknesses – particularly later in life – and defending his more down-to-earth "power of reasoning":

> I have said that in one respect my mind has changed during the last twenty or thirty years. Up to the age of thirty, or beyond it, poetry of many kinds, such as the works of Milton, Gray, Byron, Wordsworth, Coleridge, and Shelley, gave me great pleasure, and even as a schoolboy I took intense delight in Shakespeare, especially in the historical plays. I have also said that formerly pictures gave me considerable, and music very great delight. But now for many years I cannot endure to read a line of poetry: I have tried lately to read Shakespeare, and found it so intolerably dull that it nauseated me. I have also almost lost my taste for pictures or music. Music generally sets me thinking too energetically on what I have been at work on, instead of giving me pleasure..
>
> On the other hand, novels which are works of the imagination, though not of a very high order, have been for years a wonderful relief and pleasure to me, and I often bless all novelists. A surprising number have been read aloud to me, and I like all if moderately good, and if they do not end unhappily – against which a law ought to be passed. A novel, according to my taste, does not come into the first class unless it contains some person whom one can thoroughly love, and if a pretty woman all the better. This curious and lamentable loss of the higher aesthetic tastes is all the odder, as books on history, biographies, and travels (independently of any scientific facts which they may contain), and essays on all sorts of subjects interest me as much as ever they did. My mind seems to have become a kind of machine for grinding general laws out of large collections of facts, but why this should have caused the atrophy of that part of the brain alone, on which the higher tastes depend, I cannot conceive.
>
> A man with a mind more highly organized or better constituted than mine, would not, I suppose, have thus suffered; and if I had to live my life again, I would have made a rule to read some poetry and listen to some music at least once every week; for perhaps the parts of my brain now atrophied would thus have been kept active through use. The loss of these tastes is a loss of happiness, and may possibly be injurious to the intellect, and more probably to the moral character, by enfeebling the emotional part of our nature.. I have no great quickness of apprehension or wit which is so remarkable in some clever men, for instance, Huxley.. My power to follow a long and purely abstract train of thought is very limited; and therefore I could never have succeeded with metaphysics or mathematics.. Some of my critics have said, "Oh, he is a good observer, but he has no power of reasoning!" I do not think that this can be true, for the "*Origin of Species*" is one long argument from the beginning to the end, and it has convinced not a few able men. No one could have written it without having some power of reasoning. I have a fair share of invention, and of common sense or judgment, such as every fairly successful lawyer or doctor must have, but not, I believe, in any higher degree..
>
> On the favourable side of the balance, I think that I am superior to the common run of men in noticing things which easily escape attention, and in observing them carefully. My industry has been nearly as great as it could have been in the observation and collection of facts. What is far more important, my love of natural science has been steady and ardent.. Therefore my success as a man of science, whatever this may have amounted to, has been

> determined, as far as I can judge, by complex and diversified mental qualities and conditions. Of these, the most important have been – the love of science – unbounded patience in long reflecting over any subject – industry in observing and collecting facts – and a fair share of invention as well as of common sense. With such moderate abilities as I possess, it is truly surprising that I should have influenced to a considerable extent the belief of scientific men on some important points.

The last sentence is a clear example of true modesty because he has done much more than influencing "to a considerable extent the belief of scientific men on some important points." However, in overall, the fact that Darwin included these discussions about his "power of reasoning," "abstract thinking," and so on – which were written on the 1st of May, 1881, just about 1 year before he died on 19 April 1882 – in his autobiography and ended that autobiography with this humble sentence, does seem to reinforce the assertion that, in a way, Darwin's was doing a last, personal defense of his scientific legacy. The legacy of a very particular scientist, a man that wrote books combining markedly inaccurate "facts" about human evolution with amazingly brilliant, accurate, and precise observations about the life and evolution of so-many nonhuman organisms, of all types, shapes and sizes. A man of contrasts, characterized by logical inconsistencies, biases, and prejudices, as we all are.

Chapter 2
Darwin's Society and Science

Darwinism was made by Darwin and Victorian society.

(Janet Browne)

A State Funeral, the Power of Place, and "Racial Capitalism"

> *The only very marked difference between the average civilized man and the average savage is that the one is gilded and the other is painted.* (Mark Twain)

What happened when Darwin died? How did his country react to his death? The physical body of the intellectually immortal scientist was laid to rest, on April 26, 1882, in Westminster Abbey. This royal church in the center of London was, and still is – it was inscribed as a cultural World Heritage Site in 1987 –, an emblematic symbol of English power, where numerous coronations and royal weddings and funerals have occurred. This point is revealing, as are various other interesting aspects mentioned in the account provided in Desmond and Moore's 1994 *Darwin* book about what happened since Darwin's last breath, until his body was buried:

> In the Commons.. Lubbock [president of the Linnean Society] moved among his colleagues collecting signatures, and Ireland was pushed aside for a moment in favour of Darwin and English pride. Lubbock left the House with a petition stating that "it would be acceptable to a very large number of our countrymen of all classes and opinions that our illustrious countryman Mr. Darwin should be buried in Westminster Abbey." "It was very influentially signed", he told Frank Darwin.. four Fellows of the Royal Society, including the education minister, and Lyon Playfair, now the Deputy Speaker. Also down were the Under Secretary of State for Foreign Affairs, and Sir G. O. Trevelyan, Secretary of the Admiralty. There followed the Solicitor General, the Postmaster General, a Sea Lord, and the Speaker of the House.. Another signatory, Henry Campbell-Bannerman, would one day be Prime Minister.. On Saturday, The *Standard* made an emotive plea: "*Darwin.. who has brought such honour to the English name, and whose death is lamented throughout the civilized world.. should not be laid in a comparatively obscure grave. His proper place is amongst those other worthies whose reputations are landmarks in the people's history.. we owe it to posterity to place his remains in Westminster Abbey, among the illustrious dead who make that noble fane unrivalled in the world*".. Other papers also latched on to the crusade. Patriotism was

R. Diogo, *Darwin's Racism, Sexism, and Idolization*,
https://doi.org/10.1007/978-3-031-49055-2_2

> the paramount theme. Who was the Abbey's hallowed ground for, if not those who had made Britain great, extended her Empire, civilized new worlds at home and abroad?
>
> On Wednesday 26 April.. committees adjourned, judges put on mourning dress, and Parliament emptied as members trooped across the road. From embassies, scientific societies, and countless ordinary homes they came. Under leaden skies they converged on the Abbey, anticipating the awe and spectacle of a state occasion. The Darwins and Wedgwoods queued in the Jerusalem Chamber, thirty-three in all, including Galton.. In the Chapter House, where Parliament had once met, the elders of science, State, and Church, the nobility of birth and talent, stood waiting to file through the cloisters, behind the coffin. They were "the greatest gathering of intellect that was ever brought together in our country", said one. The transepts were filled with friends and guests; the south side of the nave with the bearers of black-edged admission cards. The Lord Mayor of London took his place in the sacrarium, before the altar, with still more family members. In the choir, Spencer sat forlornly among the distinguished ladies, now regretting that he was simply "one of the spectators". Finally, the doors opened to the multitudes without tickets.. Then, at midday, the moment arrived. As the Abbey bell tolled, Canon Prothero entered from the West Cloister door, leading the procession, with the choristers singing "I am the resurrection". The train of family and dignitaries shuffled past the tombs of the famous and made their way slowly through the candle-lit choir to the centre of the transept. There the coffin, draped in black velvet and covered in a spray of white blossoms, was placed beneath a lantern. A specially commissioned hymn was sung after the lesson. Composed by the Abbey's deputy organist, it was sensitive to the occasion. The lyrics were taken from the Book of Proverbs.. The Abbey interment gave tangible expression to the public feeling that Darwin, in his life and work, symbolized English success in conquering nature and civilizing the globe during Victoria's long reign.

"Who was the Abbey's hallowed ground for, if not those who had made Britain great, extended her Empire, civilized new worlds at home and abroad?" There is something odd here. This sentence, as well as many others of the above excerpt, do not match at all with the typical portrait of Darwin as a rather "private" man just obsessed in studying nature and putting forward an evolutionary theory that "triumphed" against the *status quo*. That is, against the system, the religious authority and intransigence – represented by powerful God-fearing Victorians such as Bishop Samuel Wilberforce and Richard Owen. Within such typical narratives, repeated over and over in scientific and historical publications, textbooks, movies, documentaries, and so on, we see countless times the image shown in Fig. 2.1, or similar images. That is, portraying Darwin as the lonely subject of despicable bullying, humiliation, and mockery by the God-fearing powerful Victorians: conservative politicians, religious leaders, influential journalists, and so on. However, after defending such narratives, such publications often end up by proudly describing how spectacular and pompous was Darwin's funeral at one of the major symbols of British power and religion, and how it was attended by the cream of the Victorian society, including powerful politicians and religious leaders (Fig. 2.2). This is one further example of wanting to eat the cake while having it too, because these two prevailing narratives clearly seem to contradict each other. We know that the latter narrative, about the pompous funeral at a royal church, is factually true. So might it be that the former narrative, about Darwin's fighting against the powerful religious *and* societal Victorian *status quo* is a bit exaggerated, at least part of it? Or that it has some omissions, in order to make it more appealing or creating a good story, with a

Fig. 2.1 Charles Darwin, shown as an ape, holds a mirror up to another ape: color lithograph by F. Betbeder

Fig. 2.2 The funeral ceremony of Charles Darwin at Westminster Abbey, 26 April 1882

"good" hero, a kind of "Robin Hood" scientist, fighting against the *status quo*, the well-off and the oppressors?

In his 1987 book about *Darwin*, Richards recognizes that the "war" between evolutionists such as Darwin versus creationists such as Owen is indeed often exaggerated. This is particularly important because, as you remember, Richards started that very same book with a dramatization of such a "war." However, 124 pages after that dramatized opening, Richards notes that many authors provide an "oversimplified picture of his [Darwin's] opponents, the natural theologians." It is indeed not a coincidence that Richard Owen is in fact, together with Bishop Samuel Wilberforce, one of the usual suspects that are used to represent the fight between Darwin and the *status quo*, within such dramatized scripts. Within this narrative, this was an ideological "war" not only against religion – something that, in reality, Darwin had privately said he preferred to avoid – but also against the powerful scholars that represent the old traditions of science, influenced by religion, such as natural theology. Owen was precisely one of the most important exponents of natural theology, at that time. In this regard, Richards explains that "it is commonly supposed that British natural theologians defended a position that was profoundly inimical to the theory of evolution.. since, as it is assumed, they uniformly rejected continuity of human and animal mental faculties, Darwin could have had little interest in their analyses of mind and behavior." Richards then explains that, within this erroneous narrative, Darwin's "theory of species change by natural selection must then have formed (the equally simplified) mirror image of their theological accounts." But, as he notes, the reality is that "the convictions of the natural theologians cannot.. be reduced to mere background darkness against which the light of Darwinism suddenly burst forth." In fact, as explained in Chap. 1, some of Darwin's key evolutionary ideas were actually influenced by natural theology, and he himself recognized, in the last years of his life, in his autobiography, that reading Paley's "Natural Theology" gave him "much delight."

Unfortunately, countless books, including numerous educational textbooks read and studied in school by our kids, do continue to use such incorrect quasi-religious dramatizations of a "war" between the "light" of Darwin and the darkness of religion and of natural theology. Numerous natural theologists *were* at Darwin's state funeral and profoundly revered him and, many of them were friends or had close professional ties with other members of "Darwin's Armada." For instance, as noted in Slotten's *The heretic*, even when Darwin was alive, and just 9 years after the publication of his *Origin*, in 1868, Richard Owen himself supported Wallace's application for a directorship of a museum. Later in life, Wallace continued to like Owen and respect his "immense fund of knowledge.. when their visits coincided, the two took long walks.. breathing in the scent of the magnificent conifers and discussing gardening and botany." Actually, "only seven years after the publication of the *Origin*, evolutionists now occupied positions of power in the English scientific community.. at the 1866 meeting of the British Association for the Advancement of Science.. Thomas Huxley presided over Section D, now renamed the Biology

Section.. [also,] Wallace headed the newly created Anthropology Department.” Darwin’s Armada was not “‘fighting” against power and the cream of the Victorian scientific community: they were part of it.

We can now understand why Richards probably opted to start his 1987 book in the dramatized way that he did: so later he could more effectively call the attention for a central tenet defended in his book: that contrary to such simplified “religious teleology versus scientific evolutionism” narratives, Darwin did *not* show that nature was “morally meaningless.” Richards wrote: “these characterizations of Darwin’s accomplishment control our perception of the late nineteenth and early twentieth centuries.. but are they accurate? I do not think so.” Instead, “they grievously distort historical reality.. neither Darwin nor Herbert Spencer, certainly the iconic figures of evolutionary doctrine in the nineteenth century, rendered nature ‘morally meaningless’.. on the contrary, they scientifically reconstructed nature with a moral spine.” Actually, Darwin often defined himself as an agnostic, not as an atheist. Wallace, who was also part of the so-called Darwin’s armada against the “powerful believers,” was deeply spiritual, for decades before his death. George Romanes, who was one of the most faithful and emotionally attached disciple of Darwin – being accordingly often designated as one of the first “ultra-Darwinians” – had deep religious feelings. In the last decades of his life, he believed in a mindful universe and even wrote a whole book about this topic, entitled *Thoughts on Religion*. Conwly Lloyd Morgan, who also defined himself as a Darwinian, wrote that the whole natural world gradually revealed a progressive, rational, divide plan, which he called “the world-plan.” As explained by Richards, this also applied, in a way or another, to many other Darwin followers, such as Haeckel.

In fact, there are so many of such examples that Richards declared that “after the publication of such recent historical studies on the relation of science to religion as Frank Turner’s *Between Science and Religion* and James Moore’s *Post Darwinian Controversies*, it might be thought anomalous that any scholars would still cultivate the vintage belief that Darwinism and religious conviction were fundamentally opposed.” Anomalous indeed. But this vintage *belief* is indeed still much in vogue and cultivated by numerous scholars idealizing Darwin and by most creationists demonizing him. This is in itself somewhat paradoxical, as the latter could use the argument made by Richards – that Darwin did *not* remove teleology and morality from the natural world – in their favor. But that is not the way in which dramatized idolization and demonization work: beliefs become more important than facts, and the true central questions and critical aspects are thus often omitted, neglected, manipulated, or even completely forgotten.

Another type of omission done within such simplified dramatizations is that, as noted in Chap. 1, Charles Darwin was *not* the first to have evolutionary ideas and thus to contradict the creationist belief that God created humans and other organisms, back then. Many before him had defended *transmutation ideas*, including Lamarck, his own grandfather Erasmus Darwin, and so on, as explained in Jenkins Bill’s 2019 *Evolution Before Darwin: Theories of the Transmutation of Species in*

Edinburgh, 1804–1834. This and other books clearly show that the fact that Wallace sent a manuscript describing a theory of evolution by natural selection to Darwin just 1 year before Darwin published his own book with a similar idea was not a "*striking coincidence*," contrary to what Darwin stated. As explained in Browne's *Power of Place*:

> When Darwin called this event a coincidence he was already building up and retreating behind a protective fence of his own making. He evidently found it far less stressful to characterise the situation as a "striking coincidence" than to contemplate the alternatives - alternatives that implied any number of authors might be racing towards his own personal goal or that his concepts were much less innovative than he thought. He disliked finding out that someone else could conjure up his own private brainwave. He had invested his time, his health, and his happiness in the work, and to lose this intellectual capital overnight was as cruel to him as any financial disaster.. He genuinely believed no one else held the same combination of ideas as he did..
>
> It was hard to accept that he was not the innovator he imagined he was. Along with everything else, his scientific *vanity* was badly shaken. Yet Wallace's letter was really no more of a coincidence than the invitation to travel on the Beagle had been. To start with, there were differences between the two theories.. Wallace attended far more than he did to the replacement of a parent species by an offspring variety. Wallace wrote of the way that a group of advantaged individuals, say a variety of pigeon that could fly further in times of food shortage, might in time supplant those birds that possessed less stamina. Group replaced group. His view of nature was thus less concerned with individuals than Darwin's. Second, he declared his belief that there could be no parallel between the natural process of "selection" and what went on under artificial conditions of domestication - a point diametrically opposed to Darwin. Any parallel with breeders "selecting" traits in their animals, Wallace said, was "altogether false"..
>
> Darwin had also long been blind to many of the changing currents around him. If he had been less inwardly focused on his own projects, or less preoccupied with his health and that of his family, he might not have been so hopelessly taken aback. Wallace had scattered suggestive pointers about the way his thoughts were tending in several articles published in London journals during the 1850s and deliberately raised the problem of accurate distinctions between species and varieties in letters to Darwin. Lyell had drawn these signals to Darwin's attention during a weekend visit to Down House in 1856. Other suggestive pointers were just as plain to see. *Evolution, Lyell observed, was hanging tensely in the air*. Evolution - or something very like it. If Darwin had lived in London, as Lyell did, or mixed more frequently with the intellectual avant-garde, he must surely have noticed the general swing of progressive, liberal opinion among a small circle of influential figures. Speculative developmental ideas enjoyed fairly wide currency.
>
> Mostly these ideas were loosely based on the concept of an inbuilt *advance of mankind and society*, ideas that ultimately rested on the ideologies of enlightenment and transformation disseminated by European thinkers of the late eighteenth century and revolutionary period, among them Jean-Baptiste Lamarck and Darwin's own grandfather, Dr. Erasmus Darwin, and that were now revitalised with high Victorian notions of *striding forwards*. During the 1820s and 1830s, most aspects of this *transformist philosophy of nature*, including the possibility of changing the nature of the human mind and the structure of society itself, had in Britain come to be associated with scientific rationalism and as often as not the lurking threat of political activism. But by the 1850s, intellectuals were equally liable to embrace the same motifs in the safer form of self-advance, *economic progress*, and the steady *march of civilisation* while still taking Lamarck's name as a general catch-all label for any *progressive transmutationary ideas*.

"*Evolution, Lyell observed, was hanging tensely in the air.*" So, *both* the popular culture idea that Darwin was the first to talk about evolution and thus "take God out of the equation" *and* that he then "fought" mainly alone, or just with his "bulldog" Huxley, against the "powerful theists," are not only exaggerated dramatized stories: they are historically inaccurate.

The fact that nowadays we do not hear people talking about Wallacism, and that lay people do not even know who was Wallace for that matter, clearly shows us that there is much more about the idealization of Darwin than the fact that he postulated a theory of evolution by natural selection. Maybe there was indeed a major societal and ideological "war" back then, but one in which Darwin mainly sided with, instead of against, the hierarchical Victorian *status quo* and the powerful and most "well-off" members of his society? Maybe that is, at least in part, why he became an English "'hero" that has been and continues to be revered by so many Westerns and Anglo-Saxons in particular, as opposed to the self-proclaimed "socialist" Wallace? This would help to explain why Darwin was explicitly celebrated in his funeral as a Victorian hero that helped, with his ideas and books, to make "Britain great," extend "her Empire," and civilize "new worlds at home and abroad." That is, that helped doing a lot of stuff that Wallace actually often publicly criticized. As we have seen above, we do *know* that there was a "public feeling that Darwin, in his life and work, symbolized English success in conquering nature and civilizing the globe during Victoria's long reign." We have indeed seen, in Chap. 1, that Darwin explicitly defended, and even praised, social inequalities and hierarchies and British imperialism and colonialism, again and again, in his public and private writings. For example, among many other examples that we have discussed, in *Voyage of the Beagle* he criticized the egalitarianism of the Fuegians as "primitive" and commended *hierarchy and even monarchy*:

> The perfect equality among the individuals composing the Fuegian tribes must for a long time retard their civilization. As we see those animals, whose instinct compels them to live in society and obey a chief, are most capable of improvement, so is it with the races of mankind. Whether we look at it as a cause or a consequence, the more civilized always have the most artificial governments. For instance, the inhabitants of Otaheite, who, when first discovered, were governed by hereditary kings, had arrived at a far higher grade than another branch of the same people, the New Zealanders,—who, although benefited by being compelled to turn their attention to agriculture, were republicans in the most absolute sense. In Tierra del Fuego, until some chief shall arise with power sufficient to secure any acquired advantage, such as the domesticated animals, it seems scarcely possible that the political state of the country can be improved. At present, even a piece of cloth given to one is torn into shreds and distributed; and no one individual becomes richer than another. On the other hand, it is difficult to understand how a chief can arise till there is property of some sort by which he might manifest his superiority and increase his power. I believe, in this extreme part of South America, man exists in a lower state of improvement than in any other part of the world.

In many passages of the same book, Darwin did not hide his support for colonialism – in particular, British colonialism, of course – and disgust for the supposed "cannibalism, murder, and all atrocious crimes" of the savages. For instance, when he referred to Tahiti, New Zealand, and Australia:

There are many who attack.. both the missionaries, their system, and the effects produced by it. Such reasoners never compare the present state with that of the island only twenty years ago; nor even with that of Europe at this day; but they compare it with the high standard of Gospel perfection. They expect the missionaries to effect that which the Apostles themselves failed to do. Inasmuch as the condition of the people falls short of this high standard, blame is attached to the missionary, instead of credit for that which he has effected. They forget, or will not remember, that human sacrifices, and the power of an idolatrous priesthood—a system of profligacy unparalleled in any other part of the world—infanticide a consequence of that system—bloody wars, where the conquerors spared neither women nor children – that all these have been abolished; and that dishonesty, intemperance, and licentiousness have been greatly reduced by the introduction of Christianity. In a voyager to forget these things is base ingratitude; for should he chance to be at the point of shipwreck on some unknown coast, he will most devoutly pray that the lesson of the missionary may have extended thus far..

At length we reached Waimate [in New Zealand]. After having passed over so many miles of an uninhabited useless country, the sudden appearance of an English farm-house, and its well-dressed fields, placed there as if by an enchanter's wand, was exceedingly pleasant. Mr. Williams not being at home, I received in Mr. Davies's house a cordial welcome. After drinking tea with his family party, we took a stroll about the farm. At Waimate there are three large houses, where the missionary gentlemen, Messrs. Williams, Davies, and Clarke, reside; and near them are the huts of the native labourers.. Late in the evening I went to Mr. Williams's house, where I passed the night. I found there a large party of children, collected together for Christmas Day, and all sitting round a table at tea. I never saw a nicer or more merry group; and to think that this was in the centre of the land of cannibalism, murder, and all atrocious crimes! The cordiality and happiness so plainly pictured in the faces of the little circle, appeared equally felt by the older persons of the mission..

At last we anchored within Sydney Cove. We found the little basin occupied by many large ships, and surrounded by warehouses. In the evening I walked through the town, and returned full of admiration at the whole scene. *It is a most magnificent testimony to the power of the British nation.* Here, in a less promising country, scores of years have done many more times more than an equal number of centuries have effected in South America. *My first feeling was to congratulate myself that I was born an Englishman.* Upon seeing more of the town afterwards, perhaps my admiration fell a little; but yet it is a fine town. The streets are regular, broad, clean, and kept in excellent order; the houses are of a good size, and the shops well furnished. It may be faithfully compared to the large suburbs which stretch out from London and a few other great towns in England; but not even near London or Birmingham is there an appearance of such rapid growth.

"The sudden appearance of an English farmhouse, and its well-dressed fields.. was exceedingly pleasant" and "my first feeling was to congratulate myself that I was born an Englishman": Darwin was so happy to go back to the Victorian paradise, after being for so long in the "lands of evil." Darwin's excerpt above is exceptionally revealing for two other main reasons. First, Darwin's own statements do not fit, once again, within the narrative of the "war" between Darwin the revolutionary non-believer versus the powerful backward theists. Of course, one could argue that his *Voyage* was written before Darwin changed some of his key ideas about God. However, one thing is his later ideas that God probably did not exist, another very different one is his praise for Christianity and its morals and colonialist missionaries, which he tended to praise very often. Second, the above excerpt further deconstructs the typical "everybody else was like that back then" line of defense used within Darwin's idolization. Darwin recognizes that this was not necessarily so, concerning precisely the missionaries: "there are many who attack.. both the missionaries, their system, and the effects produced by it." So, there were indeed "many" that did *not* defend, as Darwin did, the way missionaries acted during colonialism, the imperialist system in which they thrived, and the effects produced by both the missionaries and that system in general. It should however be emphasized again that despite his general praise for English colonialism, Darwin was able to recognize, occasionally, that there were at least some less positive "effects" of such colonialism. For instance, in the *Voyage* he wrote, about Australia:

> The number of aborigines is rapidly decreasing. In my whole ride, with the exception of some boys brought up by Englishmen, I saw only one other party. This decrease, no doubt, must be partly owing to the introduction of spirits, to European diseases (even the milder ones of which, such as the measles, prove very destructive), and to the gradual extinction of the wild animals. It is said that numbers of their children invariably perish in very early infancy from the effects of their wandering life; and as the difficulty of procuring food increases, so must their wandering habits increase; and hence the population, without any apparent deaths from famine, is repressed in a manner extremely sudden compared to what happens in civilized countries, where the father, though in adding to his labour he may injure himself, does not destroy his offspring.
>
> Besides the several evident causes of destruction, there appears to be some more mysterious agency generally at work. Wherever the European has trod, death seems to pursue the aboriginal. We may look to the wide extent of the Americas, Polynesia, the Cape of Good Hope, and Australia, and we find the same result. Nor is it the white man alone that thus acts the destroyer; the Polynesian of Malay extraction has in parts of the East Indian archipelago, thus driven before him the dark-coloured native. The varieties of man seem to act on each other in the same way as different species of animals—the stronger always extirpating the weaker. It was melancholy at New Zealand to hear the fine energetic natives saying that they knew the land was doomed to pass from their children. Every one has heard of the inexplicable reduction of the population in the beautiful and healthy island of Tahiti since the date of Captain Cook's voyages: although in that case we might have expected that it would have been increased; for infanticide, which formerly prevailed to so extraordinary a degree, has ceased; profligacy has greatly diminished, and the murderous wars become less frequent.

But as we can see in this last sentence, even when he was able to recognize that there were a few less positive sides about colonialism, such as the diseases brought by the colonialists, he then suggested that if it was not for those diseases or the "introduction of spirits" (drinks) or "the gradual extinction of the wild animals," it would have been "expected" that colonialism would have increased the number of indigenous peoples. As if the massacres and mass killings done by colonialists, or the huge number of slaves that lost their lives or were sent to other places by the Europeans, did somehow not even exist. There are numerous historical records showing that Europeans killed millions of indigenous people actively by forced labor and by using weapons, including horrible massacres done precisely in Australia by the English, against aborigines, as we have seen above (Fig. 1.15). To give just another example, under the rule of King Leopold II of Belgium the atrocities done by Belgians led to the killing of about 10 million Africans in the then so-called *Congo Free State* – a name that, in face of the atrocities that were done there, reflects the inhumane hypocrisy of European colonialism. In the very last two pages of his *Voyage* Darwin eloquently shows – for if there was any doubt – his praise and defense of colonialism, of white "civilization," of Christianity, of the British Empire, of the "philanthropic spirit of the British nation," of capitalism and "prosperity," and of the "British flag":

> From seeing the present state, it is impossible not to look forward with high expectations to the future progress of nearly an entire hemisphere. The march of improvement, consequent on the introduction of Christianity throughout the South Sea, probably stands by itself in the records of history. It is the more striking when we remember that only sixty years since, Cook, whose excellent judgment none will dispute, could foresee no prospect of a change. Yet these changes have now been effected by the philanthropic spirit of the British nation. In the same quarter of the globe Australia is rising, or indeed may be said to have risen, into a grand centre of civilization, which, at some not very remote period, will rule as empress over the southern hemisphere. It is impossible for an Englishman to behold these distant colonies, without a high pride and satisfaction. To hoist the British flag, seems to draw with it as a certain consequence, wealth, prosperity, and civilization.

Darwin traveled to several countries and so had the unique opportunity to see, with his own eyes, how the British Empire was using, abusing, massacring, and stealing the natural resources of the local people. But he still concluded his book with such an enthusiastic support for the "philanthropic spirit of the British nation," stating that "to hoist the British flag, seems to draw with it as a certain consequence, wealth, prosperity, and civilization." This summarizes, in a nutshell, Darwin's Victorian biases and support for the *status quo*, the well-off, the powerful, and the British Empire, and helps to understand why he was precisely praised for that during his pompous funeral at a royal church. In her 2019 book *On Fire: The (Burning) Case for a Green New Deal*, Naomi Klein argues that, apart from promoting racism, ethnocentrism, Christianity, subjugation, discrimination, oppression, and social and economic inequalities, the "*racial capitalism*"of the British Empire that was defended by authors such as Darwin also lead, ultimately, to another major problem. One that, tragically, is now faced by all of us and by countless other species: *Earth's ecological collapse*. Coming back to some of the topics discussed in her very influential *No Logo* book, she summarizes this idea as follows:

> Australia, despite its wealth, insists on massively expanding coal production in the teeth of the climate crisis; Canada has done the same with the Alberta tar sands; the United States has done the same with Bakken oil, fracked gas, and deepwater drilling, becoming the world's largest oil exporter; the United Kingdom has attempted to ram through fracking operations despite fierce opposition and evidence linking it to earthquakes..These nations led the way in forging the global supply chain that gave birth to modern capitalism, the economic system of limitless consumption and ecological depletion at the heart of the climate crisis. It's a story that begins with people stolen from Africa and lands stolen from Indigenous peoples, two practices of brutal expropriation that were so dizzyingly profitable that they generated the excess capital and power to launch the age of fossil fuel-led industrial revolution and, with it, the beginning of human-driven climate change. It was a process that required, from the start, pseudoscientific as well as *theological theories of white and Christian supremacy*, which is why the late political theorist Cedric Robinson argued that the economic system birthed by the convergence of these fires should more aptly be called "*racial capitalism*."

> Alongside the theories that rationalized treating humans as raw capitalist assets to exhaust and abuse without limit were theories that justified treating the natural world (forests, rivers, land and water animals) in precisely the same way. Millennia of accumulated human wisdom about how to safeguard and regenerate everything from forests to fish runs were swept away in favor of a new idea that there was no limit to humanity's ability to control the natural world, nor to how much wealth could be extracted from it without fear of consequence. These ideas about nature's boundlessness are not incidental to the nations of the Anglosphere; they are foundational myths, woven deep into national narratives. The huge natural wealth of the lands that would become the United States, Canada, and Australia were, from their very first contact with European ships, imagined as sort of body-double nations for colonial powers that were running out of nature to exhaust back home. No more. With the "discovery" of these seemingly limitless "new worlds", God had granted a reprieve: New England, New France, New Amsterdam, New South Wales – proof positive that Europeans would never run out of nature to exhaust. And when one swath of this new territory grew depleted or crowded, the frontier would simply advance and new "new worlds" would be named and claimed.

In this small excerpt, Klein touches on various critical topics. She summarizes, in a nutshell, the tragic economic, social, and ecological repercussions of the type of ideas so enthusiastically defended, promoted and, most importantly, "scientifically" supported and disseminated, by Darwin. She also exposes something that also applies to Darwin and that further increases the list of paradoxes and logical inconsistencies that characterize his works on human evolution and life and the way he has been idolized since then. Darwin is almost always portrayed as a pure "lover" of the natural world. However, contrary to the tone often used back then by travelers such as Wallace or Humboldt, his comments about "savage" life, be it concerning humans or other organisms, many times have a negative tone. We have seen above one of the numerous examples in which he used such a tone: "*after having passed over so many miles of an uninhabited useless country, the sudden appearance of an English farm-house, and its well-dressed fields, placed there as if by an enchanter's wand, was exceedingly pleasant.*" He called the "savage" world – including both "savage" animals and plants – as "useless country" and then got excited with the "exceedingly pleasant" "English farm-house, and its well-dressed fields." Not only that. Darwin's praise of agricultural states and their "civilization" – despite the fact that he *knew* that they were destroying the "savage" natural world – and criticism of – and often disgust toward – brutish "savages," is also at odds with someone who loves wilderness, which he clearly also loved. This is a further example of the incongruences between Darwin the biologist lover of nature and Darwin the anthropologist lover of agriculture and "civilization."

As recognized in Richards' 1987 book about Darwin, among a minority of scholars that dare to criticize Darwin more openly, one of the main criticisms about him is his use and abuse of Malthusian struggle-for-existence and Victorian capitalistic ideas and their extrapolation for the natural world as a whole, within his "scientific" theories. Such scholars make "the historical objection that Darwin unwittingly infused his theory with the political assumptions of laissez-faire English liberalism and the hedonistic selfishness of Benthamite utilitarianism and that this has hopelessly infected any evolutionary analysis of mind and morals." It is argued, Richards adds, that conceptions of rational ability and ethical choice which find root in the ideology of a particular culture – for example, nineteenth-century Victorian society – must be inherently defective or at least circumscribed thereby. As he notes, Sahlins, in his book *Use and Abuse of Biology*, maintains that "Darwinism, at first appropriated to society as 'social Darwinism', has returned to biology as a genetic capitalism." Sahlins warned us "that the current incarnation of 'social Darwinism', that is, sociobiology, 'contributes primarily to the final translation of natural selection into social exploitation'." As we have seen, later in life Darwin's continued to "scientifically" support English imperialism and the "morals" of a British Empire that caused so many atrocities, suffering, and deaths. In his *Descent*, published just 11 years before he died, he reiterated the very same ideas that he defended in the *Voyage* about these topics. With the aggravating fact that while the *Voyage* could be said to be mainly a kind of "travel account," the *Descent* was supposed to be his "scientific masterpiece" on human evolutionary history. This is an example of the kind of so-called evolutionary facts provided in the *Descent*, about these topics:

> The remarkable success of the English as colonists over other European nations, which is well illustrated by comparing the progress of the Canadians of English and French extraction, has been ascribed to their "daring and persistent energy;" but who can say how the English gained their energy. There is apparently much truth in the belief that the wonderful progress of the United States, as well as the character of the people, are the results of natural selection; the more energetic, restless, and courageous men from all parts of Europe having emigrated during the last ten or twelve generations to that great country, and having there succeeded best. Looking to the distant future, I do not think that the Rev. Mr. Zincke takes an exaggerated view when he says: "All other series of events—as that which resulted in the culture of mind in Greece, and that which resulted in the empire of Rome—only appear to have purpose and value when viewed in connection with, or rather as subsidiary to.. the great stream of Anglo-Saxon emigration to the west." Obscure as is the problem of the advance of civilisation, we can at least see that a nation which produced during a lengthened period the greatest number of highly intellectual, energetic, brave, patriotic, and benevolent men, would generally prevail over less favoured nations.

Instead of doing good and being kind to others – the very definition of *benevolent* – the most "benevolent" Anglo-Saxon men that emigrated to the west of the United States often prevailed over, oppressed, stole the lands, and massacred the "less favored." With such friendly "benevolent" men, who needs enemies?

The Other "War": Self-Preservation of the Cream of Victorian Society

> *Working men's Paris, with its Commune, will be forever celebrated as the glorious harbinger of a new society.. its martyrs are enshrined in the great heart of the working class.. its exterminators history has already nailed to that eternal pillory from which all the prayers of their priest will not avail to redeem them.* (Karl Marx)

As we have seen, one of the most interesting aspects of Darwin's autobiography is what it *does not mention*. For instance, Darwin *did not* refer to the *British flag* nor the benefits of colonialism and monarchy that he praised so much in his books, nor did he provide details about the social position and highly privileged Victorian type of life that his family enjoyed. Nor did he mention the capitalistic businesses undertaken by his own father (Box 2.1) as well as by himself. He did also not mention the Victorian political and socioeconomic context surrounding him, his scientific and societal ideas, and those that were very pleased with them and that subsequently further promoted them. Even about his grandfather Erasmus Darwin, who defended various evolutionary ideas and concepts that were very similar to the ones defended by him decades later, he only refers to him one time, in a short passage. As noted above, this does support the idea that Charles Darwin was seemingly using, at least in part, his autobiography to write a particular version of his story that he probably deemed to be more appealing, or more positively seen, by those that would read that autobiography in the future. That is, as put by Browne, in a way that was in part somewhat a camouflage, omitting the huge importance played by his *power of place* and the societal biases related to it, and paving the way for a narrative that indeed came to be the prevailing one within popular culture, focusing almost uniquely on the "good" aspects that led to his immortality. Namely, the quasi-religious tale that his ideas came from a *tabula rasa*, from him alone – the lone genius myth – in marked contrast to how his "flaws" – regarding aspects such as racism or sexism, within the very few occasions in which they are openly recognized in the literature – which are instead blamed on his society, as a *tabula inscripta*. As explained in Browne's *Voyaging* (see also Box 2.1):

> None of the influential thinkers, scientists, and philosophers of any historical period were "born" in this simple sense [as a *tabula rasa*].. Darwin was made by Darwin *and* Victorian society.. Because Darwin believed in the Victorian ethos of character - in the inbuilt advantages of mind - and unconsciously endorsed the cult of great men and public heroes that was so much a part of nineteenth-century life, he did not - could not - see that figures like himself were the product of a complex interweaving of personality and opportunity with the movements of the times. Scientific ideas and scientific fame did not come automatically to people who worked hard and collected insects, as Darwin seems to have half hoped they would. A love of natural history could not, on its own, take a governess or a mill-worker to the top of the nineteenth-century intellectual tree. Nor can it, on its own, explain Darwin.. [in this sense] his autobiography was just as much an exercise in camouflage - a disguise - as it was a methodical laying out of the bare bones of his existence.
>
> Inevitably, the things he left unsaid in this autobiography are the most revealing. Behind Darwin lay the vast unacknowledged support system of the Victorian gentry, and beyond that the farflung network of imperial, colonial Britain. He was born a wealthy child in a socially secure, well-connected family. His father was a rich physician and his mother a daughter of Josiah Wedgwood, the potter; his grandfathers on both sides were noted for their contributions to science, philosophy, and technology. He possessed many advantages in life, including an education at the best institutions Britain had to offer. The friends he made at Cambridge University proved influential figures over the years, especially in the way they generated an invitation for him to join the Beagle's expedition round the world and then eased his entrance into London's scientific circles after his return. Soon after that return, Darwin married one of his Wedgwood cousins, who provided all the secure comforts of a sedately upper-class, countrified existence..

Browne further noted:

> It is also clear that an extraordinary number of people were drawn one way or another into Darwin's scientific projects.. Darwin's work was entirely a social process in this sense, and the "facts" he collected represented a collaborative endeavour fully documented in his

extensive correspondence.. Furthermore, when the *Origin* was published, it was not Darwin but his scientific friends who took the brunt of defending the idea of evolution in public. Without Thomas Henry Huxley acting as a pugnacious bulldog to his stay-at-home Labrador, Darwin would have been lost in the ensuing conflict; and prominent figures like Joseph Hooker, Alfred Russel Wallace, Charles Lyell, and the Harvard botanist Asa Gray, who each had serious misgivings about some part of Darwin's theory or other, willingly picked up their cudgels on his behalf. Even Darwin's notorious illnesses were excused and glamorised by his contemporaries as the penalty of intellect: characteristic, it was believed, of the deeply thoughtful group of eminent men to which he belonged.. Darwin's real life, in short, belies the common view of him as an isolated recluse. There was a sliver of ice inside enabling him to make the most of all the advantages he possessed and the circumstances in which he found himself. Though he was often alone with his theories, morbidly turning over the ideas of death and struggle on which the concept of natural selection was based.. he was also, in another sense, propped up by British society. His story is the story of the era - of the different ways in which a man could emerge as a profound thinker in Victorian Britain, of the way that someone could take up and turn around the assumptions of the age and become a hero for doing so. It is the story of the transformation, in a particular time and place, of an amiable but rather aimless young man into a scientific giant.. he was no more "born a naturalist" than he was a sailor or politician.. Darwin did not simply sit in the middle of Victorian society soaking up the overriding themes of the age. On the contrary, Victorian society made him.

Box 2.1 Darwin's Father and Mother, Victorian Society, and Capitalism

Browne's *Charles Darwin voyaging* emphasizes the crucial role that contingency played in Darwins' life, work, and recognition, both during his life and after it. Here I include just a few excerpts of that book that are particularly relevant to better understand such contingencies and Darwin's *power of place*:

> Although Charles Darwin eventually became an archetypal Victorian.. his roots were embedded in a notably different context. The distinctive aura of late Georgian times left an indelible mark. Darwin's first and most formative experiences of life were politely *Regency* in spirit, seasoned with the vigorous intellectual activity of the early years of the Industrial Revolution.. The Darwins may have been sedate, but they lived through a period of extraordinary vitality and change in Britain - in politics, in social relations, in art and literature, in science, agriculture, trade, and manufacture. From the time of the American Revolution to the trial of Queen Caroline and the dramatic Cato Street conspiracy in 1820, there were hardly any other years in British history so full of contrasts, political uncertainties, and booming commercial expansion. Revolution and terror in Paris, the long-drawn-out wars with Napoleonic France, ardent radicalism at home, grain riots, crippling taxes, "gagging bills" in Parliament, and unparalleled antagonisms within English society suddenly made the prospect of civil strife - even bloody revolt - possible.
>
> It seemed as if the country was as close to internal rebellion..At the same time, new technology was transforming the national economy and landscape: factories and mills sprouted deep in the backbones of England and Scotland; iron, coal, textile, and pottery works inched through the Midlands.. With them came extraordinary riches for the privileged few. Nothing the frame-breakers could do would alter the rapid acceleration towards modern capitalism and consumer culture characterising high society in late Georgian times. Despite the fact that the rolling northern landscape revealed the dramatic effects of enclosure and the beginnings of industrial expansion, the social repercussions of these fundamental changes barely impinged on the world the affluent landed classes knew. Regency London was the brightest,

(continued)

most opulent city in Europe.. Britain was becoming the first industrial nation: a nation of shopkeepers, according to Napoleon; of manufacturers and customers; of culture, taste, and elegance at one level; of fistfights and seditious literature at another..

Few families were so well poised to take advantage of this changing world as the Darwins of Shrewsbury. Charles Robert Darwin was born on 12 February 1809, the fifth child of Susanna and Robert Waring Darwin of The Mount, a large Georgian house.. His father was a prosperous physician, one of three practising in Shrewsbury, his mother a daughter of Josiah Wedgwood, the founder of the china company and an influential Staffordshire entrepreneur.. Darwin's parents typified everything that is known about the emergent entrepreneurial society of early industrial England, a classic example of the way wealthy members of the professional and manufacturing classes created a significant niche for themselves in a changing world. But it is important to emphasise how far this social movement had already advanced by the time Darwin was born. His parents - grandparents even - considered themselves an integral part of the landed gentry: a readily recognisable upper stratum peculiar to British society firmly based on the ownership of land, usually through many generations, and accompanied by political and local commitments which counted for a great deal in the corridors of power. Not quite aristocrats, although frequently intermarrying with the aristocracy, the landed gentry of Britain considered themselves several notches higher than most industrialists or country doctors or lawyers, although, again, frequently intermarrying with them. They belonged to the ruling classes, even if they did not necessarily rule. In a culture where finely nuanced distinctions between classes governed social behaviour, these notches meant everything.

The Wedgwood family was altogether richer, more fashionable, more part of the intellectual beau monde, or as the Darwin girls sometimes put it, the "rising ton". The first Josiah Wedgwood moved in high society, courting aristocratic clients for his chinaware, running stylish showrooms in London, and creating sensational dinner services for the queen: one of the first leaders of public taste to emerge in the eighteenth century and one quick to invest his profits in the traditional finery of a large house and grounds in the countryside.. This Wedgwood connection, long-standing social respectability, and Dr. Darwin's medical success gave an indisputable upper-class gloss to the couple's position in Shrewsbury society. Dr. Darwin was also an astute financier, a new kind of private investor appearing in response to commercial expansion and changing forms of financial opportunity brought about by manufacturing and technological developments in Britain.

Charles Darwin grew up in an atmosphere imbued with the principles of financial adaptation to circumstances, of investment, returns, and profit. The doctor [his father], in fact, was one of the first capitalists of the modern era.. specialised in private money-broking. At a time when factory and mill owners were rarely themselves men of capital and Britain's banking structure was only rudimentary, individual proprietors or business partners often found it difficult to raise the circulating capital needed for large-scale works like canals and roads. They depended on investors from the professional sector, on men normally hidden from history like Dr. Darwin who were prepared to move money from fixed reserves, such as land, into the fluid stock of manufacturing companies. Dr. Darwin went further than most in the way he also acted as a financial middle man, arranging loans and raising cash for other local

(continued)

gentlemen who themselves wished to join the investors' market.. He particularly favoured mortgages, a straightforward device for people to borrow large sums on the security of land or housing. Returns took the form of quarterly interest payments or a percentage of the profit. Within a few short years, the existence of this kind of flexible paper wealth - as distinct from the fields and stately country houses of squires and aristocrats - became a crucial element in supporting the new industrial society.

Such historical accounts indeed reveal that most publications about Charles Darwin – including his own autobiography – are oversimplified and often omit that there was another major ideological "war" occurring back then, and that his ideas were influenced by and played a major role within this "war," in England. As shown in Box 2.1, Browne wrote that "as Revolution and terror in Paris, the long-drawn-out wars with Napoleonic France, ardent radicalism at home, grain riots, crippling taxes, 'gagging bills' in Parliament, and unparalleled antagonisms within English society suddenly made the prospect of civil strife – even bloody revolt – possible." This "war" was omnipresent in the minds of many English people, particularly the ones that had the most to lose: the well-off, including Charles Darwin and his family, as well as his closest friends, and powerful colleagues and allies. The "war" had involved some major events, including atrocities, in continental Europe, just at the other side of the English Channel, and there was a feeling that it could also profoundly affect England. Factual evidence, such as personal letters, show that some of Charles Darwin's family members were indeed very afraid of this ideological and societal "war": as expected, they were *not* taking the side of those that wanted to change the Victorian *status quo*. As most of the cream of Victorian society, they instead wanted to maintain their highly privileged, well-off position within the very hierarchical Victorian system. Importantly, with respect to this major "war," the "scientific" support provided in Darwin's books for Victorian racist, sexist, and ethnocentric evolutionary ladder-of-life ideas was the perfect type of "weapon" that the Victorian well-offs were desperately looking for. Such "scientifical" ideas could justify their privileged position at the very top of the ladder: hierarchies are just natural, a part of nature. Not only that, as "scientifically" defended by Darwin in his *Voyage* and *Descent* books, within human societies the more hierarchical ones are the "best," more "civilized," and more "moral" ones. As put in Desmond and Moore's 1994 *Darwin* book:

> Darwin's was a privileged life.. and it stood on a precipice in the dark days of 1848. That weekend the insurrection sweeping Italy was threatening to explode nearer home. France - in Alexis de Tocqueville's phrase - was 'sleeping on avolcano'.. The French King abdicated on the 24th and headed for exile in England. In London the rumours flew. At the Tory leader Sir Robert Peel's party on Saturday the 26th, Darwin's colleagues - Lyell, Owen, Buckland, and De la Beche - were given agraphic account of the Paris uprising by the Prussian ambassador, a member of whose staff had escaped through the revolutionaries' lines. They heard of the '30,000communists in Paris who are for property in common and no marriage, and who are much to be feared by those who have aught to lose'. After the ladies retired, there was hectic talk of the revolution, and how, with a few sensible reforms, the throne could have been saved. Peel told Lyell he feared a financial crisis in Britain as the measures of the

new Republican government - to be proclaimed the next day - frightened capitalists on both sides of the Channel. Peel's guests were the gentlemen and clergy who rubbed shoulders with Darwin at his clubs, the Geological and the Athenaeum. Some had shared his hospitality at Down the fortnight before. All were concerned to see stabilizing reforms put into place and to police the radical masses, now demanding total suffrage at home. Darwin's friends were fearful. Lyell had long deplored 'mob-rule'. Forbes, the son of a banker, was himself shortly to take up his baton against rioters..

After the initial bloodletting in Paris, when the populace took the city, Emma's aunt sighed with some relief that the revolution had become 'more a social than a political one'. She still wondered whether the new leaders would be 'able to realise their promises to the working classes,' and dreaded the 'vengeance of the monster they have unchained' if they could not. But while French workers had gained concessions, British unions were frustrated over the government's intransigence. There was nothing like the French 'Right to Work', and unemployment remained high during the 40s.. Panic spread among the wealthy. The Queen left the Palace for her own safety, and plans were drawn up to 'quell the insurrection by force'.. Everything was done to ensure that demonstrators did not occupy any buildings, as they had in Paris. The jitters were as pronounced among Darwin's colleagues, who mounted guard in the scientific institutions. At Charing Cross, Ramsay, sworn in as a special, patrolled with Forbes at the Geological Survey, shouldering his truncheon in defence of his trilobites.. The air of frantic preparation left everybody panicky, Darwin not least.. London was now up in arms, anyway. The agitators' demands - for land taxes, property taxes, wealth taxes - would have hit him hard. Here he was, a member of the despised gentry, a leisured gentleman, living, as the extremists would say, off the backbroken poor. His father was ill, he had a growing family to fend for, and, if events took a nasty turn, his investments could be wiped out.

So, *something* needed to be done, as explained by Desmond and Moore:

Herbert Spencer.. had long accepted evolution, seeing it as an accumulation of changes acquired and passed on by each individual. *Progress was a necessity*. It was a 'law underlying the whole organic creation'; civilization was 'a part of nature; all a piece with the development of the embryo or the unfolding of a flower'. It was a guarantee that evil will ultimately disappear and man 'become perfect'. By the summer of the Great Exhibition he realized that progress meant something more. George Lewes.. introduced him to the French zoologist Henri Milne-Edwards's concept of the 'physiological division of labour'. And Carpenter convinced him that evolution was a continuous change from the 'general' to the 'special'; it honed organisms to their environments, adapted each to its task, as in Milne-Edwards's *industrial analogy*. Spencer now believed that what was true for animals held for society: *progress came through specialization*. Was it not Prince Albert who said that the 'great principle of the division of labour' was the 'moving power of civilization', and that the Great Exhibition was a 'living picture of the point of development at which the whole of mankind has arrived'? However, a great stumbling block still had to be overcome, one that had damped downbelief in human perfectibility for half a century. There were always too many mouths to feed, with never enough food to go round; a painful *struggle for existence* was inevitable.

This was the Malthusian 'principle of population' and for most liberals its status was sacrosanct. Humanity's only hope lay in abstinence from sex until each couple had the means to support children. But such 'moral restraint', as the Revd Malthus called it, had never been practised by the great mass of humanity.. To those, like Holyoake, with working-class sympathies, Malthus's principle was an evil invention. It was the ideology of the workhouse, blaming poverty on the poor and blessing the rich. But to those who sided with the cotton kings and their apologists.. the Malthusian principle was a beneficent law of nature, encouraging responsibility and self-improvement..Spencer's '*Theory of Population deduced from the General Law of Animal Fertility*'.. had something for everyone - but the lion's share was for the comfortably well-off. In his view the painful Malthusian principle

> is both true and self-correcting. People who multiply beyond their means take 'the high road to extinction'.. those who remain are 'the select of their generation'. Having exercised moral restraint and foresight, they bequeath their powers of 'self-preservation'.

"Self-preservation," indeed. Self-preservation of the well-off, of the *status quo*, that was what the "privileged few" wanted. As put in one of the most detailed analyses of the history of racism, Bethencourt's book *Racisms*, the "enormous political change of the mid-nineteenth century had an impact on research.. after a long period of structural and multiple changes, the 1840s and 1850s presented turning point in which scientific research on the variety of human beings became much more assertive, ideologically aggressive, and politically engaged.. it presented a scientific effort to justify and reify divisions as well as hierarchies of races, supposed to be innate, immutable, and perpetual." In particular, "the extraordinary spread of the revolt had presented new social and political challenges: while merit still struggled to assert itself against privilege in the new system of values, the struggle for equality against inequality became an important issue in the conceptual separation of the new and old social orders.. in Europe, reflection on race and the scientific quest for the origins of human variety became a major tool for proving the supposedly inherent, rooted origins of inequality, in order to undermine the powerful movement for equality as artificial and antinatural." How wonderful it would be if a well-off, well-respected, Victorian 'white' male scientist would provide "scientific facts" to support Malthusian, "struggle-for-life" notions that would naturalize a social ladder, societal hierarchies and inequalities, and the subjugation of "others" – of the poor, the non-whites, the women, and so on (see also Box 2.2). Well, as the saying goes: so it was, Darwin's writings came at the perfect time, and the rest is history. A history that is often untold, or distorted, within the countless accounts idealizing Darwin. An often untold story that, when fully disclosed, without taboos, idealization, and demonization, can now help us to better understand the missing pieces that connect the repeatedly told narrative about Darwin – the non-believer that "triumphed" against the *status quo* of the powerful Victorian religious establishment and its natural theologian allies – to factual events such as why he had an ostentatious state funeral at a Royal Church. A story that fully goes in line to Browne's assertion "a love of natural history could not, on its own, take a governess or a millworker to the top of the nineteenth-century intellectual tree." It could not, and it did not. In the Introduction of a 2004 version of Darwin's *Descent of Man*, Moore and Desmond show, even more clearly, with whom Darwin sided, within this "war" between the Victorian well-off and the political ideas coming from the other side of the English Channel, and how this is related to the "cleansing' and "purging" of Darwin and the beginning of his idolization:

> Moral and political issues were constitutive in the *Descent of Man*, as they had been in Darwin's work from the first. He had devised his theory of natural selection as his Whigs were building the workhouses. These were made deliberately abominable to keep the able-bodied out of them and thus competing in the market place. Darwin's selection worked in the same way, with overpopulation thrusting individuals into competition, leaving only the fittest to multiply. To him nature and society were of a Malthusian piece. Thirty years on, with the population of England 50 per cent larger, society seemed to be shielding life's rightful losers. Galton and Greg feared a genetic drain, and the *Descent* prescribed Darwin's

liberal eugenic remedy: "both sexes ought to refrain from marriage if in any marked degree inferior in body or mind"; and "all ought to refrain from marriage who cannot avoid abject poverty for their children". As for the remainder, he offered his Malthusian prescription: "our natural rate of increase, though leading to many and obvious evils, must not be greatly diminished by any means". Competition must not be prevented "by laws or customs". The "best" must out-breed the rest. Rougher sorts saw Darwin sanctioning their own attacks on Christian "laws and customs". Religious decay was the goal at the secularist *Hall of Science* in London's City Road, where a member of the *International* (a federation of socialists' and workers' groups, founded in 1864, with Karl Marx effectively in control) lectured on the *Descent of Man* for ninety minutes to prolonged applause.. Conducting their own defence, they subpoenaed Darwin. Had he not liberated society from superstition and discussed sexual matters openly in the *Descent*? Darwin was appalled and pointed out the passage about not diminishing the birth rate "by any means" - that meant "any artificial means". If compelled to testify, he would denounce the defendants. So on 18 June 1877, in the Queen's Bench Court, learned legal minds debated the *Descent of Man*. Darwin was metaphorically in the dock himself now as Besant slammed the passage for "the awful amount of human misery which it accepts as the necessary condition of progress"..

While promoting a Malthusian ruthlessness, Darwin, the reclusive patriarch, cocooned with his inherited fortune, never had to compete himself.. In the *Descent of Man* he effectively justified a body of rich intellectuals, freed from daily work, gentlemen (not women) whose status gave science its imprimatur, and who could contribute to the progress of mind. The book lent evolutionary credentials to Victorian middle-class gains, not to mention sexual, ethnic and nationalistic rankings. With the *Descent* selling in London and the industrial towns for 24 shillings – over a week's wage for an average worker – one can appreciate it as a piece of self-congratulatory science for the Liberal nouveau riche and emerging agnostic patriarchs of middle England: the rising businessmen, industrialists and professionals. Darwin never intended that his evolutionary scheme, whatever its secularizing tendencies, should sanction working-class collective self-help. Unions and cooperatives, which "opposed.. competition", were, he declared in 1872, "a great evil for the future progress of mankind". On its publication, many saw the *Descent*'s science as politically loaded. But the perception did not last. Already in Darwin's lifetime, a cleansing and purging of its cultural overtones was under way. This can be seen as the Fortnightly Review's John Morley scored *The Times* for attacking Darwin "from the point of view of property" during the Paris Commune - in effect denying that science could be judged right or wrong by its political consequences. Such an "unseemly" reduction of scientific truth to a "department of daily politics" had to be exposed for the sake of a growing scientific professionalism. While evolution was tugged this way and that to support conflicting social nostrums, one figure was seen to rise above the fray. Eventually it could be claimed that the actual lack 'of unanimity among Darwinians in matters of Sociology and Politics' showed that "the principles of the Master are perfectly neutral on such questions". Parallel with this neutralizing of the "Master's" work, as contemporary political, social and moral meaning was ostensibly drained from the *Descent*, an extraneous 'Social Darwinism' was born within sociology.

It is indeed a striking historical fact that two major events symbolizing the two sides of this ideological "war" happened exactly in the same year, 1871: the Paris commune, on one side, and the publication of Darwin's *Descent*, on the other. We now know that this notable fact was far from being completely accidental. The *Descent* was exactly the type of book that so-many Victorian well-offs were waiting for. Both Darwin's idolization and the rise of Social Darwinism were only possible *because* he was part of – and provided that very influential "weapon" to – the "well-off". Darwin and other members of the scientific and societal cream of Victorian society did mainly win that "war" in the short term and continue to be partial winners in the long term. After all, an aristocracy lead by a Queen – exactly as was the case back then – as well as capitalism, systemic racism, and social hierarchies, are still part of what England is, nowadays. However, as shown in Box 2.2, many of the then so-called "radical social" demands of the "Paris Commune" are now usually accepted as a given and applied even by conservative parties in England and other countries within the Anglosphere, such as the United States, Canada, or Australia. In this sense, it is interesting to note that, today as back then, there are still some notable differences between countries such as England and in particular the United States, versus continental European countries such as France, Spain, or Portugal, which are not infrequently governed by parties that are said to have a somewhat more "Socialist tenor." The very term "socialism" continues to be often demonized by a huge number of people in the United States and to be used by many of its conservative politicians to scare a huge portion of voters in a way that somewhat resembles the type of narratives that were often used by the "well-off" Victorians when Darwin wrote his *Descent*. This while countries such as Portugal and Spain had several governments led by "socialist" parties, without that implying any type of social collapse or "degeneration" of those countries: in fact, both of them are nowadays among the favorite touristic destinations of countless US citizens.

Box 2.2 1871's Descent of Man, Paris Commune, and a Societal and Ideological "War"

In 1871, the year Darwin published the *Descent*, there was another major event that is also remembered – and celebrated by many, in a very different way in which the *Descent* is: the "Paris Commune." It refers to a local authority that is often described as "socialist" and that briefly ruled Paris from 18 March to 28 May, 1871, before many of its members were arrested and executed (Figs. 2.3 and 2.4). Knowing a bit about what happened back then allows us to better understand the links between the two events, and how interactive and dynamic human history is, including the history of science. This is because Darwin's writings – particularly the *Descent* – included societal biases and were in turn used to subsequently "scientifically" support, legitimize, and naturalize the very same societal and political ideas that led to those biases in the first place, in a self-reinforcing and self-preservation process. The following brief excerpts about the 1871 Paris commune are taken from the *New World Encyclopedia*:

> In a formal sense the Paris Commune of 1871 was simply the local authority (council of a town or district–French "commune") that exercised power in Paris for two months in the spring of 1871. But the conditions in which it was formed, its controversial decrees, and its tortured end make it one of the more important political episodes of the time.. The Commune was the result of an uprising within Paris after the

(continued)

Fig. 2.3 Destruction of the Vendôme Column during the 1871 rise of the Paris Commune; actually this picture, as others taken at that time, were later used to identify and execute Communards

Franco-Prussian War ended with France's defeat. This uprising had two root causes: on the one hand the disaster in the war, on the other the growing discontent among French workers, which can be traced to the 1830s, when the first worker uprisings took place in Lyon and Paris.. By that time hundreds of thousands of Parisians were armed members of a citizens' militia known as the "National Guard", which had been greatly expanded to help defend the city. Guard units elected their own officers, who in working-class districts included radical and socialist leaders.. Steps were being taken to form a "Central Committee" of the Guard, including patriotic republicans and socialists, both to defend Paris against a possible German attack, and also to defend the republic against a possible royalist restoration, following the election of a monarchist majority in February 1871 to the new National Assembly..

As the Central Committee of the National Guard was adopting an increasingly radical stance and steadily gaining in authority, the government felt that it could not indefinitely allow it to have four hundred cannons at its disposal.. Instead of following instructions, however, the soldiers, whose morale was in any case not high, fraternized with National Guards and local residents.. The 92 members of the Commune (or, more correctly, of the "Communal Council") included a high proportion of skilled workers and several professionals (such as doctors and journalists). Many of them were political activists, ranging from reformist republicans, through various types of socialists, to the Jacobins who tended to look back nostalgically to the Revolution of 1789.

(continued)

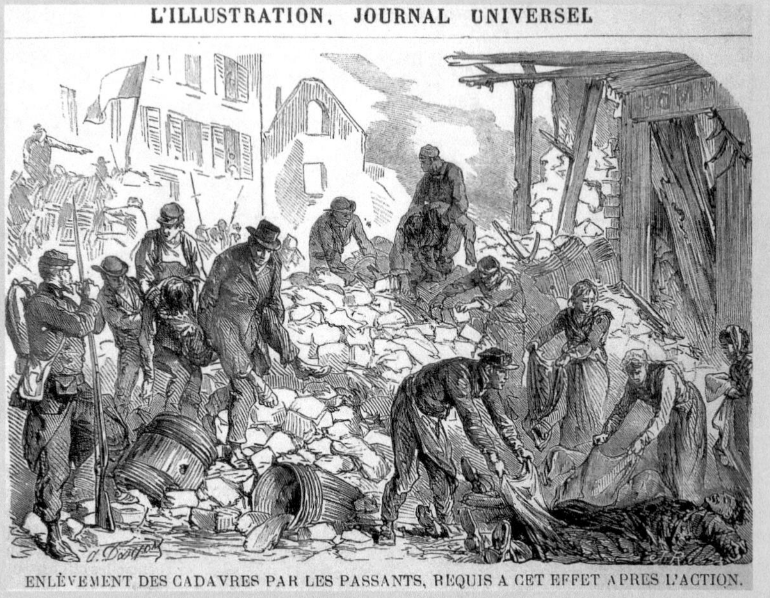

L'ILLUSTRATION, JOURNAL UNIVERSEL

ENLÈVEMENT DES CADAVRES PAR LES PASSANTS, REQUIS A CET EFFET APRES L'ACTION.

Fig. 2.4 When the 1871 Paris Commune rise was defeated, Parisians buried the bodies of the Communards in temporary mass graves. They were quickly moved to the public cemeteries, where between 6000 and 7000 Communards were buried. Painting by d'Alfred Darjou for *L'Illustration du 10 juin 1871*, displayed at the *Bibliothèque historique de la Ville de Paris*

Despite internal differences, the Council made a good start in maintaining the public services essential for a city of two million; it was also able to reach a consensus on certain policies whose content tended towards a progressive, secular and highly democratic social democracy rather than a social revolution. Lack of time (the Commune was able to meet on fewer than 60 days in all) meant that only a few decrees were actually implemented. These included the separation of church and state; the right to vote for women; the remission of rents owed for the entire period of the siege; the abolition of night work in the hundreds of Paris bakeries; the granting of pensions to the unmarried companions of National Guards killed on active service, as well as to the children if any; the free return, by the city pawnshops, of all workmen's tools and household items up to 20 francs in value, pledged during the siege as they were concerned that skilled workers had been forced to pawn their tools during the war; the postponement of commercial debt obligations, and abolition of interest on the debts; and, the right of employees to take over and run an enterprise if it were deserted by its owner, who was to receive compensation.

Some women organized a feminist movement, following on from earlier attempts in 1789 and 1848. Thus, Nathalie Lemel, a socialist bookbinder, and Élisabeth Dmitrieff, a young Russian exile and member of the Russian section of the First International, created the Union des femmes pour la défense de Paris et les soins aux blessés ("Women's Union for the Defense of Paris and Care of the Injured") on 11

(continued)

April 1871.. Believing that their struggle against patriarchy could only be followed in the frame of a global struggle against capitalism, the association demanded gender-equality, wages' equality, right of divorce for women, right to secular education and for professional education for girls. They also demanded suppression of the distinction between married women and concubines, between legitimate and natural children, the abolition of prostitution (obtaining the closing of the maisons de tolérance, or legal official brothels). The Women's Union also participated in several municipal commissions and organized cooperative workshops..

Throughout April and May, government forces, constantly increasing in number.. carried out a siege of the city's powerful defenses, and pushed the National Guards back.. By 27 May only a few pockets of resistance remained, notably the poorer eastern districts of Belleville and Ménilmontant. Fighting ended during the late afternoon or early evening of 28 May.. Marshall MacMahon issued a proclamation: "*To the inhabitants of Paris. The French army has come to save you. Paris is freed! At 4 o'clock our soldiers took the last insurgent position. Today the fight is over. Order, work and security will be reborn.*" Reprisals now began in earnest. Having supported the Commune in any way was a political crime, of which thousands could be, and were, accused. Some of the Communards were shot against what is now known as the Communards' Wall in the Père Lachaise cemetery while thousands of others were tried by summary courts martial of doubtful legality, and thousands shot.. Nearly 40,000 others were marched to Versailles for trials. For many days endless columns of men, women and children made a painful way under military escort to temporary prison quarters in Versailles. Later 12,500 were tried, and about 10,000 were found guilty: 23 men were executed; many were condemned to prison; 4,000 were deported for life to the French penal colony on the island of New Caledonia in the Pacific. The number of killed during *La Semaine Sanglante* can never be established for certain, and estimates vary from about 10,000 to 50,000.

The fact that, at that time within the Victorian society, the "well-off" were the main "winners" of the political "war" between the privileged and the "revolutionaries" allows us to understand not only Darwin's state funeral but also why those well-offs were the ones that mainly "'wrote history," back then.

Regarding this topic, one further line of evidence that the widely, and often blind, acceptation and subsequent veneration of Darwin within the academic and societal cream of the Victorian society and of many capitalistic countries was in great part due to societal and political reasons is the fact that this did not happen in socialist regimes such as the Soviet Union. One of the most emblematic examples concerns the case of soviet biology in the first decades of the twentieth century, in which Darwinism begun to be more and more criticized – as tragically symbolized by the death of Vavilov, in jail – and Lamarckism more and more dominant – as epitomized by the meteoric rise of Lysenko. As summarized by Nils Roll-Hansen in the 2011 book chapter "*Lamarckism and Lysenkoism Revisited*":

Soviet biology differed from Western biology by its tie to a philosophical worldview that was also the ideological foundation of the state. [At the beginning] 'Creative Darwinism' was part of the official world picture, a substitute for traditional religious beliefs. Ideological pressure on scientists grew as the Stalinist political system tightened its grip from the late 1920s on. Lamarckism was attractive to Marxist ideology because it promised mutual reinforcement between improvement of the social environment and improvement of the 'gene fund' of the population. New progressive scientific institutions established in the 1920s were from the start inclined toward Lamarckism. The standard argument against Mendelism was that its concept of stable genes contradicted evolutionary change and progress, biological as well as social. But toward the end of the 1920s new work on mutations helped geneticists temporarily turn the tables on Lamarckism.. The period of cultural and agricultural revolution around 1930 was marked by strong practical demands on science, guided by the ideological principle of 'unity of theory and practice'. The Lenin Academy of Agricultural Science was formed in 1929 to coordinate the numerous agricultural research institutes that had grown up during the 1920s. Its first president was the plant scientist Nikolai Vavilov, who had been the main entrepreneur in agricultural science since the early 1920s. He underlined the central role of the Lenin Academy in collectivization. It was to be 'the academy of the general staff of the agricultural revolution', the general staff being the Ministry of Agriculture.

Thus, the Lenin Academy had to take some of the responsibility for the tragic failures of forced collectivization, and the government reacted by tightening political control and planned steering in 1935. Food production remained a headache and a key issue in Soviet politics for decades. Up to the 1950s the Lenin Academy was a central scientific institution with great clout in biology, rivaling the influence of the traditional 'big' Academy of Sciences.. By early 1938 Lysenko became president {of the Lenin Academy).. Stalinist terror paved the final steps for Lysenko. Decision at the top political level, probably Stalin' s own, put Lysenko into the most influential position of Soviet biology. His social background as a son of a peasant with little formal education was a valuable asset in the turbulent 1930s. Nevertheless, his quick rise would not have been possible without widespread sympathy for Lamarckism and skepticism toward neo-Darwinism among leading Soviet plant scientists throughout the 1920s and 1930s.. Nikolai Vavilov (became) most famous as the martyr who defended genuine genetic science against Lysenko.. He was arrested in 1940, charged with espionage or collaboration with foreign powers, and sentenced to death; he perished in prison in 1943.

Among countless illustrative examples highlighting that the *power of place* also plays a huge rule within other scientific areas, one of the most enthralling concerns the development of the telephone. Particularly within the context of the present book, this example is interesting because it concerns an event that is almost totally unknown to most members of the scientific community and that took place just 1 year after the publication of Darwin's *Origin*, that is, in 1860. Let us start by the end: who invented the telephone? Probably you, as the vast majority of people, will answer by saying the name of a Victorian: Alexander Graham Bell, who patented the telephone in 1876. However, it is now known that in reality an obscure Italian, Antonio Meucci, actually delivered his telephone patent first. Even less known is the fact that 16 years before that, in 1860, an also obscure Portuguese-German, Johan Philipp Reis built the first prototype of a telephone. Reis thought electricity could be propagated through space, and undertook some experiments to show this, which he described in a manuscript entitled "*On the radiation of electricity.*" Somewhat resembling what Wallace did with Darwin in 1858, the self-taught Reis mailed the manuscript to Professor Poggendorff in 1859, so the manuscript would be published in *Annalen der Physik*. However, the manuscript was rejected, probably at least in part precisely because Reis was not part of the scientific community. But Reis did not give up, and the year after he built the first model of a telephone, which he designated as a "telephon": it covered a distance of 100 meters. In 1862, he tried again his luck with Poggendorff, but he was rejected again because the renowned Professor considered that transmission of speech was a fantasy. And, once again, the rest is history: today the vast majority of people associate the telephone to Victorian Bell, not to the Portuguese-German Reis, or to the Italian Meucci, who *did* patent it earlier than Bell. *The power of place* is critical.

This discussion about the profound links between the power of place, Darwin's Victorian biases, and the acceptance of his biased ideas by the cream of the Victorian society bring us back to the pertinent discussion provided in Lewontin's book *Biology as Ideology* mentioned in Chap. 1, and in particular to the dual process described by him:

> [There are] forces.. that .. have the power to appropriate from science ideas that are particularly suited to the maintenance and continued prosperity of the social structures of which they are a part. So other social institutions have an input into science both in what is done and how it is thought about, and they take from science concepts and ideas that then support their institutions and make them seem legitimate and natural. It is this dual process - on the one hand, of the social influence and control of what scientists do and say, and, on the other hand, the use of what scientists do and say to further support the institutions of society--that is meant when we speak of science as ideology. Regardless of one's political view, everyone must agree that we live in a world in which psychic and material welfare is very unevenly distributed. There are rich people and poor people, sick people and healthy people, people who have control over the conditions of their own lives, work, and time (like professors who are invited to give lectures on the radio and turn them into books) and those who have their tasks assigned to them, who are overseen, who have little or no control over any psychic or material aspect of their lives. There are rich countries and poor countries. Some races dominate others. Men and women have very unequal social and material power.. As such struggles occur, institutions are created whose function is to forestall violent struggle by convincing people that the society in which they live is just and fair, or if not just and fair then inevitable, and that it is quite useless to resort to violence. These are the institutions of *social legitimation*.

Social legitimation. This is indeed, in a nutshell, one of the main reasons that have led, and continue to lead in great part, to the widespread and common blind acceptation, and even idealization, of Darwin and his ideas, particularly those about human nature, social hierarchies, higher "races," and superiority of males. In fact, although Lewontin's book is mainly focused on criticizing Neo-Darwinian doctrines about the DNA such as Dawkins' concept of selfish-genes, he makes it clear that Darwin's ideas provide a critical example of the dual process mentioned above:

Despite its claims to be above society, science, like the Church before it, is a supremely social institution, reflecting and reinforcing the dominant values and views of society at each historical epoch. Sometimes the source in social experience of a scientific theory and the way in which that scientific theory is a direct translation of social experience are completely evident, even at a detailed level. The most famous case is Darwin's theory of evolution by natural selection. No scientist doubts that the organisms on earth today have evolved over billions of years from organisms that were very unlike them and that nearly all types of organisms have long since gone extinct. Moreover, we know this to be a natural process resulting from the differential survivorship of different forms. In this sense, we all accept Darwinism as true.. But Darwin's explanation for that evolution is another matter. He claimed that there was a universal struggle for existence..

He claimed that the idea for evolution by natural selection occurred to him after reading the famous "*Essay on Population*" by Thomas Malthus, a lateeighteenth-century parson and economist. The essay was an argument against the old English Poor Law, which Malthus thought too liberal, and in favor of a much stricter control of the poor so they would not breed and create social unrest. In fact, Darwin's whole theory of evolution by natural selection bears an uncanny resemblance to the political economic theory of early capitalism as developed by the Scottish economists. Darwin had some knowledge of the economic survival of the fittest because he earned his living from investment in shares he followed daily in the newspapers. What Darwin did was take early-nineteenth-century political economy and expand it to include all of natural economy. Moreover, he developed a theory of sexual selection in evolution, in which the chief force is the competition among males to be more appealing to discriminating females. This theory was meant to explain why male animals often display bright colors or complex mating dances. It is not clear that Darwin was conscious of how similar his view of sexual selection was to the standard Victorian view of the relationship between middle-class males and females. In reading Darwin's theory, one can see the proper young lady seated on her sofa while the swain on his knees before her begs for her hand, having already told her father how many hundreds a year he has in income.

Importantly, with his characteristic sharpness Lewontin points out the broader societal legacies of such biased ideas of Darwin, up to the very present, highlighting for instance that the United States provide an emblematic example where those ideas continue to be actively used for social legitimization:

We have heard over and over again in school and had it drummed into us by every organ of communication that we live in a society of free equals. The contradiction between the claimed equality of our society and the observation that great inequalities exist has been, for North Americans at least, the major social agony of the last 200 years. It has motivated an extraordinary amount of our political history. How are we to resolve the contradiction of immense inequalities in a society that claims to be founded on equality? There are two possibilities. We might say that it was all a fake, a set of slogans meant to replace a regime of aristocrats with a regime of wealth and privilege of a different sort, that inequality in our society is structural and an integral aspect of the whole of our political and social life. To say that, however, would be deeply subversive, because it would call for yet another revolution if we wanted to make good on our hopes for liberty and equality for all. It is not a popular idea among teachers, newspaper editors, college professors, successful politicians, indeed anyone who has the power to help form public consciousness. The alternative, which has been the one taken since the beginning of the nineteenth century, has been to put a new gloss on the notion of equality. Rather than equality of result, what has been meant is equality of opportunity. In this view of equality, life is a foot race. In the bad old days of the ancien régime, the aristocrats got to start at the finish line whereas all the rest of us had to start at the beginning, so the aristocrats won. In the new society, the race is fair: everyone is to begin at the starting line and everyone has an equal opportunity to finish first. Of course, some people are faster runners than others, and so some get the rewards and others don't. This is the view that the old society was characterized by artificial barriers to equality, whereas the new society allows a natural sorting process to decide who is to get the status, wealth, and power and who is not. Such a view does not threaten the status quo, but on the contrary supports it by telling those who are without power that their position is the inevitable outcome of their own innate deficiencies and that, therefore, nothing can be done about it.

Darwin, Wallace, Haeckel, Malthus, Capitalism, and Ethnocentrism

The more we claim to discriminate between cultures and customs as good and bad, the more completely do we identify ourselves with those we would condemn..by refusing to consider as human those who seem to us to be the most "savage" or "barbarous" of their representatives, we merely adopt one of their own characteristic attitudes.. the barbarian is, first and foremost, the man who believes in barbarism. (Claude Lévi-Strauss)

In *Power of Place*, Browne discusses Wallace's lack of appetite for being – and seemingly likely also a reduced ability to be – part of the "*good society*":

Underneath the welcome that Wallace received from the scientific establishment [after returning to England for good], however, there lurked a hint of rapacity. These men-about-town fell on him with alarming speed, seemingly impatient to suck him dry. They wanted to know what he could offer, which sword he could rattle. In actual fact Wallace found it hard to identify any suitable role for himself in Britain in the post-*Origin* years. At times it may have appeared to him that he was hardly needed. Much of the evolutionary flare-up had already passed him by. Darwin seemed fully in command of spreading the word at home and overseas, and the public business of defence lay in the hands of individuals like Huxley, who dominated nearly every corner of the controversy and appeared unwilling to share the limelight. Furthermore, Darwin's name was becoming an acknowledged synonym for the theory of natural selection. Any chance of an evolutionary movement called Wallacism - even if Wallace wished for such a thing - had probably disappeared. Several years afterwards when reflecting on his return to England, Wallace said quietly of the theory that Darwin "had already made it his own".

He may have been content to let it go. For a while he concentrated on distributing and classifying his Malaysian collections.. He contributed important papers to the *Zoological*, *Entomological*, and *Linnean* Societies. Nor did he lose sight of natural selection. In private, he began to probe some of the most enigmatic aspects of his and Darwin's evolutionary scheme, thinking hard about the protective colours of animals, the mental capacities of mankind, and the emergence of early human societies. But he was unsettled by London's superficial existence. With a start, he remembered exactly why he had abandoned the metropolis for the impenetrable green of Malaysia. "Talking without having anything to say", he later wrote in his memoirs, "and merely for politeness or to pass the time, was most difficult and disagreeable". Lady Lyell's impression when she met him in 1863 for the first time, at a London lunch party hosted by Lyell in their home in Harley Street, was of a very

> retiring figure, "shy, awkward and quite unused to *good society*". If truth be told, Wallace was overawed by the Lyells' style of living, with its daunting array of *silver forks and high-table conversation*.. All in all, Wallace found it increasingly more convenient to shut himself up with his bird skins than to accept dinner invitations.

Perhaps while Wallace was dreaming about shutting himself up from Victorian society, he was remembering the small hut in which he lived "pretty comfortably" for 6 weeks in Waigiou Island, far away from "civilization" (Fig. 2.5). However, it is important to mention that the fact that Wallace did not have Darwin's *Power of Place* does not mean that Wallace was as unknown to the general public in England, back then, as it is today globally. As noted above, later in life, particularly between the end of his world travels and the moment he began to "go astray" with some rather unscientific ideas, he became a prominent member within the scientific community and achieved public recognition within the Victorian society. There are accounts indicating that well-connected Victorians proposed, during Wallace's life, that his death should be followed by a ceremony in Westminster Abbey – likely less pompous than that of Darwin – and that it was Wallace who refused that idea. Not unexpectedly, one would say. One interesting aspect connecting Wallace and Darwin is that, despite the way their scientific ideas, and career, diverged later in their lives, and particularly after they were dead, during life they always remained connected, in a way. Namely, Darwin always acknowledged, and was very thankful – both privately and publicly – to Wallace's humble attitude concerning the "priority" issue regarding the theory of evolution by natural selection. For instance, Darwin helped Wallace to obtain a pension. So, Wallace actually benefited from the congeniality

Fig. 2.5 A sketch of Wallace working under his "dwarfs house" – a small hut probably constructed from dried vegetation, in Waigeo Island, Indonesia – from his book *The Malay Archipelago*

between the two men even after Darwin died, as pointed out in McCalman's *Darwin's armada*:

> Darwin was not to know [after he died].. that the pension he'd obtained for Alfred Wallace would sustain his friend for thirty more years of campaigning for social justice and writing daring books on every facet of science, even including the biology of other planets. We can imagine how satisfying Darwin would have found the scene at the Linnean Society in 1908, when on the commemoration of their joint publication Wallace was issued with a gold Darwin-Wallace medal stamped with their bearded faces. At the ceremony, Sir Joseph Hooker gave a gracious speech on being presented with a silver copy of the same medal. By this time, Wallace's eccentricities had been forgotten not only by Hooker, but by British science as a whole. The tireless and good humoured old collector had become a national treasure, and was awarded the Copley Medal and an Order of Merit in the same year.
>
> At the golden jubilee of the publication of *Origin* in 1909, Wallace gave the official Royal Institution lecture. Though his voice was feeble, his passion was undimmed. Darwin, he said, had given the world a theory which surpassed all others in its understanding of our planet: "*its persistence in ever-changing but unchecked development throughout the geological ages, the exact adaptations of every species to its actual environment both inorganic and organic, and the exquisite forms of beauty and harmony in flower and fruit, in mammal and bird, in mollusk and in the infinitude of the insect-tribes; all of which had been brought into existence through the unknown but supremely marvellous powers of Life, in strict relation to that great law of Usefulness, which constitutes the fundamental principle of Darwinism.*" There was something apt, too, about the fact that these two tough southern voyagers, Alfred Wallace and Joseph Hooker, both died peacefully in their sleep after long lives: Hooker in December 1911, Wallace two years later. Hooker was buried at Kew Gardens and Wallace at the local cemetery in Broadstone, the last town he'd lived in. Both had resisted efforts to book them a place in Westminster Abbey, though they could not escape the trappings of fame altogether. On 1 November 1915, amid the carnage of a different and bloodier war, they were commemorated with a plaque in the Abbey.

This precious little story about how Darwin helped Wallace to get a pension stresses, once again, that one should avoid demonizing either Darwin or Wallace, or idealizing any of them, or both, and over-dramatizing the tensions between them. Another story that is often untold, and that is quite related to the contingencies of life and how they influence science, is that apart from being very enthusiastic about capitalistic ideas, Charles Darwin was directly engaged in capitalistic businesses, as noted above. One of the reasons that might explain why this is a mainly untold story, within the innumerable voluminous books and documentaries focusing specifically on Darwin's life, is that this story goes a bit against the constructed image of Darwin as a "pure scientist." Within that construction, Darwin is mostly portrayed as someone that just happened to be born in a well-off family and that then merely passively took advantage of this contingency to undertake his "only true passion" every time he could, almost obsessively: to study nature. But Darwin was actively engaged in capitalistic investments and businesses – he seemingly dedicated a significant amount of time doing so, as stated in Desmond and Moore's 1994 *Darwin* book:

> Money was the root of the matter. On paper Charles and Emma were doing nicely. Their joint annual income had just passed £3000, putting them in the top few per cent of rentiers

nationwide. It came from shrewd investment of their inherited capital. Besides the Beesby farm in Lincolnshire, valued at some £14,000, Charles had acquired assets worth about £40,000 on his father's death. The Doctor left him another Lincolnshire farm at Sutterton Fen, a £13,000 mortgage to the Earl of Powis, and sufficient other funds to enable him to invest massively in British and American industry. Emma, too, had profited handsomely from her father's will. Her assets were held in trust by brother Josiah and Erasmus Darwin, a contemporary caution to shield a wife from her husband's creditors. This Wedgwood wealth came to something over £25,000, including an interest in the family firm, canal and railway shares, and a big mortgage to the son of a Shropshire squire. In all she and Charles had more than £80,000 in investments. A large and growing proportion was in securities, mostly railway stock. Charles was financing the new age of iron and steam. Britain in 1851 had 6800 miles of track, seven times the length of all the rail networks of the five countries of western Europe put together. This achievement, like the building of the pyramids or the Great Wall of China, had cost dear. The railway boom had been a free-for-all, fuelled by mad speculation and condoned by laissez-faire. Fortunes were made overnight, then lost - and so were lives..Charles.. and Emma came into their money as the boom was subsiding, but he still bided his time, bought when the market was sluggish, and went in for low-risk loans. In 1847, after the spring panic and the rise in interest rates, he began pouring thousands into the Leeds and Bradford Railway.

A few years later he got out of London and North Western shares too late and lost about £800, but in 1854 he put £20,000 into the Great Northern Railway and lived on the proceeds for years. He had become an astute financier. '*The present low prices of Guaranteed Railway shares*', he reasoned with his Lincolnshire land agent.. '*for supposing I could get nearly 5 per cent, the extra interest beyond what I shd. get from land, would during the rest of my life (as I do not spend my whole income) make up a depreciation fund at compound interest, to compensate for any fall in Gold*'. That year his income went up to £4600 in all, half of which was reinvested. But Charles did not come by this financial know-how easily. For years he had read the news, watched the markets, taken expert advice.. '*Though I am a rich man*', he wrote to his old ship's servant Covington, long since relocated down-under, '*when I think of the future I very often ardently wish I was settled in one of our Colonies.. Tell me how far you think a gentleman with capital would get on in New South Wales.. What interest can you get for money in a safe investment? How dear is food.. How much land have you?*' Actually, the place Charles said '*I fancy most*' was '*the middle States of N. America*'. The Empire also held out promise, a bolthole if the economy plunged. '*The English certainly are a noble race*', he added, '*and a grand thing it is that we have got securely hold of Australia and New Zealand*'..Charles wanted to hear about the gold.. There was another reason to emigrate. On 2 December 1851 Louis Napoleon seized power in a coup d'état. The spectre of a new Napoleonic empire sent the patriotic British into a spin. 'The fall of France seems decreed by Heaven', Emma's Aunt Jessie mused. 'Now I think everything may be possible, even an invasion'. Over dinner at Downe Capt. Sulivan frightened everybody with his scenario for a French landing and occupation of the home counties. It left Charles with a recurrent nightmare of the French army sweeping up 'the Westerham & Sevenoaks roads' in a pincer movement and surrounding Downe. Seeing the village cut off by Napoleon - the 'Beast' as Jessie called him (from personal acquaintance) - was not total paranoia. Charles's friends were equally vehement. The anti-Catholic Lyell cursed the dictator 'and his pretorian guards and Jesuits'.

Darwin was interested in gold businesses, interested in living in the English colonies despite having witnessed with his own eyes the atrocities that Europeans did to "others," and was doing capitalist investments. There is obviously nothing wrong with that per se, but these are indeed the type of facts that are almost never discussed within the vast literature idealizing Darwin, and that do contrast with the type of "pure naturalist"'tales often constructed about him. The part about him saying that "the English certainly are a noble race" and that "a grand thing it is that we have got securely hold of Australia and New Zealand" is often also not discussed at length in most accounts about Darwin but is at least recognized by some scholars. Even authors of books that idealize Darwin, such as Bowler in his *Darwin Deleted*, recognize that: "I must concede that Darwinism did become involved with the culture of imperialism, providing a source of extremely effective rhetoric.. the imperialists certainly used Darwinian terminology.. and in a few cases they were (even) genuine scientific Darwinists." In this sense, another interesting aspect of Desmond and Moore's more nuanced 1994 book is that they highlight the types of processes involved in Darwin's circular reasoning. That is, in the way in which he incorporated Victorian imperialist and Malthusian concepts and ideas about industrial progress and the mechanization of the workforce into his evolutionary theories, and how he then constructed "scientific facts" to support those theories:

> Just as his Malthusian insight had come from population theory, so his mechanism for creating diversity looked like a blueprint for industrial progress. Darwin was a heavy investor in industry. His Wedgwood cousins were among the pioneers of factory organization. They created a production-line mentality with a marked division of labour among the work force, pushing up productivity by giving each operative a single, specialized function. This mechanisation of the labour force, and its effect on output, was totally familiar to Darwin. Endless trips to Uncle Jos's house as a young man had ensured that, and the Darwin library was stocked with books on economy and manufacture. Every gentleman living off his industrial shares understood 'division of labour'. It was synonymous with specialization and speed in a steam-powered society. It promised wealth and booming markets, and the industrial metaphor seemed to stretch to nature herself. It was the catch-phrase of the age.. Herbert Spencer, that erstwhile railway surveyor, thundering down his own branch line, was not alone in importing it into science. Darwin recognized that, just as industry expanded when the workers specialized, so did life. But Nature had the 'more efficient workshops'. He argued that natural selection would automatically increase the 'physiological division of labour' among animals caught in competitive situations. Stressful competition in overcrowded areas - what he called Nature's 'manufactory of species' - favoured variants that could exploit free niches. These individuals would seize on new opportunities, exploit the available openings on that spot.. Just as a crowded metropolis like London could accommodate all manner of skilled trades, each working next to one another, yet without any direct competition, so species escaped the pressure by finding unoccupied niches in Nature's market place..
>
> The metaphoric extension was complete. Nature was a self-improving 'workshop', evolution the dynamic economy of life. The creation of wealth and the production of species obeyed similar laws. Division of labour was nature's way as well as man's. But economic doctrine was promiscuously mixed with partisan politics in Britain, and Darwin introduced the subject by citing the zoologist Milne-Edwards's use of the term 'division of labour' rather than an economist's. A political taint would have made natural selection too much of a target. To be successful, evolution had to be seen standing on the solid rock of science. With progress guaranteed in Nature's workshop, as much as it was in Uncle Jos's, Darwin's self-evolving Nature was like the expanding, diversifying empires of the Dissenting cotton kings and pottery patriarchs. And indeed, the 50s were a decade of accelerating production; the boom years when economic laws seemed as iron-clad as Nature's own. But the masters' views were very unlike those from the shop-floor. Darwin was siding with the factory bosses and his fellow investors, and he would have no truck with any other. He dismissed critics who saw mechanization beggar the workforce, and those included Emma's uncle, the economist Jean Sismondi. Brutal competition, a dehumanizing division of labour, the

> 'unjust' distribution of profits: Sismondi had slated them all as 'scourges.' So Darwin did have alternative models available. But he rejected them. He accepted that Nature's struggle.. This was the price of progress and diversification, and there was no bucking it.. Economists had called for a specialized work force, free markets, and a rail network to reduce transport costs. Their utilitarian ethos had led to the railway mania as much as a laissez-faire Nature. Darwin put his mouth where his money was. He spent tens of thousands of pounds on railway companies, and twenty years of his life revealing the competitive, specialized, and labour-intensive aspect of Nature's 'workshops'. He was placing Nature on industry's side.

In a nutshell, Desmond and Moore deconstructed the most used line of defense – which was even used by them, in books such as the *Sacred Cause* – to "cleanse" Darwin from his sexist and ethnocentric writings: "Darwin did have alternative models available.. but he rejected them.. he [sided] with the factory bosses and his fellow investors." That is, Darwin's ideas were not merely a passive byproduct of Victorian ideas: *he actively* reinforced – to the point of constructing "facts" – *specific* societal ideas that *he praised* and that were crucial to legitimize that people *like him* had the type of privileges that *he had*. Victorian society was obviously not monochromatic. There were "alternative models" around, but Darwin consistently preferred to side with the "factory bosses," the well-off, the privileged, the elites, the colonialists, the societal oppressors. Another of the countless times in which Darwin revealed his patronizing, ethnocentric way of seeing "others" and how for him "progress" was essentially inevitably leading to Western-like civilizations, Western-like clothes, and Western-like "working" culture, was in a passage of the *Voyage of the Beagle*. We wrote that "many Indians of pure blood [that] reside.. on the outskirts of the town [a town in Brazil].. are considered civilized; but what their character may have gained by a lesser degree of ferocity, is almost counterbalanced by their entire immorality." But, he noted, "some of the younger men are, however, improving; they are willing to labour, and a short time since a party went on a sealing-voyage, and behaved very well.. they were now enjoying the fruits of their labour, by being dressed in very gay, clean clothes, and by being very idle." The taste they showed in their dress, he added, "was admirable; if you could have turned one of these young Indians into a statue of bronze, his drapery would have been perfectly graceful." In this sense, Darwin was so socially conservative that in some parts of that book he does literally seem to be an old-fashioned Pope talking. For instance, when he described "sensuality" and "mockery of all religion" as "vices" and equated them to corruption. He stated: "the character of the higher and more educated classes who reside in the towns, partakes, but perhaps in a lesser degree, of the good parts of the Gaucho, but is, I fear, stained by many vices of which he is free.. sensuality, mockery of all religion, and the grossest corruption, are far from uncommon."

When we take all these pieces of information into account, it is not difficult to understand why Darwin sided with authors such as Haeckel and why they sided with him, nor why subsequent socially conservative authors such as the leaders of the KKK were so willing to cite him to justify their white supremacist, racist, sexist, ethnocentric, and conservative ideas. As recognized even by Bowler, in *Darwin Deleted*, in the "absence [at that time] of fossil hominids, modern savages were treated as equivalent of these primitive ancestors, and the physical anthropologists' alleged evidence of small brains and apelike features in the 'lowest' races was called in to confirm the link.. Darwin certainly contributed to this process in his *Descent of man*." And, as Bowler added, "in Germany Ernst Haeckel"– often associated with the Nazi ideology – "built the idea that the human race show different levels of development firmly into his Darwinism." Strikingly, many of the countless scientists and historians that continue to idealize Darwin, when confronted with this direct link between Darwin and someone with such a noticeable racist reputation such as Haeckel, quickly try to "cleanse" Darwin's name from it.

However, once again, by doing this what those Darwin enthusiasts are doing is to deny what was written by their own idol, Darwin. For instance, Darwin explicitly stated, in a letter about his 1871 *Descent* book, that "*this last naturalist [Haeckel].. has recently.. published his 'Naturliche Schopfungs-geschichte,' in which he fully discusses the genealogy of man.. if this work had appeared before my [Descent] essay had been written, I should probably never have completed it.*" This is because, he added, "almost all the conclusions at which I have arrived, I find confirmed by this naturalist, whose knowledge on many points is much fuller than mine." So, Darwin explicitly recognized that the ideas written in a profoundly racist book by a profoundly racist author such as Haeckel – who is directly related to Nazi ideology by many scholars nowadays – were a confirmation of "almost all the conclusions at which" he "arrived" in his *Descent*.

Newton's Mechanicism, Externalism, Adaptationism, and Social Legitimization

> *The best and safest way of philosophising seems to be, first to enquire diligently into the properties of things, and to establish those properties by experiences [experiments] and then to proceed slowly to hypotheses for the explanation of them.. for hypotheses should be employed only in explaining the properties of things, but not assumed in determining them; unless so far as they may furnish experiments.* (Isaac Newton)

In the last page of *Power of Place*, Browne wrote: "Darwin's grave was in the nave, near Sir John Herschel and Isaac Newton.. The stone was inscribed with his name and dates, and only later did the Royal Society organise a fund to add a bronze tablet that described his contribution to science." She added that, as "the *Pall Mall Gazette* declared, well-wishers believed they were laying there "the greatest Englishman since Newton," one who had given "the same stir, the same direction to all that is most characteristic in the intellectual energy of the nineteenth century, as did Locke and Newton in the eighteenth." No one, said the *Times*, has "wielded a power over men and their intelligences more complete than that which for the last twenty-three years has emanated from a simple country house in Kent." This *post-mortem* recurrent comparison of the immortal giants Darwin and Newton surely would have pleased Darwin: ultimately, he *did* become recognized as the Newton of biology. In a way, this was predictable, in light of everything we have seen so far. Newton's writings were obviously not completely immune to societal biases – all scientific works are affected by biases, in one way or another. But some are less so, and because many of the specific ideas of Newton's work did not concern our own species, they were much less biased, in what concerns societal prejudices, than those of Darwin's *Descent* and *Expression*. Gravity, for example, affects all humans in the same way: there are no chosen ones, no special ones, it does not make some of them more "moral" or superior to others.

Related to that, there are other crucial differences between how people perceive Darwin and Newton, globally. Darwin is almost as famous within those groups that tend to idolize him, such as Western male scientists in particular, and those that tend to demonize him, such as numerous creationists in the United States. Moreover, most people in the planet do not agree with the assertion that humans evolved from other primates. In contrast, Newton's concept of gravity is accepted by most people, and Newton is not at all so commonly demonized by U.S. creationists as Darwin is, for instance. This is because accepting that gravity exists does not put their *beliefs* and, in particular, the idea that *God created humans in His own image,* directly into question, although Newtonian mechanics did ultimately contribute to the more materialistic way of seeing the world that became increasingly in vogue within later scholars, such as Darwin. In this sense, one can say that the fact that Newton's ideas did not concern so directly societal topics nor the origin of our species explains why his name is not so often related to quasi-religious tales of both demonization and idolization as it occurs with Darwin. As discussed in Chap. 1, there are no scientists today – at least that I know of – defending the idea that what Newton said was "enough" to understand the cosmos. Or defending that different ideas about the cosmos, such as those of Albert Einstein or Stephen Hawking, should be just disregarded or are in a way "heretic" or "dangerous" because they can be used by "others," such as creationists, to attack the way in which "we," scientists, see the world.

In contrast, as noted above many current scholars that are part of the "cream" of evolutionary biology – such as Wray and colleagues, in 2014, Coyne, in 2016, Stoltzfuz, in 2017, Futuyma in, 2017, and Gupta and colleagues, in 2017, among many others – explicitly defended the position that what we have from Darwin, or from the subsequent "Modern Synthesis," is enough to understand biological evolution in the overall. Apart from this idolization and the fact that many scholars that defend such views clearly feel "identified" with Darwin's ingroup – they are mostly Western scientists, and most of them are males – there are other reasons that have likely also contributed to such a wide, and many times uncritical, acceptance of Darwin's evolutionary ideas within Western scientists in general. For instance, maybe many biologists are more interested in studying specific details than in testing and discussing broader evolutionary ideas and concepts. Also, the publish or perish pressure current scientific system, in which scientists have a huge pressure to publish more and more papers, as well as the demands to undertake grant projects that are mostly focused on very specific issues, makes it easier to study and publish about "details" and avoid "negative results" that would contradict the prevailing Darwinian paradigms.

One of the goals of my 2017 book *Evolution Driven by Organismal Behavior* was to provide a discussion of some of the key Darwinian paradigms that are often accepted a priori as the intellectual framework of many papers and grant proposals written within fields such as biology and anthropology. As explained in that book, concerning the topic of Darwin and Newton, it is not at all an historical accident that Darwin referred to gravity, and to Newton's mechanics, in the last sentence of his most prominent work, his 1859's *Origin*. He wrote: "there is grandeur in this view of life, with its several powers.. that, whilst this planet has gone cycling on according to *the fixed law of gravity*, from so simple a beginning endless forms most beautiful and most wonderful have been, and are being, evolved." As put in Ruse's 2018 *On Purpose*: "what spurred the (Darwin's) move to natural selection was the strongly felt need to be the Newton of biology - to find a cause for the change."

Well before Newton or Darwin, some ancient philosophies had already been developed around the idea that the cosmos could be *reducible* to mechanical principles, such as those related to the motion and collision of *matter*. That is why one says that those old philosophies were related to *reductionism* and *materialism*. For reasons discussed in detail in my 2021 book *Meaning of life*, those ideas were not so prominent within Western thought and science during the European Middle Ages – also called medieval period, from about the fifth century to the fifteenth century. Mechanicism began to be particularly in vogue with the so-called scientific revolution of the seventeenth century: more and more scholars started to defend the idea that all natural phenomena could eventually be explained in terms of mechanical laws, an idea that had a deep influence in Darwin's ideas about the natural world. In this sense, one can say that although Darwin the brilliant biologist was much less influenced by biases than Darwin the less successful anthropologist, the former was obviously also influenced by them. Such biases refer not only to the Malthusian "struggle-for-existence" ideas as explained above but also to Newton's mechanist ideas and in particular to the parallel that Darwin constructed between the force of gravity exerted onto inanimate objects and the "force" exerted by natural selection onto living beings.

For instance, in his second transmutation notebook, Darwin asked: "Why is thought being a secretion of brain, more wonderful than gravity a property of matter?" As Hoffmeyer put it in a 2013 book chapter, "Darwin created a perfectly *externalist* theory, a theory that seeks to explain the internal properties of organisms, their adaptations, exclusively in terms of properties of their external environments, natural selection pressures." However, as emphasized by Hoffmeyer, Darwin "was not a fundamentalist in his externalism, as were most of his Neo-Darwinist followers in the twentieth century, who thought they could get rid of organismic agency by enthroning the gene and seeing organisms as passive derivatives of genotypes." This is actually a view that is so much in vogue today, when the media and general public – and even some scientists – suggest that there is for instance an "obesity gene," or an "alcoholism gene," and so on, as if we are merely passive slaves of genes. In the vast majority of the cases that we know of so far, at the maximum the mutation of a single gene can lead to some predisposition to undertake a certain behavior A or B, within an often much broader context in which there are many other factors that are in play in what concerns the eventual undertaking of that behavior.

The observations of indigenous nomadic hunter-gatherer groups by scientists such as Darwin, Wallace, and Huxley actually show us how such narratives about an "obesity" gene are not only oversimplistic but completely wrong. For instance, to my knowledge, they did not refer to nomadic hunter-gatherers that, back then, were morbidly obese. Obviously such so-called "obesity" genes did not appear suddenly only in the last centuries within our species and did not appear specifically in countries such as the United States, in which a vast proportion of the population is sedentary and eats huge amounts of fat and sugar. That is, such genes were very likely also found in at least some people within the hunter-gatherer groups described by those scholars back then, but there is almost always a very complex interplay between genes and an enormous number of epigenetic factors – that is, things that are not related to specific changes in the genotype – such as cultural habits, personal decisions and lifestyle, and so on. So, no, we are not mere slaves, or servers of our genes. This is another example of how the idolization of Darwin leads to cases of Darwinists or Neo-Darwinists that are more "papist than the pope," defending extreme versions of Darwin's ideas. This is precisely the case of Dawkins and his 1976 book *Selfish Gene*, a title that in point of fact gives the idea, particularly to those among the broader public and the media that did not read the whole book, that the bodies of organisms are mainly passive vehicles for the replication – and thus indeed *servers* – of selfish genes. As explained in Noble's 2017 book *Dance to the Tune of Life*:

> The language of Neo-Darwinism and twentieth-century biology reflects highly reductionist philosophical and scientific viewpoints, the concepts of which are not required by the scientific discoveries themselves. In fact, it can be shown that, in the case of some of the central concepts of Neo-Darwinism, such as 'selfish genes'.. no biological experiment could possibly distinguish even between completely opposite conceptual interpretations of the same experimental findings. There is no biological experiment that could distinguish between the selfish gene theory and its opposites, such as 'imprisoned' or 'cooperative' genes. This point was implicitly conceded long ago by Richard Dawkins in his 1982 book *The Extended Phenotype*, where he wrote 'I doubt that there is any experiment that could prove my claim'. [Such Neo-Darwinist] concepts therefore form a biased interpretive veneer that can hide those discoveries in a web of interpretation. I refer to a web of interpretation since it is the whole conceptual scheme of Neo-Darwinism that creates the difficulty. Each concept and metaphor reinforces the overall mind-set until it is almost impossible to stand outside it and to appreciate how beguiling it is. Since Neo-Darwinism has dominated

> biological science for over half a century, its viewpoint is now so embedded in the scientific literature, including standard school and university textbooks, that many biological scientists may themselves not recognise its conceptual nature, let alone question incoherencies or identify flaws.

This is a further example highlighting how many scientific ideas, including those that are become to be seen as dogmas by some Neo-Darwinists, are not only non-falsifiable, but even non testable, as is the case with the notion of "selfish genes." From Darwin's notion that natural selection was mainly an "external" force and organisms mainly passive when they face that "force," such ideas go all the way to suggest that the bodies of living beings are *in general* passive, in face of things like selfish genes. Apart from being non-falsifiable and very likely inaccurate, these latter ideas do a disservice to Darwin and his more nuanced theories about biological evolution. As put by Delisle in *The Darwinian Tradition in Context*, Darwinism and Neo-Darwinism comprise multiple ideas and ideologies, many of them being markedly different from what Darwin actually wrote. For instance, Darwin was extremely careful to distinguish his notion of ("external") natural selection and his concept of "sexual selection" associated with the behavioral choices made by organisms of the very same species being selected. So, Darwin's sexual selection corresponded, in a way, to a subset of "organic selection" sensu Baldwin, in the sense that in both of them organisms – and not the external environment – were the main selectors. This also happened, in a different way, with Darwin's "artificial selection," which as we have seen mainly refers to the behavioral choices of humans concerning traits of other taxa, for instance, when they choose the traits of animals and plants during domestication. So, in two of the types of selection to which Darwin referred – sexual and artificial – organisms were seen as active players of evolution, as *the selectors*. And, as we have seen, in the *Descent* Darwin stated that those two types of selection had played a particularly important role in human evolution, probably in part because he wanted to distinguish his ideas from Wallace's ones but also because his Victorian way of seeing the world would somewhat conflict with the idea that humans are just mainly passive evolutionary players within this planet.

Having said this, it is important to point out that in his opus magnum, the *Origin*, Darwin the biologist paid much more attention to (external) "natural selection," when he referred to the natural world as a whole, and not specifically to humans. Also, in the *Origin* he often grouped, and established parallels between, that type of selection and "artificial selection," contrasting them to "sexual selection." That is, in the *Origin* he *mainly* emphasized the passive role of organisms in biological evolution. This is because both in his "natural selection" and "artificial selection" the organisms *being selected* are mainly passive players, powerless in face of the "selectors." That is, in face of the external environment and of humans, respectively. Actually, if we think about this more deeply, Darwin's grouping contrasts with a much more logical assembling: "external" natural selection sensu Darwin versus organic selection sensu Baldwin, the latter including both "artificial" and "sexual" selection sensu Darwin.

Critically, as pointed out previously, in a letter from Darwin to Lyell – 18 June 1958, excerpt from Wetherington's 2011 book *Readings in the History of Evolutionary Theory* – Darwin made it very clear that his notion of "struggle-for-existence" was not just a crucial part of his theory of evolution by natural selection. The whole theory *depended* on it: "I explained to you here very briefly my views of 'natural selection' depending on the Struggle for existence." By establishing a parallel between planets impotently moved by the force of gravity and passive organisms selected by the external environment, and by combining this idea with an emphasis on the Malthusian notion of "struggle-for-existence," Darwin was indeed attributing a particular powerful strength to his "external" natural selection within the natural world as a whole. The vast majority of biologists nowadays – including myself – accept the crucial role played by Darwin's "external" natural selection, no doubt about that. But this acceptance does not mean that cases of natural selection are always, or even often, particularly strong phenomena that tend to lead to an optimal, or at least a suboptimal, current "design" due to a continuous, suffocating, struggle for life. Depending on the specificities of when, where, and how it occurs, natural selection sensu Darwin can be very strong, or more relaxed, or very relaxed, as indicated by the frequent, and usually much neglected, occurrence of phenomena such as "maladaptive behavior syndromes" and morpho-etho-ecological mismatches. These latter phenomena are mismatches between the anatomy of an organism and its behavior or ecology, or between the latter two. I provided many examples of such mismatches in my 2017 book. Similarly, within "artificial selection," there are also cases in which the selection is quite intense – for instance occurring at each generation, as in the famous Siberian domesticated foxes described in Dugatkin and Trut's outstanding 2017 book *How to tame a fox* –, while in others it is much more relaxed – for example, within some stray dogs or cats of villages of many countries of the globe in which there is plenty of food available, such as rats or items from garbages.

The discussion between the influence of Newton's writings on Darwin's mostly externalist ideas about evolution and the natural world, and the subsequent acceptation, and even exaggeration, of such Darwin's ideas within academia and many other layers of society, leads us again to Lewontin's book *Biology as Ideology*. This is because this discussion provides another emblematic example of Lewontin's concept of dual process, involving the indoctrination of scientists with dominant societal ideas and the subsequent use of the biased writings of such scientists to further support and legitimize those very same societal ideas, in a vicious cycle. As put by Lewontin, Darwin's mostly externalist view of a natural world governed by a "struggle-for-existence" was, and continues to be, so easily and often blindly accepted by those defending the *status quo* in great part because it supports, and helps to legitimize, that *status quo*, including the notion that humans are separated from, and thus can use as they please, the "natural world":

The separation between nature and nurture, between the organism and the environment, goes back to Charles Darwin, who finally brought biology into the modern mechanistic world view. Before Darwin, it was the general view that what was outside and what was inside were part of the same whole system and one could influence the other. The most famous theory of evolution before Darwin was that of Jean Baptiste Lamarck, who believed in the inheritance of acquired characteristics. Changes occurred in the environment that caused changes in the body or behavior of organisms, and it was believed that the changes induced by the environment would enter into the hereditary structure of the organisms and would be passed on to the next generation. In this view, nothing separates what is outside from what is inside because external alterations would enter into the organism and be perpetuated in future generations. Darwin completely rejected this world view and replaced it with one in which organisms and environment were totally separated. The external world had its own laws, its own mechanisms of operation. Organisms confronted these and experienced them and either successfully adapted to them or failed. The rule of life, according to Darwin, is "adapt or die." Those organisms whose properties enabled them to cope with the problems set by the external world would survive and leave offspring, and the others would fail to do so. The species would change, not because the environment directly caused physical and body changes in organisms, but because those organisms smart enough to be able to handle the problems thrown at them by nature would leave more offspring, who would resemble them. The deep point of Darwinism was the separation between the forces of the environment that create the problems and the internal forces of the organism that throw up solutions to problems more or less at random, the correct solutions being preserved. The external and internal forces of the world behave independently. The only connection between them is a passive one.

The organisms who happen to be lucky enough to find a match between what is going on inside themselves and what was going on outside themselves survive. Darwin's view was essential to our successful unraveling of evolution. Lamarck was simply wrong about the way the environment influences heredity, and Darwin's alienation of the organism from the environment was an essential first step in a correct description of the way the forces of nature act on each other. The problem is that it was only a first step, and we have become frozen there. Modern biology has become completely committed to the view that organisms are nothing but the battle grounds between the outside forces and the inside forces. Organisms are the passive consequences of external and internal activities beyond their control. *This view has important political reverberations. It implies that the world is outside our control*, that we must take it as we find it and do the best we can to make our way through the mine field of life using whatever equipment our genes have provided to us to get to the other side in one piece. What is so extraordinary about the view of an external environment set for us by nature, and essentially unchangeable except in the sense that we might ruin it and destroy the delicate balance that nature has created in our absence, is that it is completely in contradiction to what we know about organisms and environment.. We must replace the adaptationist view of life with a constructionist one. It is not that organisms find environments and either adapt themselves to the environments or die. They actually construct their environment out of bits and pieces.

Importantly, Lewontin noted that:

In Darwin's view, organisms were acted upon by the environment; they were the passive objects and the external world was the active subject. This alienation of the organism from its outside world means that the outside world has its own laws that are independent of the organisms and so cannot be changed by those organisms. Organisms find the world as it is, and they must either adapt or die. "Nature - love it or leave it" - it is the natural analog of the old saw that you can't fight city hall.. this is an impoverished and incorrect view of the actual relationship between organisms and the world they occupy, a world that living organisms by and large create by their own living activities. So, the ideology of modern science, including modern biology, makes the atom or individual the causal source of all the properties of larger collections. It prescribes a way of studying the world, which is to cut it up into the individual bits that cause it and to study the properties of these isolated bits. It breaks the world down into independent autonomous domains, the internal and the external. Causes are either internal or external, and there is no mutual dependency between them. For biology, this world view has resulted in a particular picture of organisms and their total life activity.. The world outside us poses certain problems, which we do not create but only experience as objects. The problems are to find a mate, to find food, to win out in competition over others, to acquire a large part of the world resources as our own, and if we have the right kinds of genes we will be able to solve the problems and leave more offspring. So in this view, it is really our genes that are propagating themselves through us. We are only their instruments, their temporary vehicles through which the self-replicating molecules that make us up either succeed or fail to spread through the world. In the words of Richard Dawkins, one of the leading proponents of this biological view, we are "lumbering robots" whose genes "created us body and mind".

Some years ago, in his 2014 book *The Meaning of Human Existence*, Edward Wilson wrote a very interesting personal story about these issues, and in particular about how Neo-Darwinists such as Dawkins tend to not only construct biased tales but also to become authoritarian – sometimes in a very dogmatic, fundamentalist way – when other scholars dare to not blindly accept those tales. The story is also worthy of note because it shows the still prevailing obsession of some evolutionary biologists to focus almost entirely – and erroneously – on concepts such as inclusive fitness and kin-selection to explain things such as social behavior, cooperation, and altruism, therefore further emphasizing the links between such biased biological tales and their broader societal repercussions. As we have seen in Chap. 1, such concepts and explanations are in great part based on the type of struggle-for-existence ideas defended by Darwin. They show how some of Darwin's ideas that were originally based on Victorian tales do indeed continue to lead to rather incomplete or even factually inaccurate evolutionary explanations that are still accepted by many current biologists and anthropologists. About this topic, Wilson wrote:

At first I found the theory of inclusive fitness, winnowed down to a few cases of kin selection that might be studied in nature, enchanting. In 1965, a year after Hamilton's article, I defended the theory at a meeting.. Hamilton himself was at my side that evening. In my two books formulating the new discipline of sociobiology, *The Insect Societies* (in 1971) and *Sociobiology: The New Synthesis* (in 1975), I promoted kin selection as a key part of the genetic explanation of advanced social behavior, treating it as equal in importance to caste, communication, and the other principal subjects that make up sociobiology. In 1976 the eloquent science journalist Richard Dawkins explained the idea to the general public in his best-selling book *The Selfish Gene*. Soon kin selection and some version of inclusive fitness were installed in textbooks and popular articles on social evolution. During the following three decades a large volume of general and abstract extensions of the theory of kin selection was tested, especially in ants and extensions of the theory of kin selection was tested, especially in ants and other social insects, and purportedly found proof in studies on rank orders, conflict, and gender investment. By 2000 the central role of kin selection and its extensive inclusive fitness had approached the stature of *dogma*. It was a common practice for writers of technical papers to acknowledge the truth of the theory, even if the content of the data to be presented were only distantly relevant to it. Academic careers had been built upon it by then, and international prizes awarded.

Yet the theory of inclusive fitness was not just wrong, but fundamentally wrong. Looking back today, it is apparent that by the 1990s two seismic flaws had already appeared and begun to widen. Extensions of the theory itself were growing increasingly abstract, hence remote from the empirical work that continued to flourish elsewhere in sociobiology. At the same time the empirical research devoted to the theory remained limited to a small number of measurable phenomena. Writings on the theory mostly in the social insects were repetitive. They offered more and more about proportionately fewer topics. The grand patterns of ecology, phylogeny, division of labor, neurobiology, communication, and social physiology remained virtually untouched by the asseverations of the inclusive theorists. Much of the popular writing devoted to it was not new but affirmative in tone, declaring how great the theory was yet to become.

In 2010, the dominance of inclusive fitness theory was finally broken. After struggling as a member of the small but still muted contrarian school for a decade, I joined two Harvard mathematicians and theoretical biologists, Martin Nowak and Corina Tarnita, for a top-to-bottom analysis of inclusive fitness. Nowak and Tarnita had independently discovered that the foundational assumptions of inclusive fitness theory were unsound, while I had demonstrated that the field data used to support the theory could be explained equally well, or better, with direct natural selection - as in the sex-allocation case of ants just described. No fewer than 137 biologists committed to inclusive fitness theory in their research or teaching signed a protest in a Nature article published the following year. When I repeated part of my argument as a chapter in the 2012 book *The Social Conquest of Earth*, Richard Dawkins responded with the indignant fervor of a true believer. In his review for the British magazine Prospect, he urged others not to read what I had written, but instead to cast the entire book away, "*with great force*", no less.

It is effectively remarkable how well into the twenty-first century influential scholars such as Dawkins – who is moreover now particularly known for his criticism of religion and religious fundamentalists – behave in ways that resemble so much the way in which such fundamentalists have operated for millennia. A scientist urging people to cast books away, as if they were "dangerous blasphemies," is indeed rather odd. Moreover, by doing so, such prominent scientists are using their *power of place* to try to assure that their *biased narratives* will be the ones that will be disseminated to and accepted by their peers, the media, and the general public, and thus thought to the next generations, so they become indoctrinated to keep *believing* in them in the future.

As we can see in this story, this is not only about Dawkins, or about using him as a "straw-man": instead, such "straw-men" and "straw-women" are very much present – and at least some of them are very prominent – within the scientific community, because as stated by Wilson, at least 137 biologists signed that letter. But perhaps the most fascinating part of the story, within the context of the present book, is that even authors such as Edward Wilson, who now state that they no longer subscribe to concepts such as Dawkins' selfish genes, continue to somewhat follow the seventeenth-century mechanicist and reductionist traditions that influenced Darwin. For instance, as astutely noted in the excellent 2013 book edited by Henning and Scarfe, *Beyond Mechanism: Putting Life Back into Biology*, Wilson often described insect colonies using terms such as "growth-maximizing machines" formed by "cellular automata" whose operations can be portrayed using language of physical or computer science. Similarly, the notion of evolutionary passivity of organisms within a Darwinian suffocating "struggle-for-existence" continues to be commonly emphasized to this very day by the word "survival" in the most common current definition of natural selection: "the differential *survival* and reproduction of individuals due to the differences in phenotype." Such notions also continue to be emphasized in the still prevailing Neo-Darwinist definition of evolution: "changes in allele frequencies within populations."

Darwinian Fundamentalism, Capitalism, Individualism, and Social Darwinism

> *Social Darwinism had continued to flourish in German. Together with Mendelian genetics, it was widely thought to provide a scientific basis for the eugenic 'Racial Hygiene' movement.* (Jonathan Glover)

In a 2017 book chapter, Depew astutely recognizes that in Bowler's 2013 book *Darwin Deleted* "what counts as Darwinism is not far removed from what Gould called "Darwinian fundamentalism." It is true, he wrote, "that in recent decades gene-by-gene, trait-by-trait adaptationism, especially applied to animal and human behavior, passes as Darwinism's highest achievement, final justification, and hence defining mark." It is important to note that, historically, what Stephen Jay Gould called *Darwinian Fundamentalism* started as soon, or even before, Darwin died in 1882, being promoted by the "ultra-Darwinians," as scholars such as Romanes called them. Basically, then as now, there were always those that were more Papist than Pope Darwin. For instance, some of them defended that natural selection is the *only* truly relevant mechanism of biological evolution. *Darwin Deleted* is a powerful example of how both adaptationism and the quasi-religious idealization of Darwin are indeed still so prevalent today.

Within the typical "God-good versus Devil-bad" narrative, within Bowler's book – which, it should be noted, is otherwise a fascinating work – the "positive" aspects of Darwin are assigned to his unique genius: "no one else, not even Wallace, was in a position to duplicate Darwin's complete theory of evolution by natural selection." What about the "negative" aspects of Darwin's ideas? Bowler does recognize that some scholars do associate some of Darwin's ideas to "an outgrowth of Victorian cutthroat capitalism - social Darwinism was possible because the selection theory was actually modeled on the ideology of competitive individualism." He also recognizes that Darwin saw Malthus' struggle-for-existence as "the driving force of selection" and admits that this war-like notion "was a product of the individualistic utilitarian ideology.. more individuals are born that can be fed, so many must die, and the result is competition for scarce resources." Bowler even acknowledges that "Darwin drew upon the Malthusian image of a world ruled by scarcity and struggle to promote his theory.. he certainly modified that image by making struggle a creative force." And he notes that some scholars have related "social Darwinism"and in particular a strong version of the notion of "struggle-for-existence" with "militarism, racism, or eugenics." He moreover also acknowledges that Darwin "may have highlighted the harsh implications of this image of nature" and that "Darwinism *was* involved, certainly in the promotion of the heartless individualism of the mid-19th century middle classes, and less directly in the promotion of the later, very different models of 'progress through struggle' – Darwin himself shared some of the concerns that drove social Darwinism."

So, after discussing all these "negative" items, and admitting that at least some of them were indeed accurate, how does Bowler react to them? Does he also attribute them to Darwin's unique mind, as he did for the "positive" items? He did not. Instead, after explaining that his aim was not "to absolve Darwinian from all responsibility" Bowler puzzling concluded, at the end of his book, that "most of the effects that have been labeled as 'social Darwinism' could have emerged" without Darwin. He added that "some of these effects, most notably scientific racism, might well have been even more strident in the absence of the Darwinian theory." This latter statement if of course non-testable and seems more to be mainly related to an attempt to end the book with a kind of feel-good tone than with anything discussed in the book itself, or anything based on any kind of plausible evidence to support it, because Bowler does not provide any type of factual information that would support such a "might well have been" statement.

Contrary to Bowler's more idealized volume, Moore and Desmond provide a much more sober and realistic account about the direct links between Darwin, Darwinism, and Social Darwinism, in the introduction of their 2004 version of Darwin's *Descent of Man*. In particular, they recognize a critical point – that capitalism, racism, ethnocentrism, and other key ideas of Social Darwinism obviously "already existed in Victorian culture.. what Darwinism did was to give [it].. a scientific cloak":

> As contemporary political, social and moral meaning was ostensibly drained from the *Descent*, an extraneous 'Social Darwinism' was born within sociology. This was Darwinian cut-throat competition applied directly to society, rather than nature. The term 'Social Darwinism' first appeared in English around 1900. Until then, all Darwin's doctrines, as well as his allies' beliefs and critics' fears, ran under the rubric of 'Darwinism'. This included the social. Indeed, the distinction between Darwinism and Social Darwinism would have been lost on the author of the *Descent of Man*. However, the purging of his theories by scientists left Darwinism with quite a different ring by the 1890s - depoliticized, and of use to the new professionals, who by definition had to stand on neutral ground. At the same time, a strike-torn socialist revival was under way. Capitalism needed a sociology that could help contain the demands of labour as economies restructured; sociology needed to guard its own professional patch from encroachment by biological theories.
>
> Thus from mixed motives a discrete 'Social Darwinism' emerged. By 1900 a distinction could be made between the politically neutral theories of the *Descent*, rendered useful to the swelling ranks of professional biologists, and the charged Social Darwinism of sociologists and of politicians who needed this justification of competitive capitalism to stem the rising tide of socialism, cooperation and labour unrest. The historical picture of Darwin's day was further muddied when, fifty years later, American scholars projected Social Darwinism back into the mid-nineteenth century. They were trying to contrast an ugly laissezfaire past

> ruled by cut-throat competition with the beneficial stabilizing effects of state intervention in the 1930s Depression. They tarred Victorian capitalists as 'Social Darwinists' and then blamed this philosophical perversion of Darwin's science on his contemporary, the laissez-faire philosopher Herbert Spencer (who did indeed advocate deregulation and unhampered individual competition).
>
> Unfettered rivalry in the marketplace, among men, races and nations, the belief that the mighty inherit the earth and progress depends on reformers and their governments letting them get on with it - none of this, it was claimed, could be deduced from Darwin's works. Darwinism was science, Social Darwinism ideology, and never the twain should meet. The persistence of this view is evidence that the *Descent of Man* remains Darwin's greatest unread book. This shielding of Darwin's corpus only increased in the wake of the eugenics horrors of the Second World War; indeed, the desire to divorce Darwin's 'pure' science from any supposed perverted consequences in the early twentieth century explains the vehemence of many debunkings of Social Darwinism's alleged successor, sociobiology. Social Darwinism is decried as a sullying of pure Darwinism: its 'prejudices' were superimposed on Darwin's science by racists, sexists and eugenists.
>
> Yet a contextual understanding of Darwin's process of creation shows how issues of race, gender and class were integral to his thought - indeed, one cannot explain the origins and development of the *Descent of Man* without them. Science is a messy, socially embedded business, Darwin's particularly so; and while hagiographers may venerate the founding documents of their professions, the historian's task is to trace the contingent influences in the production of such works. In Darwin's case, race, Malthusian insights and middle-class mores were central to his theorizing. Even sensitive historians of science, recognizing the *Descent*'s racial and class doctrines of progress, have labelled it 'deviationist', for departing from the scientific purity of the *Origin of Species*. Others have declared it perverse to deny that Darwin was a 'Social Darwinist', but they too are making a mistake, for they graft an anachronistic concept onto Darwin himself. Perhaps a fresh view from an ethnic minority perspective is needed to see (in Kenan Malik's words) that "Darwinian Man.. was not manufactured by Darwinian theory.. he already existed in Victorian culture, whether in the theories of Herbert Spencer and Henry Maine, or in the novels of Emile Zola and George Eliot.. What Darwinism did was to give him a scientific cloak".

This is indeed one of the most critical points that is too often ignored, omitted, or confused, by those that idolize Darwin, when they use the "all people back then were like that" line of defense. As stressed before, one thing is to have Tom at a bar saying racist, sexist, or ethnocentric stuff to anyone that might be drunk enough to listen to what he says. Another very different thing is to have Darwin – who in 1871 was already much revered – publishing a book including such racist, sexist, and ethnocentric ideas as if they were "evolutionary fact", thus giving them a "scientific cloak." This allowed people like David Duke and many other white supremacist and racist prominent figures and politicians to proclaim that their ideas were backed by science: even the most famous biologist that ever lived said those things, so clearly they had to be *true*.

About the war-like notion of struggle-for-existence, Wetherington noted, in *Readings in the History of Evolutionary Theory*, that the repeated and inflated way in which this notion was used in Darwin's evolutionary works was profoundly influenced by what Darwin read – for instance, Malthus and Smith – and what he perceived as "normal" in England, at that specific epoch. Namely, the "London Charles [Darwin].. settled in [after his travels] had added a million souls - numbering about 2.3 million.. lighted factories could employ more people for longer hours.. poverty increased." The "unbelievable density of humanity - over four hundred people per acre in Greater London - brought the rampant disease, increased mortality, and accelerated reproduction so starkly described by Adam Smith and enumerated by Thomas Malthus." In other words, some of the most crucial aspects of Darwin's evolutionary ideas were in part the result of *extrapolating to the whole natural world what he saw in a specific city – London – at a specific epoch – the Victorian era*. Another interesting, although more subjective, point about this topic is made in Quammen's *The Reluctant Mr. Darwin*. He argues that Darwin's fixation with applying Malthusian ideas to the natural world as a whole reflected not only what was happening in, and the prevailing biases within, Victorian society but also the generally "dark" way in which he perceived life, overall. According to Quammen, this might be related to specific events that occurred in Darwin's own life, such as his constant health problems, which somewhat mirrored the type of constant, omnipresent struggle-for-existence that Darwin applied so harshly to his evolutionary theories:

> On the first page Malthus paraphrased Benjamin Franklin, of all people, to the effect that every species has a tendency to proliferate beyond its available resources, and that nothing limits the total number of individuals except "their crowding and interfering with each other's means of subsistence". Empty the planet of life, Franklin had posited, seed it anew with just one or two forms - fennel plants, say, or Englishmen - and within a relatively short time Earth will be overrun with nothing but Englishmen and fennel. The inherent rate of population growth is geometric - that is, any population can multiply itself by some factor, not just add to itself, with each generation. For humans, Malthus calculated, the inherent rate amounts to doubling a population every twenty five years.. Under normal circumstances, on a teeming planet as opposed to an empty one, runaway population growth is prevented by what Malthus called "checks". The ultimate check is starvation. For humans it results from the fact that, while population is increasing geometrically, ever-intensified efforts at increase the food supply arithmetically.. But food supply directly limits population numbers only during famine. Another kind of check is voluntary: the decision to refrain from marrying, to marry late in life, or to practice birth control (of which Malthus, a wholesome parson of pre-Victorian views, didn't approve). Still other checks operate continually: *overcrowding, unwholesome work, extreme poverty, bad care of children, endemic disease, epidemic, war, and anything else that might contribute to sterility, sexual abstinence, or early death.*
>
> Generally speaking, Malthus wrote, you could boil them all down into "moral restraint, vice, and misery". Darwin read this and something went click. He was less interested in moral restraint and vice than in what "misery" might mean to a mockingbird, a tortoise, an ape, or a stalk of fennel. He ruminated in his "D" notebook about "the warring of the species as inference from Malthus". The geometric population increase of animals, as of humans, is prevented by such Malthusian checks, he wrote. He imagined it all freshly. Take the birds of Europe. They are well known to naturalists and their populations are (or were in his time, anyway) relatively stable. Every year, each species suffers a steady rate of death from hawk predation, from cold, from other causes, roughly maintaining its net population level against the rate of increase from fledglings. Food supply remains limited, nesting space remains limited, but breeding, laying, and hatching continue to push against those limits. Everything is interconnected and uneasily balanced. If the hawks decrease in number, the bird populations they prey upon will be affected, somehow. With new clarity Darwin saw predation, competition, excess reproduction, death - and their consequences. "One may say there is a force like a hundred thousand wedges", he wrote, and that it's trying to "force every kind of adapted structure into the gaps in the *oeconomy of Nature*, or

rather forming gaps by thrusting out weaker ones". The final result of all this wedging, Darwin added, "must be to sort out proper structure & adapt it to change". In shorthand scrawl, he had his big idea. Years later he would articulate the details and call it "natural selection".

Aside discussions on what is pessimism and optimism and which of the two might apply or not to Darwin's view of life, the critical point here is that such a harsh struggle-for-existence view of the natural world is in great part factually inaccurate. Yes, many animals might get sick, die an early death, or starve, and so on. But this does not happen to *most* animals in all places, particularly in places with a high abundance of natural resources, such as many tropical forests. I have been myself in many of them – and there are moreover very detailed ecological studies done by other scholars, about them – and I have not seen *vice, misery, starvation, famine, overcrowding, extreme poverty, bad care of children, endemic disease, epidemic, war, sterility, early death* in all, or even most, of them. Many of these concepts clearly do not apply to the vast majority of nonhuman animals: do tigers take care of their progeny "badly", or have wars between them, or a high prevalence of sterility, of overcrowding, for instance? Moreover, contrary to Malthus, Darwin did have the opportunity to see, with *his own eyes*, during his trips, that this was not the case, neither with most non-human organisms nor with most non-European human societies that he encountered. Indeed, these 'dark' items do not apply to about 99.8% of human existence, because as noted above items such as wars, slavery, and so on mainly apply to sedentary/agricultural societies. In his travels, Darwin obviously saw that things such as overcrowding and unwholesome work do not apply, in general, to human hunter-gatherers. Contrary to the terror stories created by Europeans to justify the enslavement and colonization of "others" – which were often *believed* by Darwin as noted in Chap. 1 – most nomadic hunter-gatherer groups are not constantly in a state of vice, misery, starvation, famine, or extreme poverty.

To give just an example, among countless others that are very well documented, in his 2017 book *Affluence Without Abundance*, Suzman discusses this subject based on the observations that he did of the southern Africa's San peoples for nearly a quarter of a century, as well as on an extensive literature review about the lifeways of hunter-gatherers. He explains that one of the most detailed measures of these items within these groups was made by Richard Lee, who, "struck by the apparent lack of effort that went into the food quest" in the!Kung – which often call themselves Ju/'hoansi"(Fig. 2.6) – decided to analyze how much time they dedicated to get food. Lee "established that on average.. healthy adults worked 17.1 hours per week on food colleting, with that number skewed upward by hunting trips, which almost always took up much more time than gathering excursions.. for women, the workweek rarely exceeded 12 hours." Lee's survey "also revealed that the Ju/'hoansi ate well.. adults consumed on average over 2300 calories of food each day.. this is more or less the recommended caloric intake for adults according to the World Health Organization." This idea of "affluence without abundance" was also supported by Suzman's own observations of the San. These numbers are impressive, because they mean that the *total average* – including both women (12 hours per week) and men (17.1 hours per week of men) – of 'work' time spent is 14.55 hours per week, so basically just 2 hours per day for the sampled !Kung adult, healthy

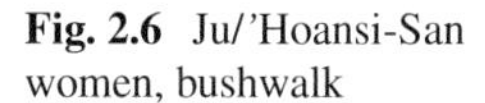

Fig. 2.6 Ju/'Hoansi-San women, bushwalk

population. That is, much less than half the *average* of most countries nowadays, particularly those so-called "highly developed" countries such as Japan.

Todes' 1989 book *Darwin Without Malthus* also discusses in detail some of Darwin's Malthusian fallacies and the related obsession that he had with the notion of a "struggle-for-existence." Todes explains that Russian biologists in general had no problem accepting Darwin's transformism and natural selection but did have a huge problem with his "struggle-for-existence" metaphor and the related capitalistic notions of individual selfishness, due to obvious political – including socialism – and geographical circumstances. As noted by Todes, in many parts of Siberia one is lucky to see even a single animal, for hours and hours – as I can attest myself, after spending several days traveling in and out of a Tran Siberian train. Personally, what I found particularly striking, after another trip I did, namely to the Galapagos islands, is how the reality of Galapagos' nature contradicted the view shared by many scholars and lay people that Darwin's key evolutionary ideas where chiefly based on what he observed in those islands. Some were, but others clearly were not, because among all the numerous regions I have traveled to in the globe, those islands are actually one of the places that conform *less* to Darwin's notion of a "struggle-for-existence." Numerous scholars that have studied the fauna and flora of those islands in detail – see, for instance, Jackson's 2016 book *Galapagos: A Natural History* – pointed that those islands have in general many natural resources and moreover many of their large animals have no natural predators at all. Therefore, as such large animals were also not predated by humans until relatively recently, they are in general very relaxed and tame and often do not display obvious signs of being afraid of human and nonhuman animals and or of being in a constant life-and-death struggle for existence, well on the contrary. In fact, more and more researchers are recognizing the fact that many of the stories on how Darwin's observation of Galapagos' animals played a seminal role in his ideas about evolution are, indeed, often a myth. This applies even, to a certain extent, to the famous Galapagos finches, as explained for instance in Stephen Moss' recent book *Ten Birds That Changed the World*.

To end this section, I will quote a few lines of Ruse's *Darwinism as a Religion*, which summarize how Darwin's ideas of the natural world were highly influenced by theories of Malthus and Adam Smith, and how those ideas then often became dogmatically accepted in Western countries as if they were part of a secular religion, not only by scientists but also by poets and novelists and therefore within popular culture in general:

> It is important to note that, for Darwin, change was not random. It was a matter of having feature that lead to success.. Natural selection gives a scientific explanation of final causes. Darwin always stressed that it was the influence of Paley's *Natural Theology*.. As also was Adam Smith. The Scotsman's theory of the division of labor - you get much more done if you divide the jobs among specialists – fed right into what Darwin was to call his "principle of divergence." We have many different forms of organisms, because they do better in the struggle for existence if they are specialized for certain niches and lifestyles and do not try to do everything.. For Darwin, selection is always a matter of one organism against another - we are all self-regarding if you like. No one puts themselves out for anyone else unless there is return - like help or passing on one's heredity (as with children), and so forth. Here Darwin differed from Wallace. The latter, much influenced by the socialism of the 1840s reformer Robert Owen, always found a place for selection aiding the group even at the expense of the individual.. Do note that Darwin was fully committed to biological progress and this was a reflection of his belief in social and cultural Progress - what other philosophy would be embraced by the grandchild of one of the greatest successes in the Industrial Revolution?
>
> How or why did Darwin think we are special? Obviously in one way because of the role of sexual selection, but we can put this on one side for a moment. We will return to it. For the moment, focus just on the fact that selection of one sort or another was crucial in our making. The point for Darwin, however, is that we weren't just made. We won. Darwin does.. think we have the biggest onboard computers, that this makes us superior, and that getting big brains was something predictable given selection. [He wrote that] "If we look at the differentiation and specialisation of the several organs of each being when adult (and this will include the advancement of the brain for intellectual purposes) as the best standard of highness of organisation, natural selection clearly leads towards highness". [Subsequently] Darwinian evolutionary popular science [became] one of the public domain. Do note, though, unlike pre-*Origin* evolutionary thinking, it was not rejected and despised and found threatening by the scientific elite. They may not have wanted to use selection as a tool, but they certainly accepted it for what they thought it was. Do note also that as something in the public domain, although evolutionary thinking was no longer simply the epiphenomenon of a particular philosophy or world view (Progress), it was still perfectly legitimate to introduce social and other values and link them to the discussion. That is what a popular science is all about. Darwin himself realized all this and in respects took what was on offer. Certainly, compared to the *Origin of Species*.. the Descent of Man is written in a more popular way with greater willingness to introduce social values. Dealing with humans, of course, this is bound to happen to a certain extent, but Darwin saw the opportunity and embraced it.

Just So Stories for Little Children, Evolutionary Psychology, and Evolutionary Medicine

> *You find that people cooperate, you say, 'Yeah, that contributes to their genes' perpetuating'..you find that they fight, you say, 'Sure, that's obvious, because it means that their genes perpetuate and not somebody else's'..in fact, just about anything you find, you can make up some story for it.* (Noam Chomsky)

Saying that most Neo-Darwinists, and in particular "ultra-adaptationists," were more Papist than Pope Darwin himself does not mean that Darwin was not an adaptationist. Darwin's adaptationism is made clear in one of the concepts that ultra-adaptationists accept as a dogma and, often, exaggerate: his "law of utility." As he wrote in his private *Old and Useless Notes*: "nothing but that which has beneficial tendency through many ages could be acquired, & we are certain from our reason, that all which (as we must admit) has been acquired, does possess the beneficial tendency.. it is probable that becomes instinctive which is repeated under many generations ..& only that which is beneficial to race, will have reoccurred." This view about nature is very similar to Aristotle's 'nature does nothing in vain' (see Boxes 1.1 and 3.1). Darwin's statement is partially inaccurate, because it is well known that many traits evolved without being directly beneficial to the organisms in which they evolved. Some of them evolved by chance, for instance, due to genetic drift. Many traits are originally mainly neutral or even detrimental and evolve because they are linked to – an evolutionary package – or as a result of – an evolutionary consequence – the evolution of other features that were beneficial to the organisms.

Having said that, it is important to emphasize two important points. First, Darwin did not subscribe to such an ultra-adaptationist view of nature in his later writings, including those about human evolution. The *Descent*'s first part about atavisms is a powerful, and brilliant, example of that: he lists a huge number of features present in humans that seem mainly to be related to evolutionary or developmental constraints, as they apparently have no "utility" to us. The tail that we have when we are embryos is an emblematic example of that. As noted in Richards' 1987 book about *Darwin*, in *The expression of the emotions in man and animals* Darwin went as far as to argue that the same happened to various behavioral traits. He noted that Darwin "would use these very examples to propose that anomalous and useless instincts would be maintained in a species if they had once been important - and therefore deeply impressed into its heritable substance - and if no significant changes in the environment made them harmful." As put by Richards, "such cases indicated the kinds of incongruity, the failures of fit between animal instincts and environmental requirements, that caught Fleming's eye - and, of course, Darwin's - but escaped the dogmatically blinkered" scientists since then. Even earlier than that, in his 1859 *On the Origin of Species*, Darwin had clearly stated that vestigial and rudimentary organs, "or parts in this strange condition, bearing the stamp of inutility, are extremely common throughout nature."

The second, related, point is that there is indeed a key difference between even some of the more adaptationist ideas earlier defended by Darwin in his *Old and Useless Notes* and the more extreme type of adaptationism that has been and continues to be so popular among so many Neo-Darwinists. Particularly, those that obsessively try to find a "benefit" not only for the moment when a certain trait *appears* but *also* for *each and every moment* in which that trait persisted during evolution. Basically, within such an extreme adaptionist framework, it is not only that "nature does nothing in vain" but that "nature preserves nothing in vain," not even by "inertia" or due to developmental or behavioral constraints. Basically, within this extreme view of life, literally "*nothing exists in vain*." So, if something exists now, either if it originated a long time ago – as the tail that we keep when we are embryos – or more recently – such as the huge current number of suicides, or deaths by overdoses, or deaths or people that decide to risk their lives by going alone to climb a mountain or explore a cave or jump from very high places to a river or sea – this has to always bring an evolutionary *advantage* of some kind. Evolutionary psychology, behavioral ecology, and evolutionary medicine continue to be among the most "dogmatically blinkered" areas of science, being often plagued by this "nothing exists in vain" extremist type of adaptationism. Contrary to the narrative that nobody today follows such an extremist view and that talking about this topic is just using a "straw-man" example, there are indeed an alarming number of such adaptationist just-so-stories within the works published in these fields, as well as in many other ones, as we shall see (see also Boxes 2.3 and 2.4). The term just-so-stories refers to the fairytale-like creations of Kipling's *Just So Stories for Little Children*, which were typically unfalsifiable ad hoc tales based on little or no empirical evidence: they were instead stories to be told to, and amuse, kids.

Box 2.3 Cultural and Human Behavioral Ecology, Anthropology, and Adaptationism

In Kelly's 2013 book *The Lifeways of Hunter-Gatherers*, he provides a pertinent criticism of adaptationism as applied to the study of hunter-gatherers in particular, and anthropology in general. This despite the fact that his book does not completely escape from falling into the trap of adaptationism in some passages and arguments. The criticism he does is made during the brief account that he provides on the history of the disciplines commonly known as cultural ecology and human behavioral ecology. He explains that "cultural ecological studies tried to account for behaviors by showing how they were functionally linked to the acquisition of food in a particular region - example, how they improved foraging efficiency, reduced risk, or netted the highest returns." Adaptationists are very uncomfortable with the existence of evolutionary mismatches and by consequence with human behaviors that might actually be irrational and/or detrimental to the people displaying them. However, things are usually much more complex than what is often asserted

(continued)

in the over simplistic and often unrealistic stories proposed by adaptationists, which too often focus on how a certain trait – such as having more lateral eyes, or knuckle-walking – has provided an "evolutionary advantage" to a certain group of animals. One of Kelly's criticisms of adaptationism is related to one of the most significant theoretical flaws of cultural ecology, which is:

> Neofunctionalist concept of adaptation. By "neofunctionalist" we mean that cultural ecologists assumed that the "function" of behavior was to keep their society in balance with the environment. The term "adaptation" consequently came to refer to any behavior that seemed a reasonable way to maintain the *status quo*. Adaptation was seen as a state of being rather than what it is: a continual process of becoming. This led to an important tautology: behavior is adaptive because it exists - otherwise, it would not exist. But this Panglossian view of life held an important contradiction, for it assumed that if a behavior exists because it accomplishes a goal more effectively than other techniques or strategies, then, presumably, at some time those former techniques or strategies had existed. In this regard, cultural ecologists were like culture area theoreticians, in that they assumed that societies went through changes in the past but were, at the time of study, "best" adapted to their environment. It requires an unwarranted level of confidence to assume that societies had finally figured out adaptation just as anthropologists arrived on the scene.. Cultural ecology.. was a "theory of consequences", in which the end result, the consequence of adaptation, defined the process rather than vice versa. Like the culture area concept, cultural ecology did not specify how adaptive change occurs. When external circumstances changed, people seemed to decide that this or that way of doing things was better for the group. But the way in which these decisions were made was nebulous; and it was not clear what was meant by better (avoid extinction? increase tribal size? more offspring? stronger offspring? psychological satisfaction?). This produced some important paradoxes.

However, in reality, human behavioral ecology is also plagued by adaptationism. This can be seen by the very definition given by Kelly: "human behavioral ecology is less concerned with biology and more concerned with understanding how different human behaviors are adaptive within a particular environmental and social context." That is, it does not seem concerned so much in investigating *if* certain behaviors are adaptive – because it assumes a priori that they *are* adaptive – but rather makes a huge effort explaining *how* they are adaptive. As noted by Kelly, "the majority of behavioral ecologists, therefore, adhere to a 'weak sociobiological thesis', in which "people tend to select behaviors from a range of variants whose net effect, on average, in a given social and ecological context is to maximize individual reproductive or inclusive fitness.. it does assume that humans subconsciously evaluate the reproductive consequences of behaviors." I am always surprised on how so many researchers within a certain field of science can make such a priori assumptions. Are people that suicide themselves, that take drugs and are killed by them in overdoses, that drink alcoholic drinks continuously, or that decide to risk their lives by going alone to climb a mountain or explore a cave

(continued)

alone, really "subconsciously evaluating the reproductive consequences of their behaviors," all the time?

Jared Diamond also emphasized, in *The world until today*, that one cannot reduce discussions about the life-ways of hunter-gatherers – or of any human or nonhuman group, for that matter – to an adaptationist approach. He stressed that one needs to at least consider other types of approaches, namely:

> A second approach, lying at the opposite pole from that first approach [adaptationism], views each society as unique because of its particular history, and considers cultural beliefs and practices as largely independent variables not dictated by environmental conditions.. The Kaulong people, one of dozens of small populations living along the southern watershed of the island of New Britain just east of New Guinea, formerly practised the ritualized strangling of widows. When a man died, his widow called upon her brothers to strangle her. She was not murderously strangled against her will, nor was she pressured into this ritualized form of suicide by other members of her society. Instead, she had grown up observing it as the custom, followed the custom when she became widowed herself, strongly urged her brothers (or else her son if she had no brothers) to fulfill their solemn obligation to strangle her despite their natural reluctance, and sat cooperatively as they did strangle her. No scholar has claimed that Kaulong widow strangling was in any way beneficial to Kaulong society or to the long-term (posthumous) genetic interests of the strangled widow or her relatives. No environmental scientist has recognized any feature of the Kaulong environment tending to make widow strangling more beneficial or understandable there than on New Britain's northern watershed, or further east or west along New Britain's southern watershed. I don't know of other societies practising ritualized widow strangling on New Britain or New Guinea, except for the related Sengseng people neighboring the Kaulong.
>
> Instead, it seems necessary to view Kaulong widow strangling as an independent historical cultural trait that arose for some unknown reason in that particular area of New Britain, and that might eventually have been eliminated by natural selection among societies (i.e., through other New Britain societies not practising widow strangling thereby gaining advantages over the Kaulong), but that persisted for some considerable time until outside pressure and contact caused it to be abandoned after about 1957. Anyone familiar with any other society will be able to think of less extreme traits that characterize that society, that may lack obvious benefits or may even appear harmful to that society, and that aren't clearly an outcome of local conditions. Yet another approach towards understanding differences among societies is to recognize cultural beliefs and practices that have a wide regional distribution, and that spread historically over that region without being clearly related to the local conditions.
>
> Familiar examples are the near-ubiquity of monotheistic religions and non-tonal languages in Europe, contrasting with the frequency of non-monotheistic religions and tonal languages in China and adjacent parts of Southeast Asia. We know a lot about the origins and historical spreads of each type of religion and language in each region. However, I am not aware of convincing reasons why tonal languages would work less well in European environments, nor why monotheistic religions would be intrinsically unsuitable in Chinese and Southeast Asian environments. Religions, languages, and other beliefs and practices may spread in either of two ways. One way is by people expanding and taking their culture with them, as illustrated by European emigrants to the Americas and Australia establishing European languages and European-like societies there. The other way is as the result of people adopting beliefs and practices of other cultures: for example, modern Japanese people adopting Western clothing styles, and modern Americans adopting the habit of eating sushi, without Western emigrants having overrun Japan or Japanese emigrants having overrun the U.S.

Box 2.4 Design, Postpartum Depression, and Evolutionary Psychology and Psychiatry

The voluminous 2012 book *Origin(s) of Design in Nature: A Fresh, Interdisciplinary Look at How Design Emerges in Complex Systems, Especially Life*, edited by Swan and colleagues, is a very interesting one. For example, it includes a highly informative chapter by Adriaens, titled "*Design and disorder - Gould, adaptationism and evolutionary psychiatry.*" In it, Adriaens shows that many – not all, of course – of the narratives circulating within the field of evolutionary psychiatry might well be "just-so stories." This is because they often assume a priori the dogmatic adaptationist tale that if a certain trait exists it has necessarily to be an advantageous adaptation. Even if the trait occurs only within a very small subset of the human population and is, moreover, medically considered to be a mental disorder. Adriaens explains:

> Most evolutionary psychiatrists.. are tried and tested in the adaptationist tradition. Generally, there are two ways to be an adaptationist about mental disorders. First of all, some evolutionary psychiatrists disagree with mainstream psychiatry in suggesting that mental disorders are not disorders or dysfunctions at all, but adaptations. Thus, they have been spread over the population by natural selection because they confer some reproductive advantage to their bearers. The idea that some mental disorders may have some functional significance may seem outrageous, but such adaptationist hypotheses have been and are still being defended in the literature today, particularly in relation to depressive disorders. Hagen (in 1999), for example, has hypothesized that women affected by postpartum depression may signal that they are suffering an important fitness cost, either because they lack paternal or social support or because their newborn baby is in bad health. In this view, postpartum depression would be a bargaining strategy, enabling women to negotiate greater levels of investment from others.
>
> Similar hypotheses suggest that typical depressive symptoms, such as a loss of appetite and excessive ruminating, may have been designed by natural selection to signal yielding in a fierce social competition that cannot be won and to reconsider unfeasible ambitions and investments. The gist of these hypotheses is that depression is not a disorder, but a useful psychological mechanism that enables us to cope with the inevitable adversities of life, much like how fever enables us to fight bacterial infections and how coughing and sneezing help us to keep our airways clear. To my knowledge, [Stephen Jay] Gould has barely written anything about [mental] disorders. On a rare occasion, however, he does discuss the evolution of mental disorders, in a book review of Sigmund Freud's posthumously published *A Phylogenetic fantasy*. Freud's text is a rather rumbling attempt to examine 'how much the phylogenetic disposition can contribute to the understanding of the neuroses', particularly by linking up our ancestor's vicissitudes during and immediately after the last Ice Age with man's present day vulnerability to a series of mental illnesses. In Freud's view, for example, the disposition to phobia derives from our progenitors' useful fears when confronted with the privations of the Ice Age. Freud's just-so story confirms Gould's earlier claim that psychoanalysis is a textbook example of the pervasive influence of recapitulationism and Lamarckism. But there is more. In one of the last paragraphs of his review, Gould notes: 'I also deplore the overly adaptationist premise that any evolved feature not making sense in our present life must have arisen long ago for a good reason rooted in past conditions now altered. In our tough, complex, and partly random world, many features just don't make functional sense, period'.

(continued)

[There is] a second way of being an adaptationist about mental disorders. For convenience's sake, I will refer to such explanations as mismatch explanations. Mismatch explanations of mental disorders build on one of the central ideas in evolutionary psychology - another recent evolutionary discipline crucified by Gould for being 'ultra-adaptationist'. Evolutionary psychologists claim that our ancestral environment, i.e. the environment in which most of the evolution of our species took place, differs substantially from our modern cultural environment. Or, in the words of Tooby and Cosmides, 'our modern skulls house a stone age mind'. As a result, we are much better at solving the problems faced by our hunter-gatherer ancestors than the problems we encounter in modern cities. Evolutionary psychiatrists often consider this mismatch to be the hotbed of many of today's mental disorders. Continuing Freud's example of phobia, they hold that such disorders mostly involve natural threats, such as snakes, spiders and heights. These threats were probably common in our ancestral environment, but they certainly aren't the most dangerous things in our contemporary environment. We do not fear guns the way we fear snakes, for example, even though guns pose a much greater threat to our fitness today than snakes do. In Gould's view, however, there is no need to assume that currently maladaptive traits were once adaptive. Mental disorders, he suggests, may not have an evolutionary history at all, let alone a functional one: 'we need not view schizophrenia, paranoia, and depression as postglacial adaptations gone awry: perhaps these illnesses are immediate pathologies, with remediable medical causes, pure and simple'.

In evolutionary psychiatry, Gould's solution is known as a breakdown explanation or medical explanation - a third category of evolutionary explanations of mental disorders. As Murphy notes, for example, both adaptationist and mismatch explanations seem to assume that 'none of our psychopathology involves something going wrong with our minds', while 'nobody should deny that our evolved nature suffers from a variety of malfunctions and other pathologies'. [Furthermore, many] claim that most evolutionary explanations of mental disorders, including trade-off explanations, smell of panglossianism: 'evolutionarily oriented mental health researchers, such as Darwinian psychiatrists and evolutionary psychologists, often go to torturous lengths to find hidden adaptive benefits that could explain the evolutionary persistence of profoundly harmful mental disorders such as schizophrenia or anorexia, but these accounts are often frustratingly implausible or hard to test'. When charging biologists (and, later on, philosophers of biology) with panglossianism, Gould and Lewontin did not only criticize their overly optimistic view of life but also their laziness in testing the predictions that follow from their hypotheses. Anyone can easily come up with stories about the function of, say, male baldness, being homesick or athletic skills, but there is an important difference between *just-so stories* and real science.

However, it is important to point out that there is a main difference between those that defend such extremist views within evolutionary medicine, versus within evolutionary psychology and behavioral ecology. The former try to find a benefit for the features that our ancestors acquired a long time ago, but then concede that the dramatic changes of lifestyle brought by let's say bipedalism, sedentism, agriculture, or industrialization created evolutionary mismatches that led, or are starting to lead, to a plethora of current pathologies. So, something like the rising cases of anxiety, or suicides, in a fast-paced so-called "modern" country such as Japan could be seen, by those doing research in evolutionary, as something "negative" that resulted from such evolutionary mismatches. Such a view contrasts with the much more extremist type of assertions that continue to be made by many evolutionary psychologists and behavioral ecologists. A 2017 book chapter by Pigliucci discusses and provides several examples of this: "behavioral biologists.. are still clinging to simplistic notions from sociobiology and evolutionary biology, which have long since been debunked." It is "not the basic idea that behaviors, and especially human behaviors, evolve by natural selection and other means that is problematic.. the problem, rather, lies with some of the specific claims made, and methods used, by evolutionary psychologists."

It is indeed striking – and troubling, for science in general – that there is probably not even a single human behavioral trait that at least some scholars have listed as a crutial "adaptation" – that is, as evolutionarily advantageous –, including phenomena such as suicide, postpartum depression, or deaths by drug overdose (see Boxes 2.3 and 2.4). It does not matter if suicides or deaths by overdose, lead to the *complete termination* of the two most important factors related to evolution by natural selection according to Darwin's theories: survival and reproduction of the individual. For such ultra-adaptationists, this does not matter because such cases *have to be adaptative anyways: everything has to be*, so they just need to "find" the "true" adaptative functions of suicide, death by overdose, and so on, and if they are not able to do so, they will just state that more research needs to be done until we find those "real" functions. This is indeed extremely similar to what happens with 'intelligent design' creationists, who would never put in question the basic premise of their beliefs – that God created the natural world - and that therefore just try to "find" signs of that creation within nature.

For example, in Aubin et al.'s 2013 paper "*The evolutionary puzzle of suicide,*" the abstract summarizes this type of circular reasoning. First, the authors recognize that "mechanisms of self-destruction are difficult to reconcile with evolution's first rule of thumb: survive and reproduce." But, they note, "evolutionary success ultimately depends on inclusive fitness.. the altruistic suicide hypothesis posits that the presence of low reproductive potential and burdensomeness toward kin can increase the inclusive fitness payoff of self-removal" In fact, this is not the only "hypothesis" used to explain why "suicide" might be evolutionarily "good." For instance, there is also the "bargaining suicide hypothesis," which "assumes that suicide attempts could function as an honest signal of need.. the payoff may be positive if the suicidal person has a low reproductive potential." And these are just two of the countless current adaptationist hypotheses about suicide, the A and B of the series, there are also C, D, E, F, G, H, and so on (see Box 2.4).

Of course, such adaptationist hypotheses do not apply to at least a substantial part of the 800 hundred thousand or so suicides that happen in agricultural societies every year. First, the family members or loved ones of the people that kill themselves are often those that suffer the most after the suicides take place, leading many of them to suffer a plethora of psychological problems or to having harsher lives than they had before. Second, empirical data from clinical and psychological studies show that such suicides are in at least some cases related to physiological phenomena such as hormonal unbalances or to particularly stressful or damaging previous experiences suffered by those that committed suicide, like being raped or being bullied, feeling very lonely, or other kinds of traumatic experiences. Therefore, saying that suicides are a way to increase group selection is not only directly ignoring the data available about those cases, but also at least in part a lack of respect toward the persons that took their lives away and their families and loved ones that suffered so much with that. Such narratives exaggerate the type of externalism that Darwin did when he compared natural selection to gravity and take the agency from those individual human beings that, for many complex reasons, ended up by deciding to commit suicide. As if they were merely passive automata or brainless robots that blindly obeyed to the external "force" of natural selection for the sake of "improving" group selection.

Third, we know that societies that tend to be more individualistic and to promote individual selfishness and have a more hierarchical structure and a more capitalistic "work" culture – all being items that Darwin praised – tend to have more suicides than for instance egalitarian nomadic groups. Not the other way round. So, it makes no sense to argue that suicides are in general something done for the well-being of the "group" or the community. The young people in Japan that work too much, or are sleep deprived, or exhausted, or feel that they are not – or are seen by society as not being – "productive" or "successful" enough, do they kill themselves for the sake of their kin or closer group? Most of them do not even live with their families, and very often not even with loved ones, many of them are socially isolated, living mainly alone in big cities such as Tokyo. As we will see below, the dramatic increase of sleep deprivation in "modern societies" seems to be one important factor within the dramatic increase of suicides in such societies: part of the huge price that those societies pay for their obsession with "work," productivity and capitalistic "success." Apart from being very rare in nomadic hunter-gatherer societies, suicides are almost – or completely, some scientists argue, although this is still subject of controversy – inexistent within highly social nonhuman animals such as baboons. Do those animals not have kin selection? For ultra-adaptationists, such animals do have kin selection – one of their favorite biological terms – so what is their explanation for the absence, or almost absence, of suicide in those species? All the tales that are created to argue that suicide is adaptative, such as the altruistic suicide hypothesis or the bargaining suicide hypothesis would in theory also apply in groups such as bonobos or common chimpanzees, after all.

For those that would quickly try to quickly absolve Darwin's ideas regarding the construction of such factually inaccurate tales, I should mention here a sentence that Aubin and colleagues used immediately after their paper's abstract. The sentence clearly, and plainly, reveals the main theoretical framework that such authors use as the a priori assumptions for their works: *"natural selection will never produce in a being any structure more injurious than beneficial to that being, for natural selection acts solely by and for the good of each.. no organ will be formed for the purpose of causing pain or for doing an injury to its possessor - Charles Darwin."* Stephen Jay Gould – the scientist that coined the term "just-so-stories" from Kipling's *Just So Stories for Little Children* to refer to such adaptationist tales - recognized that such tales are indeed characteristic of both hardcore Neo-Darwinist ultra-adaptationists and less hard-core Darwinians, including Darwin himself. As noted by Gould, when an adaptationist wants to "find out" the adaptative function of a certain biological feature, she or he hypothesizes that the function is A or B or C or D. Then, even if all those hypotheses are contradicted, she or he would just state that this simply means that the "function" is not yet known and would then try E, F, G, H, and so on, instead of being at least open to the hypothesis that maybe this all means that there is simply no current adaptative function at all. As astutely put by Landau in his 1991 book *Narratives on Human Evolution*: "like the hand of Providence in the biblical account, natural selection justifies even where it fails to explain.. what happens is not always 'right' or well understood, but it is 'fit'."

Such circular reasoning is deeply related to a common human feature associated with our obsession to create just-so-stories – we are the "storyteller animal": our profoundly embedded tendency to seek for a "purpose." As explained in Chap. 1, and discussed in much detail in Ruse's 1996 book *Monad to Man: The Concept of Progress in Evolutionary Biology* and 2003 book *Darwin and Design: Does Evolution Have a Purpose?*, this is indeed one of the most profound paradoxes concerning Darwin. He tried so hard to avoid applying teleology in his theories, but was not successful in doing so, and even recognized this, openly, in some of his letters. Unfortunately, the teleological narratives that he defended and promoted, or the way in which they were subsequently used by other authors, political leaders, and so on, were not mere historical anecdotes. They have strongly influenced not only biologists but also scholars from various other areas of science such as anthropology (see Box 2.3), psychology, and psychiatry, and this is one of the major reasons why so many evolutionary psychologists and psychiatrists are still so influenced by such teleological notions (see Box 2.4). Freud was just one of the very influential psychologists that used such adaptationist ideas and further created countless others that were particularly far from reality, including several misogynistic, racist, and ethnocentric inaccurate tales. Freud's ideas, and those of other prominent psychologists that used similar factually wrong adaptative just-so-stories, continue to influence many clinical psychologists, and therefore affect, sometimes in very tragic ways, their millions of patients. More details about Freud and this particular topic are given in Frederick Crews' 2017 book *Freud: The Making of an Illusion*, which I will further discuss below.

Another example of how such adaptationist ideas are also prominent within authors that work with other groups of organisms, such as plants, concerns a monograph that I otherwise much enjoyed reading, Mancuso and Viola's 2015 *Brilliant Green: The Surprising History and Science of Plant Intelligence*. The authors stated: "in the plant world as in the animals, no one does anything for nothing" – a direct reference to, and almost a copy of, Aristotle's teleological notion of "nature does nothing in vain" that so much influenced Darwin. Another example is Walker's 2017 otherwise superb book *Why We Sleep*, in which he wrote "why would *Mother Nature* design this strange equation of oscillatory phases of sleep?" Apart from being highly teleological – talking, as Darwin did, as if Mother Nature was somewhat an individual agent with a masterplan – this sentence typifies an inaccurate use of the scientific methodology, in which the *adaptationist dogma* is considered to be a self-evident truth that does not even need to be tested. This is indeed further demonstrated by Walker's answer to his own teleological why-question: "we have not achieved [yet] a scientific consensus to justify why [sleep is advantageous], although there are some theories." Yes, theories A, B, C, D, or if they are shown to be wrong, one can easily order more stories – E, F, and so on – from the menu. I am not saying that the existence of oscillatory phases of sleep might not have a current "function" or did not bring an "advantage" to at least some of our primate, mammalian, and/or tetrapod ancestors at a moment in time, or when it evolved in the first place. This might well be so. But this should be treated as a scientific hypothesis that needs to be *tested*, not as a scientific self-evident dogma that *just needs to be explained*, similarly to what creationists do when they blindly accept that organisms were created by God and merely then try to *explain* how or when this happened. This is *not* how the scientific method is supposed to be employed.

Another – and even more teleological and ultra-Darwinian, and in a way quasi-religious – example is the 2018 book *The Science of Sin: Why We Do the Things We Know We Shouldn't*. Published by neurobiologist, writer, and broadcaster Jack Lewis, the book has been widely discussed by the media and social media, including various TV shows, public talks, and other sources that Lewis used to disseminate his ideas. Each of its chapters is dedicated to one of the "*seven capital sins*." And, in each and every one of them, Lewis repeats this circular reasoning. For instance, in the chapter about "anger," he states: "based on the familial logic that we have used throughout this book.. there is always *something* positive to be said about such horrible and malevolent behaviors, an aspect without which we could not be." Within this reasoning, he further adds that "if there were no benefits, anger would have been eliminated from our human genetic heritage since a long time ago." There are many layers of factually inaccurate evolutionary statements in this very short excerpt. First, "anger" is not part of our genetic heritage. There are no genes for "anger," as "anger" can result from the interplay of a plethora of internal biological factors, related to both genetic and epigenetic phenomena, and of external factors concerning the actions of other people, the external environment, and so on.

Second, such teleological ideas completely exclude the possibility that at least some traits, even behavioral ones, that arose in the evolution of a certain group of organisms might just be neutral or the byproduct of other evolved features. Third, such adaptationist just-so-stories also totally leave out the prospect that at least some behavioral features that might have been advantageous at a certain period might have become neutral or even detrimental later in time and remained so for millions of years, for instance, due to social/cultural transmission. Contrary to ultra-adaptationists that repeatedly used such tales, Darwin recognized this latter point in the *Expression*, as we have seen. Fourth, after repeating such a circular reasoning in each chapter of his book, Lewis then refers to cases of "brain dysfunctions," "mal-functions," or "injuries" that can be related to each of the "sins" discussed by him. For instance, Lewis admits that some of the behavioral traits he discussed could be related to brain dysfunctions that are in turn linked to phenomena such as brain tumors. So, this case shows that a certain behavioral trait can be mainly related to such a pathology, instead of being associated with a direct previous or current evolutionary advantage. Moreover, contrary to the tale that something that happens in the natural world for millions of years necessarily needs to be adaptive, tumors and other similar pathological phenomena that are often deadly have occurred in humans and many other animals since times immemorial. Unless adaptationists want to go as far as to argue that having a brain tumor is "advantageous" for humans or that having tumors in general is "adaptative."

Darwin never went so far. He had no problem defending that at least some diseases or other conditions such as being attacked and killed by parasites were not related at all to evolutionary advantages for the group of individuals being killed or attacked, nor to any kind of evolutionary "endgame" or "purpose." He famously explained this point in a letter to Asa Gray about the existence of God: "*with respect to the theological view of the question.. this is always painful to me.. I am bewildered.. I had no intention to write atheistically, but I own that I cannot see as plainly as others do, and as I should wish to do, evidence of design and beneficence on all sides of us.. there seems to me too much misery in the world.*" He added: "*I cannot persuade myself that a beneficent and omnipotent God would have designedly created the Ichneumonidae [a parasitoid wasp family] with the express intention of their feeding within the living bodies of caterpillars.*" That is, according to Darwin the parasites benefit with this, but he would never go as far as to say that such a painful death of the caterpillars also benefited the caterpillars or their kin – this would be as absurd as saying that humans, as a species, benefit from killing themselves with a gun, or with a drug overdose, or from being killed by tumors.

Another example of how such adaptationist tales are widespread in numerous fields of science concerns Max Cooper, an outstanding scientist and one of the most renowned immunologists. He stated that, if our immune systems did not need both T and B cells, one of them would not exist, because "we do not keep what is not useful." In the 2010 book, *Evolutionary Behavioral Ecology* Fox and Westneat provided a particularly passionate defense of adaptationism. They stated, for example, that "Gould and Lewontin objected to the adaptationist paradigm adhered to by most behavioral ecologists..[which] remains dominant in behavioral ecology because, in case after case, the focus on adaptationist explanations has led to new insights." They argued that "a perfect example" of such insights was "David Lack's hypothesis that clutch size in birds would be optimized to balance the number of offspring produced with the parent's ability to feed those offspring well enough to survive." As they admited, "experimental studies on multiple species of birds revealed that clutch sizes were close to, but did not match exactly, what Lack predicted." However, they argued that, because "Lack was *invested* in the adaptationist paradigm," "despite the possibility that many nonadaptive hypotheses could be proposed to explain the disparity between data and theory, Lack chose instead to hypothesize that other factors affected selection on clutch size." Specifically, "this search for adaptive explanations led to a diversity of new adaptive hypotheses," for instance "many studies show that parental work load is indeed important in lifetime reproductive success."

They then concluded that "no doubt some nonadaptive processes also affect clutch size in birds, but Lack's focus on adaptive processes nonetheless led to substantial new insights." I need to emphasize that this is not an attack on Fox or on Westneat, as many of their works are otherwise excellent. However, "leading to new insights" clearly does not necessarily mean that one is following a correct scientific path. Non-scientific reasoning also leads to "new insights," including those of the biased "studies" of supporters of *Intelligent Design*. This is evidenced by the huge number of books and papers published about intelligent design, every year, precisely because their authors are *invested* in blindly pursuing a path and type of reasoning that is impossible to falsify. Ironically, in a nutshell one can summarize this discussion by saying that the obsessive search for evolutionary *advantages* everywhere, by scientists in general and adaptationists in particular, is actually almost always a *non-advantageous* research methodology and thus a *maladaptive* trait for science as a whole.

Importantly, apart from using such a circular reasoning and questionable methodology in their own works, many ultra-adaptationists often try to *force* others to think like they do and to apply those same methodological flaws in their studies or to prevent them from publishing their works if they do not do so. Due to their still prevailing *power of place* within many fields of science, ultra-adaptationists unfortunately frequently succeed in doing so. This happens, for instance, during the process of peer reviewing, when scientists review the studies of other scholars before those studies can be officially published in peer-reviewed journals. Often, a paper is sent by the editor of the journal, or by someone else from its editorial team, to two, three, or more scientists that supposedly have a wide knowledge about the topic being discussed in the paper, or at least are considered prominent scholars within the respective scientific area. Such a peer-reviewing methodology sounds good in theory, and many times works very well, no doubt about that. But sometimes reviewers do abuse their power during the reviewing process, particularly when they have a prominent *power of place* within that scientific area. That is precisely one of the many factors why sometimes factually inaccurate "facts," ideas, views, or theories can remain so prevalent in science, and ultimately in society in general, for a long time, as we have pointed out above. This can last for decades or even longer – as long as the old guard continues to occupy positions of power and to be extremely influential, and to use such power and influence to interfere in the peer-reviewing process, or to hire people with similar views to occupy other prominent positions in academia or editorial teams of top journals.

Among numerous disconcerting examples of such practices, an illustrative one was provided in Prum's 2017 book *The Evolution of Beauty*. It concerns a case that happened to him when he submitted a paper to a peer-reviewed journal about bird behavior. The data that he collected for that paper indicated that a certain feature evolved "through arbitrary mate choice." As he notes, "the reviewers.. argued.. that I had not specifically rejected each of the many adaptative hypotheses that they could imagine.. of course, this made it impossible to 'prove' my point, and I ultimately cut this section out of the manuscript in order to publish the paper." He thus asks: "how many of these adaptative hypotheses.. would I have to test before I could conclude that any given display trait was arbitrary? When should I ever be done with this task? Even if I were able to test every adaptative explanation they could think of.. their reasoning implied that I would have to test other hypotheses in order to satisfy other skeptical reviewers, and then others, ad infinitum." He concludes: "I was trapped.. the prevailing standard of evidence meant it would be impossible for me to ever conclude that any trait had evolved.. arbitrarily."

In the same book, Prum discusses the numerous cases of quasi-religious, unfalsifiable adaptationist ideas that are published every single year and are still so prevalent within various scientific fields. As he notes, some of the most striking examples come from "contemporary *evolutionary psychology*," which, as he explained, "has a profound, constitutive, often fanatical commitment to the universal efficacy of adaptation by natural selection.. [which] is *the organizing principle* of the field." Using an exaggerated tone to emphasize his message, Prum writes that "there is never any doubt what the conclusion of any evolutionary psychology study will be.. the only question is how far the study will have to go to get there.. this is how a faith-based scientific discipline operates - looking for new reasons, however inadequate, to maintain belief in a theory that has failed." He asks "where's the harm in this intellectual mission? What concerns me most is not merely that so much of evolutionary psychology is bad science.. what's worse is that evolutionary psychology is beginning to influence how we think about our sexual desires, behavior, and attitudes." Sadly, the term "beginning to influence" does not reflect the whole truth. As we have seen, evolutionary psychology has *already* done a huge societal damage in the sense that it has influenced the way in which numerous scholars, psychologists, their patients, and a vast part of the broader public in general, think about human evolution and the human mind. This includes the way women continue to be seen as "more passive," "less innovative," and "less sexual" than men.

A 2019 paper by Brady and colleagues alludes to this problem. They correctly pointed out that "evolutionary biologists tend to approach the study of the natural world within a framework of adaptation, inspired perhaps by the power of natural selection to produce fitness advantages that drive population persistence and biological diversity." "In contrast," they noted, "evolution has rarely been studied through the lens of adaptation's complement, maladaptation.. this contrast is surprising because maladaptation is a prevalent feature of evolution." That is, "population trait values are rarely distributed optimally; local populations often have lower fitness than imported ones; populations decline; and local and global extinctions are common."

The current Darwinian adaptationist *status quo* is also recognized by some prominent historians of science, such as Ruse. In his 2018 book *On Purpose*, he states: "*so where are we today in evolutionary thinking? Don't go away with the message that.. biologists today are now questioning seriously what was labeled.. the design-like nature of the world.*" According to him, "*in the world of organisms, adaptation is the norm - the hugely well-justified null hypothesis - and it is your task to make the contrary case if you wish.. purpose thinking rules, and it is cherished.. today's biologists use end-directed thinking and language when they are dealing with organisms.*" It is indeed striking that since Darwin published the *Origin* in 1859 – and in great part *because* of the overall adaptationist tone of that book – adaptationism has been, and continues to be, so widespread despite having so many theoretical flaws and being contradicted by so many empirical case studies.

I discussed these issues in detail in my 2017 book *Evolution Driven by Organismal Behavior* and, focusing more on a broader historical and societal perspective, in my 2021 book *Meaning of Life, Human Nature, and Delusions*. As pointed out in those books, fortunately there have been some changes in the last decades, although very slowly. One of the main reasons for this change was the strong criticism that Gould and his colleagues did of adaptationist just-so-stories, from the 1970s on. That criticism has led to an increasing number of evolutionary biologists – particularly evolutionary developmental biologists, also called Evo-Devoists – to either explicitly criticize some of the key ideas of the Neo-Darwinian Modern Synthesis or to say that such a synthesis needs to be expanded. Despite being still a minority, and of often being criticized by Darwinian adaptationists as discussed above, such scholars have provided very pertinent arguments in favor of a change of mindset toward a less biased and more encompassing, comprehensive, and realistic "Extended Evolutionary Synthesis." One interesting historical aspect about this issue is that many of the ideas defended by such scholars, and the empirical experimental data provided in numerous recent studies published by them as well as by other scientists, have revived the ideas of authors that have been labeled as – and thus mostly ignored because they were – "'Darwinian heretics." For instance, of one of the scientists that has often been seen by Darwinists and Neo-Darwinists as a "heretic," whose ideas are often still taught in schools and universities, and used in documentaries and TV shows, as the antithesis of Darwin's ideas: Lamarck (Fig. 2.7). Of course, such a war-like depiction of Darwin's versus Lamarck's ideas is completely untrue, historically, because Darwin used many ideas from Lamarck in his own works, including one that is nowadays portrayed as one of the most "heretic" of all: Lamarck's "use-disuse idea." As explained in Chap. 1, in a simplified way – such as it is so often taught in schools – under this idea a given giraffe could, over a long time of straining to reach high branches, develop an elongated neck, and then this feature could be passed to the next generations. That is, this phenomenon would involve the evolutionary transmission of *characters acquired during the life of the giraffe*. Darwin repeatedly used this notion in numerous passages of his writings, including of his major masterpiece, the *Origin*.

Such an idea was very successful for a long time, and after Darwin died there was a time in which Lamarck's way of seeing biological evolution actually became

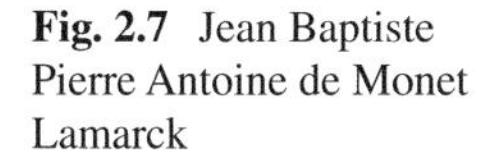

Fig. 2.7 Jean Baptiste Pierre Antoine de Monet Lamarck

to be accepted by a huge number of scholars. However, with the rise of genetics, and in particular of the gene-centric view of evolution that became, and still is, prevalent in various fields of science, including medicine, Lamarck's ideas began to be seen as *the* example of how biological evolution does *not* occur. This is because, under such a gene-centric view, changes that do *not* occur in our DNA – or in the RNA of organisms such as viruses, including those that led to new variants of COVID-19 – *cannot* be transmitted evolutionarily. This view has however been contradicted by a huge number of molecular, developmental, behavioral, and ecological works that show that there are in fact various types of extra-genetic inheritance. These include, for instance, behavioral inheritance associated with niche construction, which is deeply related to cultural evolution in humans. Moreover, we now know that there are many examples of epigenetic inheritance occurring inside our own bodies. That is, changes that do not directly affect the order of the nucleotides of our DNA – adenine (often abbreviated as "A"), thymine ("T"), guanine ("G"), and cytosine ("C") – *can* be transmitted to the *bodies* of our children and even grandchildren.

Regarding these topics, it has been fascinating – both scientifically and sociologically – to assist to confrontations between scholars that write papers arguing that, because of such new data, "*Lamarck rises from his grave*" – a 2017 paper by Wang and colleagues – versus others that publish works titled "*We should not use the term Lamarckian*" – a 2019 paper by Speijer 2019. The latter paper literally states that "discussing examples of inheritance of acquired characteristics.. is all fine as long as it is clear they do not embody alternatives to Darwinism, but illustrate the incredible versatility of natural evolution working in accordance with its basic assumptions.. to paraphrase the most celebrated words from Darwin's magnum opus: 'by so simple a model endless forms most beautiful and most wonderful have been, and are being, explained'.. so, stop using the 'L-word!'" "I don't know if there is a more illustrative example to show how there is indeed still often a quasi-religious veneration of Darwin and Darwinism by many scientists, who can literally went so far as telling other scholars to not commit the "heresy" of using the name of the "heretic" Lamarck. To the point that, in their papers, they try to not even pronounce the name of that "heretic," using instead just the capital letter L. This brings to mind a commonly used religious expression from the European Middle Ages: "don't speak of the devil.. or he shall appear." As noted in the first chapter of the 2011 book *Transformations of Lamarckism* by Gabriel Motzkin, before Lamarck started to rise from the grave in the last decades, Lamarck was indeed one of the most used "heretics" within this quasi-religious worship of Darwin: "like all of you, I took an undergraduate course (in my case, some forty-six yearsago) in which I learned that Lamarck was bad and Darwin was good." This point was reiterated by Pietro Corsi in the second chapter of the same book: "Lamarck's name has often been evoked to contrast his insistence on acquired individual variations with the mainstream interpretation of the *Darwinian doctrine*.. as a consequence, reference to Lamarck has often been made with polemical intent, and has rarely been based on firsthand acquaintance with his works and the biological, philosophical, and even political debates of which he was part."

As I also discussed in *Evolution Driven by Organismal Behavior*, two other prominent Darwinian "heretics" have been rising from their graves in the last decades due to a plethora of new developments and empirical studies published within various fields of science: Baldwin and Goldschmidt. Baldwin's idea of "organic selection" has been increasingly cited in recent years because of the growing empirical evidence showing that organisms are *not* – contrary to what Darwin's unfortunate gravity-natural selection metaphor suggested – mainly passive evolutionary players. They are very often crucial *active* players in their own evolutionary history, for instance, in cases of niche construction. Regarding Goldschmidt, more and more authors are now showing that many of his ideas are also supported by empirical data. This includes Goldschmidt "hopeful monsters" concept, which was so often discredited, and even ridiculed, by Neo-Darwinists because within the Darwinian "gravity-natural selection" and "struggle-for-existence" frameworks, organisms – including their anatomical features – should in theory optimally "fit" the habitats in which they live. If organism are mainly passive players within the natural selection exerted by their external environment – often portrayed by Darwin as a kind of omnipresent and omniscient *Mother Nature* –, under an intense struggle for existence their selected traits should optimally fit the habitats in which they live. However, as we have seen, this is often not the case in biological evolution: anatomical and behavioral traits are instead often mainly "good enough" sketches within the context of the environments inhabited by the organisms that possess them, rather than "optimal designs."

Within the evolutionary framework resulting from the combination of the Darwinian "gravity-natural selection" and "struggle-for-existence" metaphors with Darwin's view that evolution was a slow and gradual process, Goldschmidt's "hopeful monsters" concept involving evolutionary "jumps" would in theory not be feasible. This is because, in a simplified way, the "macromutations" postulated by Goldschmidt would in theory result in at least some "monstrosities" that would be far from being the most optimal "fit" within the external environment occupied by the macromutated organisms. That is, those features would not lead to an *immediate* increase of the survival and/or reproduction of those organisms, and most likely would actually decrease one or both of them, thus ultimately leading those macromutated organisms to be purged from existence. However, there are clear empirical examples of many organisms that were and are "hopeful monsters," such as chameleons. As explained in a paper published by two of my closest colleagues and me – Diogo et al. (2017b) – chameleons are "monsters" sensu Goldschmidt because they have features that were and still are described as "monstrosities" – often named, in medicine, as "severe congenital malformations" – in humans. These include syndactyly – having some digits, or at least their skin, fused into "superdigits" – and zygodactyly – being able to oppose such "superdigits" against one another, for instance, to grasp tree branches (Fig. 2.8). Chameleons are also "hopeful" sensu Goldschmidt because the ancestors of today's chameleons that first had such "monstrosities" were able to survive *despite* having them. That is, they were "good enough" to survive back them and, at later evolutionary stages, their descendants were even able to use those "monstrosities" in a way that allowed them to occupy specific niches very efficientely. For instance, within arboreal habitats, being nowadays quite "successful" and widespread as a group – about 202 species of chameleons are often recognized, most of them still living today.

Fig. 2.8 A "hopeful monster" living among us: the flap-necked chameleon

The fact that Goldschmidt, as well as Lamarck and Baldwin, are rising from their graves is the result of – and, in turn, is further contributing to – a change in mindset about how the evolution of life is being perceived by more and more scholars, although they are still mostly a minority. As noted above, in the 1970s and 1980s Gould and his colleagues put in question, based on empirical evidence that they gathered from the fossil record, that evolution is always, or even mostly, gradual. Eldredge and Gould proposed their notion of punctuated equilibrium in a 1972 paper and provided further evidence since then, as summarized in Gould's 2002 *opus magnum*, *The Structure of Evolutionary Theory*, and more recently in Eldredge's 2014 book *Extinction and Evolution: What Fossils Reveal About the History of Life*. Such works were crucial to call into question the almost dogmatic acceptance of Darwin's evolutionary ideas, and although the change of mindset catalyzed by those works is still far from being consensual within the scientific community, it is slowly occurring even in fields that have been typically dominated by ultra-adaptationist just-so-stories, such as evolutionary medicine. For instance, Randolph Nesse, who is one of the most prominent and influential figures within that field, wrote in his 2019 book *Good Reasons for Bad Feelings*: "During the first several months of our work, we made a fundamental mistake: we tried to find evolutionary explanations for diseases. Why, we asked, did natural selection shape coronary artery disease? Why did it shape breast cancer? Why did it shape schizophrenia? Finally we recognized our mistake." He recognized that he and his colleagues were "viewing diseases as adaptations.. a serious error that remains common in evolutionary medicine.. but," as he explains, "diseases are not adaptations.. they do not have evolutionary explanations.. they were not shaped by natural selection."

This error is indeed still far too common in evolutionary medicine, and even more in fields such as evolutionary psychology, no doubt about that. But the fact that influential scholars such as Nesse are now explicitly and publicly recognizing that they committed such errors in the past is a very good sign that things are indeed starting to change. It is time to start actively recognizing, in a broader, active way, that such factually inaccurate adaptationist and teleological evolutionary just-stories have plagued our view of evolution, of medicine, of science in general, and of our daily lives, for too long. Instead of blindly accepting a priori the idea that evolution is a gradual, optimized, struggle-for-existence "war"-like process, we need to accurately apply the scientific method and empirical test if this is truly so, for each and every case study available. Often the truth is somewhere in the middle of extreme views of the natural world: the evolution of each group of organisms depends on numerous specific local, temporal, contingent, and random factors. So, in some cases where there are more resources available, the evolutionary history of those groups can be more of the "good enough" type and allow some type of evolutionary jumps leading to hopeful monsters, while if there are less resources it can be more an intense type of "struggle-for-existence" gradually leading to traits that are more matched to the external environment. Within the evolutionary history of the countless numbers of organisms that have lived in this vast planet, each of these and many other types of items – gradual versus jumps, optimized versus good enough, struggle-for-existence versus evolutionarily relaxed, as well as neutral versus non-neutral, and so on – probably combined with other ones in completely different ways. As explained in a 2019 paper entitled "*Many roads lead to Rome: neutral phenotypes in microorganisms*" by Nanjundiah, "more than one physiological profile is consistent with the normal development of the group in a given environment; the alternatives are neutral.. an unintended consequence of overlooking phenotypic heterogeneity [that is, variation] is that one can fall into the trap of accepting a seemingly plausible, but possibly erroneous, adaptive explanation for a 'normal' wild-type phenotype."

Social Darwinism, Selfishness, and Today's Popular Culture

> *The world says: 'You have needs - satisfy them. You have as much right as the rich and the mighty. Don't hesitate to satisfy your needs; indeed, expand your needs and demand more'.. this is the worldly doctrine of today.. and they believe that this is freedom.. the result for the rich is isolation and suicide, for the poor, envy and murder.* (Fyodor Dostoyevsky)

One of the fascinating aspects of the widespread acceptance, and even idealization, of Darwin's biased views about human societies and their "progress," is that those views, and the so-called Social Darwinist ideas related to them, continue to be in vogue within social conservative creationists of Western countries such as the U.S.. This highlights the fact that, apart from the "war" between Darwin's evolutionary ideas versus creationist ideas, there was and still is indeed another "war," that between conservative, well-off groups, or dominant groups that want to keep the *status quo* and use Darwin's ideas to legitimize it, versus those that want to change or end that *status quo*. Social Darwinist ideas are very often used by CEO's and other "well-offs" within not only Western countries but also within many other so-called "developed" countries that have adopted a more Western type of capitalistic socioeconomic framework, such as Japan and South Korean. A recent, 2021, article by Jeremy Lent, "*The evolutionary idea that all organisms are singularly wired to ruthlessly pass on their genes is flawed,*" stresses that influential well-off people that were crucial in the history of the U.S. and within the development of modern Western culture and economic ideas in general, such as Andrew Carnegie and John Rockfeller, did commonly cite ideas disseminated by Darwin, and Social Darwinists, to support their views:

> There's an unforgettable moment in the movie "*Wall Street*" when financier Gordon Gekko tells the shareholders of Teldar Paper why his buyout proposal, incorporating massive lay-offs, is not only profitable, but morally legitimate. With his slicked-back hair and custom-tailored suit, he struts to the front of the hall and proclaims that there is a "new law of evolution in corporate America." It's a simple law, he explains: "The point is, ladies and

gentlemen, that greed - for lack of a better word - is good.. greed is right.. greed works.. greed clarifies, cuts through, and captures the essence of the evolutionary spirit". Greed, Gekko is declaring, is the basis of evolution and all that's arisen from it - including human supremacy. Gekko's speech was unleashed on moviegoers in 1987 as the world was reeling from an early encounter with the excesses arising from global financial deregulation. His signature claim – "Greed is good!" - has since become the stuff of legend, strikingly capturing the ethos of unrestrained, free market capitalism that has come to dominate mainstream thinking.

The idea that selfishness and greed are drivers of evolution, and therefore possess underlying virtue, has been around for over a century, ever since Charles Darwin's theory of evolution became widely accepted. The archetypal robber barons, Andrew Carnegie and John D. Rockefeller, both argued that the "survival of the fittest" principle [coined by Darwin from Spencer] justified their cutthroat tactics. But the publication in 1976 of [Neo-Darwinist] Richard Dawkins's bestseller, "*The Selfish Gene*", adroitly repackaged the notion for modern times, reducing the complexities of evolution to a brutally elemental simplicity. As Dawkins summarized it: "*The argument of this book is that we, and all other animals, are machines created by our genes. Like successful Chicago gangsters, our genes have survived, in some cases for millions of years, in a highly competitive world. This entitles us to expect certain qualities in our genes. I shall argue that a predominant quality to be expected in a successful gene is ruthless selfishness. This gene selfishness will usually give rise to selfishness in individual behavior.. Much as we might wish to believe otherwise, universal love and the welfare of the species as a whole are concepts that simply do not make evolutionary sense*".

With the notion of the "selfish gene" as the ultimate driver of evolution, Dawkins helped forge the moral framework of his age. Influential thought leaders have since infused this supposed biological truth into economics, politics, and business. "The economy of nature is competitive from beginning to end", writes sociobiologist M. T. Ghiselin, coeditor of the *Journal of Bioeconomics*. It's difficult to overstate the pervasiveness of Dawkins's selfish gene theory in popular culture. In a nutshell, the underlying story goes something like this: All organisms in nature are simply vessels for the replication of the selfish genes that control us. As such, all living entities - including humans - are driven to compete ruthlessly to pass on their genes. This *struggle for reproduction* is the underlying engine of evolution, as occasional positive random mutations in genes give an entity a competitive edge to beat out weaker rivals. Any apparently altruistic behavior is merely a convenient tactic for a concealed selfish goal. Since nature works most effectively based on selfishness, human society should be similarly organized, which is why free market capitalism has been so successful in dominating all other socioeconomic models.

After explaining this, Lent discusses in some detail how these highly exaggerated and sometimes even completely erroneous concepts about biological evolution have been contradicted by a plethora of recent scientific studies. That second part of the article is critical in the sense that it shows that more and more people are starting to openly recognize the fallacies of such evolutionary concepts. We have already briefly discussed some of the studies and ideas that have been contributing to this change of paradigm among biologists, including the elaboration of an *extended evolutionary synthesis*. As pointed out by Lent:

Pervasive as it has become throughout our culture, the story of the selfish gene is based on fundamental misconceptions. In recent decades, researchers in evolutionary biology have overturned virtually every significant assumption in the selfish gene account. In its place, they have developed a far more sophisticated conception of how evolution works, revealing the rich tapestry of nature's dynamic interconnectedness. Rather than evolution being driven by competition, it turns out that cooperation has played a far more important role in producing the great transitions that led to Earth's current breathtaking state of diversity and beauty. The trouble with the selfish gene story is not just that it is scientifically flawed; it's also that it presents such an impoverished view of life's dazzling magnificence. The discoveries of modern researchers showing how life evolved to its current state of lavish abundance reveal a spectacle of awe-inspiring complexity, mind-boggling dynamic feedback loops, and infinitely subtle interconnections..

Organisms.. play a crucial role in looking after themselves and their offspring in sophisticated ways. But how about the "selfish" question? Have we just relocated the "selfishness" of the gene to that of an individual organism? In fact, one of the most important findings in modern biology has been that cooperation, not selfish competition, has been the foremost driving force in each of life's major evolutionary transitions since it began on Earth billions of years ago.. Even as species differentiated, they developed ways to trade their own specialized skills for the unique skills of other species that could help them thrive. This process, known as mutually beneficial symbiosis (or mutualism) is so widespread throughout nature that it forms a bedrock of every ecology on Earth. The prevalence of mutualism means that life is rarely a zero-sum game, where a species can only gain at the expense of another. On the contrary, by working together, species have co-created ecosystems everywhere in which the whole is far greater than the sum of the parts. These deeply intimate symbioses are everywhere in nature, forming the foundation of the living world.

It's impossible take a walk in the woods, eat a meal, or dip in the ocean, without participating in the deep symbioses that have nourished life's plenitude. On the most fundamental level, plants have specialized in transforming sunlight into chemical energy that provides food for other creatures, whose waste then fertilizes the soil that the plants rely on. If you hike in the woods, you may notice how the trees provide shade that maintains moisture for creatures on the ground. Below you, mycorrhizal fungi maintain underground networks allowing "guilds" of trees to exchange carbon and nutrients among each other in a sophisticated interplay of resources that's been dubbed the "wood-wide web." These symbiotic relationships are frequently so intimate that we rely on them without even knowing about it. We share our bodies with a vast multitude of bacteria - more than the number of cells we call our own. We need them to help us perform biochemical tricks that we can't do ourselves, such as producing enzymes to digest food that our own enzymes can't manage. These symbionts are so important to us that, after birth, a mother's milk contains special sugars that the baby can't digest but provide nutrition for the newborn's symbiotic bacteria.. In countless instances, over hundreds of millions of years, life has decided time and again, that things work better together.

What is particularly striking is that, even in the 2020s, one needs to write such articles or whole books to explain not only to the general public but even to so many scholars, that cooperation is *not* at all an exception, but instead an extremely common occurrence in nature. Apart from relationships between different species such as those mentioned by Lent, there are obviously billions of organisms of social species that could not exist if there was no cooperation: that is what social organisms do, they cooperate. How do termites build a nest? Can a single individual termite do it? How can it be that we keep seeing so many specialized papers stating, as they a priori dogmatic framework, that "*the origin of human cooperation poses an evolutionary problem,*" as done in a 2021 paper entitled "*The evolutionary origins of*

Fig. 2.9 Not so long ago: indoctrination can not only naturalize racism, but also make the general public to not even know about such atrocious racist events, or to think that they happened a "long, long time ago." But this is not so: the Tulsa massacre was real, and today's centenarians were already alive when it happened, in 1921

cooperation in the hominin lineage"published in the *Journal of Philosophy*, by the very prestigious Chicago University Press. To put this in perspective, this paper was published exactly a century after a massacre that highlighted so clearly the tragic legacies of the Social Darwinist selfishness, "war"-like, "struggle-for existence, ethnocentric, discriminatory way of seeing the world and how the dominant groups use it for their own benefit: the so-called Tulsa race massacre, which happened in the U.S., May 31 and June 1, 1921 (Fig. 2.9). Mobs of European descendants, including many that were deputized and received weapons from city officials, attacked African-Americans both on the ground and by using aircrafts. This resulted in the destruction of more than 35 square blocks of what was at that time the wealthiest African American community in the U.S., the admission in hospitals of more than 800 people, and the death of likely 75 to 300 people, according to a 2001 state commission.

Using a typical argument employed by those that deny the existence of systemic racism, one could argue that things are better now, and indeed they are in some way. But, unfortunately, many of the biased just-so-stories that have led to such massacres, or at least to justify them, are still very popular among not only the media but also numerous academics, as we have seen. Such narratives are for instance used, by many, to justify why, many decades after the Tulsa massacre, in 2006–2010, the median income in Tulsa County among the households of

European descendants was $50,842, while for African American households it was $25,979. Such narratives continue to be used to legitimize a *system* that continues to lead to such tragical inequities. For instance, many, including academics that have won the Nobel Prize such as James Watson, continue to argue that such huge disparities are not the result of systemic racism, but instead evidence that African Americans are in average intellectually inferior to European descendants. Within such a fallacious argument, there is no "excuse" when a poor family, despite having one, or more jobs, has no money to send its kids to the best schools. Such kids will never have the *choice* of being born in well-off, privileged, well-connected families and to be able to study in the most expensive schools. But according to such deniers that is still the individual "fault" of those kids, or of their families, or of their specific "racial" group. It is not the fault of the system, of the *status quo*, it is exclusively their fault, they are just inferior, nothing can be done about that, that is just the way nature is.

The links between Social Darwinism, indoctrination, popular culture, political ideologies, and scientific biases and their legacies were brilliantly discussed in a 2019 book by Edgar Cabanas and Eva Illouz, entitled *Manufacturing Happy Citizens: How the Science and Industry of Happiness Control Our Lives*. As explained by these authors, in numerous Hollywood movies and books the "hero" is:

> Depicted as the quintessential self-made individual.. and his[/her] life as a sort of Social Darwinist struggle for upward mobility which ends with the key message that meritocracy works because persistence and personal effort are always rewarded.. Irrespective of how good or bad as science the science of happiness might indeed be, it is essential to interrogate and examine which social agents find the notion of happiness useful, what and whose interests and ideological assumptions it serves, and what the economic and political consequences of its broad social implementation are. In this regard, it is noteworthy that the scientific approach to happiness and the happiness industry that emerges and expands around it contribute significantly to legitimizing the assumption that wealth and poverty, success and failure, health and illness are of our own making. This also lends legitimacy to the idea that there are no structural problems but only psychological shortages; that, in sum, there is no such thing as society but only individuals, to use Margaret Thatcher's phrase inspired by Friedrich Hayek..
>
> Throughout the 1990s, *psychologists and economists collaborated* to develop new questionnaires, scales and methodologies in the attempt to objectively measure concepts such as happiness, subjective well-being and the hedonic balance between positive and negative affect.. In his most important and influential book on the relationship between happiness and politics, *Happiness: Lessons from a New Science*.. [Richard Layard, an economist] would claim that happiness is not only measurable, but self-evidently good. Layard coincided with positive psychologists in stating that happiness should be understood as a natural, objective goal that all human beings inherently pursue: "*We naturally look for the ultimate goal that enables us to judge other goals by how they contribute to it.. happiness is that ultimate goal because, unlike other goals, it is self-evidently good..if we are asked why happiness matters, we can give no further, external reason.. it just obviously does matter.. as the American Declaration of Independence says, it is a 'self-evident' objective*". It is worth noting, though, that this claim should be understood as posited rather than proven, as ideological rather than scientific; a tautological affirmation that, as Layard himself men-

> tions, *lacks further, external reasons that can justify it*. The absence of solid theoretical underpinnings notwithstanding, self-reliance in the accurate and unbiased measurement of happiness has indeed been one of the most significant charms with which the scientific discourse of happiness has wormed itself within the individualistic, technocratic and utilitarian soul of neoliberal politics.

Importantly, many of the factually inaccurate ideas that are now being "proven" – or better said accepted as a priori as dogmas – by numerous happiness scientists and economists come precisely from the very same type of utilitarian, individualistic, and Malthusian societal ideas that Darwin believed in, and that were promoted and *naturalized* in his evolutionary works. As noted by Cabanas and Illouz:

> According to happiness economists, Bentham's .. utilitarianism has ceased to be an abstract utopia of social engineering, to instead become a scientific reality in which the good life is amenable to technocracy by integrating moods and feelings, meanings, development and even the most intimate nooks and crannies of the psyche into a mass-scale calculus of consumption, efficiency, productivity and national progress.. Inequality is one of the latest and most striking examples. According to recent studies, and contrary to the claims of many other economists that the idea of a social floor, redistribution and equality are indispensable for social prosperity, dignity, recognition and welfare, research on large databases seems now to prove that income inequality and capital concentration have a positive relationship to happiness and economic progress, especially in developing countries. Apparently, inequality is accompanied not by resentment, but by a '*hope factor*' according to which the poor perceive the success of the rich as a harbinger of opportunity, thus raising hope and happiness related to a higher motivation to thrive..
>
> In conclusion, some have argued that movements such as positive psychology would improve scientifically if they actually acknowledged *their historical and cultural background*, as well as their *ideological and individualist biases and preferences*. We would concur with this argument, but we do not think that this will happen. The main reason is that the *strength* of positive psychology lies, precisely, in the *denial of these backgrounds and biases*: it is by being presented as apolitical that it is truly effective as an ideological tool. This statement is well applied to positive psychology as well as to happiness economics, which draw much of their cultural power, scientific authority and social influence from supporting and practising as valid and universal what the individualistic, utilitarian and therapeutic worldview of neoliberalism has already presumed to be true and desirable for individuals and societies alike.

Indeed, what a coincidence that scholars that live in Western neoliberal individualistic hierarchical societies have "scientifically concluded" that the happiest people in the whole globe are precisely from.. those societies. Similarly, Darwin "concluded" that Europeans were the "highest" human beings, James Watson defends that "whites" are "superior," those that defend the most the caste system in India are mostly the privileged ones, white supremacists are almost exclusively "white," those that defend that Judaism, Islam or Hinduism are the best religion are almost all Jews, Muslims, or Hindus, respectively, and so on. Scientists such as Watson should know better: they should easily realize that this is not a coincidence at all, that they are just "seeing" what they want to *believe* in, within their biased world in which their "group" is always seen as the "superior" one.

This discussion thus makes us come full circle to the beginning of this chapter and the commemoration of Darwin, in a state funeral at a Royal Church, as an exponent and promoter of Victorian society, British imperialism, and their superior "morality." Why did Darwin *not* see the coincidence that the evolutionary "facts" that he wrote in his books, such as that women are inferior, Europeans are superior, social hierarchies are good, and so on, were exactly the same things that he already took for granted before he boarded the Beagle? This type of self-reinforcing, circular reasoning is in great part why a huge number of happiness scientists now argue that their ideas are based not only on "facts," but specifically on "evolutionary facts" – that is, that their ideas are just an "elementary" part of the natural world. We, scientists, can and *should* do much better than this.

Chapter 3
Racism and Its Societal Repercussions

To see what is in front of one's nose needs a constant struggle.
(George Orwell)

Putting Things in Context: Scientific Biases and Racism Before Darwin

> *Ignorance and prejudice are the handmaidens of propaganda.. our mission, therefore, is to confront ignorance with knowledge, bigotry with tolerance, and isolation with the outstretched hand of generosity.. Racism can, will, and must be defeated.* (Kofi Annan)

Racism, especially scientific racism, did not begin with Darwin. Since the very first writings about "other people," there have been racist assertions and ideas. I have provided a detailed account of the history of racist just-so-stories in my last book *Meaning of Life, Human Nature, and Delusions*, so, here, I will just briefly summarize some key concepts and events. As noted above, such stories are often related to the notion of a *scala naturae* or a "chain of being" (Fig. 1.7), which, in turn, is deeply related to our quest to understand humankind and its place in nature, including its "subgroups" or imaginary "races." This notion was prominent in ancient Greece and was promoted by Aristotle and, since then, has been adopted by most Western societies – and by many Christian and Muslim theologians. A wonderful book that I highly recommend to those interested in knowing more about Aristotle's life, his works and ideas, and their influence on biology and on Western science and society in general is Leroi's 2004 book *The Lagoon: How Aristotle Invented Science*. As pointed out by Leroi, together with his notion of a *scala naturae*, some other key societal concepts of Aristotle, such as his thoughts about women being inferior and nature "doing nothing in vain," significantly influenced subsequent scholars, including Darwin, as seen above:

> The history of Western thought is littered with teleologists. From fourth-century Attica to twenty-first-century Kansas, the *Argument from Design* has never lost its appeal. Aristotle

R. Diogo, *Darwin's Racism, Sexism, and Idolization*,
https://doi.org/10.1007/978-3-031-49055-2_3

> and Darwin, however, share the more unusual conviction that though the organic world is filled with design there is no designer. But if the designer is dead for whose benefit is the design? It's the prosecutor's question: *cui bono*? Darwin answered that individuals benefit. Biologists have batted the question about ever since. The answers that they've essayed are: memes, genes, individuals, groups, species, some combination or all of the above. Aristotle, however, generally appears to agree with Darwin: organs exist for the sake of the survival and reproduction of individual animals. This is why so much of his biology seems so familiar. Yet there is a deep difference between Aristotle's teleology and Darwin's adaptationism, one that appears when we follow the chain of explanation that any theory of organic design invites. Why does the elephant have a trunk? To snorkel. Why must it snorkel? Because it's slow and lives in swamps. Why is it slow? Because it's big. Why is it big? To defend itself. Why must it defend itself? Because it wants to survive and reproduce. Why does it want to survive and reproduce? Because.. Because natural selection has designed the elephant to reproduce itself.
>
> Darwin gave teleology a mechanistic explanation. He halted the march of whys. It is for this reason that Ogle celebrated Darwin as Democritus reincarnated.. When Aristotle speaks of the divine he is not – the point must be made again – invoking a divine craftsman for none exists; rather, he is telling us that immortality is a property of divine things and that reproduction makes animals a little bit divine.. his teleology is riddled with such value judgments. He says that the position of the heart in the middle of the body is dictated by its embryonic origins. But it is also located more *above* than *below* and more *before* than *behind*, 'For nature when allocating places puts more honourable things in more honourable positions, unless something more important prevents this' – the language suggests the seating plan at a dinner. One may wonder why, then, the human heart (actually its apex) is located on the inferior left, but Aristotle has inserted a caveat – '*when nature does nothing (in vain)..*' – and gives a patently *ad hoc* explanation that it's needed there to 'balance the cooling of things on the left.' He thinks, of course, that the right-hand side of the body, being more honourable, is hotter than the left, and that this is especially so in humans, and so the heart has to shift to compensate for the left's relative coolness. Plato's influence is most obvious when Aristotle considers man. He is explicit: man is his model not only because he's the animal we know best, but also because he is the most perfect animal of all.

The ancient Greeks and Romans, as well as other societies such as the ancient Egyptians, defended that "other" peoples – for instance, "blacks" – or societies were "inferior," but the explanations for why this was so were in general related to epigenetic factors (see also Box 1.3). That is, such "inferiority" was not defined by birth but instead because the "others" lived in a different way and at a different place, inhaled a different kind of air, ate different types of food, and so on. I call this type of "cultural" racism, which was the most prevalent and widespread until just a few centuries ago, type A/epigenetic racism. That is why the first Portuguese or Spanish Christians, or the Middle Eastern Muslims, who started to dominate some regions of Africa and of Central and South America, repeatedly tried to "civilize" – or, as they would often argue, "save the savage souls" of – the Indigenous people of those regions. Sometimes they literally took some indigenous people in boats to European or Middle Eastern cities with the *hope* that they could be *changed*, go up the 'chain of life', and become '*civilized*'. Although any type of racism is unfortunate and based on factually inaccurate narratives, type A racism is clearly much less "hard-core" and damaging than the more recent type B/innate racism, which began to be widely accepted, in particular by Western scholars, and then by many Western and non-Western lay people, since the end of the seventeenth century. This is because although in type A things can be changed, at least in theory – for instance, the "inferior" ones can become as "civilized" as the "superior" ones if they attend their schools – in type B, there is no way that things can be changed, that is, the "inferior" ones are naturally deemed to always be "inferior."

This innate type of racism has led to some of the most atrocious acts committed by humans in the last few centuries, including the genocide of Armenians by the Turks, of Tutsis and Batwa people by the Hutus in Rwanda, and of Jews and Gypsies by the Nazis (Fig. 3.1; see also Box 3.5). The type of white supremacy ideas promoted by the Nazis, which are tragically still in vogue in a relatively significant portion of the population of various countries – including many neo-Nazi and white supremacy groups that have become more prominent and publicly visible in the U.S. in recent years – are indeed an emblematic case of such an innate type of racism. The difference between these two types of racism is widely consensual among scholars that focus on such issues – see, for example, the excellent 1981 book by Gould, *The Mismeasure of Man*, or Bethencourt's 2013 book *Racisms* – but, unfortunately, continues to be too often neglected in discussions about racism in the media and in educational materials, and therefore among the general public. As noted above, a similar, and related, mismatch between the current scholarly knowledge and general public perception also applies to the very notion of "race," which

Fig. 3.1 What type B/innate racism can lead to: Einsatzgruppe – Schutzstaffel (SS) paramilitary death squads of Nazi Germany – shooting a woman and a child, near Ivangorod, Ukraine, 1942. Literally, thousands of such profoundly disturbing pictures showing merciless horrendous acts could be shown to highlight the types of atrocities that can be undertaken and seen as "morally" justified by a group of humans when it *believes* that it is innately superior to "others"

is obviously mostly used in tales constructed under the framework of an innate type of racism. Although the vast majority of biologists and anthropologists recognize that there are *no* different living human "races" – and the American Anthropological Association has officially stated this fact – most lay people, and even governmental documents and questionnaires such as those used in the U.S., often still refer to "races."

There were various critical reasons that led to the rise of type B racism at the end of the seventeenth century, including economic ones. Scientifically, a major reason – which is also mainly unknown to most lay people and even to most scholars – concerns the first detailed description of the bones and soft tissues of a great ape – likely a chimpanzee – by a Western scientist, Edward Tyson, in 1699. This was just about half a century after Tulpius – a physician and anatomist immortalized in Rembrandt's painting *The Anatomical Lesson* – published the first relatively realistic external anatomical depiction of a great ape (Fig. 3.2), which was likely a common chimpanzee, a bonobo, or, according to some scholars, possibly an orangutan. A critical point emphasized in Fig. 3.2 is how great apes were generally depicted in a rather more "positive" and "docile" – and, thus, in a general in a more realistic – way in the seventeenth century, in contrast to the much more aggressive way in which they began to be represented after the rise of scientific type B racism in the eighteenth and nineteenth centuries.

Fig. 3.2 Depiction of a great ape in Tulpius' 1641 *Observationes medicae*

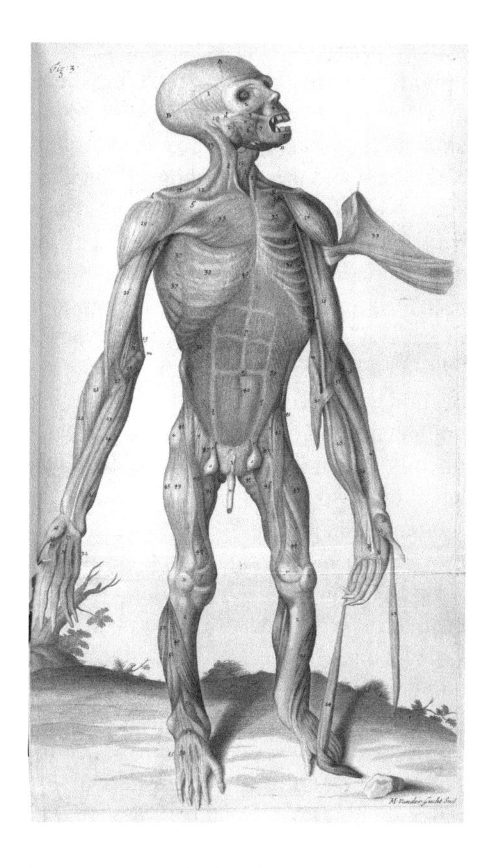

Fig. 3.3 One of the depictions of the muscles of the chimpanzee dissected in Tyson's 1699 work

This marked change of attitude toward apes was precisely related, in part, to the publication of Tyson's landmark work and its pertinent observations of how human-like apes were. Tyson's 1699 monograph, which was entitled *Orang-Outang, sive Homo sylvestris, or the Anatomy of a Pygmie Compared to That of a Monkey, an Ape and a Man* and included a series of strikingly detailed and beautiful anatomical drawings (Fig. 3.3), contributed to change the scientific perception of humanity and its place in nature. That monograph was the culmination of a trend that emerged in the fourteenth century among Western scholars and artists, which started to depict nonhuman primates in a more naturalistic, realistic, and positive way. In the preface of the 1943 volume entitled *Edward Tyson, M.D., F.R.S., 1650–1708*, Montagu noted that Tyson "did not discover the theory of evolution.. but he accomplished in a modest and honest way a goodly share of the (anatomical) analytical work without which the scientific formulation of that theory would have remained impossible," being, in this sense, "a forerunner.. of [Darwin's] the *Origin of species* (1859) and the *Descent of man* (1871)."

Despite a few specific anatomical errors, the central conclusion of Tyson's work is correct and indeed represents a key moment not only in biology but also in science in general: he found more anatomical features shared by chimpanzees and humans (48 according to him) than by chimpanzees and monkeys (34 according to his comparison). That is, that work was the *first* known scientific publication to provide empirical data contradicting the tales constructed by many religions as well as many prominent ancient writers about a dichotomy between humans and other animals. Tyson's comparative data empirically contradicted the existence of such a dichotomy and therefore showed that humans are not so special and that there is no "ladder of life." This is because the data showed not only that there is a continuity between humans and other animals but also that humans are actually anatomically – and behaviorally, as we now know – more similar to other animals such as chimpanzees than those animals are to other nonhuman animals such as monkeys. As noted in Lovejoy's 1936 book, in the eighteenth century, decades after Tyson's 1699 work, "the sense of the separation between man from the rest of the animal creation was beginning to break down." Rousseau asserted in 1753 that humans and great apes – orangutans and chimpanzees, as gorillas were seemingly still not described scientifically at that time – should be included in the same species and that language is not natural to humans but is instead "an art which one variety of this species (humans) has gradually developed." In 1781, Bonnet stated that great apes have the size, members, carriage, and "upright posture" of humans, given that they have a "true face," are "entirely destitute of a tail," are "susceptible to education", and even have a sort of "politeness," irrespective of whether we compare their minds or bodies with ours. "We are astonished to see how slight and how few are the differences, and how manifold and how marked are the resemblances," he wrote.

It was in part the discomfort caused by such striking human–chimpanzee similarities that led several thinkers and scholars to create a new type of savage-civilized fictional dichotomy, in which the "true" gap was no longer between humans and other animals but instead between "civilized" European humans and "savage" non-European humans plus other primates. For instance, in 1714, just 15 years after Tyson's work, Blackmore and Hughes, noting how "surprising and delightful it is" to trace "the scale or gradual ascent from minerals to man," placed the African Hottentots (see Fig. 3.4) between "humans" – including people mainly from "civilized" countries – and apes. They wrote: "the ape or the monkey that bears the greatest similitude to man, is the next order of animals below him.. as the Hottentot, or stupid native of Nova Zembla."

Linnaeus' 1735 *Systema Naturae* is a landmark example of the influence of Tyson's 1699 work and the reactionary responses to it by some prominent European thinkers and researchers. Linnaeus' work was profoundly influenced by the ideas of Blumenbach and Camper, who are usually considered the "father of physical anthropology" and "the father of racial anatomical studies," respectively. On the one hand, Linnaeus went one step further than Tyson did because in the great scheme of classification of the living world that he proposed in the tenth edition of *Systema Naturae* (1758), he further developed his previous classifications and listed two *Homo* species: *Homo sapiens* ("*Homo Diurnus*") and *Homo troglodytes* ("*Homo nocturnus*") (see Fig. 3.5). The latter species included the great apes, namely, *Homo sylvestris Orang-Outang*, which is both Bontius' orangutan and Tulpius' and Tyson's chimpanzees – as noted above, gorillas had not been officially described at that time.

Fig. 3.4 "A Pair of Broad Bottoms," a caricature of Sarah Baartman – the "Hottentot Venus" – by William Heath from 1810. She was a South African Khoi women who, due to the Western objectification of her buttocks, was exhibited in "freak shows" in Europe, in the nineteenth century. Hottentot was the name that the Europeans then used to designate the Khoi people, who many Europeans saw as a direct link – and "evolutionary relict" – between nonhuman apes and monkeys and "superior" humans: themselves, obviously

However, this suggested human–animal continuity was accompanied, on the other hand, by the reactionary division of humans into innately different groups defined by both social/moral *and* anatomical traits. In this sense, Linnaeus' work became a landmark publication that changed Western science in a very dark – and factually inaccurate – way that was completely the opposite of what Tyson's revolutionary, open-minded, empirical work revealed.

That is, Linnaeus' *Systema Naturae* was the *first* influential scientific work by a renowned biologist to imply that different "groups" of humans were defined by *fixed, unchangeable, morphological, and mental traits*. He constructed, based on factually incorrect data, the framework for scientific type B/innate racism. Specifically, he defined four different human "groups" or "variants," which later came to be commonly known as "races" and prevail till today, as attested by the still common use of the terms "blacks" – "Africans" and their descendants; "whites" – "Europeans" and their descendants; "yellows" or "pales" – "Asians;" and

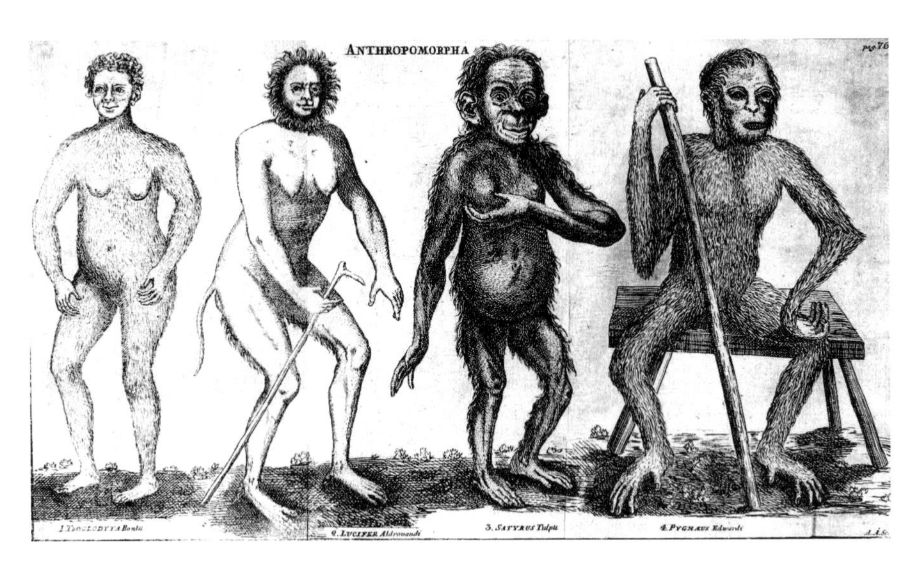

Fig. 3.5 The “Anthropomorpha” of Linnaeus (1935): *Troglodyta*, *Lucifer*, *Satyrus*, and *Pygmaeus*, which are based on a mixture between scientific descriptions such as those of Tyson (1699) and imaginary human-like creatures mentioned by earlier authors

“reds” – “Native Americans.” According to him, these four groups were characterized as follows: (1) (Native) Americans: red, bilious, straight – governed by customs; (2) Europeans: white, sanguine, muscular – governed by customs; (3) Asians: sallow (pale), melancholic, stiff – governed by opinion; and (4) Africans: black, phlegmatic, stiff – governed by chance. Using Linnaeus’ definitions, a “black” person is *always* governed by “chance.” There is no “escape” from it: “black” people were born like that and would remain that way their whole lives; this was as “innate” as the color of their skin – using current terms, it was a genetically imprinted characteristic of that “race.” These are exactly the types of ideas – that “blacks” are, were, and always will be mentally and morally “inferior,” so one has the “moral” right to expel, segregate, enslave them, or even lynch/kill them – that Nazis, neo-Nazis, White supremacists, and others continue to defend even today, almost three centuries later (Fig. 5.6).

How did racist scientists back then justify the existence of “innate” irremediable “racial” differences, particularly in Linnaeus’s epoch when most of them were *not* transmutationists – that is, when most of them defended that humans were created by God? At the beginning, they could use just-so-stories based on religious ideas. For example, many natural theologians defended, as did many religious leaders, that such differences were the result of a “degeneration” from an “ideal,” “superior’ type” – such as from Adam and Eve, who were supposedly “white” in these tales. That is, according to such a monogenesist conception of humanity, because all humans are part of the same species as they descended from a single man and a single woman, the existence of and the differences between human “races” can only be due to such a “degeneration.” Later, when more and more Western scientists started to accept transmutation, most of them did what humans are very good at doing: creating, and importantly *believing* in, equally racist and factually inaccurate stories. In particular, after the publication of Darwin's *Origin* and *Descent*, many scholars started to defend the polygenesist conception of humanity, which remained very popular until the first decades of the twentieth century. According to this idea, the different color of skin or other traits existing within *living* human individuals are due to the fact that such individuals belong to different human species: we are not all *Homo sapiens*. Many scholars started to defend that "blacks" are part of a "species" that in many ways is more similar, and *evolutionarily* more closely related, to chimpanzees than to the "*sapiens*" species that include "whites." Both these explanations of human "differences" are equally racist, but the polygenesist one is even more inaccurate in the sense that comparative genetic studies have confirmed that all living humans belong to a single species and, obviously, no single living human is more closely related to any nonhuman primate than to other living humans.

In a further example of how many Darwin followers that were originally inspired by Darwin's writings then became more extreme papists than their Pope is that many of those Western scholars defending such polygenesist ideas often cited Darwin's racist theories, particularly those of his *Descent* book. However, in that book, Darwin makes it clear that he defended the monogenesist view of evolution. In a way, this might seem paradoxical because Darwin also made clear, in that book, that for him the differences between some human "races" could be appalling, to a point that made it difficult for him to accept that they could indeed belong to the same species. However, basically, despite such racist ideas, he *needed* to defend that all humans, and all *organisms* for that matter, descend from a single, *continuous* evolutionary lineage: this point was crucial to provide a stronger basis and theoretical framework for his whole theory of evolution by natural selection. When we consider such logical inconsistent ideas and statements about human evolution that, as noted above, are so typical of Darwin's *Descent*, one can understand why even polygenesists would feel that citing this book could support their arguments, within a type B/innate racist framework. This is because Darwin often mixed, in a highly confusing and incongruent way, this type of racism with cultural A type racism, for instance, when he suggested that some differences between "races" were so well-defined and evolutionarily old that they likely could not be overcome, as we have seen in Chap. 1. In that chapter, we have seen that Darwin's *beliefs* even led him to apply such an innate type of racism to socially constructed items such as "morals," to the point that he even explicitly defended that morals can become innate – one of the factually most inaccurate evolutionary statements published by Darwin.

Similarly, Darwin also applied such erroneous tales to genders, for instance, when he suggested that women were less prominent in science because they were, in general, innately mentally inferior to men. In this sense, there is no doubt that Darwin's beliefs, biases, erroneous ideas, and the "facts" that he constructed to support such fictional tales played a huge part in the rise of evolutionary scientific racism at the end of the nineteenth century and in the first decades of the twentieth century – a new, critical subset of the scientific racism that started much earlier. The fictional evolutionary racist and sexist tales he defended in the *Descent* and the *Expression* did indeed significantly contribute to the "biologization of the social" – a term used in a chapter of Reynaud-Paligot' book *The Invention of Race*, – or, I would say, to the anatomization of the social/moral. This is because Darwin's tales allowed to justify, evolutionarily, what Linnaeus had postulated much earlier in a non-evolutionary context: the innate, irreversible differences – not only concerning anatomical traits but also mental ones – between human "races" and genders. As explained by Reynaud-Paligot, such racial and gender determinism is deeply related to the idea of evolutionary biological transmission, through blood and heredity, of the anatomical *as well as* the intellectual and moral attributes of a given people.

Kendi's 2016 brilliant book *Stamped from the Beginning* shows how the innate and cultural types of racism described above are respectively associated with two main types of political views about how to "deal" with "others": the segregationist view and the integrationist or assimilationist view. One of the most original and remarkable aspects of Kendi's book is that it reveals that even those scholars, politicians, and activists that history and educational books, or today's newspapers, tend to praise for being more "progressive" and "humanist" in their effort to integrate or assimilate – rather than segregate – "others" into "our" culture and societies actually tend to follow, at least in some respects, racist narratives. In other words, they are not able to escape from the more ancestral type of racism, namely, cultural racism:

> The title *Stamped from the Beginning* comes from a speech that Mississippi senator Jefferson Davis gave on the floor of the U.S. Senate on April 12, 1860. This future president of the Confederacy objected to a bill funding Black education in Washington, DC. "This Government was not founded by negroes nor for negroes", but "by white men for white men", Davis lectured his colleagues. The bill was based on the false notion of racial equality, he declared. The "inequality of the white and black races" was "stamped from the beginning". It may not be surprising that Jefferson Davis regarded Black people as biologically distinct and inferior to White people – and Black skin as an ugly stamp on the beautiful White canvas of normal human skin – and this Black stamp as a signifier of the Negro's everlasting inferiority. This kind of segregationist thinking is perhaps easier to identify – and easier to condemn – as obviously racist.. Historically, there have been three sides to this heated argument. A group we can call segregationists has blamed Black people themselves for the racial disparities. A group we can call antiracists has pointed to racial discrimination. A group we can call assimilationists has tried to argue for both, saying that Black people and racial discrimination were to blame for racial disparities.
>
> And yet so many prominent Americans, many of whom we celebrate for their progressive ideas and activism, many of whom had very good intentions, subscribed to assimilationist thinking that also served up racist beliefs about Black inferiority. We have remembered assimilationists' glorious struggle against racial discrimination, and tucked away their inglorious partial blaming of inferior Black behavior for racial disparities. In embracing biological racial equality, assimilationists point to environment – hot climates, discrimination, culture, and poverty – as the creators of inferior Black behaviors. For solutions, they maintain that the ugly Black stamp can be erased – that inferior Black behaviors can be developed, given the proper environment. As such, assimilationists constantly encourage Black adoption of White cultural traits and/or physical ideals. In his landmark 1944 study of race relations, a study widely regarded as one of the instigators of the civil rights movement, Swedish economist and Nobel Laureate Gunnar Myrdal wrote, "It is to the advantage of American Negroes as individuals and as a group to become assimilated into American culture, to acquire the traits held in esteem by the dominant white Americans".

He had also claimed, in *An American Dilemma*, that "in practically all its divergences, American Negro culture is.. a distorted development, or a pathological condition, of the general American culture."

Kendi's book also shows that racist narratives have been so prevalent in society and in science that even those "immortals" that Darwin so much admired and wanted to join, such as Isaac Newton, believed in and used them in their "scientific" writings:

One of the early leaders of the Royal Society was one of England's most celebrated young scholars, the author of *The Sceptical Chymist* (1661) and the father of English chemistry – Robert Boyle. In 1665, Boyle urged his European peers to compile more 'natural' histories of foreign lands and peoples, with Richard Ligon's *Historie of Barbados* serving as the racist prototype. The year before, Boyle had jumped into the ring of the racial debate with *Of the Nature of Whiteness and Blackness*. He rejected both curse and climate theorists and knocked up a foundational antiracist idea: 'The Seat' of human pigmentation 'seems to be but the thin Epidermes, or outward Skin,' he wrote. And yet, this antiracist idea of skin color being only skin deep did not stop Boyle from judging different colors. Black skin, he maintained, was an 'ugly' deformity of normal Whiteness. The physics of light, Boyle argued, showed that Whiteness was 'the chiefest color.' He claimed to have ignored his personal 'opinions' and 'clearly and faithfully' presented the truth, as his Royal Society deeded. As Boyle and the Royal Society promoted the innovation and circulation of racist ideas, they promoted objectivity in all their writings. Intellectuals from Geneva to Boston, including Richard Mather's youngest son, Increase Mather, carefully read and loudly hailed Boyle's work in 1664.

A twenty-two-year-old unremarkable Cambridge student from a farming family copied full quotations. As he rose in stature over the next forty years to become one of the most influential scientists of all time, Isaac Newton took it upon himself to substantiate Boyle's color law: light is white is standard. In 1704, a year after he assumed the presidency of the Royal Society, Newton released one of the most eminent books of the modern era, *Opticks*. "Whiteness is produced by the Convention of all Colors", he wrote. Newton created a color wheel to illustrate his thesis. "The center" was "white of the first order", and all the other colors were positioned in relation to their "distance from Whiteness". In one of the foundational books of the upcoming European intellectual renaissance, Newton imaged "perfect whiteness." Thanks to this malleable concept in Western Europe, the British were free to lump the multiethnic Native Americans and the multiethnic Africans into the same racial groups. [Centuries later] assimilationists first used and defined and popularized the term "racism" during the 1940s. All the while, they refused to define their own assimilationist ideas of Black behavioral inferiority as racist. These assimilationists defined only segregationist ideas of Black biological inferiority as racist. And segregationists, too, have always resisted the label of "racist". They have claimed instead that they were merely articulating God's word, nature's design, science's plan, or plain old common sense.

However, it should be noted that there is a huge difference between Darwin and Newton, concerning racism. Contrary to Darwin, Newton did *not* actively engage in constructing *a plethora of new "scientific" facts* that clearly contradicted the factual evidence to which he was directly exposed in his travels around the globe, in order to support such racist narratives and disseminate and popularize them. This is why, today, white supremacists such as David Duke and many neo-Nazis directly quote Darwin's writings in their books or on their websites to support their racist ideas – not Newton's works.

Darwin, Indoctrination, and the Repercussions of Scientific Evolutionary Racism

> *White people don't need a law against rape, but if you fill this room up with your normal black bucks, you would, because niggers are basically primitive animals.* (David Duke)

Scientific evolutionary racism has permeated a huge number of societal layers, including education, the media, the art industry, politics, sports, and so on. As stated by Corbey in his 2005 book *The Metaphysics of Apes*, "the Darwinian perception of nature as competition provided new support to the age-old icon of a beastly, human-like, and now preferably apish Other." It was not a coincidence that particularly at the end of the nineteenth century, many ethnographical books and anthropological narratives started to portray non-European men carrying weapons, to emphasize tales about the "aggressive practices of primitive savages," such as warfare, hunting, and raping. Often directly quoting Darwin's writings about how evolution was mainly a selfish, continuous war-like struggle for existence, many scholars began to argue that human "races" were naturally condemned to compete against one another until ultimately only one of them prevailed. As stated by Andreassen in *The Invention of Race*, at the end of the nineteenth century "indigenous people were literally being exterminated by white colonizers in Australia, but their extermination was not understood as a result of the atrocities being committed against them but rather as a result of biological determinism that mandated that the stronger (white) race survive while the weaker race (of color) disappeared." As we have seen above, in the *Descent*, Darwin specifically stated, over and over, the "scientific fact" that "others" – including Australian aborigines – were indeed *naturally* doomed to extinction or extermination. So, those that colonized and killed them were able to easily justify themselves by saying that they were just helping out Darwin's omniscient *Mother Nature*.

One example of how the "apish other" concept became quickly disseminated and normalized in popular culture is the 1933 movie King Kong. This movie became extremely popular among Westerners at a time when both non-European humans and great apes were increasingly depicted as evolutionarily primitive, aggressive brutes in Western media, literature, and science. A very interesting book about this movie and its links to and influence on popular culture is Erb's 1998 *Tracking King Kong: A Hollywood Icon of World Culture.* The message of the widely popular image of King Kong shown in Fig. 3.6 is simple: the apish other – King Kong represents both African apes and humans – who cannot just be ignored, or even segregated into a separate place far from European descendants because it will not just stay "there," in that part of the city, country, or continent. No, the apish other will instead aggressively come to "your" continent, country, and cities, to "your" neighborhoods and streets, and destroy "your" civilization – represented here by a plane, one of the main symbols of Western "civilization" in the 1930s. Not only that: the apish other will also take – and likely rape – "your" women – represented here not simply by a Western woman but by a *Western blonde* woman, in case there were still some doubts about the key take-home message.

Fig. 3.6 An Austrian poster to advertise the 1933 movie King Kong

Importantly, the very aggressive apish King Kong shown in that figure clearly contrasts with both the earlier and far more realistic scientific illustrations of apes provided by Western scholars such Tulp in 1641 and Tyson in 1699 (Fig. 3.2) and the views of earlier writers such as Rousseau who tended to emphasize the peaceful behavior of both apes and "noble human savages" (see also Box 1.3). This and many other similar examples clearly highlight how the scientific narratives about apes and the apish others changed since Tyson's 1699 work. On the one hand, the huge discomfort created by Tyson's 1699 factual deconstruction of the erroneous old concept of human–animal dichotomy affected not only religious leaders or lay people in general but also numerous scholars. They would not accept the fact that humans are mainly just dressed apes, so they created a fictitious 'apish other' that contrasted with the truly civilized humans – which obviously were Westerners, as they were. On the other hand, Darwin's racist "scientific" statements about how repulsive non-European groups such as the Fuegians were and how Europeans were not only more evolved but also mentally and morally superior was critical to then construct the notion of an evolutionarily primitive brutish amoral apish other. In a nutshell, while Darwin needed to support a monogenesist ladder-of-life evolutionary framework in which Europeans derived from "apish others," his *Descent* and *Expression* suggested that those who stayed at the apish other level would very likely never be able to replicate the astonishing evolutionary and moral steps that led to the Western "civilization."

In *The metaphysics of Apes*, Corbey makes a pertinent point concerning the way in which Western science constructed, in the nineteenth century and the first decades of the twentieth century, this notion that both apes and "primitive human races" were "powerful personifications of wildernesses to be fought heroically and conquered by civilized Westerners." He explains that apart from the most famous biologist – Darwin – the most famous psychologist – the psychoanalyst Sigmund Freud, influenced by Darwin's writings – also played a critical role in the widespread dissemination of the "beast-in-man stereotype" in both scientific and popular cultural discourses. As other European males, Freud had the biases and personal motivation to actively contribute to the construction of the notion of the evolutionarily primitive apish others. For more details on how Freud's ideas about human evolution were deeply influenced by Darwin's writings, particularly by Darwin's *Descent*, I strongly recommend Lucille Ritvo's 1990 volume *Darwin's Influence on Freud: A Tale of Two Sciences*.

Like Darwin, Freud was biased, racist, ethnocentric, and misogynist and also subscribed to fictitious teleological tales about "progress" in human evolution. However, there is a huge difference between Darwin and Freud. As noted above, while Darwin's writings about human evolution – as well as his general evolutionary ideas such as the exaggerated war-like, selfish struggle for existence - were affected by his Victorian biases, many of his specific observations about most non-human organisms were amazingly accurate. So, many of the books of Darwin that do not deal with humans are full of excellent science, which very few biologists, and scholars in general, were able to parallel. This contrasts with Freud, who basically wrote about, and worked on, humans, basing many of his theories on Darwin's biased writings about human evolution, or, even worse, on even more fictitious just-so-stories. Accordingly, more and more scholars, including numerous psychologists, now openly dare to state that saying that Freud's "psychoanalytic theories" were scientific is too far-stretched: they were mostly fictional, non-testable tales. This tendency also began to permeate the media and broader public in the last few decades, as summarized for instance in a short 2002 article published in *The Guardian*, entitled "Scientist or storyteller?" This article referred to the publication, by the publishing company Penguin, of the major translations of Freud's work under the general editorship of Adam Phillips, which prompted "serious questions about the nature of Freud's contribution and his legacy."

As correctly stated in the last paragraph of the article, Freud's "psychoanalytic theories" are not only chiefly based on, but are in themselves, basically fictional tales: "Philosophies that capture the imagination never wholly fade.. from Animism to Zoarastrianism, every view known to man retains at least a few devotees.. there might always be Freudians, and there will always be admirers of Freud's great imaginative and literary powers." "These two," the article added, "as the foregoing remarks suggest, are intimately linked.. but as to Freud's claims upon truth, the judgment of time seems to be running against him." Among the various recent books emphasizing this point, the one that I would recommend is Crews' 2017 book *Freud: The Making of an Illusion*. This is because this book directly connects these topics with a key issue discussed in the present volume, namely, that Freud's huge success and influence among Westerners, particularly Western male scholars, was not achieved *despite* his use of teleological, racist, ethnocentric, and sexist tales but in great part *because* of that. In this sense, Freud – a Darwin – provides an emblematic example of how the vicious, self-reinforcing cycle of indoctrination and systemic racism and sexism is propagated. In Box 3.1, I provide a powerful example – among countless ones – of the types of horrendous practices that occurred within the framework of that systematic racism, which were specifically justified by scientific evolutionary racist and ethnocentric

Fig. 3.7 A poster for the 1904 Louisiana Purchase Exposition, informally known as the St. Louis World's Fair, painted by Alphonse Mucha. Such "fairs" often promoted racist, ethnocentric, colonialist, and imperialist Western narratives

Fig. 3.8 A disturbing example of how racist, ethnocentric, colonialist, and imperialist Western narratives were promoted and justified, by showcasing how "inferior" the "others" were. Ota Benga (second from the left, with monkey), a widower from what is now the Democratic Republic of Congo, was exhibited in New York's Bronx Zoo in 1910 – here he is shown with other African men who were exhibited to the public and studied by psychologists at the 1904 St. Louis World's Fair

Box 3.1 Circus Africanus, Human Zoos, World's Fairs, Racist Scholars, and Western "Progress"

The horrific case of Ota Benda (Fig. 3.8) as well as other horrendous cases that should have never happened in human history are described in detail in the chapter entitled "Circus Africanus" in Washington's outstanding – and profoundly disturbing – 2006 book *Medical Apartheid*:

> By 1904, swashbuckling missionary-explorer Samuel Phillips Verner had acquired a veritable Noah's Ark of exotic fauna during three trips to the interior of the Dark Continent. The last expedition was commissioned in 1903 by the St. Louis Exposition Company, which paid the South Carolina–born Verner to hunt men instead of monkeys: He was to bring African Pygmies to America for display at the St. Louis World's Fair." [Fig. 3.7]. "Upon his return to America, Verner found himself romanticized as a reincarnation of Dr. David Livingstone, whom he claimed as his "posthumous mentor". As an ordained minister in the Presbyterian Church, Verner was also lionized in church circles as an imparter of morality to the Congo natives he doggedly hectored at the Southern Presbyterian Missionary House in Luebo, chiding

(continued)

them for their immodest dress and sexual behavior. His American admirers did not know that between 1895 and 1899, Verner had fathered a daughter and son on an African orphan girl there. By 1906, the World's Fair was over and the cash-strapped Verner was selling off his animals, artifacts, and more. Upon the receipt of a financial gift, he bestowed a prized equatorial specimen upon William T. Hornaday, director of the Bronx Zoological Gardens.

Verner's present was twenty-three-year old Ota Benga, an Mbuti widower from southern Africa, in what is now the Democratic Republic of Congo. Around 1903, Benga had returned from a hunting trip, only to find his village in smoking ruins and his wife, children, and entire tribe slaughtered by Force Publique thugs supported by the Belgian government. Benga himself was seized and sold into Verner's hands. Hornaday's views about the natives of sub-Saharan Africa mirrored Verner's own, conscripting Darwin in the service of racism: He told the New York Times that there exists "a close analogy of the African savage to the apes". *Scientific American* agreed: "The Congo pygmies [are] small, apelike, elfish creatures, furtive and mischievous, they closely parallel the brownies and goblins of our fairy tales. They live in the dense tangled forests in absolute savagery.. while they exhibit many ape-like features in their bodies".

But Hornaday espoused a more progressive vision as a scientific artist, and we have him to thank for the modern American zoo. As chief taxidermist of the National Museum (the Smithsonian), a position he held until 1890, he had inherited a static mausoleum of tatty taxidermy enshrined on plaster pedestals with only laconic placards to suggest what the animal had been like in life. In 1888, Hornaday persuaded the museum to add a wing of living animals in lifelike settings, which proved so popular a revolution that it became the National Zoological Gardens. He resigned over differences of vision, but in 1896 he reemerged as the first director of the New York Zoological Gardens (known as the Bronx Zoo), the world's largest, lushest, and most varied zoo. Hornaday's passion was for colorful verisimilitude in the re-creation of his animals' natural habitats. With a verdant Bronx park as his canvas, Hormaday installed colorful exotic animals of every genus grouped with their natural companions amid native vegetation. So when Benga was locked in the monkey house, before the staring crowd and with keepers always nearby, he was given a bow and arrow to brandish, his cage was littered with bones, and his two cage mates were Dinah, a gorilla, and an orangutan called Dohung.

The placard on Benga's enclosure read, "The African Pygmy, 'Ota Benga'.. height 4 feet 11 inches.. weight 103 pounds.. brought from the Kasai River, Congo Free State, South Central Africa by Dr. Samuel P. Verner.. exhibited each afternoon during September". *New York Times* headline trumpeted, "*BUSHMAN SHARES A CAGE WITH THE BRONX PARK APES*". Black New Yorkers were incensed, and representatives of the clergy, led by the Reverend Dr. MacArthur, pressed Mayor George B. McClellan to withdraw the city's support from the exhibit. As another minister, a Reverend Gordon, told the New York Times, "Our race.. depressed enough without exhibiting one of us with the apes.. we think we are worthy of being considered human beings, with souls". *The Times* turned an unsympathetic ear to African American objections: "One reverend colored brother objects to the curious exhibition on the grounds that it is an impious effort to lend credibility to Darwin's dreadful theories.. the reverend colored brother should be told that evolution.. is now taught in the textbooks of all the schools, and that it is no more debatable than the multiplication table".

(continued)

The swipe at creationism did not address Gordon's immediate concerns but did hit a nerve among many whites who shared Gordon's outrage. Some were angered by this inhumane insult to blacks, and others, who opposed the teaching of Darwin's theory of evolution, were afraid that Benga's dramatic presence would offer a powerful plebeian argument for the theory of evolution. The entertainment of a "monkey-man" might persuade people who were untouched by the theory's scientific merits. Mayor McClellan snubbed the black delegation, referring them to the Parks Department, and another *Times* account hinted that Benga differed little from the zoo's animals: "Ota Benga.. is a normal specimen of his race or tribe, with a brain as much developed as are those of its other members.. and can be studied with profit.. the pygmies are an efficient people in their native forests.. but they are very low in the human scale, and the suggestion that Benga should be in a school instead of a cage ignores the high probability that school would be a place of torture to him and one from which he could draw no advantage whatever".

A lively epistolatory debate ensued in the pages of the *Times*, heavily weighted in favor of retaining Benga, and many of the letters were signed by respondents with M.D. and Ph.D. degrees. One doctor suggested, "It is a pity that Dr. Hornaday does not introduce the system of short lectures or talks in connection with such exhibitions.. to] help our clergymen to familiarize themselves with the scientific point of view so foreign to many of them". *Times* journalists agreed that Benga provided a valuable tool for illustrating basic evolutionary precepts: To oppose his internment was to oppose science. These precepts included physical similarities to the lower primates that scientific racism was beginning to popularize widely. Anthropometric portraits of blacks and apes demonstrated how blacks' facial angles, stature, stance, and gait resembled those of monkeys, chimpanzees, and orangutans. Blacks' hair, or "wool", was compared to animal pelts. Such uncomplimentary images were published in scientific journals and would soon adorn children's textbooks. Scientists alleged that apes preferred to mate with black women, just as black men lusted after white women, their own evolutionary "betters".

At the zoo, the *Times* revealed that Benga's situation was escalating: "There were 40,000 visitors to the park on Sunday.. nearly every man, woman and child of this crowd made for the monkey house to see the star attraction in the park, the wild man from Africa.. they chased him about the grounds all day, howling, jeering, and yelling.. some of them poked him in the ribs, others tripped him up, all laughed at him". Finally, Benga retaliated by attacking visitors with a knife and a bow and arrows, and the zoo ejected him. Black New Yorkers organized a collection, which was insufficient to return him home, as he wished, but provided enough to cap his filed teeth and send him to the Virginia Theological Seminary and College, where he proved himself an able student. Benga then found work in a Lynchburg, Virginia, tobacco factory, where he fit in well as an efficient worker and a beloved Pied Piper who taught local children to fish and hunt. But he spoke often and tearfully of wishing to return home to the Congo, and when he realized he could never save enough for passage, his depression became profound. In 1916, Benga committed suicide with that ubiquitous icon of Western technological achievement, a handgun.

tales, at the end of the nineteenth century and in the first decades of the twentieth century – precisely the time when scientific eugenics was rising in popularity across various Western countries, as we will see below.

For example, Ota Benda's tragic example discussed in Box 3.1 shows how such tales naturalize, legitimize, and "morally" and "scientifically" justify his "acquisition" by the Bronx Zoological Gardens in 1906, after being first "displayed" at the 1904 St. Louis World's Fair (Figs. 3.7 and 3.8). Yes, you read correctly: a human being was "acquired" by a zoo. And yes, this happened in the U.S., less than 12 decades ago. Think about this: the grandparents and even the parents of some people living today were alive back then. Not only that, but the *international* "fair" – held in St. Louis, Missouri – in which he was "displayed" – received local, state, *and* federal funds, was attended by nearly 20 million people, and had exhibition spaces from more than 60 countries and 43 U.S. states. This is another powerful example of how systemic racism and indoctrination are part of a huge, *active*, societal, and bureaucratical machine – a machine that ultimately led to a situation in which a huge number of Western people that saw Oto Benda's "display" at that fair (Fig. 3.8) did not think that there was any problem with that. Clearly, if those people were from Africa, or had not been indoctrinated from a young age with such racist tales, then that would very likely not have been the case: cultural indoctrination is key. This point stresses that we should *all* make an active effort to critically think about and question the things that we are told/taught at school or at home or see on the television about what is "normal." We always should take a step back and think deeply before accepting the tales constructed by our societies, for instance, about how our "group" or "culture" is the best, the most beautiful, or the most "moral."

It is not at all a mere historical coincidence that the very same year in which Oto Benda was displayed as an apish other at an "international fair," the Germans started the Herero–Nama genocide of numerous "apish others" in Africa. Various thousands of Herero and Nama men, women, and children were shot, tortured, or driven

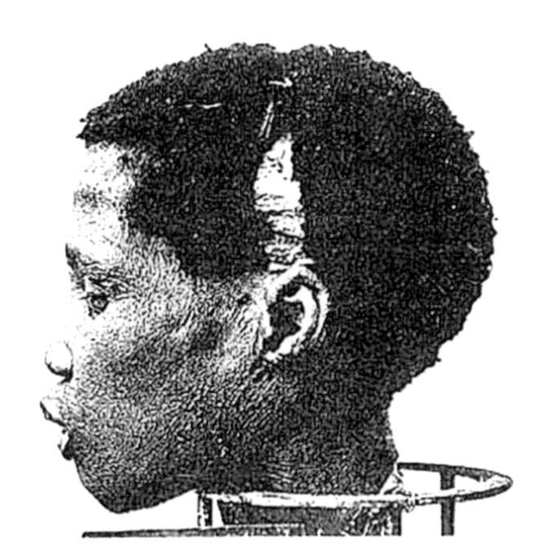

Fig. 3.9 At the very same time that Western countries were organizing "World's Fairs" to show their 'superiority' and higher "civilizations" and "morality," they were undertaking genocides such as this one, in Africa. According to numerous historians, from 1904 to 1908, more than 80% of Namibia's Herero and 50% of its Nama people died – many due to inhumane conditions and starvation – in a genocide carried out by German forces in concentration camps. Here, a picture from the Shark Island Death Camp

into the Kalahari Desert to starve, by German troops, between 1904 and 1908 (Fig. 3.9). As explained above, genocides were one of the logical consequences of the construction of the aggressive, brutish, "savage" amoral apish others. As made very clear in the subtitle of Darwin's 1859 opus magnum, in his evolutionary concept of a war-like struggle for existence between "races," only some of them would ultimately be "favored." Within the various "solutions" proposed by many followers of the eugenics movement that were influenced by that war-like "struggle-for-existence" evolutionary notion, a critical one was sterilization of the "others." Another, even more radical, "solution" was mass killings and even genocides, as carried out by the Germans between 1904 and 1906 in Africa and just a few decades later in Europe. Obviously, the Germans were not the only ones killing numerous "less evolved or civilized others." The machine of system racism and ethnocentrism led, for instance, to the killing of about 10 million Africans under the rule of King Leopold II of Belgium in the then so-called Congo Free State, and the English committed many massacres in Australia and many other regions of the globe, as we have seen (Fig. 1.15).

An illustrative and upsetting example of how the scientific evolutionary racist construction of 'apish others' also applied to African apes, in parallel to what was being done to African people, concerns the colonial propaganda film made in the 1950s in the Belgian Congo. This film was made on behalf of the Belgian government and circulated broadly in Belgian cinemas, programmed on Sunday afternoons for families with children. As described by Corbey in *The Metaphysics of Apes*, "the footage shows in great and, by present-day standards, shocking detail how scientists of the Royal Belgian Institute of Natural Sciences shoot and kill and adult female gorilla carrying young; subsequently the body is skinned and washed in a nearby stream, with the distressed youngster sitting next to it; the adult's skeleton, skin and other body parts were collected for scientific [anatomical] study and conservation, while the live young gorilla was sent to the Antwerp zoo."

What is revealing about such an indoctrination process, and shows that it is far from over, is that most Westerners – and actually most people across the planet – including the vast majority of people in the very countries that promoted such mass killings and genocides, had never even heard them. In fact, the Herero–Nama genocide was only officially recognized by Germany a few weeks before I wrote these lines, in May 2021. That is, 116 years after it happened. Such omissions and lack of historical knowledge are, again, not at all a coincidence. Among the countless things we learn in school, including the "good" things done by European leaders such as King Leopold II of Belgium – who is still shown as a hero in numerous statues in Europe – we do not learn anything about such genocides and mass killings of Africans, Australians, or Native Americans. The major genocide we learn about, in detail, in school, is also a very atrocious one – the *Holocaust* – in which the victims were mostly Europeans, that is, European Jews, Gypsies, homosexuals, and so on. Why are the mass killings and genocides of Africans, Australians, Asians, and Native Americans much less discussed in Western countries? Why do many educational materials continue to dedicate several pages idolizing Linnaeus, Darwin, Columbus, and so on, including the writings in which they defended racist ideas, such as the *Descent*, while not referring at all to the Herero–Nama genocide or to the 10 million Africans killed in the "Congo Free State"?

Fortunately, while such an indoctrination process continues to be done until this very day, it needs to be said that the way in which it is done in various Western countries nowadays is much less biased, or less extreme, than it was before the 1950s. Since then, there has been a change in the mindset within parts of the Western scientific community and the broader public towards a vision of continuity and unity both between all extant human groups and between them and other primates. This change was in great part due to World War II, which provided the most horrible, and direct, contradiction of the Western belief that Westerners were leading the whole globe to an overall societal and "moral" progress. The Holocaust, together with many other genocides, mass killings, and other atrocities committed by Western countries in the decades that preceded it, clearly showed that this was not the case. For instance, the UNESCO Statement on Race – an official declaration against racism that attempted to break the connection between "race" and biological determinism – was published in 1950. In fact, it should be noted that in various countries, eugenics was already losing some ground as early as the 1930s because geneticists could not scientifically support any of the key conceptual tenets defended by eugenicists and also because of the socioeconomic crisis that affected many of these countries in that decade, as explained in Kevles' 1998 book, *In the Name of Eugenics*. We will further discuss that book, and eugenics, in the next sections.

This change of attitude also led to – and was then interactively further expanded by – new and less biased comparative anatomical, behavioral, and genetic works focusing on nonhuman primates, such as the groundbreaking behavioral works about apes carried out by researchers such as Dian Fossey, Birute Galdikas, and Jane Goodall in the second half of the twentieth century. It is not a coincidence that these groundbreaking and less biased behavior works that contributed to deconstruct some key racist and sexist inaccurate tales about our evolutionary history were done by female scientists. The very fact that the scientific community was willing to accept the importance and innovative character of the studies conducted by these women was in itself another sign that the times were indeed changing, in a way. One crucial consequence of such scientific and societal changes is that, nowadays, many researchers – such as De Waal, as emphasized in his popular 2016 book *Are We Smart Enough to Know How Smart Animals Are?* – are now recognizing that the main problem of previous behavioral studies was not that they were anthropomorphizing nonhuman primates but instead the widespread acceptance of the dogmatic idea that nonhuman primates had to be very different from us because we are "special" and "unique" within the animal kingdom. Contrary to that dogmatic idea, we now know that apes also use tools, display highly complex behaviors, including some related to "fairness," "morality," and "altruism," have similar emotions and display similar facial expressions, plan tasks in advance, are able to deal with abstract concepts, and so on. A broader, related reflection of this change of mindset is The Great Ape Project, which calls for great apes to be accorded the same basic rights as humans. In addition, more and more authors began to defend the idea that humans and chimpanzees should be placed in the same genus. This recognition of the huge biological proximity of humans and chimpanzees, which was first empirically shown three centuries earlier by Tyson, in 1699, was, for instance, reflected in the 1990s by the provocative title of Jared Diamond's excellent 1994 book, *The Rise and Fall of the Third Chimpanzee*.

As has always been happening in the course of human history, such societal and scientific changes are not welcomed by all members of the societies in which they occur, particularly by those that are from the "group," or "race," or "gender" that was the most favored before those changes. We have seen above that the idolization of Darwin started precisely in great part due to the backlash by well-off males of the English society to the huge societal changes that were happening in continental European countries such as France in the nineteenth century. A similar backlash is happening right now: not only among many lay people such as those subscribing to the slogan "Make America Great Again" in the United States, who are obviously mostly European descendants, but also among European descendant scholars, particularly males such as Steven Pinker. In fact, the huge mediatic attention given to authors such as Pinker, and his ideas that Westerners are leading the "progress" of the planet through an expanding moral circle, is precisely related to the fact that so many European descendants are becoming highly uncomfortable with the change of the *status quo* that is slowly happening in our society. Once again, this shows us the profound links between the biases and power of place of scientists, politics, the media, and popular culture (see also Box 3.2).

Box 3.2 Racism, Science, Sociobiology, and Prevailing Societal Inequities

In his 2016 book *Stamped from the Beginning*, Kendi recognizes that in the last few decades, there have been significant changes and positive aspects concerning the fight against racism and discrimination in the United States, as reflected, for example, by Obama's presidential victory. However, he points out that, at the same time, and precisely mainly as a reaction to those changes, there were also very negative aspects:

> As the economic and racial disparities grew and middle-class incomes became more unstable in the late 1970s and early 1980s, old segregationist fields – like evolutionary psychology, preaching genetic intellectual hierarchies, and physical anthropology, preaching biological racial distinctions – and new fields, like sociobiology, all seemed to grow in popularity. After all, new racist ideas were needed to rationalize the newly growing disparities. Harvard biologist Edward Osborne Wilson, who was trained in the dual-evolution theory, published *Sociobiology: The New Synthesis* in 1975. Wilson more or less called on American scholars to find 'the biological basis of all forms of social behavior in all kinds of organisms, including man.' Though most sociobiologists did not apply sociobiology directly to race, the unproven theory underlying sociobiology itself allowed believers to apply the field's principles to racial disparities and arrive at racist ideas that blamed Blacks' social behavior for their plight. It was the first great academic theory in the post-1960s era whose producers tried to avoid the label 'racist.' Intellectuals and politicians were producing

(continued)

theories – like welfare recipients are lazy, or inner cities are dangerous, or poor people are ignorant, or one-parent households are immoral – that allowed Americans to call Black people lazy, dangerous, and immoral without ever saying 'Black people,' which allowed them to deflect charges of racism.

Assimilationists and antiracists, realizing the implications of Sociobiology, mounted a spirited reproach, which led to a spirited academic and popular debate over its merits and political significance during the late 1970s and early 1980s. Harvard evolutionary biologist Stephen Jay Gould, who released *The Mismeasure of Man* in 1981, led the reproach in the biological sciences against segregationist ideas. Edward Osborne Wilson, not to be deterred, emerged as a public intellectual. He no doubt enjoyed hearing Americans say unproven statements that showed how popular his theories had become, such as when someone quips that a particular behavior 'is in my DNA.' He no doubt enjoyed, as well, taking home two Pulitzer Prizes for his books and a National Medal of Science from President Jimmy Carter. Wilson's sociobiology promoted but never proved the existence of genes for behaviors like meanness, aggression, conformity, homosexuality, and even xenophobia and racism..

Weeks after passing the most antiracist bill of the decade over Reagan's veto – the Comprehensive Anti-Apartheid Act with its strict economic sanctions – Congress passed the most racist bill of the decade. On October 27, 1986, Reagan, "with great pleasure", signed the Anti-Drug Abuse Act, supported by both Republicans and Democrats. "The American people want their government to get tough and to go on the offensive", Reagan commented. By signing the bill, he put the presidential seal on the "Just say no" campaign and on the "tough laws" that would now supposedly deter drug abuse. While the AntiDrug Abuse Act prescribed a minimum five-year sentence for a dealer or user caught with five grams of crack, the amount typically handled by Blacks and poor people, the mostly White and rich users and dealers of powder cocaine – who operated in neighborhoods with fewer police – had to be caught with five hundred grams to receive the same five-year minimum sentence. Racist ideas then defended this racist and elitist policy. The bipartisan act led to the mass incarceration of Americans. The prison population quadrupled between 1980 and 2000 due entirely to stiffer sentencing policies, not more crime. Between 1985 and 2000, drug offenses accounted for two-thirds of the spike in the inmate population. By 2000, Blacks comprised 62.7 percent and Whites 36.7 percent of all drug offenders in state prisons – and not because they were selling or using more drugs. That year, the National Household Survey on Drug Abuse reported that 6.4 percent of Whites and 6.4 percent of Blacks were using illegal drugs.

Racial studies on drug dealers usually found similar rates. One 2012 analysis, the National Survey on Drug Use and Health, found that White youths (6.6 percent) were 32 percent more likely than Black youths (5 percent) to sell drugs. But Black youths were far more likely to get arrested for it. During the crack craze in the late 1980s and early 1990s, the situation was the same. Whites and Blacks were selling and consuming illegal drugs at similar rates, but the Black users and dealers were getting arrested and convicted much more. In 1996, when two-thirds of the crack users were White or Latina/o, 84.5 percent of the defendants convicted of crack possession were Black. Even without the crucial factor of racial profiling of Blacks as drug dealers and users by the police, a general rule applied that still applies today: wherever there are more police, there are more arrests, and wherever there are more arrests, people perceive there is more crime, which then justifies more police, and more arrests, and supposedly more crime.. After all, African Americans possessed 1 percent of the national wealth in 1990, after holding 0.5 percent in 1865, even as the Black population remained at around 10 to 14 percent during that period.

Another example concerns the backlash against the recent Black Lives Matter (BLM) movement within the United States and its spread across many other countries. This backlash was mainly undertaken by European descendants, particularly those in positions of power, either in science or within the society more generally. As explained above, that is, by definition, what reactionaries do: they react against societal changes, demanding the continuation of the prevailing *status quo* that favors them. For instance, they argue that they "feel" that we are already mainly in a "post-racial" time. First of all, it is easy to "feel" this way when you have never felt oppressed or discriminated against because of your skin color. Second, this "gut feeling" is factually wrong. For instance, as noted in a report published by the Brookings Institution on February 27, 2020, based on a detailed analysis of empirical data, "a close examination of wealth in the U.S. finds evidence of staggering racial disparities.. at $171,000, the net worth of a typical white family is nearly ten times greater than that of a Black family ($17,150) in 2016" (Fig. 3.10). The report states that such "gaps in wealth between black and white households reveal the effects of accumulated inequality and discrimination, as well as differences in power and opportunity that can be traced back to this nation's inception.. the Black-white wealth gap reflects a society that has not and does not afford equality of opportunity to all its citizens." Moreover, as reported in Sorenson's 2009 book *Ape*, detailed archival content analyses reveal that news articles keep on creating implicit associations between "black criminals" and apes and that those identified as more "ape-like" are more likely to be condemned to life sentences or to be executed. In face of such empirical data, it is also not a coincidence that in the very same year that Sorenson published his book, there was widespread dissemination and discussion in the media of the graphic, violent cartoon mentioned above suggesting that Barack Obama was an ape that was precisely destined to be executed by police officers (Fig. 1.19). Despite significant changes, the tales about evolutionarily 'primitive apish others' remain highly present in our society.

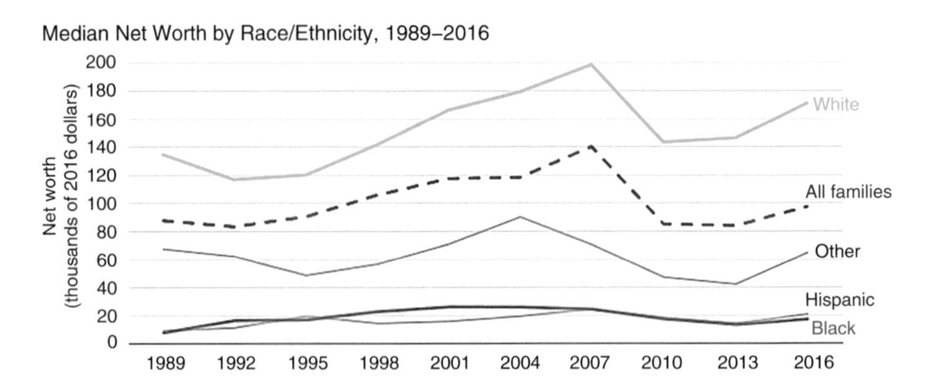

Fig. 3.10 Median net worth by so called "race", or ethnicity, between 1989 and 2016, in the U.S.

Darwin's Encounters with Non-Europeans, "Civilization," and Western Atrocities

The more I see of what you call civilization, the more highly I think of what you call savagery. (Robert Howard)

We can now return to the analysis of a critical question: if in Darwin's idea of natural selection, organisms are mainly adapted to their *local* habitats, at a *specific time* in history, then why did he still refer so often to *general* "progress," "favored or preferred races," and to morally or biologically "higher" human groups? In Chap. 1, we saw that these notions were not minor details in his work, as indicated by the very subtitle of his *opus magnum*, the *Origin*: "the preservation of favored races in the struggle for life." As noted in Ruse's *On Purpose*, Darwin did recognize such logical inconsistencies in some parts of his books.

One case in which Darwin explicitly did so, and directly related some of his ideas to teleological narratives in general and religious beliefs in particular, is a sentence from his autobiography, where he confessed that when he was writing his *Origin*, he felt the "extreme difficulty or rather impossibility of conceiving this immense and wonderful universe, including man with his capacity of looking far backwards and far into futurity, as the result of blind chance or necessity." Thus, he added, "when thus reflecting I feel compelled to look to a *First Cause* having an *intelligent mind* in some degree analogous to that of man; and I deserve to be called a *Theist*." As put by Ruse, "one feels a little as if Darwin is like Moses – he led his children to the Promised Land but never got there himself." In recent years, some scholars have started to publicly recognize that Darwin's writings are indeed characterized by such major logical incongruities. For instance, in his 2017 book *The Darwinian Tradition in Context* and the 2019 book *Charles Darwin's Incomplete Revolution*, Delisle emphasized that "the first to challenge Darwin was Darwin himself in the *Origin*.. the reader willing to go beyond Darwin's rhetoric encounters a book displaying at least five independent sets of issues or pictures.. while some are squarely incompatible with one another, others are less than clearly related to each other" (see also Box 3.3).

Box 3.3 Enlightenment, Faith in Science, "Progress," Darwin, and Racism

The third chapter of Gray's *Seven Types of Atheism*, entitled "A strange faith in science", provides illustrative examples of how belief and teleological narratives, particularly about "progress," have been deeply embedded in the ideas and works of numerous renowned scientists, including Darwin. A major problem is that such beliefs and narratives are often deeply related to, or are used to justify, racist and ethnocentric views, as noted by Gray:

(continued)

In 1929, the Thinker's Library, a series of books published by the Rationalist Press Association in London to counter the influence of religion in Britain, produced an English translation of the German biologist Ernst Haeckel's 1899 book *The riddle of the universe*.. strongly hostile to Jewish and Christian traditions, Haeckel founded a new religion called Monism, which spread widely among intellectuals in central Europe. Among Monist tenets was a 'scientific anthropology' according to which the human species was composed of a hierarchy of racial groups, with white Europeans at the top. At the time 'scientific racism' was not unusual in books promoting rationalism. Like Haeckel a proponent of a 'religion of science', Julian Huxley [grandson of Thomas Huxley] joined Haeckel in promoting theories of innate racial inequality. In 1931, he wrote that there was 'a certain amount of evidence that the Negro is an earlier product of human evolution than the Mongolian or the European, and as such might be expected to have advanced less, both in body and in mind'. In the early twentieth century such attitudes were commonplace among rationalists. In his best-selling book *Anticipations* (in 1901) H. G. Wells, also a contributor to the Thinker's Library, wrote of a new world order ruled by a scientific elite drawn from the most advanced peoples of the world.

Regarding the fate of 'backward' or 'inefficient' peoples, he wrote: "And for the rest, those swarms of black and brown, and dirty-white, and yellow people, who do not come into the needs of efficiency? Well, the world is a world, not a charitable institution, and I take it that they will have to go.. it is their portion to die out and disappear." Huxley's 'evolutionary humanism' asserted that if humankind was to ascend to a higher level, evolution would have to be consciously planned. Some religious thinkers followed Huxley in thinking in this way. A. N. Whitehead (1861–1947) and Samuel Alexander (1859–1938) developed a type of 'evolutionary theology' in which the universe was becoming more conscious of itself – a process that would culminate in the emergence of a Supreme Being much like the God of monotheistic religion. The French Jesuit theologian Pierre Teilhard de Chardin (1881–1955) developed a similar view in which the universe was evolving towards an 'Omega Point' of maximal consciousness.

Much of the Enlightenment was an attempt to demonstrate the superiority of one section of humankind – that of Europe and its colonial outposts – over all the rest. Evangelists for the Enlightenment will say this was a departure from the 'true' Enlightenment, which is innocent of all evil. Just as religious believers will tell you that 'true' Christianity played no part in the Inquisition, secular humanists insist that the Enlightenment had no responsibility for the rise of modern racism. This is demonstrably false. *Modern racist ideology is an Enlightenment project.* Racism and anti-Semitism are not incidental defects in Enlightenment thinking. They flow from some of the Enlightenment's central beliefs. For Voltaire, Hume and Kant, European civilization was not only the highest there had ever been. It was the model for a civilization that would replace all others. The 'scientific racism' of the nineteenth and early twentieth centuries continued a view of humankind promoted by some of the greatest Enlightenment thinkers.

All these philosophies rely on an idea of evolution. There is a problem, however. As understood in Darwin's theory, the universe is not in any sense evolving towards a higher level. Thinking of evolution as a movement towards greater consciousness misses Darwin's achievement, which was to expel teleology – explaining things in terms of the purposes they may serve rather than the causes that produced them – from science. As he wrote in his *Autobiography*, "There seems to be no more design in the

(continued)

variability of organic beings, and in the action of natural selection, than in the course in which the wind blows." As Darwin makes clear in this passage, natural selection is a purposeless process. He did not always stick with this view, however. On the last page of *On the origin of species*, he wrote: "we may be certain that the ordinary succession by generations has never once been broken, and that no cataclysm has desolated the whole world.. hence we may look with some confidence to a future of great length.. and as natural selection works solely for the good of each being, all corporeal and mental endowments will tend to *progress to perfection*." In fact the theory of natural selection contains no idea of progress or of perfection. Darwin's inability always to accept the logic of his own theory is revealing. An eminent Victorian, he could not help believing that natural selection favoured 'progress to perfection'. Many less scrupulous thinkers have followed Darwin in this belief.

Some of the arguments used by Darwin, Darwinians, and neo-Darwinians to deny the incongruities between Darwin's theory of evolution by natural selection concerning local adaptations to local environments and teleological tales about "progress" and "higher" organisms involved the use of the notion of an "arms race." Let us say that if predator A and its prey B are coevolving, and they respectively acquire "better" evolutionary "weapons" to hunt and to not be hunted, then we could explain how *both* A and B could get "better" with time. However, this argument is not convincing. First, even if that was the case, A and B would just be better at hunting B and not be hunted by A, respectively: they would not be "better" animals *as a whole*, evolutionarily. Life is much more than predator A being able to hunt or not hunt B – predators often hunt many different types of prey, each prey is often eaten by many types of predators, and, in addition, both prey and predators have many other types of ecological interactions, including competition, cooperation, having sex with others, taking care of their progeny, building nests, and so on. It would be like saying that humans that lived one million years ago were "better" as a whole just because they were "better" at escaping from a single predator, let us say leopards. Many other crucial aspects were relevant for humans at that time, such as using fire, making and using stone tools, cooperating, migrating, and so on.

What makes the use of the "arms race"' argument to justify such teleological tales even less convincing is that in most of the cases in which Darwin refers to "progress," "higher clades," or "favored races," he is not even referring at all to being the "better" prey or predator. For instance, although in some passages Darwin states that humans are not really special animals within the natural world, in many others he clearly suggests that they are. As stated by Ruse in his book *On Purpose*, Darwin was "deeply committed to the cultural ideology of progress and to the belief in biological progress, something that ends not just with human beings but with Europeans, preferably English capitalists." As discussed above, there are indeed profound historical links between not only the Age of Enlightenment and the notion of progress but also the Industrial Revolution, capitalism, and the concept of individual selfishness that was so prominent in Darwin's scientific ideas (see also Box 3.3). Ruse wrote that "Adam Smith was important [regarding the notion of progress], with his ideas of the importance of a division of labor and of the Invisible Hand making a virtue of individual selfishness."

This topic was astutely analyzed in Landau's book *Narratives on Human Evolution*, which highlights that many elements of Darwin's works – particularly *The Descent of Man* – are indeed similar to imaginary narratives and tales typically used in folklore and myth. For instance, "the principle of natural selection, or 'struggle-for-existence', remains the chief agent of [Darwinian] evolution.. it also explains events according to their consequences or 'final causes'.. [it] may appear to operate in a teleological fashion, as though directed toward some overall design or purpose.. Darwin.. confesses that he does believe human evolution has been toward a preferred and higher state." As she further noted, in Darwin's *Descent*, as in other writings on human evolutionary history such as those of Haeckel's, "like most narratives, the story of human evolution is subject to an intrinsic 'teleological determinism': elements are present not as they occur but as they contribute to the outcome of the story."

Let us thus examine in further detail an issue mentioned in Chap. 1 that is deeply linked to these topics: Darwin's personal encounters with, and constructed ideas about, non-Europeans. One of the evolutionary "facts" constructed by Darwin in the *Descent*, concerning such encounters, was that "nomadic habits, whether over wide plains, or through the dense forests of the tropics, or along the shores of the sea, have in every case been highly detrimental." This "fact" is inaccurate because for 99.8% of our evolutionary history, since humans and chimpanzees diverged, our ancestors were mainly nomads. Not only that: almost all nonhuman animals are also nomadic. So, how can it be that "in every case" being nomadic is "highly detrimental," when Darwin suggested that organisms with highly detrimental traits are often removed from the natural world by *Mother Nature*'s broom? The answer is that, as we have seen in Chaps. 1 and 2, what Darwin "saw" – or better said the "facts" he said he "saw" in his travels around the globe – *concerning humans* was markedly different from what was truly before his eyes and what he had no problems in seeing when it concerned the observation of other organisms.

An illustrative example mentioned in Chap. 1 concerns his encounters with the nomadic Fuegians that repulsed him so much (Figs. 1.13 and 1.14). Before Darwin's travels, Malthus, looking for an example of the world's most downtrodden "savages," had already written about "the wretched inhabitants of Tierra del Fuego," who had been said by some earlier European travelers to be "at the bottom of the scale of human beings." When Darwin arrived at Tierra del Fuego, he *agreed* with Malthus and *confirmed his own a priori biases*, stating: "I believe if the world was searched, no lower grade of man could be found" (see also Box 3.3). As we have seen, Captain Robert Fitzroy of the Beagle had picked up three Fuegians (Fig. 1.14) on an earlier voyage and took them to England to introduce them to the "highest" of civilizations, before returning them to Tierra del Fuego so that they could serve as missionaries. However, just a year later, in 1934, the huts and gardens that the

British sailors built for those three Fuegians to help them "civilize" the other Fuegians were empty. Jemmy Button, one of the three, later told the crew that he and the other two had reverted to their former way of living (see Fig. 1.14). A shocked Darwin wrote about this in his journal – he had never seen "so complete and grievous a change.. it was painful to behold him." Fitzroy then proposed something to Jemmy that, according to the type of ethnocentric tales constructed by Darwin, should quickly be accepted by any 'non-civilized' person: he told Jemmy that he could take him back to 'civilization', if Jemmy wanted to. Not only to civilization but to its very pinnacle: England. Surprisingly – for Fitzroy and Darwin – Jemmy answered that he had "not the least wish to return to England" because he was "happy and contented" with "plenty fruits," "plenty fish," and "plenty birdies."

As a fascinating example of the lack of understanding of and empathy towards the way of life, aspirations, and priorities of the Indigenous people typically manifested by biased Europeans, Darwin was particularly puzzled and could not comprehend how someone from the "lower grade of man" did not want to live a "highly civilized" life in London. In fact, contrary to people like Fitzroy, for Darwin this was not just a mere incident or a mere personal anecdote: Jemmy's decision went against the very "facts" about the evolution of humanity, civilization, and morality that he had written in his notebooks and diary back then. This is truly the crux of the matter, which most scholars fail – or do not want – to recognize when they use the "everybody was racist and ethnocentric back then" line of defense to absolve Darwin: for basically "everybody else," things were completely different because such ideas were not a fundamental part of their work and careers. It is exactly because of this that it is obviously much more alarming and damaging to the society as a whole and to science in particular if a Nobel laureate scientist such as James Watson (Fig. 3.11) publicly uses his scientific *power of place* to assert that it is an empirical "fact" that women or Africans are mentally "inferior", than if this is said by a drunk lay person at a bar (see also Box 3.6).

Accordingly, contrary to "everybody else," Darwin had to further construct new evolutionary "facts" in a biased, self-reinforcing, "confirmation bias" way, as a snowball, in order to justify Jemmy's "puzzling" refusal to go back to London, the epicenter of the pinnacle of evolution and morality. Specifically, Darwin wrote that Jemmy's decision to not go back to London was very likely related to Jemmy's "young and nice-looking [Fuegian] wife." Like many adaptationists continue to create just-so-stories to support *a priori* assumptions based on wrong scientific ideas, Darwin created a just-so-story to support his *a priori* assumptions. That is, in the mind of someone like Darwin, who was *sure* that the "whites" living in England were the highest, more advanced, and civilized group that was "favored" by evolution, what other reason could a "savage" such as Jemmy have to not want to go back to London, other than something as "primitive" as his bodily desire for a "savage" young and beautiful woman? This is what Darwin wrote in his *Voyage of the Beagle*, a book that was widely sold, read, and used not as a fictional account or a compilation of personal biased ideas but as a nonfictional collection of fascinating *true facts* about the natural world and "other," non-European people.

Fig. 3.11 James Watson, a Nobel laureate, is an emblematic example of scientific biases: he shows how scholars that are capable of making important scientific discoveries are often also not immune to the heavy chains of indoctrination and societal, inaccurate, racist and misogynistic imaginary tales. Such people are particularly dangerous because they precisely use their *power of place* to give the appearance that they have a deep knowledge about societal issues they often know little about, in order to "scientifically" support their *a priori* racist and sexist beliefs, dragging countless scholars and lay people down with them into the spiral of indoctrination and discrimination

Apart from such "nonfictional" books, Darwin also asserted such evolutionary "facts" in his private letters, which provide clear evidence that his attitude toward Jemmy and other "others" was both biased and patronizing. For instance, in a letter to his sister Catherine, he starts by saying that Jemmy was very "civilized" when he started the trip back from England with him but now was a "thin," "squalid savage" – a "poor savage" who "had nothing." About Jemmy's "puzzling" refusal to go back to England, he told her:

> After visiting some of the southern islands, we beat up through the magnificent scenery of the Beagle Channel to Jemmy Button's country [Jemmy Button, York Minster, and Fuegia Basket, were natives of Tierra del Fuego, brought to England by Captain Fitz-Roy in his former voyage, and restored to their country by him in 1832]. We could hardly recognise poor Jemmy. Instead of the clean, welldressed stout lad we left him, we found him a naked, thin, squalid savage. York and Fuegia had moved to their own country some months ago, the former having stolen all Jemmy's clothes. Now he had nothing except a bit of blanket round his waist. Poor Jemmy was very glad to see us, and, with his usual good feeling, brought several presents (otter-skins, which are most valuable to themselves) for his old friends. The Captain offered to take him to England, but this, to our surprise, he at once refused. In the evening his young wife came alongside and showed us the reason. He was quite contented. Last year, in the height of his indignation, he said "his country people no sabe nothing – damned fools" – now they were very good people, with TOO much to eat, and all the luxuries of life. Jemmy and his wife paddled away in their canoe loaded with presents, and very

happy. The most curious thing is, that Jemmy, instead of recovering his own language, has taught all his friends a little English. "J. Button's canoe" and "Jemmy's wife come," "Give me knife," etc., was said by several of them.

As can be seen in this letter, Jemmy did tell Darwin and the other Europeans the true reasons why he refused to go back to England. It was simply because "he was quite contented" with the life he had in Tierra Del Fuego with the other Fuegians. However, Darwin's *a priori* biases and *a posteriori* constructed evolutionary "facts" did not allow him to take Jemmy's *point of view* into consideration. Instead, seemingly infuriated, Darwin put in question Jemmy's explanation and wrote that "last year, in the height of his indignation, he [Jemmy] said '*his country people no sabe nothing – damned fools*' – now they were very good people, with too much to eat, and all the luxuries of life." This is clearly a Victorian man in denial mode. A man that could not accept that "others" did not dream to be like him. A man that concluded his thoughts about this "puzzle" by asserting that, after all, "instead of recovering his own language," Jemmy "has taught all his friends a little English." A sign that, after all, the "superiority" of the English culture and language proclaimed as a "fact" by Darwin was vindicated: at least in part, "civilized" good did prevail over brutish savagery (see also Box 3.4).

Box 3.4 Darwin, Fuegians, Biases, Animality, and Devildom

In *Charles Darwin: Voyaging*, Browne discusses, with her typical sharpness, the broader context concerning Darwin's encounters with, and reactions to, the Fuegians:

The shock of seeing genuine wild men as the Beagle ran down the eastern tip of the continent never thereafter left him [Darwin; he wrote] "I shall never forget how savage & wild one group was.. four or five men suddenly appeared on a cliff near to us, they were absolutely naked & with long streaming hair; springing from the ground & waving their arms around their heads, they sent forth most hideous yells.. their appearance was so strange, that it was scarcely like that of earthly inhabitants". From the very first sighting, Darwin was dazed; the absolute primitiveness sent him reeling. "They are as savage as the most curious person would desire" he announced to Fox in disbelief. As he grasped for metaphors, the only remotely appropriate pictures that came to mind were of *animality* and *devildom*: images entirely outside the human realm and indicative of his inability to adjust to what he was seeing.

Smeared with red and white paint, hair lankily hanging, their only garment a dirty guanaco skin thrown over the shoulders, gesticulating wildly in the mist, the Fuegians made him think of the devils in *Der Freischutz*: "like the troubled spirits of another world", he declared to Henslow. Fires blazing on each and every headland as they passed – whether to attract the ship's attention or to spread the news of their arrival, Darwin could not guess – added to the otherworldly effect. Most of the men they could see were running: running so fast that their noses bled and their mouths frothed, which mingling with the red and white paint made them seem "like so many demoniacs who had been fighting". He was truly in Magellan's "land of fire", he told his diary. "The sight of a naked savage in his native land is an event which can never be forgotten".

(continued)

> He thrilled to the barbaric glamour of it all, the "surrounding savage magnificence" of the country matching what he felt to be the raw brutishness of the inhabitants.. A menacing air – not unwelcome to Darwin at this stage of the voyage – hung over them as the Beagle progressed. Wary tribesmen perched on a "wild peak overhanging the sea", while men on the hills inexplicably kept up a "loud sonorous shout". Taking his turn on the night-watch, Darwin revelled in desolate aboriginal nature: "*There is something very solemn in such scenes; the consciousness rushes on the mind in how remote a corner of the globe you are then in; all tends to this end, the quiet of the night is only interrupted by the heavy breathing of the men & the cry of the night birds, the occasional distant bark of a dog reminds one that the Fuegians may be prowling, close to the tents, ready for a fatal rush*"..
>
> To Darwin, the episode was gripping in a different way. Naïvely, he recounted his surprise that Jemmy was almost another species of man compared with those who were his literal relatives: "I could not have believed how wide was the difference, between savage and civilised man". Where was the noble savage of his grandfather's [Erasmus Darwin] day or the unclothed Adam of the Bible? Expostulating to Caroline, he found it hard to acknowledge blood relationship with such wild people: "an untamed savage is I really think one of the most extraordinary spectacles in the world.. in the naked barbarian, with his body coated with paint, whose very gestures, whether they may be peacible or hostile are unintelligible, with difficulty we see a fellow creature". Thinking over Jemmy's good qualities, he found it remarkable that he was a member of the same race, and once doubtless possessing the same character as the "miserable, degraded savages whom we first met here".

There are many other examples highlighting how Darwin's racist and ethnocentric biases prevented him from truly understanding the point of view of "others" and their lifeways, culture, and even adaptations to their local environments. For instance, he did not try to understand, or asked the Fuegians, why they were often mainly naked in such a windy and often very cold region of the globe such as Patagonia. Instead, he *chose* to just repeat the typical things that biased Westerners would say about naked "savages," for instance stating how he was repulsed by the Fuegians' nakedness and how Western clothing and civilization would do them well. In *Voyage of the Beagle*, he wrote:

> While going one day on shore near Wollaston Island, we pulled alongside a canoe with six Fuegians. These were the most abject and miserable creatures I anywhere beheld. On the east coast the natives, as we have seen, have guanaco cloaks, and on the west they possess seal-skins. Amongst these central tribes the men generally have an otter-skin, or some small scrap about as large as a pocket-handkerchief, which is barely sufficient to cover their backs as low down as their loins. It is laced across the breast by strings, and according as the wind blows, it is shifted from side to side. But these Fuegians in the canoe were quite naked, and even one full-grown woman was absolutely so. It was raining heavily, and the fresh water, together with the spray, trickled down her body. In another harbour not far distant, a woman, who was suckling a recently-born child, came one day alongside the vessel, and remained there out of mere curiosity, whilst the sleet fell and thawed on her naked bosom, and on the skin of her naked baby!

> These poor wretches were stunted in their growth, their hideous faces bedaubed with white paint, their skins filthy and greasy, their hair entangled, their voices discordant, and their gestures violent. Viewing such men, one can hardly make one's self believe that they are fellow-creatures, and inhabitants of the same world. It is a common subject of conjecture what pleasure in life some of the lower animals can enjoy: how much more reasonably the same question may be asked with respect to these barbarians! At night, five or six human beings, naked and scarcely protected from the wind and rain of this tempestuous climate, sleep on the wet ground coiled up like animals. Whenever it is low water, winter or summer, night or day, they must rise to pick shell-fish from the rocks; and the women either dive to collect sea-eggs, or sit patiently in their canoes, and with a baited hair-line without any hook, jerk out little fish. If a seal is killed, or the floating carcass of a putrid whale is discovered, it is a feast; and such miserable food is assisted by a few tasteless berries and fungi.

From this excerpt, it is clear that Darwin knew that the Fuegians were perfectly capable of making clothes from either guanacos or seals (Fig. 1.13). If the Fuegians were not of our species, then Darwin the biologist would probably use his brilliant naturalist skills to try to understand why they did not use more clothing in such cold conditions and to analyze whether such a behavior could actually be adaptative to those local conditions. However, as it often happens when it concerns humans, the brilliant observational skills of Darwin the biologist were eclipsed by the ethnocentric biases of Darwin the anthropologist: he just *assumed* that this was one more detrimental behavior – as were the "nomadic habits" displayed by "savages." Is this assumption scientifically accurate? Very likely, it is not. One of the major reasons that led to the demise of the Fuegians is the typical "kindness" brought by the "superior civilized whites": guns, violence, germs, steel, and imposed Western behaviors. Various scholars argue that one of such forced behaviors that seems to have resulted in increased diseases and deaths of the Fuegians was precisely the imposition to wear Western clothes. As explained by Blanco-Wells and colleagues in a recent 2021 paper entitled "Plagues, past, and futures for the Yagan canoe people of Cape Horn, southern Chile":

> For Hyades and Deniker (1891).. the harsh climate is admirably endured, from a physiologic point of view, by the savage Fuegians with whom we lived, and we could not find any dead among them because of the phthisis [pulmonary tuberculosis or a similar wasting disease].. The correlation between settlement and disease is historically controversial. According to Martin Gusinde, who visited the Yagan at different times in the 1920s, the deaths were mainly the result of the introduction of tuberculosis and measles that took place after European and European-American colonization. The role played by smallpox, whooping cough, typhus, flu, and syphilis was minor, according to the Catholic priest. "Sad gifts given by Europeanism to the Yámanas! The families remaining in the forests did not have contact with the Europeans and kept themselves safe until they shared the sad fate of their tribe's companions". Thus, "the primitive population of about 2500 members of the tribe had descended, by late 1945, to less than 50". Citing news received from Patagonia, he claimed (in 1951), "there will be no Yámana soon" Gusinde considered European clothes to be the sources of contagion: "Those who sent clothes for the so poorly dressed savages did not know that the Indians were magnificently accustomed to their extraordinary climate and that nude they had the advantage of receiving the heat from their bonfires directly and abundantly".

> According to Penelope Dransart, who wrote about the history of clothing in the Tierra del Fuego, "at the time Gusinde did his fieldwork, people were virtually all wearing Western-style garments. Imported garments were proving to be remarkably unsuitable for their needs.. they provided a home for fleas, which were introduced by Europeans along with the clothing". Increased commercial exploitation of pinniped populations (especially fur seals) by Europeans and U.S. navigators also produced, around the same time, a crisis in the Yagan's "main source of food", forcing them "to survive with less nourishing and scarcer resources" while "infectious diseases for which they had no immunity ravaged populations" [according to Orquera and Piana]. In less than 20 years, the economic and demographic collapse induced the dispersed family groups to settle close to the missions and new sheep-farming stations. Summarizing, "disease, extermination, and the activities of gold seekers and sheep raisers.. brought about a swift decline" [as stated by Valory in 1967], in the condition of the Yagan people, forcing the survivors to adjust to their new surroundings.

I was at Tierra del Fuego a few years ago. Many of its areas are agglomerations of relatively small parcels of land surrounded by water: there is *a lot* of water. The Fuegians, who were mainly adapted to such windy, cold, and wet conditions, as well as to hunting various types of fish and other aquatic animals such as seals, probably got wet very often. So, it is indeed likely that, as suggested by various scholars, one of the major reasons for them to not wear more clothing is that by doing so they could get dry faster by "receiving the heat from their bonfires directly and abundantly." This idea is also supported by the fact that historians and priests such as Martín Gusinde – who became famous for his ethnological and anthropological works – noted that the Fuegians were seemingly *not* in decline before they met the Westerners. That is, it does seem that apart from the guns, violence, germs, and steel that the Westerners brought, the practice of enforcing the use of Western clothing did also contribute to the Fuegians' decline. Those clothes would become wet for long periods of time and moreover tend to become full of fleas, both things leading to diseases. In this sense, Darwin's ethnocentric evolutionary assumptions were likely wrong in two major ways. First, he was not able to understand the behavioral and ecological traits related to the Fuegians' common nakedness. Second, he could not perceive that under such ecological conditions, it was probably the adoption of the Westerners' "higher" habits and "superior" clothing that further contributed to the Fuegians' sickening, death, and ultimate decline. This reminds us, once again, of George Orwell's quote: "to see what is in front of one's nose needs a constant struggle." Darwin failed to engage in such a struggle to truly try to comprehend "other" peoples' lifeways and adaptations: in at least some cases, he apparently did not even try to do so. Unfortunately, until this very day, there have been many cases similar to that of Jemmy, in which following the type A cultural–epigenetic–integrationist type of racism, often mixed with type B innate racism, racist people take "others" to their "civilizations" in order to try to "tame" and "civilize" them.

There are also cases in which "others" went to "civilized" countries for other reasons that were not directly related to such racist tales, including love, as reported in a 2013 BBC News article entitled "Return to the rainforest: a son's search for his Amazonian mother." This article refers to the story of an anthropologist named Kenneth and a Yonomamo woman named Yarima: they met in the Amazonia and then got married, had children, and went to live in the United States. However, they did not live happily ever after in the "civilized" world, just like in most fairytales. Instead, Yarima could not take it anymore and went as far as to leave her own children in the United States and return, alone, to Amazonia:

> Life in New Jersey was not working out for Yarima. It wasn't the weather, food or modern technology but the absence of *close human relations*. The Yanomami day begins and ends in the shapono, open to relatives, friends, neighbours and enemies. But Yarima's day in the U.S. began and ended in a closed box, cut off from society. Other than Kenneth, no-one could communicate with Yarima in her own language and she had no means of speaking with her family back home. In Hasupuweteri, the men disappeared for a few hours in the day to go hunting, but husbands did not disappear all day, every day. Yarima would spend the day at home or roaming the shopping malls. Good also gave her video and sound recordings from Hasupuweteri that she would listen to over and over. Together with a co-writer, David Chanoff, Kenneth wrote his memoir, which was reviewed well, sold well and was translated into nine languages. He and Yarima became minor celebrities, appearing in People magazine three times.
>
> Articles appeared in newspapers with titles like *Americanization of a stone age woman* and *Two worlds: one love*. A 1992 film with National Geographic charted the family's first visit back to the jungle for almost four years. A five year-old David is seen squabbling with Vanessa over a heavy bunch of plantains, while baby Daniel is carried on Yarima's back in a sling attached to a headband, in the traditional Yanomami style. The film contains some joyful moments of Yarima showing off her children to her sister and going crab hunting again in the creeks, but it also captures her despondency. "They say I have become a nabuh," Yarima's translated voiceover tells us. "I live in a place where I do not gather wood and no-one hunts.. the women do not call me to go kill fish.. sometimes I get tired of being in the house, so I get angry with my husband.. I go to the stores and look at clothing.. it isn't like in the jungle.. people are separate and alone.. it must be that they do not like their mothers". A few months after the making of the film, on another return trip to Hasupuweteri, Yarima decided to stay.

I will end this section with a topic I briefly referred to in Chap. 1 and that was covered in the news while I was writing these lines, as it is related to the issues discussed above and demonstrates how the racist and ethnocentric legacies of the past continue to be so critical even today. A *BuzzFeed.News* article entitled "Canada is mourning 215 indigenous children after their remains were found at a residential school" explained:

> On Monday [May, 31, 2021], Perry Bellegarde, national chief of the Assembly of First Nations, said the find should be a "catalyst" for further work uncovering these graves at school sites throughout the country. "It's a prime opportunity to do this very, very important work in all the residential school grounds," he told reporters. "This has to be further researched and investigated. Kamloops is one school. There were over 130 residential schools that were operating across Canada". For more than 100 years, thousands of Indigenous children were sent to residential schools across Canada, taken from their parents in an attempt to assimilate them. These schools, many run by religious institutions, operated as recently as 1996 and were rife with abuse. The Truth and Reconciliation Commission of Canada, which was formed to examine the legacy of the schools, has identified at least 4,100 children who died of disease or by accident in these schools. It's estimated that the actual number of deaths could be more than 6,000. The U.S. government used similar boarding schools, forcing Native American children to live and learn away from their families, and systematically cutting them off from their communities and culture [
> Fig. 3.12].

Fig. 3.12 A direct legacy of the types of racist, ethnocentric, colonialist, imperialist narratives about "civilization" and "progress" that were prevalent in Western countries in the nineteenth century and that were believed, defended, and scientifically supported and widely disseminated to the broader public as "empirical facts" by Darwin. Native American boarding schools became widely common in the United States and Canada during the nineteenth and the first half of the twentieth centuries, in order to vilify and erase Native American culture and "civilize" them by integrating them into the "superior" Euro-American culture. This picture shows Native American students at the Carlisle Indian Industrial School, Pennsylvania, around 1900

> In its 2015 report, the Canadian government's Truth and Reconciliation Commission called the schools an act of "cultural genocide". The report also came up with 94 calls to action for reconciliation, including six in regard to the children who died in residential schools. Recommended actions include creating an online registry of school cemeteries and making sure descendants know where their family members were buried.. "This is the reality of the genocide that was, and is, inflicted upon us as Indigenous peoples by the colonial state. Today we honour the lives of those children, and hold prayers that they, and their families, may finally be at peace". Chief Clarence Louie of the Syilx Okanagan Nation in British Columbia said that Indigenous people still feel the impact of the schools today.

This tragic example highlights several important points discussed above. One has to do with the still mainly ethnocentric attitude of many prominent current scholars such as Steven Pinker and his ideas of an "expanding moral circle." As we have seen, the moral "progress" to which Pinker refers in his books is similar to the type of "moral" progress described by Darwin in his *Descent*, about 150 years ago. For both scholars, the pinnacle of this "progress" is the "superior" morality of Western countries. When they say that "others" are becoming more moral, or that the "moral circle" is being expanded to include them, this ultimately means that *they* are becoming more like *us* Westerners. Darwin explicitly praised Indigenous people that dressed like Westerners, was disgusted by those that did not or that were mainly naked, and applauded the "civilization" of Indigenous people by Christian missionaries. Such a "civilizing" process is exactly the type of thing that was supposedly being done at those Canadian and U.S. "residential schools" – a euphemism for what were in reality mainly centers of cultural genocide (see also Box 3.5). As Darwin explicitly defended in his scientific works, either the brutish "savages" would become "civilized" or they would become "naturally extinct" – another euphemism because this often included being massacred by Westerners, as recognized by him.

Such examples of cultural or physical genocide are directly connected, at a theoretical level at least, with the evolutionary "facts" stated by Darwin and the racist and ethnocentric ideas of current scholars such as Pinker and his notion of an "expanding moral circle." This is because, if there is an expanding moral circle, then those that are not part of it are basically amoral or morally subhuman. Even those of us that criticize such ethnocentric and racist narratives tend to fall into their trap in our daily lives, due to the things we have learned and the way we were and continue to be indoctrinated. For instance, we often say things such as: since the rise of agriculture, humans have tended to use more milk in their diets. However, this is not only ethnocentric but also factually wrong. A more correct way of saying this, historically and factually, would be: within *those human groups* that formed agricultural societies, independently within several regions of the globe, people have tended to use more milk in their diets. This is because many societies never became agricultural, and many of those that did were forced to do so either by force or through indoctrination. Moreover, there are several known cases of societies that adopted an agricultural lifestyle, either voluntarily or involuntarily, and that later reverted to a nonagricultural lifestyle. For instance, according to a 2005 paper by Oota and colleagues, this seems to have been the case with the Mlabri, a group of people who nowadays range across the Nan, Phrae, and Phayao provinces of north and northeastern Thailand and the Sayaburi province of western Laos.

By saying things such as "since the rise of agriculture," or of "industrialization," humanity did this or that, we are excluding all nonagricultural or nonindustrialized groups from "humanity." Coincidently, when I was writing part of this section, an interesting paper emphasizing these very same points, entitled "People of the book: empire and social science in the Islamic Commonwealth period," was published by Musa al-Gharbi. This paper shows how important it is to have an insight of the "others" and to take in their points of views about these topics – in this case, of a Muslim scholar – so that one can see how others perceive our Western scientific biases and narratives. Importantly, this paper highlights how the ideas now defended by authors such as Pinker are indeed based on the teleological evolutionary ideas of authors such as Darwin – it refers, for instance, to people being "unfit for survival" and to the idea defended in the *Descent* that those that are not "civilized" are "naturally" doomed to cease existing. It also emphasizes how such ideas are therefore also related, tragically, to "imperialism, colonialism, eugenics, and genocide":

> Modernity came to be viewed as a universal process which, while birthed in Europe, would soon swallow all other societies and cultures as well. Every religion would undergo some version of a Reformation and Enlightenment (or be lost to history). Every society would become secular, industrial, urban and capitalist. While this process would generally be as

> brutal, destructive and disorienting for others as it was for Europe, these birthing pangs were predicted to eventually give way to an age of unprecedented worldwide peace and prosperity. Looking to other historical and cultural contexts increasingly became a means through which people in the West demonstrated their superiority over others and justified their eschatology of progress. 'The West' was thereby placed, in a sense, beyond good and evil. There could be no blame in destroying communities and cultures which were already destined for extinction. While perhaps tragic to observe, it was ultimately a kindness to efficiently eliminate socio-cultural phenotypes which were unfit for survival in the 'modern' world, or to accelerate mankind's evolution by reconstructing the exotic and primitive 'other' in the image of the 'modern' West. Operating under such auspices, [19th century Western] social science would go on to play a critical role in justifying imperialism, colonialism, eugenics and genocide.
>
> Many have argued that the colonial and imperial origins of [19th century] Western social science continue to structure its theory and practice today. Territoriality around scholarly disciplines is held to reflect an imperial mindset. Researchers continue to value forms of knowledge useful for administration, surveillance or dominance of subaltern populations – with 'social problems' defined largely in terms of these populations. 'Insider' and 'outsider' groups continue to be studied and discussed in asymmetrical ways – with the sociological 'lens' turned primarily towards the latter. Social research continues to be largely extractive: data is taken from populations to be utilized and analyzed by outsiders. The preferences, priorities and perspectives of those from whom information is collected are often not taken seriously, or in any case, tend to be subordinated to the needs and desires of researchers. Aspirations towards empiricism, objectivity, rationality, universality and positivism are held to privilege values and worldviews prevalent among those *who dominate the social order*. Meanwhile, work by and for people from historically marginalized and disadvantaged groups tends to be viewed as less rigorous or significant. These and other unsavory features are held to be consequences of the imperial origins of Western social science.

In a nutshell, in his paper, al-Gharbi summarizes the circular reasoning, self-reinforcing narratives, and links between ethnocentrism, politics, and societal and scientific biases, not only within social sciences but also within sciences as a whole because, as we have seen in this book, this clearly has also applied, and continues to apply, to natural sciences as well as to other fields of science and knowledge and, therefore ultimately to popular culture.

Not Everybody Was Necessarily "Like That" Back Then: Wallace

> *I could show fight on natural selection having done and doing more for the progress of civilization than you seem inclined to admit. Remember what risk the nations of Europe ran, not so many centuries ago of being overwhelmed by the Turks, and how ridiculous such an idea now is! The more civilised so-called Caucasian races have beaten the Turkish hollow in the struggle for existence. Looking to the world at no very distant date, what an endless number of the lower races will have been eliminated by the higher civilized races throughout the world.* (Charles Darwin)

In *Readings in the History of Evolutionary Theory*, Wetherington noted that "the idea that progress is a natural condition of the social order did not await Darwin for its expression.. it was present at the Enlightenment.. (Darwin's) natural selection simply gave it a sense of scientific authenticity." Even in *Darwin Deleted*, which, as noted above, mainly defends narratives often used to idolize Darwin, Bowler did recognize that "evolutionism.. offered a plausible (scientific) explanation of why some races might not have advanced as far as others up the scale leading from the ancestral ape.. most of the Darwinians endorsed this way of thinking.. with the notable exception of Wallace." This brings us full circle to the discussion of Chap. 1 about how the writings of naturalists such as Wallace and Humboldt show the fallacies of the "everybody was like that back then" typical narrative used to defend the racist and sexist "facts" constructed in Darwin's writings. Nobody is arguing here that such naturalists did not defend some racist or sexist ideas in private and even in certain books and specialized papers.

For instance, regarding Humboldt, Bethencourt wrote in his books *Racisms*:

> Alexander von Humboldt (1769–1859).. one of the major scientific authorities of the first half of the nineteenth century.. considered language as intimately associated with the affinity of races, implying the similarity of linguistic structures. He advocated the unity of humankind, citing the many intermediate gradations in skin color and skull forms – gradations that would make it impossible to establish a clear distinction between races.. According to Humboldt, comparisons of the black populations of Africa, South India, and the west Australian archipelagos had shown no connection between skin color, woolly hair, and cast of countenance. Humboldt reasserted Buffon's old assertion concerning fertility and reproduction: different races were variations within a single species, not different species of a genus, since in the latter case hybrid descendants would remain sterile. Humboldt went further than previous authors, denouncing the lack of clear definition in the noun races and instead proposing the use of the expression varieties of human beings. He mentioned the five races identified by Blumenbach.. but stated that "we fail to recognise any typical sharpness of definition, or any general or well-established principle, in the division of these groups." Humboldt also observed that several groups could not be included in any category, and that geographic areas could not serve as points of departure for races in any precise way, since several regions had been inhabited at different periods by different groups. In his view, to search for the "cradle of the human race" was to pursue a myth. Finally, Humboldt explicitly refused "the depressing assumption of superior and inferior races of men," and the unhappy Aristotelian doctrine of slavery as an institution condoned by a nature that bestowed unequal rights to freedom on human beings. Humboldt considered that all nations were destined for freedom and (quoting his brother) denounced the erection of barriers among humans to prevent natural perfectibility – the result of prejudice.

That is, in general, Humboldt had not only a much more positive reaction to and was more emphatic toward "others" than Darwin was but also explicitly criticized a crucial idea that was defended in Darwin's books *Descent* and *Expression* as a scientific "fact": that "whites" or Europeans were superior to other groups.

Regarding Wallace, as explained in Chap. 1 and in books such as Brantlinger's *Dark Vanishings: Discourse on the Extinction of Primitive Races*, it is a historical fact that in general his writings about his encounters with non-Europeans had a very different tone when compared to those of Darwin. Brantlinger astutely compares the remarkable differences concerning what can be described, in a simplified way obviously, as Darwin's more Hobbesian views versus Wallace's more Rousseauistic views about "savages" and "civilization." Previously, we mainly referred to

Wallace's encounters with Native Americans. So, now, let us explore in a deeper way this issue and see what Brantlinger had to say about Wallace's encounters with and views about Polynesians and how this contrasted with Darwin's encounters with and ideas about "others":

> The same elegiac message is evident in Wallace's 1869 [volume] *Malay Archipelago* [as in his accounts of his South American travels].. If the tide of colonization should be turned to New Guinea, there can be little doubt of the early extinction of the Papuan race. A warlike and energetic people, who will not submit to national slavery or to domestic servitude, must disappear before the white man as surely as do the wolf and the tiger. Despite his romantic inclinations, in 1864 Wallace declared, quoting the subtitle of *Origin of Species*, that the "great law of 'the preservation of favoured races in the struggle for life'.. leads to the inevitable extinction of all those low and mentally undeveloped populations with which Europeans come in contact" (*Origin of Human Races*). Darwin had his "Malthusian moment", and so did Wallace. Both were led to their insights about the mechanism of evolution by Malthus's account of savage customs and the "struggle for survival" among human populations.
>
> As James Moore puts it: "Reading Malthus, [Wallace] grasped that living nature was in effect the workhouse world writ large.. ruthless struggle was everywhere the law, not just among London's starving poor.. adaptation comes through competition.. progress costs lives". Nevertheless, Wallace was more humane or at least more conflicted in his political outlook than either Darwin or Huxley. Despite or perhaps because of his application of the "great law" of survival of the fittest to savages and barbarians, [Crook noted that] Wallace became both a socialist and "a spirited critic of modern war and imperialism". Thus Wallace could write: "*naturalists need not be bound by the same rule as politicians, and may be permitted to recognize the just claims of the more ancient inhabitants, and to raise up fallen nationalities.. the aborigines and not the invaders must be looked upon as the rightful owners of the soil*".

Brantlinger further notes that:

> In contrast to most Social Darwinists, who agreed with Huxley, Spencer, and Francis Galton in opposing state interference in the operations of the capitalist marketplace, Wallace added so-called free trade to the causes of the extinction of primitive races. Wallace reverses Huxley's identification of colonies with gardens and natives with weeds. This reversal expresses the division in Wallace's thinking between romantic primitivism and evolutionary theory, and between his anti-imperialist and anti-capitalist political inclinations and social Darwinism. Throughout his writings, Wallace attempted to combine these positions, just as he also sought to reconcile science and spiritualism. In the conclusion to *The Malay Archipelago*, Wallace contends that, if the progress of civilization is at all a meaningful concept, then that progress must be leading humanity toward an "ideally perfect social state". In such a state, every individual would be able to maximize his or her freedom, "intellectual organization", and morality, and to do so only by following "the free impulses of [human] nature". Sounding like the Lévi-Strauss of *Tristes Tropiques* or, for that matter, like Rousseau, Wallace continues: "*Now it is very remarkable that among people in a very low stage of civilization we find some approach to such a perfect social state.. I have lived with communities of savages in South America and in the East, who have no laws or law courts but the public opinion of the village freely expressed.. each man scrupulously respects the rights of his fellow, and any infraction of those rights rarely or never takes place.. In such a community, all are nearly equal. There are none of those wide distinctions, of education and ignorance, wealth and poverty, master and servant, which are the product of our civilization*".
>
> There is not even, Wallace continues, "that severe competition and struggle for existence, or for wealth, which the dense population of civilized countries inevitably creates".

> [In contrast] Wallace's *Wonderful Century* (1898) is mainly a paean to progress on many fronts of 19th century European civilization, though he also considers its "failures". Among these are "the plunder of the earth", by which Wallace means both imperialism and capitalism. Here the operation of civilization on the non-civilized parts of the world is not merely the neutral action of natural selection or survival of the fittest but the *immoral action of power and greed*. In the chapter on "plunder of the earth", Wallace condemns the scramble for Africa that began in earnest after the Berlin Conference of 1884: "*The result, so far, has been the sale of vast quantities of rum and gunpowder; much bloodshed, owing to the objection of the natives to the seizure of their lands and their cattle; great demoralization both of black and white; and the both of black and white; and the condemnation of the conquered tribes to a modified form of slavery*". So, too, in India and Ireland the results of British domination have been, says Wallace, "*rebellion, recurrent famines, and plague in India; discontent, chronic want and misery; famines more or less severe, and continuous depopulation in our sister-island at home – these must surely be reckoned among the most terrible and most disastrous failures of the 19th Century*". What is more, "if the Spaniards exterminated the natives of the West Indies, we have done the same thing in Tasmania, and almost the same in temperate Australia". Both his socialist sympathies for the working class at home and his lengthy sojourns among "savage" peoples in South America and Indonesia helped Wallace approximate "the perceptive modes of modern anthropology", including its cultural relativism.

Once again, we see a marked contrast between Wallace's – even when, later in life, he came to praise the notion of "progress" more – and Darwin's ideas. In fact, an important point that needs to be made is that Darwin defended biased ideas about "races" and women in a rather constant way throughout his whole life, that is, from his very first "repulsive" encounters with non-Europeans when he was still young until the racist and sexist statements that he made in the *Descent* and even in major English newspapers such as *The Times*, just a few years before he died, as we will see in Chap. 4. Such constancy contrasts, for example, with the case of Thomas Huxley, who seems to have been more empathetic to "others" during his youth trips around the globe and then began to defend more conservative ethnocentric ideas later in life. In the case of Darwin, it is difficult to argue that there was a moment in his life – including in his youth – in which his ideas were truly idealistic or at least markedly less conservative, ethnocentric, or misogynistic. As a matter of fact, as we have just seen, even the old Wallace, who praised the concept of "progress" a bit more in an 1898 book, was markedly less intellectually conservative that the young Darwin that published, 59 years before, the *Voyage of the Beagle*. The old Wallace said things that Darwin was never able to recognize. This includes not only the recognition of how harmful the English empire truly was, in places such as Australia – a region that Darwin had explicitly stated, as a "fact" in his scientific books, to have benefited a lot from English imperialism – but also various other points concerning a notion that was at the very center of Darwin's books and that he accordingly exaggerated, as we have seen above: the notion of struggle for existence.

Wallace was capable of recognizing, in 1898, what many anthropologists, historians, and other scholars have only started to recognize in recent decades: that what Malthus and Darwin called the "struggle for existence" in human societies mainly applies – although there are exceptions, as always – to a "dense population of civilized countries," that is, mostly to sedentary and in particular agricultural societies.

As Brantlinger perceptively recognized, Wallace's ideas about human evolution and history were in a way very similar to "the perceptive modes of modern anthropology" – something that cannot be said about Darwin's sexist and ethnocentric views. It is not a mere historical coincidence that Western anthropologists and scholars in general did not embrace Wallace's more "modern" and less racist and sexist "perceptive modes" for a long time and instead commonly embraced Darwin's views. As explained above, this was in great part precisely *because* Wallace's ideas were against the narratives that most Western scholars wanted so much to *believe* in: for them, Darwin's writings were music to their ears.

Having said this, Wallace should not be idealized, or be used to defend a biased, simplified narrative about the "good" Wallace *versus* the "bad" Darwin. Idealizations and biases are just wrong. For instance, we know that, despite being in general much more empathic toward "others" than Darwin was, Wallace did *decide* to make some rather racist public statements later in life, as explained in Slotten's *The Heretic in Darwin's Court*:

> On March 1, 1864, Wallace read his essay '*The Origin of Human Races and the Antiquity of Man Deduced from the Theory of Natural Selection*'. The subject matter of his essay was undoubtedly calculated to raise the hackles of the more conservative members of the Anthropological Society.. [He argued that] while other animals evolved better physical characteristics for pursuing prey or eluding capture, developing more versatile organs to cope with the vicissitudes of climate, man, "by the mere capacity of clothing himself, and making weapons and tools", took away from nature "that power of changing external form and structure which she exercises over all other animals". From that point on, the human mind was subjected to the same influences from which the human body had escaped. Every slight variation in mental and moral nature that guarded against adversity and enhanced mutual comfort and protection was preserved and therefore accumulated. The better and higher specimens of humanity increased and spread, while the lower and more brutal dwindled and eventually died out. Even the very lowest races of man were elevated far above the "brutes".
>
> The culmination of this evolution was the "wonderful intellect of the Germanic races". "And is it not the fact that in all ages, and in every quarter of the globe, the inhabitants of temperate have been superior to those of tropical countries?" Wallace asked. "All the great invasions and displacements of races have been from North to South". Quoting Darwin, Wallace said that it was the same great law of "preservation of favoured races in the struggle for life" that led to the inevitable extinction of weaker and inferior peoples. Just as the weeds of Europe overran North America and Australia, extinguishing the less adaptable native varieties, so the morally and intellectually superior Europeans had overrun, outbred, and extinguished the less adaptable aboriginal peoples. Foreshadowing social Darwinism, Wallace united European imperialism with the law of natural selection in his essay on the human species.

However, as noted by Slotten, even those cases in which Wallace made such racist proclamations basically concerned type A/cultural racism, and not the type B innate racism that Darwin so often flirted with in his writings and that later became so prevalent among eugenicists such as Darwin's own son, Leonard Darwin. For instance, as explained by Slotten, just 2 weeks after Wallace made the statements quoted in the above excerpt, he answered in the following way to the criticisms made about his paper by James Hunt, who defended that human "races" were innately different:

> Hunt was.. angered by what he perceived to be Wallace's defamation of the English national character. Based on the observations of the Romans, at one point Wallace claimed that the indigenous Britons had been savages when Julius Caesar conquered the British Isles. This was not a fact, Hunt said, but a tradition, based on the "barest" of historical evidence.. Wallace again clashed with Hunt at the next meeting, on March 15, when Wallace objected to a paper entitled "*Notes on the Capabilities of the Negro for Civilisation*". Although he agreed that the intellect of the black African was currently inferior to that of the white European, the black African was not the lowest grade of humanity. Blacks were energetic and intelligent, he believed, and with encouragement from a superior civilization they would rise higher than had ever been seen in the past. It was unfair to compare a group of people only recently freed from slavery with the Hindu or Chinese people, who belonged to the oldest civilizations on earth. Blacks were not lazier than anyone else – all people were lazy and required prodding to work. The European had never seen the black man under favorable circumstances, as he had in the Papuan regions of the Malay Archipelago, where he lived and worked with people who he believed were closely related to the tribal peoples of Africa. Hunt expressed outrage at Wallace's suggestion that with the help of Europeans blacks might rise to their level. There was an unbridgeable world of difference between the two races, Hunt said, and "it is the duty of anthropologists to oppose the opinion attempted to be established of the equality of the Negro and White Man".
>
> [In stark contrast, Wallace] questioned the European notion of progress, which to him was more like retrogression. It was true that civilized people had surpassed savages in intellectual and material achievements, he said, but they had not advanced equally in the moral sphere. Without according its less fortunate citizens a greater share of influence in legislation, commerce, and social organization, European civilization would never attain any real superiority even over the "better class" of savage. In Wallace's opinion, the European system of government remained in a state of barbarism.

Eugenics, Leonard Darwin, and Tragic Societal Legacies of Science

> *Of all the problems which will have to be faced in the future, in my opinion, the most difficult will be those concerning the treatment of the inferior races of mankind.. my firm conviction is that if wide-spread Eugenic reforms are not adopted during the next hundred years or so, our Western Civilization is inevitably destined to such a slow and gradual decay as that which has been experienced in the past by every great ancient civilization.. the size and the importance of the United States throws on you a special responsibility in your endeavours to safeguard the future of our race.* (Leonard Darwin)

In the last 150 years, particularly since the rise of eugenics, there have been several "us" *versus* "them" atrocities, cultural genocides, and mass killings (Figs. 1.15, 2.9, 3.1, 3.8, and 3.9). Tragically, and very sadly for a scientist as myself, there is a *considerable amount of evidence that shows*, beyond any doubt, that some of the evolutionary racist and ethnocentric "facts" disseminated in Darwin's books, and of the simplistic and often erroneous metaphors that he used to disseminate his evolutionary ideas – such as those about a *struggle for existence* and the notion of *favored races* – were *directly and actively used by ideologists that promoted or justified at least some of those atrocities*. Among the many examples that could be provided here, I will start with one that concerns one of the most emblematic and influential white supremacist groups in the United States, the Ku Klux Klan (KKK). This group formed 6 years after the publication of Darwin's 1859 *On the Origin of Species* and became progressively more prominent and influential, having committed numerous "us" *versus* "them" atrocities across the United States. David Duke, who as noted above was a former "Grand Wizard" of the KKK, explicitly wrote, in his 1998 book *My Awakening: A Path to Racial Understanding*:

> Charles Darwin.. demonstrated that principles of heredity combined with what he called, *Natural Selection*, had developed the exceptional abilities of mankind itself.. his masterpiece, *Origin of species* has a subtitle that expresses his whole idea in a nutshell: *The Preservation of Favoured Races in the Struggle for Life* – preserving the Caucasian race is but a precondition for continuing its evolution to a higher level.

I can understand that scholars can engage in never-ending discussions about whether Darwin was or not careless, or very careless, or just a bit careless, or not careless at all when he wrote such simplistic catchy metaphors and racist and ethnocentric evolutionary "facts" in his works. But it is odd that so many scholars continue to deny that such metaphors and 'facts' did provide easy ammunition for white supremacists such as Duke and for eugenicists, or that they did not use Darwin's works to promote or justify their racist and eugenic ideas, as argued for instance by Bowler in his *Darwin Deleted*. There is no doubt that Darwin's writings provided a unique opportunity for people that had such ideologies: they could cite his name and works in order to 'scientifically' legitimize them. We know that this was so because ideologists such as Duke and many other white supremacists and eugenicists *did* use Darwin's name and writings to justify what they promoted or did. We cannot deny this tragic reality or bury our heads into the sand and pretend that this never happen. It did, and it continues to happen, as can be easily attested by an analysis of the "scientific" ideas that are commonly used on the websites of many current white supremacist groups in the United States or Europe to justify their ideologies.

Are such groups wrong in doing so? Are they misusing or distorting Darwin's ideas? I would say that their *practices* are very likely going against Darwin's *intentions*, in the sense that I do not think – but, again, this cannot be tested – that Darwin would approve the lynching of African-Americans, or other similar atrocities; that was not his style at all, mainly because he was a pacifist and opposed violence. However, regarding Darwin's *ideas* per se – not his intentions or practices – one needs to admit, without taboos or idealizations, that the racist and ethnocentric ideas disseminated in Darwin's writings, in particular his *Descent*, are in fact in many ways very similar to the ideas of groups such as the KKK concerning human "races," their struggle for existence, and the "superiority" and "higher morality" of European descendants. The saying goes: do as I say, not as I do. Fortunately, what Darwin said was much worse than what he did and seemingly than what he indented to promote, but, unfortunately, what he said was used over and over by ideologists to defend or justify what he did not do or intended to do. It is indeed very important, and disconcerting, to note that Duke precisely referred to the part of the title of Darwin's 1859 book that most scholars omit today – "the preservation of favored races in the struggle for life." Darwin himself deleted this subtitle in later versions of the *Origin*, perhaps because he recognized how wrong and dangerous it was to have used it in the first place. As noted above, in contrast to scholarly books that mainly focus on Darwin, which tend to idealize him or at least to omit such historical facts, academic books that deal with broader topics such as the history of racism or eugenics are more likely to refer to these darker historical repercussions of Darwin's writings. For instance, in *Stamped from the Beginning*, Kendi has explicitly discussed such historical facts:

> [In the 1860s] it looked as if polygenesis had finally become mainstream. In actuality, the days of the notion of separately created human species were numbered. Another pernicious theory of the human species was about to take hold, one that would be used by racist apologists for the next one hundred years. In August 1860, polygenesist Josiah C. Nott took some time away from raising Alabama's first medical school (now in Birmingham). He skimmed through a five-hundred-page tome published the previous November in England. It had a long title, *On the origin of species by means of natural selection, or the preservation of favoured races in the struggle for life*. Nott probably knew the author: the eminent, antislavery British.. biologist Charles Darwin. "The view which most naturalists entertain, and which I formerly entertained – namely, that each species has been independently created – is erroneous", Darwin famously declared. "I am fully convinced that species are not immutable". Recent discoveries were showing, he explained, that humans had originated much earlier than a few thousand years ago.
>
> Darwin effectively declared war on biblical chronology and the ruling conception of polygenesis, offering a new ruling idea: natural selection. In the "recurring struggle for existence", he wrote, "all corporeal and mental endowments will tend to progress towards perfection". Darwin did not explicitly claim that the white race had been naturally selected to evolve toward perfection. He hardly spent any writing time on humans in *The Origin of Species*. He had a grander purpose: proving that all living things the world over were struggling, evolving, spreading, and facing extinction or perfection. Darwin did, however, open the door for bigots to use his theory by referring to "civilized" states, the "savage races of man", and "half-civilized man", and calling the natives of southern Africa and their descendants "the lowest savages". Over the course of the 1860s, the Western reception of Darwin transformed from opposition to skepticism to approval to hailing praise. The sensitive, private, and sickly Darwin let his many friends develop his ideas and engage his critics. The mind of English polymath Herbert Spencer became the ultimate womb for Darwin's ideas, his writings the amplifier of what came to be known as Social Darwinism. In *Principles of Biology* in 1864, Spencer.. [used] the iconic phrase "survival of the fittest". He religiously believed that human behavior was inherited. Superior hereditary traits made the "dominant races" better fit to survive than the "inferior races". Spencer spent the rest of his life calling for governments to get out of the way of the struggle for existence. In his quest to limit government, Spencer ignored the discriminators, probably knowing they were rigging the struggle for existence. Longing for *ideas to justify the nation's growing inequities*, American elites firmly embraced Charles Darwin and fell head over heels for Herbert Spencer.
>
> Charles Darwin's scholarly circle grew immeasurably over the 1860s, encircling the entire Western world. The *Origin of Species* even changed the life of Darwin's cousin, Sir Francis Galton. The father of modern statistics, Galton created the concepts of correlation and regression toward the mean and blazed the trail for the use of questionnaires and surveys to collect data. In *Hereditary Genius* (1869), he used his data to popularize the myth that parents passed on hereditary traits like intelligence that environment could not alter. "The average intellectual standard of the negro race is some two grades below our own",

Galton wrote. He coined the phrase "nature *versus* nurture", claiming that nature was undefeated. Galton urged governments to rid the world of all naturally unselected peoples, or at least stop them from reproducing, a social policy he called "eugenics" in 1883. Darwin did not stop his adherents from applying the principles of natural selection to humans. However, the largely unknown codiscoverer of natural selection did. By 1869, British naturalist Alfred Russel Wallace professed that human spirituality and the equal capacity of healthy brains took humans outside of natural selection. Then again, as Wallace made a name for himself as the most egalitarian English scientist of his generation, he still professed European culture to be superior to any other.

Darwin attempted to prove once and for all that natural selection applied to humans in *Descent of Man*, released in 1871. In the book, he was all over the place as he related race and intelligence. He spoke about the "mental similarity between the most distinct races of man", and then claimed that "the American aborigines, Negroes and Europeans differ as much from each other in mind as any three races that can be named". He noted that he was "incessantly struck" by some South Americans and "a full-blood negro" acquaintance who impressed him with "how similar their minds were to ours". On racial evolution, he said that the "civilized races" had "extended, and are now everywhere extending, their range, so as to take the place of the lower races". A future evolutionary break would occur between "civilized" Whites and "some ape" – unlike like the present break "between the negro or Australian and the gorilla". Both assimilationists and segregationists hailed *Descent of Man*. Assimilationists read Darwin as saying Blacks could one day evolve into White civilization; segregationists read him as saying Blacks were bound for extinction.

In fact, the Eurocentric evolutionary ideas of Darwin, and of many of his Darwinian and neo-Darwinian followers, became so prominent that even Bowler recognizes, in *Darwin Deleted*:

I must concede that Darwinism did become involved with the culture of imperialism, providing a source of extremely effective rhetoric.. the imperialists certainly used Darwinian terminology.. and in a few cases they were (even) genuine scientific Darwinists.. In the absence (as yet) of fossil hominids, modern savages were treated as equivalent of these primitive ancestors, and the physical anthropologists' alleged evidence of small brains and apelike features in the 'lowest' races was called in to confirm the link.. Darwin certainly contributed to this process in his *Descent of man*, and in Germany Ernst Haeckel built the idea that the human race show different levels of development firmly into his Darwinism.

As explained in Chap. 1, Haeckel is often associated with Nazi ideology, so the direct links between him and Darwin and between the ideas disseminated in their writings often make those scientists and historians that idealize Darwin particularly uncomfortable. Two typical reactions are often seen among such scientists and historians. One, less common, as it goes against a huge plethora of historical facts, is to try to distance Haeckel from Nazi ideology. We will discuss one of such cases in some detail in the next sections. Another more common reaction is to distance Haeckel's ideas from Darwin's ones, or vice versa, despite the fact that both these options also go against unequivocal historical evidence. For instance, typical narratives that are often used in such cases are that Haeckel misused Darwin's ideas, that Darwin did not support Haeckel's ideas, or that Haeckel's ethnocentric and racist ideas had nothing to do with Darwin's ones. Of course, such narratives are themselves contradictory. First of all, if Haeckel followed and used Darwin's ideas to justify his racist and ethnocentric ideas, it was because he felt that the latter were very similar to those defended by Darwin. Moreover, as we have seen, this was actually recognized by Darwin himself, in a letter about his 1871 *Descent* book, in which he explicitly wrote that "this last naturalist [Haeckel].. has recently.. published his '*Naturliche Schopfungs-geschichte*,' in which he fully discusses the genealogy of man.. if this work had appeared before my [*Descent*] essay had been written, I should probably never have completed it." This is because "almost all the conclusions at which I have arrived, I find confirmed by this naturalist, whose knowledge on many points is much fuller than mine."

So, unless Darwin's idolizers build a narrative that their idol was lying in this letter, they would need to recognize this historical fact. Many of them instead opt for recurrently using a third option: they just omit what their idol said, in those cases in which what he said does not fit within their idolized construction of him. This means that they often have to omit a *lot* of things that Darwin wrote, not only in his personal letters but also in his books, including one of the books that they tend to idealize the most, the *Descent*. This is because, apart from the numerous racist, sexist, and ethnocentric excerpts of the *Descent* that we have already mentioned throughout this book, there are many others. As we obviously have no space to include all of them in this book, I will just refer to some of the most extreme or relevant ones, particularly those that are almost never mentioned in most publications about Darwin, to make this point. For instance, in the following excerpt of the *Descent*, Darwin provides disturbing scientifically inaccurate "facts" about "others", other than women and non-Westerners, such as people with congenital malformations and disabilities:

> The strong tendency in our nearest allies, the monkeys, in microcephalous idiots, and in the barbarous races of mankind, to imitate whatever they hear deserves notice.. Judging from the hideous ornaments and the equally hideous music admired by most savages, it might be urged that their aesthetic faculty was not so highly developed as in certain animals, for instance, in birds. Obviously no animal would be capable of admiring such scenes as the heavens at night, a beautiful landscape, or refined music; but such high tastes, depending as they do on culture and complex associations, are not enjoyed by barbarians or by uneducated persons.. Sympathy beyond the confines of man, that is humanity to the lower animals, seems to be one of the latest moral acquisitions. It is apparently unfelt by savages, except towards their pets. How little the old Romans knew of it is shewn by their abhorrent gladiatorial exhibitions. The very idea of humanity, as far as I could observe, was new to most of the Gauchos of the Pampas.. It seems at first sight a monstrous supposition that the jet-blackness of the negro should have been gained through sexual selection; but this view is supported by various analogies, and we know that negroes admire their own colour. With mammals, when the sexes differ in colour, the male is often black or much darker than the female; and it depends merely on the form of inheritance whether this or any other tint is transmitted to both sexes or to one alone. The resemblance to a negro in miniature of *Pithecia satanas* [a dark New-World monkey] with his jet black skin, white rolling eyeballs, and hair parted on the top of the head, is almost ludicrous.
>
> Human song is generally admitted to be the basis or origin of instrumental music. As neither the enjoyment nor the capacity of producing musical notes are faculties of the least use to man in reference to his daily habits of life, they must be ranked amongst the most mysterious with which he is endowed. They are present, though in a very rude condition, in men of all races, even the most savage; but so different is the taste of the several races, that our music gives no pleasure to savages, and their music is to us in most cases hideous and unmeaning.. We see that the musical faculties, which are not wholly deficient in any race, are capable of prompt and high development, for Hottentots and Negroes have become

> excellent musicians, although in their native countries they rarely practise anything that we should consider music. Schweinfurth, however, was pleased with some of the simple melodies which he heard in the interior of Africa. But there is nothing anomalous in the musical faculties lying dormant in man: some species of birds which never naturally sing, can without much difficulty be taught to do so.

When one reads such excerpts, it seems difficult to argue that they were just the result of Darwin's unconscious biases. For instance, for someone that traveled around the planet and encountered so many groups of Indigenous peoples, it seems unlikely that he was being honest when he wrote that Indigenous people in their "native countries" rarely "practice anything that we should consider music." The vast majority of Indigenous peoples, everywhere in the globe, obviously practice something that almost every European descendant – including even extremist white supremacists such as KKK's David Duke – would consider *to be music*. Some Indigenous groups from the Congo basin, such as the Baka, perform polyphonic songs that have almost no parallel, in terms of their complexity, anywhere else in the globe. It seems very implausible that Darwin never heard Indigenous people playing music during his Beagle voyage. Or that he would hear them doing so and would truly consider that this was *not* "what we should consider music." There is ample empirical evidence that people such as the Fuegians, with whom he interacted so closely, did perform music, as described, for instance, in Erich von Hornbostel's work *The Music of the Fuegians*. So, either Darwin was not being honest, in the above excerpt, or, if he was, his racist and ethnocentric biases were so extreme that Darwin the anthropologist was unable to both see what was truly before his eyes *and hear* the sounds around him.

As wisely noted in Landau's *Narratives on Human Evolution*, one particularly striking aspect that is too often neglected in the literature is how "the real struggle in [Darwin's] *Descent of man* occurs not between animals and men but between humans of varying intellects." Darwin explicitly wrote: "for my own part I would as soon be descended from that heroic little monkey.. or from that old baboon.. – as from a savage who delights to torture his enemies, offers up bloody sacrifices, practices infanticide without remorse, treats his wives like slaves, knows no decency, and is haunted by the grossest superstitions." As Landau points out, "next to savages who are cruel and false, the European appears kindest and most faithful" in Darwin's writings. Furthermore, as she explains, the problem is *not only* that Darwin had indirect links, and *direct* familiar connections, with the eugenics movement – including his cousin and his own son, as we will see below – but also that Darwin's own writings resemble in a disturbing way those used by authors within that movement. For example, in *The Descent* he wrote:

> With savages, the weak in body or mind are soon eliminated; and those that survive commonly exhibit a vigorous state of health. We civilised men, on the other hand, do our utmost to check the process of elimination; we build asylums for the imbecile, the maimed, and the sick; we institute poor-laws; and our medical men exert their utmost skill to save the life of every one to the last moment. There is reason to believe that vaccination has preserved thousands, who from a weak constitution would formerly have succumbed to small-pox. Thus the weak members of civilised societies propagate their kind. No one who has attended to the breeding of domestic animals will doubt that this must be highly injurious to the race

> of man. It is surprising how soon a want of care, or care wrongly directed, leads to the degeneration of a domestic race; but excepting in the case of man himself, hardly any one is so ignorant as to allow his worst animals to breed. The aid which we feel impelled to give to the helpless is mainly an incidental result of the instinct of sympathy, which was originally acquired as part of the social instincts, but subsequently rendered, in the manner previously indicated, more tender and more widely diffused.
>
> Nor could we check our sympathy, if so urged by hard reason, without deterioration in the noblest part of our nature. The surgeon may harden himself whilst performing an operation, for he knows that he is acting for the good of his patient; but if we were intentionally to neglect the weak and helpless, it could only be for a contingent benefit, with a certain and great present evil. Hence we must bear without complaining the undoubtedly bad effects of the weak surviving and propagating their kind; but there appears to be at least one check in steady action, namely the weaker and inferior members of society not marrying so freely as the sound; and this check might be indefinitely increased, though this is more to be hoped for than expected, by the weak in body or mind refraining from marriage. In all civilised countries man accumulates property and bequeaths it to his children. So that the children in the same country do not by any means start fair in the race for success. But this is far from an unmixed evil; for without the accumulation of capital the arts could not progress; and it is chiefly through their power that the civilised races have extended, and are now everywhere extending, their range, so as to take the place of the lower races.

As a few scholars have dared to state in their works, there are so many layers of racism, ethnocentrism, hierarchism, eugenic ideas, and scientific inaccuracies within this single passage of the *Descent* that, if someone would say that this was an excerpt of let us say Hitler's *Mein Kampf*, this could sound plausible. However, tragically, this passage was published in a book that is currently still considered by countless scholars, including both biologists and anthropologists, as a "masterpiece" of the most renowned, and idolized, biologist of all times. This is what is truly concerning: not only that Darwin published these things as "facts" in a scientific book about human evolution but also that so many people continue to idolize Darwin and that book, which more than 150 years later continues to be openly celebrated by so many Western historians and scientists as well as the media and the broader public in general. A book that includes several other disturbing passages, such as an excerpt to which we have referred in a previous chapter. It is worthy to come back to this excerpt because it contextualizes the topics discussed in the present section, and, importantly, it is the type of passage that is almost always neglected in the countless books idolizing Darwin:

> In regard to the moral qualities, some elimination of the worst dispositions is always in progress even in the most civilised nations. Malefactors are executed, or imprisoned for long periods, so that they cannot freely transmit their bad qualities. Melancholic and insane persons are confined, or commit suicide. Violent and quarrelsome men often come to a bloody end. Restless men who will not follow any steady occupation—and this relic of barbarism is a great check to civilisation—emigrate to newly-settled countries, where they prove useful pioneers. Intemperance is so highly destructive, that the expectation of life of the intemperate, at the age, for instance, of thirty, is only 13·8 years; whilst for the rural labourers of England at the same age it is 40·59 years. Profligate women bear few children, and profligate men rarely marry; both suffer from disease. In the breeding of domestic animals, the elimination of those individuals, though few in number, which are in any marked manner inferior, is by no means an unimportant element towards success. This especially holds good with injurious characters which tend to reappear through reversion, such as

blackness in sheep; and with mankind some of the worst dispositions, which occasionally without any assignable cause make their appearance in families, may perhaps be reversions to a savage state, from which we are not removed by very many generations. This view seems indeed recognised in the common expression that such men are the black sheep of the family.

A most important obstacle in civilised countries to an increase in the number of men of a superior class has been strongly urged by Mr. Greg and Mr. Galton, namely, the fact that the very poor and reckless, who are often degraded by vice, almost invariably marry early, whilst the careful and frugal, who are generally otherwise virtuous, marry late in life, so that they may be able to support themselves and their children in comfort. Those who marry early produce within a given period not only a greater number of generations, but, as shewn by Dr. Duncan, they produce many more children. The children, moreover, that are born by mothers during the prime of life are heavier and larger, and therefore probably more vigorous, than those born at other periods. Thus the reckless, degraded, and often vicious members of society, tend to increase at a quicker rate than the provident and generally virtuous members. Or as Mr. Greg puts the case: '*The careless, squalid, unaspiring Irishman multiplies like rabbits: the frugal, foreseeing, self-respecting, ambitious Scot, stern in his morality, spiritual in his faith, sagacious and disciplined in his intelligence, passes his best years in struggle and in celibacy, marries late, and leaves few behind him. Given a land originally peopled by a thousand Saxons and a thousand Celts—and in a dozen generations five-sixths of the population would be Celts, but five-sixths of the property, of the power, of the intellect, would belong to the one-sixth of Saxons that remained. In the eternal 'struggle for existence,' it would be the inferior and less favoured race that had prevailed—and prevailed by virtue not of its good qualities but of its faults.*'

In this excerpt, Darwin directly contradicts what many of those idolizing him have defended: that the link between Galton – a particularly prominent eugenicist who was Darwin's cousin – and Darwin was merely a familiar one. This is because there is clearly a profound conceptual, interactive link between the ideas of the two men. Darwin cited his cousin's eugenicist ideas in the *Descent* to support his scientific "facts" about human evolution and humanity, and Galton cited Darwin's books to support his eugenicist, racist, and ethnocentric ideas. Related to this topic, which we have briefly covered in Chap. 1, there is a story that is also almost always left untold in the countless accounts about Darwin: an ever stronger link between Darwin and another renowned eugenicist. Someone who was actually *directly raised by him*: his own son, Leonard Darwin. The fact that both Charles Darwin's cousin – who Darwin cited in his books – and his own son – who he raised – became very prominent actors within the eugenics movement is as coincidental as was the fact that Darwin had a state funeral in a Royal church. After all, Leonard was raised, lived, and heard private comments on a daily basis from a man that cited Galton and wrote passages such as the ones we saw above, as well as the following one, which was also included in the *Descent*:

Man scans with scrupulous care the character and pedigree of his horses, cattle, and dogs before he matches them; but when he comes to his own marriage he rarely, or never, takes any such care. He is impelled by nearly the same motives as the lower animals, when they are left to their own free choice, though he is in so far superior to them that he highly values mental charms and virtues. On the other hand he is strongly attracted by mere wealth or rank. Yet he might by selection do something not only for the bodily constitution and frame of his offspring, but for their intellectual and moral qualities. Both sexes ought to refrain from marriage if they are in any marked degree inferior in body or mind; but such hopes are

Utopian and will never be even partially realised until the laws of inheritance are thoroughly known. Everyone does good service, who aids towards this end. When the principles of breeding and inheritance are better understood, we shall not hear ignorant members of our legislature rejecting with scorn a plan for ascertaining whether or not consanguineous marriages are injurious to man.

The advancement of the welfare of mankind is a most intricate problem: all ought to refrain from marriage who cannot avoid abject poverty for their children; for poverty is not only a great evil, but tends to its own increase by leading to recklessness in marriage. On the other hand, as Mr. Galton has remarked, if the prudent avoid marriage, whilst the reckless marry, the inferior members tend to supplant the better members of society. Man, like every other animal, has no doubt advanced to his present high condition through a struggle for existence consequent on his rapid multiplication; and if he is to advance still higher, it is to be feared that he must remain subject to a severe struggle. Otherwise he would sink into indolence, and the more gifted men would not be more successful in the battle of life than the less gifted. Hence our natural rate of increase, though leading to many and obvious evils, must not be greatly diminished by any means. There should be open competition for all men; and the most able should not be prevented by laws or customs from succeeding best and rearing the largest number of offspring. Important as the struggle for existence has been and even still is, yet as far as the highest part of man's nature is concerned there are other agencies more important. For the moral qualities are advanced, either directly or indirectly, much more through the effects of habit, the reasoning powers, instruction, religion, etc., than through natural selection; though to this latter agency may be safely attributed the social instincts, which afforded the basis for the development of the moral sense.

It is essential to note that, in a very wide sense, "eugenics" can be broadly defined as a set of beliefs and/or practices that aim at improving the genetic quality of a human population, so not all eugenicists were necessarily racist. It is also important to emphasize that among those eugenicists that were racist, many did not go all the way to approve measures such as the sterilization or murder of "others" and were instead more focused on using science to "improve" qualities via, for instance, the production of "positive traits." However, numerous eugenicists did support such horrible practices. This is why it is so crucial to underline the often untold story about how one of Darwin's sons, Leonard Darwin, became so predominantly involved in one of the most *negative versions of eugenics*. One of the few authors that had no problem to openly discuss, without taboos, this often neglected story was Edwin Black, in his book *Against the Weak*:

By 1912, America's negative eugenics had been purveyed to like-minded social engineers throughout Europe.. hence the First International Congress of Eugenics attracted several hundred delegates and speakers from the United States, Belgium, England, France, Germany, Italy, Japan, Spain and Norway. Major Leonard Darwin, son of Charles Darwin and head of the EES [UK's Eugenics Education Society], was appointed congress president.. Leonard Darwin revealed his true feelings in a speech to the adjunct *Cambridge University Eugenics Society*. "The first step to be taken", he explained, "ought to be to establish some system by which all children at school reported by their instructors to be specially stupid, all juvenile offenders awaiting trial, all ins-and-outs at workhouses, and all convicted prisoners should be examined by trained experts in mental defects in order to place on a register the names of all those thus ascertained to be definitely abnormal". Like his colleagues in America, [Leonard] Darwin wanted to identify not just the so-called unfit, but their entire families as well. [Leonard] Darwin emphasized, "From the Eugenic standpoint this method would no doubt be insufficient, for the defects of relatives are only second in importance to the defects of the individuals themselves-indeed, in some cases [the defects of relatives] are of far greater importance". British eugenicists were convinced that just seeming normal was not enough – the unfit were ancestrally flawed. Even if an individual appeared normal and begat normal children, he or she could still be a "carrier" who needed to be sterilized.

Even Bowler, while trying to minimize as much as he could the links between Darwin and Darwinism and eugenics in his *Darwin Deleted*, had to recognize that it was Darwin's cousin, Francis Galton, who began to argue for a "eugenic program – in effect a call to impose a mechanism of artificial selection on the human race." As Bowler noted, "in a civilized society, we do not restrict the ability of people to have children, which means that even those with the lowest mental and moral capacities continue to breed.. both [Charles] Darwin and Galton worried that this might lead to degeneration." This program became popular in Britain, "and sterilization programs were introduced in a number of American states.. the movement became particularly active in Germany, where the Nazis went beyond mere sterilization and began to exterminate those elements of society they wished to suppress." In Todes' *Darwin Without Malthus*, he explains that Darwin's notion of "struggle for existence" was said to be the "most severe between the individuals of the same species, for they frequent the same districts, require the same food, and are exposed to the same dangers.. [Charles] Darwin used the words 'struggle' and 'competition' interchangeably.. the metaphor 'struggle for existence,' and in such phrases as 'the great battle for life' and the 'war of nature' contributed a certain rhetorical power to his argument." According to Todes, by sacrificing accuracy for eloquence and proposing that within such a struggle "death is generally prompt, and that the vigorous, the healthy and the happy survive and multiply," Darwin *did* provide *easy and powerful ammunition* to eugenicists around the globe. Similarly, in one of the chapters of his 2014 book *The invention of Race*, Andreassen wrote that "Darwin's arguments about the survival of the fittest became central to theories about racial hierarchies and human development.. many scientists began to see the different races competing against one another; the stronger and more intelligent would thrive, while the weaker and less intelligent races declined.. [in other words] racial Darwinism" (see also Box 3.5).

A succinct summary of just a few of the numerous atrocious repercussions – many of them prevalent even today – of the influential ideas of Galton and other eugenicists that were not only supported but also directly quoted by Charles Darwin in his *Descent* has been provided in a recent 2021 paper by Adam Rutherford:

Galton was Charles Darwin's half cousin, and greatly admired him. Much of his rationale for eugenics was drawn from Darwin's work on evolution by both natural and artificial selection. Galton's proposed application of selective breeding of humans for the general improvement of a people was reliant on a 19th century and thus simplistic notion that characteristics such as intellectual and physical capabilities or psychological disorders are biologically heritable, and therefore could be modified or eradicated from a population. Global political support for eugenics came from the height of power. In his years in government in the UK until he enlisted in the army in 1915, Winston Churchill frequently spoke and wrote positively about eugenics. In 1910, he asked the British Home Office to look into sterilization of the 'feeble-minded,' in the terminology of the time, based on Indiana's eugenics legislation, that had come into force in 1907, the first in the world. In 1912, Churchill spoke

> of using 'Röntgen Rays'—X-rays—to sterilize men and women, and he included sterilization in early drafts of what became the *1913 Mental Deficiencies Act*. The sterilization laws were removed at the third and final reading of the bill, but legalization to institutionalize British people remained, under four categories of undesirability: Idiots, Imbeciles, Feeble minded people and Moral Imbeciles, as was the Edwardian parlance for all manner of psychological, cognitive and mental health conditions.
>
> Similar views were held by the most senior politicians in the United States. 'Society has no business to permit degenerates to reproduce their kind' wrote Theodore Roosevelt in a letter to the eugenicist Charles Davenport. 'Some day we will realize that the prime duty, the inescapable duty of the good citizen of the right type is to leave his blood behind him in the world, and that we have no business to perpetuate citizens of the wrong type.' Galton's work was taken up in many countries, and most obviously enacted under the deranged policies of the Nazis during the Third Reich and the Holocaust. In the United States, and a few other countries, the forced, involuntary and often secret sterilization of 'undesirables' was embraced enthusiastically. From 1907, when Indiana passed the first state mandate, until 1963, forced sterilization was legally administered in 31 states, with California the most vigorous adopter. In the 20th century, more than 60,000 men and women, though mostly women, were sterilized for a variety of so-called undesirable traits – men frequently to curtail the propagation of criminal behaviors. Though eugenics and scientific racism are distinct, eugenics policies disproportionally affected minority groups in many countries. Native American women were forcibly sterilized in their thousands, and as late as the 1970s, black women with multiple children were being sterilized under the threat of withheld welfare, or in some cases without their knowledge. The legacy of these eugenics programs persists: in 2020, there were credible allegations that women detained in U.S. Immigration and Customs Enforcement facilities had undergone unnecessary gynecological procedures, including hysterectomies.

Interestingly, the title of Rutherford's paper is "Race, eugenics, and the canceling of great scientists." As we have seen in Chap. 1, being the author of a book entitled *How to Argue with a Racist*, Rutherford's arguments are overall somewhat subjective and confusing, particularly about Darwin, as when Rutherford refers to the now so typically misused term "canceling":

> Darwin himself is not exempt from historical reassessment. Though expressing essentially humanist views throughout the *Descent of Man* and arguing against racial categorizations, that book also contains passages that today jar as being both scientifically specious and politically outmoded. "We may also infer" he writes on page 361, "from the law of the deviation from averages, so well illustrated by Mr. Galton, in his work on '*Hereditary Genius*', that if men are capable of a decided pre-eminence over women in many subjects, the average of mental power in man must be above that of woman". On race, he speaks of how the "civilised races of man will almost certainly exterminate and replace throughout the world the savage races". These, and other descriptions by Darwin of differences between the sexes and various human populations, are well documented. However, it is also clear that these views, neither atypical or extreme for his time, were not a major part of his overall body of work, nor a central thesis in the *Descent of Man*.. We cannot and should not abandon nor trash the scientific works of Galton, Fisher and others on whose shoulders we stand. Their techniques are in constant use, for the betterment of science and all humans, and these are formidable legacies. But we can choose not to honor their names.

In other words, authors such as Rutherford have the merit, on the one hand, to recognize that the works of scholars such as Darwin and Galton had societal repercussions that are felt even today, which is something to be applauded and shows how things are indeed starting to change, slowly. However, on the other hand, they still fail to recognize that such repercussions are also affecting writers such as themselves. That is, they fail to recognize that the existence of systemic racism, sexism, and ethnocentrism within science today is in great part precisely the legacy of people such as Darwin and Galton and the indoctrination process that "sells" their works about human evolution and human "races" as "formidable legacies." This topic is related to the confusion created by authors such as Rutherford regarding the way in which they use the term "canceling" – in this case, in the phrase "canceling of great scientists." What truly contributes to the "canceling" of history is to describe works such as Darwin's *Descent* and Galton's writings as "formidable legacies." They were not. As noted above, there are parts of Darwin's *Descent* that are scientifically very strong, including those about atavisms in humans and sexual selection in other animals. However, the parts about human evolution include many racist, sexist, ethnocentric, ladder-of-life notions that are factually inaccurate and that therefore cannot be considered to be scientifically "formidable" at all (see also Box 3.5).

Box 3.5 Darwin's Ideas, Capitalism, Inequalities, Racism, and the Notion of "Progress"

In Bethencourt's 2013 book *Racisms*, he discusses the links between Darwin's ideas, capitalism, inequalities, racism, and the notion of "progress" and emphasizes the inconsistencies between the writings of Darwin the anthropologist about these subjects and some of the key evolutionary ideas defended by Darwin the biologist:

> Darwin's.. remark on the social system of the Fuegians shows again how reflections on the different stages of humankind and prevailing racial constructions were linked to the issue of inequality, which at the turn of the century was dealt with in a debate between William Godwin and Thomas Robert Malthus that is reflected here through reference to Malthus's assertion that only self-interest motivates humankind. For Darwin, the perfect equality among the individuals composing the Fuegian tribes had retarded their civilization. Peoples governed by hereditary kings were considered most capable of improvement, and among races, the more civilized ones had the more sophisticated governments. *Darwin equated equality with baseness*: pieces of cloth given to the Fuegians were torn into shreds and distributed; no individual would be richer than the others. Individual property, the notion of superiority, and an accumulation of power were unthinkable in this tribal regime, yet for Darwin they were the sinews of improvement. The comparison between the "savages" and "barbarians" that Darwin met during his voyage around the world highlights his hierarchy.
>
> The Fuegians were placed at the bottom of the scale, along with the war-like cannibals and murderous New Zealanders (or Maori), Australian aborigines (skillful with the boomerang, spear, and throwing stick in climbing trees and methods of hunting, but feeble in mental capacity), and "wretched" South African tribes prowling the land in search of roots. They were all contrasted unfavorably with the relatively civilized South Sea islanders – the manners and even tattoos of the Tahitians were praised – and proficient Eskimos, with their subterranean huts and fully equipped canoes. Darwin possessed an independent and acute mind, although for

(continued)

some of his observations he was indebted to Captain Cook's journals. These observations reveal continuities in the descriptions of the peoples of the world, reminding us of the early accounts of Native Americans by Columbus, Vespucci, or Caminha, even though the detachment (and repugnance) concerning savages sounds even more pronounced after centuries of contact. The divergence was reinforced by the eighteenth century notion of civilization, and enhanced by the industrial revolution and progress in the comfort of daily life as well as the quality of transportation. The voyage of the Beagle was certainly more comfortable and safe than previous circumnavigations of the world, although the second trip made by Cook (1772–75) had been particularly successful in terms of a radical reduction in the loss of human lives. The filthiness of the native body along with the scanty clothes, diabolic body paintings and tattoos, constant warfare driven by revenge, cannibalism, cruelty, and absence of justice as well as the inferior local languages were not new topics; vehement disgust was an expression of the Europeans' projection of their own self-perception. Darwin's descriptions, however, represent the highest level then reached by travel accounts. They were attentive to habitat, housing, material culture, family structure, division of labor, and political specialization.

The claim that inequality was a source of social improvement lay at the core of contemporary debates between socialists and liberals; it shows that Darwin was aware of the major social and political discussions of his time. Nevertheless, Darwin's lack of empathy concerning the savages did not shake his abolitionist convictions. Darwin expressed his indignation when confronted with the daily cruelty toward slaves in Rio de Janeiro, where he saw instruments for their torture, heard the cries of slaves being punished, and intervened on various occasions to stop further suffering. He equated slavery with the moral debasement of a whole society; he protested against the idea of slavery as a tolerable evil, denouncing the way in which people were "blind[ed] by the constitutional gaiety of the negro"; and he refused the attempt to "palliate slavery by comparing the state of slaves with our poorer countrymen". Darwin raised a crucial issue that could be related to many other situations of oppression: "those who look tenderly at the slave owner and with a cold heart at the slave, never seem to put themselves into the position of the latter", concluding emotionally, "it makes one's blood boil, yet heart tremble, to think that we Englishmen and our American descendants, with their boastful cry of liberty, have been and are so guilty: but it is a consolation to reflect, that we at least have made a greater sacrifice, than ever made by any nation, to expiate our sins".. Scholars James Moore and Adrian Desmond attribute Darwin's more conservative stance in [his later book *The Descent of Man*].. to the hardening of attitudes in the 1860s.

But if we take one significant example, Darwin's remark concerning the immorality of savages, used to counter the Irish historian William Lecky's benevolent arguments, was in line with his observations of savages during his voyage on the Beagle. In terms of eugenics, though, I would agree with Moore and Desmond: Darwin quoted William Greg, Wallace, and Francis Galton on the failure of natural selection in civilized nations, as a result of vaccinations, poor laws, and asylums – medical care and social assistance for the less fortunate, which promoted the survival and propagation of the weaker members of society, leading to a "deterioration in the noblest part of our nature". Darwin blended eugenics with an essentialist approach to nations. He drew attention to another process of negative selection, produced by the Spanish Inquisition over centuries, which systematically excluded those people most ambitious in thought and action, and thus was responsible for long-term decline, while the emigration of the most energetic people to British America had produced the opposite outcome. But Darwin acknowledged that all

(continued)

civilized nations descended from barbarians, showing the possible improvement of savages through independent steps along the scale of civilization. He quoted anthropologist Edward Tylor, who in 1865 had published *Researches into the Early History of Mankind and Development of Civilization*, based on the idea of intellectual abilities shared by all groups of people and differences in social evolution resulting from education. Darwin explicitly rejected the idea of human being's decline: "to believe that man was aboriginally civilized and then suffered utter degradation in so many regions, is to take a pitiably low view of human nature", maintaining instead "that progress has been much more general than retrogression; that man has risen, though by slow and interrupted steps, from a lowly condition to the highest standards as yet attained by him in knowledge, morals and religion".

When I discuss the history of eugenics in my talks around the globe, lay people, journalists, and even scholars often tell me things such as: eugenics is a thing of the past; it mainly ended with the Holocaust; there has been a huge progress since then, so why are you talking about such past ideas? However, as we have seen throughout this book, human history, and biological evolution in general, is not at all a linear line toward "progress" but is instead a spiral with circles that are often strikingly similar – not equal, as history can obviously never repeat itself – to each other. Right now, while I write these lines, there are many countries applying political ideologies that are in a way somewhat similar to eugenic ideologies. This includes the most populous country on the planet, China. For example, recently China has started to actively promote the narrative that Chinese families should try to have three children because otherwise the country will face huge problems, such as a lack of young people, and thus a burden on the social system to support the life and health problems of older people and so on. We have heard many times the same narrative recently in other regions of the planet, such as many European countries, accompanied by political policies to motivate the "native" people from those countries to have more children while making it more and more difficult for migrants from Africa and other continents to become citizens of such countries. Basically, the message is: we need young people but from "our" country; we need our "native" people, not "other" people from "other" continents. They are essentially doing the same thing as China, but in a way China is taking this to another level, as it is doing this even at an *internal* level, with its "own" people, by actively promoting the reproduction of the "best" Chinese people – that is, from the Han ethnicity – and trying to reduce as much as possible the reproduction of the "worst" Chinese people – for instance, Xinjiang's ethnic Uyghur population (see Box 3.6).

Box 3.6 Unpunished Eugenicist Ideologies and Practices in 2021

A recent study has revealed some numbers that plainly show how humans are indeed still following, producing, facing, and millions of them suffering with, eugenic ideologies, policies, and practices even in the 2020s. As summarized in a *Reuters* news article published on June 7, 2021, entitled "China policies could cut millions of Uyghur births in Xinjiang":

(continued)

Chinese birth control policies could cut between 2.6 to 4.5 million births of the Uyghur and other ethnic minorities in southern Xinjiang within 20 years, up to a third of the region's projected minority population, according to a new analysis by a German researcher. The report.. also includes a previously unreported cache of research produced by Chinese academics and officials on Beijing's intent behind the birth control policies in Xinjiang, where official data shows birth-rates have already dropped by 48.7% between 2017 and 2019. Adrian Zenz's research comes amid growing calls among some western countries for an investigation into whether China's actions in Xinjiang amount to genocide, a charge Beijing vehemently denies. The research by Zenz is the first such peer reviewed analysis of the long-term population impact of Beijing's multi-year crackdown in the western region. Rights groups, researchers and some residents say the policies include newly enforced birth limits on Uyghur and other mainly Muslim ethnic minorities, the transfers of workers to other regions and the internment of an estimated one million Uyghurs and other ethnic minorities in a network of camps. "This (research and analysis) really shows the intent behind the Chinese government's long-term plan for the Uyghur population", Zenz told Reuters.

The Chinese government has not made public any official target for reducing the proportion of Uyghur and other ethnic minorities in Xinjiang. But based on analysis of official birth data, demographic projections and ethnic ratios proposed by Chinese academics and officials, Zenz estimates Beijing's policies could increase the predominant Han Chinese population in southern Xinjiang to around 25% from 8.4% currently. "This goal is only achievable if they do what they have been doing, which is drastically suppressing (Uyghur) birth rates", Zenz said. China has previously said the current drop in ethnic minority birth rates is due to the full implementation of the region's existing birth quotas as well as development factors, including an increase in per capita income and wider access to family planning services.. The new research compares a population projection done by Xinjiang-based researchers for the government-run Chinese Academy of Sciences based on data predating the crackdown, to official data on birth-rates and what Beijing describes as "*population optimization*" measures for Xinjiang's ethnic minorities introduced since 2017. It found the population of ethnic minorities in Uyghur-dominated southern Xinjiang would reach between 8.6–10.5 million by 2040 under the new birth prevention policies. That compares to 13.14 million projected by Chinese researchers using data predating the implemented birth policies and a current population of around 9.47 million..

The move to prevent births among Uyghur and other minorities is in sharp contrast with China's wider birth policies. Last week, Beijing announced married couples can have three children, up from two, the largest such policy shift since the one child policy was scrapped in 2016 in response to China's rapidly ageing population. The announcement contained no reference to any specific ethnic groups. Before then, measures officially limited the country's majority Han ethnic group and minority groups including Uyghur to two children – three in rural areas. However, Uyghurs and other ethnic minorities had historically been partially excluded from those birth limits as part of preferential policies designed to benefit the minority communities. Some residents, researchers and rights groups say the newly enforced rules now disproportionately impact Islamic minorities, who face detention for exceeding birth quotas, rather than fines as elsewhere in China. In a Communist Party record leaked in 2020, also reported by Zenz, a re-education camp in southern Xinjiang's Karakax county listed birth violations as the reason for internment in 149 cases out of 484 detailed in the list. China has called the list a "fabrication". Birth quotas for ethnic minorities have become strictly enforced in Xinjiang since 2017, including though

(continued)

the separation of married couples, and the use of *sterilisation procedures, intrauterine devices* (IUDs) and abortions, three Uyghur people and one health official inside Xinjiang told Reuters.. In Xinjiang counties where Uyghurs are the majority ethnic group, birth rates dropped 50.1% in 2019, for example, compared to a 19.7% drop in majority ethnic Han counties, according to official data compiled by Zenz.

Zenz's report says analyses published by state funded academics and officials between 2014 and 2020 show the strict implementation of the policies are driven by national security concerns, and are motivated by a desire to dilute the Uyghur population, increase Han migration and boost loyalty to the ruling Communist Party. For example, 15 documents created by state funded academics and officials showcased in the Zenz report include comments from Xinjiang officials and state-affiliated academics referencing the need to increase the proportion of Han residents and decrease the ratio of Uyghurs or described the high concentration of Uyghurs as a threat to social stability. "The problem in southern Xinjiang is mainly the unbalanced population structure.. the proportion of the Han population is too low", Liu Yilei, an academic and the deputy secretary general of the Communist Party committee of the Xinjiang Production and Construction Corps, a government body with administrative authority in the region, told a July 2020 symposium, published on the Xinjiang University website. Xinjiang must "end the dominance of the Uyghur group", said Liao Zhaoyu, dean of the institute of frontier history and geography at Xinjiang's Tarim University at an academic event in 2015, shortly before the birth policies and broader internment programme were enforced in full..

Zenz and other experts point to the *1948 Convention on the Prevention and Punishment of the Crime of Genocide*, which lists birth prevention targeting an ethnic group as one act that could qualify as genocide. The United States government and parliaments in countries including Britain and Canada have described China's birth prevention and mass detention policies in Xinjiang as genocide. However, some academics and politicians say there is insufficient evidence of intent by Beijing to destroy an ethnic population in part or full to meet the threshold for a genocide determination. No such formal criminal charges have been laid against Chinese or Xinjiang officials because of a lack of available evidence on and insight into the policies in the region. Prosecuting officials would also be complex and require a high bar of proof. Additionally, China is not party to the *International Criminal Court* (ICC), the top international court that prosecutes genocide and other serious crimes, and which can only bring action against states within its jurisdiction.

Hitler's Struggle and the Arian Struggle for Existence

> *Truly, this earth is a trophy cup for the industrious man.. and this rightly so, in the service of natural selection.. he who does not possess the force to secure his Lebensraum in this world, and, if necessary, to enlarge it, does not deserve to possess the necessities of life.. he must step aside and allow stronger peoples to pass him by.* (Adolf Hitler)

Black's 2003 book *War Against the Weak* provides a superb historical context to understand the rise of eugenics and its profound and disturbing – and often untold – direct links to the subsequent rise of the area of biology that is now known as "genetics." As we have seen, James Watson, the Nobel laureate biologist that coauthored with Francis Crick the 1953 paper proposing the double helix structure of

DNA, is one of the most famous living geneticists and is an emblematic example of how this dark racist past of eugenics still affects the field of genetics. Watson has a remarkably long history of making sexist and racist comments and *continues* to proudly defend scientifically inaccurate "facts" such as that "Africans" are mentally inferior because of the genes they have, and so on. It is striking how a Nobel laureate biologist does not even seem to know – or does not want to acknowledge – that people from Africa do not constitute a true biological – monophyletic – group, as we have seen before. That is, Watson's assertions have nothing to do with real genetic facts and are instead just a "modern" repetition of the old scientific "innate type of racism" that emerged in the eighteenth century, proclaiming that even a single drop of "blood" – in this case, updated to "genes" – of an "African" would lead to biological and mental inferiority, no matter what, when, and where (see Box 3.7). There is nothing new under the sun, in this respect, well into the twenty-first century.

Black's 2003 book shows that not only a huge number of biologists but also a vast number of famous U.S. philanthropists – such as Carnegie and Rockefeller – as well as the enormously renowned and influential institutions that they created – including the Carnegie Institution and Rockefeller Foundation – were deeply involved in the eugenic enterprise. In other words, it is not only "giant" scholars such as Watson, Darwin, and Aristotle but also "discoverers" such as Christopher Columbus, "thinkers" such as Kant, "philanthropists" such as Carnegie and Rockefeller, and physicians such as Marion Sims (Fig. 3.13), that contributed to either systemic racism, or systemic sexism, or, often, to both. While we were indoctrinated to *believe* that Western "civilizations" are mostly built "on the shoulders of noble giants," and of "humanists," the reality is very different. Almost all those "giants" that are idolized and displayed as "heroes" in thousands of statues around the globe and in an endless number of books, documentaries, and TV series – including those made to be read, seen, or heard by kids – were and are often racist or misogynistic or commonly both. This is precisely how indoctrination works. What our kids learn about them in schools is just part of their true story – a tale that only shows the "noble," "humanist" parts but omits anything "negative" concerning their racism, sexism, ethnocentrism, and, in many cases, the atrocities that they indirectly or directly – as was the case with Columbus and Sims, as we will see below – promoted or justified. Black explained:

> The victims of eugenics [in the first decades of the 20th century within the USA] were poor urban dwellers and rural 'white trash' from New England to California, immigrants from across Europe, Blacks, Jews, Mexicans, Native Americans, epileptics, alcoholics, petty criminals, the mentally ill and anyone else who did not resemble the blond and blue-eyed Nordic ideal the eugenics movement glorified. Eugenics contaminated many otherwise worthy social, medical and educational causes from the birth control movement to the development of psychology to urban sanitation. Psychologists persecuted their patients. Teachers stigmatized their students. Charitable associations clamored to send those in need of help to lethal chambers they hoped would be constructed. Immigration assistance bureaus connived to send the most needy to sterilization mills. Leaders of the ophthalmology profession conducted a long and chilling political campaign to round up and coercively sterilize every relative of every American with a vision problem. All of this churned throughout

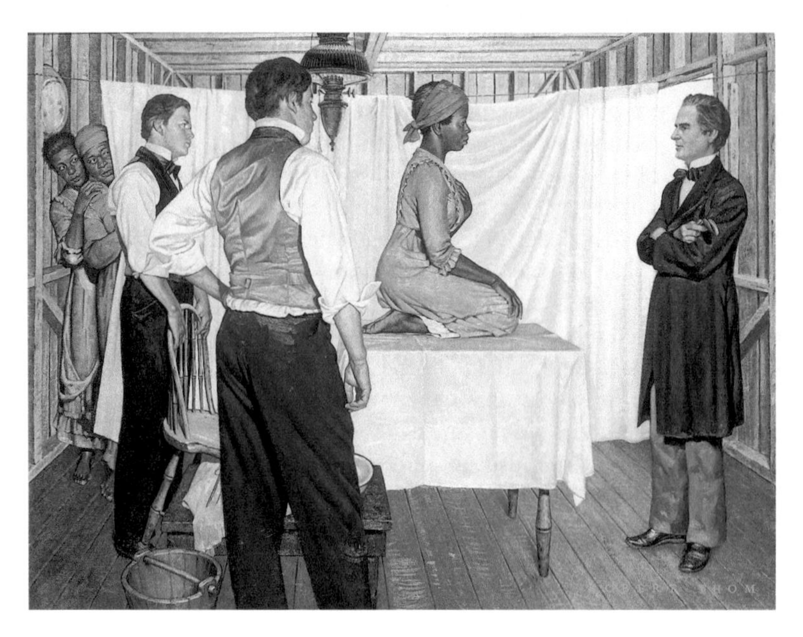

Fig. 3.13 Robert Thom's "J. Marion Sims: Gynecologic Surgeon," an oil representation of an experimental surgery upon a powerless slave, from Thom's *The History of Medicine*, circa 1952, archived at the University of Michigan

America years before the Third Reich rose in Germany. Eugenics targeted all mankind, so of course its scope was global. American eugenic evangelists spawned similar movements and practices throughout Europe, Latin America and Asia. Forced sterilization laws and regimens took root on every continent. Each local American eugenic ordinance or statute – from Virginia to Oregon – was promoted internationally as yet another precedent to be emulated by the international movement. A tightly-knit network of mainstream medical and eugenical journals, international meetings and conferences kept the generals and soldiers of eugenics up to date and armed for their nation's next legislative opportunity.

Eventually, America's eugenic movement spread to Germany as well, where it caught the fascination of Adolf Hitler and the Nazi movement. Under Hitler, eugenics careened beyond any American eugenicist's dream. National Socialism transduced America's quest for a 'superior Nordic race' into Hitler's drive for an 'Aryan master race.' The Nazis were fond of saying 'National Socialism is nothing but applied biology,' and in 1934 the *Richmond Times-Dispatch* quoted a prominent American eugenicist as saying, 'the Germans are beating us at our own game.' Nazi eugenics quickly outpaced American eugenics in both velocity and ferocity. In the 1930s, Germany assumed the lead in the international movement. Hitler's eugenics was backed by brutal decrees, custom-designed IBM data processing machines, eugenical courts, mass sterilization mills, concentration camps, and virulent biological anti-Semitism – all of which enjoyed the open approval of leading American eugenicists and their institutions. The cheering quieted, but only reluctantly, when the United States entered the war in December of 1941. Then, out of sight of the world, Germany's eugenic warriors operated extermination centers.

Eventually, Germany's eugenic madness led to the Holocaust, the destruction of the Gypsies, the rape of Poland and the decimation of all Europe. But none of America's far-reaching scientific racism would have risen above ignorant rants without the backing of corporate philanthropic largess. Within these pages you will discover the sad truth of how

> the scientific rationales that drove killer doctors at Auschwitz were first concocted on Long Island at the Carnegie Institution's eugenic enterprise at Cold Spring Harbor.. during the prewar Hitler regime, the Carnegie Institution, through its Cold Spring Harbor complex, enthusiastically propagandized for the Nazi regime and even distributed anti-Semitic Nazi Party films to American high schools.. [there were] links between the Rockefeller Foundation's massive financial grants and the German scientific establishment that began the eugenic programs that were finished by Mengele at Auschwitz. Only after the truth about Nazi extermination became known did the American eugenics movement fade. American eugenic institutions rushed to change their names from eugenics to genetics. With its new identity, the remnant eugenics movement reinvented itself and helped establish the modem, enlightened human genetic revolution. Although the rhetoric and the organizational names had changed, the laws and mindsets were left in place. So for decades after Nuremberg labeled eugenic methods genocide and crimes against humanity, America continued to forcibly sterilize and prohibit eugenically undesirable marriages.

Among the powerful, wealthy, and influential figures that supported the eugenic enterprise, there are even people that were, or had been, U.S. presidents, as explained by Black:

> Eventually, the eugenics movement and its supporters began to speak a common language that crept into the general mindset of many of America's most influential thinkers. On January 3, 1913, former President Theodore Roosevelt wrote Davenport, "I agree with you.. that society has no business to permit degenerates to reproduce their kind.. some day, we will realize that the prime duty, the inescapable duty, of the good citizen of the right type, is to leave his or her blood behind him in the world; and that we have no business to permit the perpetuation of citizens of the wrong type". Episcopalian Bishop John T. Dallas of Concord, New Hampshire, issued a public statement: "eugenics is one of the very most important subjects that the present generation has to consider". Episcopalian Bishop Thomas F. Gailor of Memphis, Tennessee, issued a similar statement: "The science of eugenics.. by devising methods for the prevention of the propagation of the feebleminded, criminal and unfit members of the community, is.. one of the most important and valuable contributions to civilization". Dr. Ada Comstock, president of Radcliffe College, declared publicly, "eugenics is 'the greatest concern of the human race'. The development of civilization depends upon it". Dr. Albert Wiggam, an author and a leading member of the *American Association for the Advancement of Science*, pronounced his belief: "Had Jesus been among us, he would have been president of the First Eugenic Congress".

However, it is important to stress that some of the things defended by the most hard-core U.S. eugenicists were much harsher than what was defended originally by many of the followers of Galton in England, as noted by Black:

> While many of America's elite exalted eugenics, the original Galtonian eugenicists in Britain were horrified by the sham science they saw thriving in the United States and taking root in their own country. In a merciless 1913 scientific paper written on behalf of the Galton Laboratory, British scientist David Heron publicly excoriated the American eugenics of Davenport, Laughlin, and the Eugenics Record Office. Using the harshest possible language, Heron warned against "certain recent American work which has been welcomed in this country as of first-class importance, but the teaching of which we hold to be fallacious and indeed actually dangerous to social welfare". His accusations: "careless presentation of data, inaccurate methods of analysis, irresponsible expression of conclusions, and rapid change of opinion". Heron lamented further, "those of us who have the highest hopes for the new science of Eugenics in the future are not a little alarmed by many of the recent contributions to the subject which threaten to place Eugenics.. entirely outside the pale of true science.. when we find such teaching-based on the flimsiest of theories and on the most

superficial of inquiries-proclaimed in the name of Eugenics, and spoken of as 'entirely splendid work', we feel that it is not possible to use criticism too harsh, nor words too strong in repudiation of advice which, if accepted, must mean the death of Eugenics as a science". Heron emphasized "that the material has been collected in a most unsatisfactory manner, that the data have been tabled in a most slipshod fashion, and that the Mendelian conclusions drawn have no justification whatever.." He went so far as to say the data had been deliberately skewed. As an example, he observed that "a family containing a large number of defectives is more likely to be recorded than a family containing a small number of defectives". In sum, he called American eugenics rubbish.

Unfortunately, the more extreme, "rubbish" version of eugenics thriving in the United States was highly successful in exporting itself to many other countries, including the United Kingdom – backed by politicians who have otherwise accomplished highly praiseworthy things, such as Winston Churchill – and Germany, contributing in a way to the ideology of Hitler's Third Reich:

By 1912, America's negative eugenics had been purveyed to like-minded social engineers throughout Europe, especially in Germany and the Scandinavian nations, where theories of Nordic superiority were well received. Hence the *First International Congress of Eugenics* attracted several hundred delegates and speakers from the United States, Belgium, England, France, Germany, Italy, Japan, Spain and Norway.. [for instance] the ambitious British eugenic plans encompassed not just those who seemed mentally inferior, but also criminals, debtors, paupers, alcoholics, recipients of charity and "other parasites". In 1909 and 1910, other so-called welfare societies for the feebleminded, such as the *Cambridge Association for the Care of the Feebleminded*, contacted the Eugenics Education Society to urge more joint lobbying of the government to sanction forced sterilization. Mass letter-writing campaigns began. Every candidate for Parliament was sent a letter demanding they "support measures.. that tend to discourage parenthood on the part of the feebleminded and other degenerate types". As in America, sterilization advocacy focused first and foremost on the most obviously impaired, in this case, the feeble-minded, but then escalated to include "other degenerate types". Seeking support for the Mental Deficiency Act, society members mailed letters to every sitting member of Parliament, long lists of social welfare officials, and virtually every education committee in England. When preliminary governmental committees shrank from support, the society simply redoubled its letter-writing campaign. Finally the government agreed to consider the legislation. Home Secretary Winston Churchill, an enthusiastic supporter of eugenics, reassured one group of eugenicists that Britain's 120,000 feebleminded persons "should, if possible, be segregated under proper conditions so that their curse died with them and was not transmitted to future generations". The plan called for the creation of vast colonies. Thousands of Britain's unfit would be moved into these colonies to live out their days..

Germany was no exception. German eugenicists had formed academic and personal relationships with Davenport and the American eugenic establishment from the turn of the century. Even after World War I, when Germany would not cooperate with the International Federation of Eugenic Organizations because of French, English and Belgian involvement, its bonds with Davenport and the rest of the U.S. movement remained strong. American foundations such as the Carnegie Institution and the Rockefeller Foundation generously funded German race biology with hundreds of thousands of dollars, even as Americans stood in breadlines. Germany had certainly developed its own body of eugenic knowledge and library of publications. Yet German readers still closely followed American eugenic accomplishments as the model: biological courts, forced sterilization, detention for the socially inadequate, debates on euthanasia. As America's elite were describing the socially worthless and the ancestrally unfit as "bacteria", "vermin", "mongrels" and "subhuman", a superior race of Nordics was increasingly seen as the final solution to the globe's eugenic problems. America had established the value of race and blood. In Germany, the concept was known as *Rasse und Blut*. U.S. proposals, laws, eugenic investigations and ideology were not undertaken quietly out of sight of Gennan activists.

They became inspirational blueprints for Germany's rising tide of race biologists and race-based hatemongers, be they white-coated doctors studying *Eugenical News* and attending congresses in New York, or brown-shirted agitators waving banners and screaming for social upheaval in the streets of Munich. One such agitator was a disgruntled corporal in the German army. He was an extreme nationalist who also considered himself a race biologist and an advocate of a master race. He was willing to use force to achieve his nationalist racial goals. His inner circle included Germany's most prominent eugenic publisher. In 1924, he was serving time in prison for mob action. While in prison, he spent his time poring over eugenic textbooks, which extensively quoted Davenport, Popenoe and other American raceo-logical stalwarts. Moreover, he closely followed the writings of Leon Whitney, president of the American Eugenics Society, and Madison Grant, who extolled the Nordic race and bemoaned its corruption by Jews, Negroes, Slavs and others who did not possess blond hair and blue eyes. The young Gennan corporal even wrote one of them fan mail.. [he] would soon burn and gas his name into the blackest corner of history. He would duplicate the American eugenic program-both that which was legislated and that which was only brashly advocated-and his group would consistently point to the United States as setting the precedents for Germany's actions. And then this man would go further than any American eugenicist ever dreamed, further than the world would ever tolerate, further than humanity will ever forget. The man who sent those letters was Adolf Hitler.

As also pointed out by Black, the links between the eugenics movement in the United States, Hitler, and teleological narratives are indeed disturbing:

Where did Hitler develop his racist and anti-Semitic views? Certainly not from anything he read or heard from America. Hitler became a mad racist dictator based solely on his own inner monstrosity, with no assistance from anything written or spoken in English. But like many rabid racists, from Plecker in Virginia to Rentoul in England, Hitler preferred to legitimize his race hatred by medicalizing it, and wrapping it in a more palatable pseudoscientific facade-eugenics. Indeed, Hitler was able to recruit more followers among reasonable Germans by claiming that science was on his side. The intellectual outlines of the eugenics Hitler adopted in 1924 were strictly American. He merely compounded all the virulence of long established American race science with his fanatic anti-Jewish rage. Hitler's extremist eugenic science, which in many ways seemed like the logical extension of America's own entrenched programs and advocacy, eventually helped shape the institutions and even the machinery of the Third Reich's genocide.

By the time Hitler's concept of Aryan superiority emerged, his politics had completely fused into a biological and eugenic mindset. When Hitler used the term master race, he meant just that, a biological "master race". America crusaded for a biologically superior race, which would gradually wipe away the existence of all inferior strains. Hitler would crusade for a master race to quickly dominate all others. In Hitler's view, eugenically inferior groups, such as Poles and Russians, would be permitted to exist but were destined to serve Germany's master race. Hitler demonized the Jewish community as social, political and racial poison, that is, a biological menace. He vowed that the Jewish community would be neutralized, dismantled and removed from Europe. Nazi eugenics would ultimately dictate who would be persecuted, how people would live, and how they would die. Nazi doctors would become the unseen generals in Hitler's war against the Jews and other Europeans deemed inferior. Doctors would create the science, devise the eugenic formulas, write the legislation, and even hand-select the victims for sterilization, euthanasia and mass extermination. Hitler's deputy, Rudolf Hess, coined a popular adage in the Reich, "National Socialism is nothing but applied biology".

In page after page of Mein Kampf's rantings, Hitler recited social Darwinian imperatives, condemned the concept of charity, and praised the policies of the United States and its quest for Nordic purity. Perhaps no passage better summarized Hitler's views than this from chapter 11: "the Germanic inhabitant of the American continent, who has remained racially pure and unmixed, rose to be master of the continent; he will remain the master as long as he does not fall a victim to defilement of the blood".. Moreover, as Hitler's knowledge of American pedigree techniques broadened, he came to realize that even he might have been eugenically excluded. In later years, he conceded at a dinner engagement, "I was shown a questionnaire drawn up by the Ministry of the Interior, which it was proposed to put to people whom it was deemed desirable to sterilize. At least three-quarters of the questions asked would have defeated my own good mother. If this system had been introduced before my birth, I am pretty sure I should never have been born at all".. On January 30, 1933, Adolf Hitler seized power following an inconclusive election.

During the twelve-year Reich, he never varied from the eugenic doctrines of identification, segregation, sterilization, euthanasia, eugenic courts and eventually mass termination of germ plasm in lethal chambers. During the Reich's first ten years, eugenicists across America welcomed Hitler's plans as the logical fulfillment of their own decades of research and effort. Indeed, they were envious as Hitler rapidly began sterilizing hundreds of thousands and systematically eliminating non-Aryans from German society. This included the Jews. Ten years after Virginia passed its 1924 sterilization act, Joseph Dejarnette, superintendent of Virginia's Western State Hospital, complained in the *Richmond TimesDispatch*, "the Germans are beating us at our own game". Most of all, American raceologists were intensely proud to have inspired the purely eugenic state the Nazis were constructing. In those early years of the Third Reich, Hitler and his race hygienists carefully crafted eugenic legislation modeled on laws already introduced across America, upheld by the Supreme Court and routinely enforced. Nazi doctors and even Hitler himself regularly communicated with American eugenicists from New York to California, ensuring that Germany would scrupulously follow the path blazed by the United States. American eugenicists were eager to assist.

A crucial point of Black's *War Against the Weak* concerns the resurgence, in the last few decades, of many of the troubling aspects of eugenics, now often rebranded under the name of "genetics" as noted above, or, as Black calls it, "newgenics." This is a clear example of how the so-called racist and sexist "scientific facts" that are still being defended by prominent evolutionary scientists nowadays do have a clear, direct link to eugenics and therefore to ideas defended by Darwin's cousin Galton and son Leonard Darwin and explicitly defended and quoted by Charles himself, including in his *Descent* (see Box 3.7).

Box 3.7 Eugenics, Newgenics, Genetics, Scientific Racism, and James Watson

In *War Against the Weak*, Black discusses the disturbing links between eugenics and what he defines as "newgenics," that is, a kind of rebranded eugenics under the name of "genetics":

> Insurance companies vigorously claim they do not seek ancestral or genetic information. This is not true. In fact, the international insurance field considers ancestral and genetic information its newest high priority. The industry is now grappling with the notion of underwriting not only the individual applicant, but his family history as well. Insurers increasingly consider genetic traits "preexisting conditions" that

(continued)

should either be excluded or factored into premiums. A healthy individual may be without symptoms, or asymptomatic, but descend from a family with a history of a disease. In the industry's view, that individual presumably knows his family history; the insurance company doesn't. Insurers call this disparity "asymmetrical information", and it is hotly discussed at numerous industry symposiums and in professional papers.

Governments and privacy groups worldwide want to prohibit the acquisition and use of genetic testing. Many in the insurance world, however, argue that their industry cannot survive without such information, and the resulting coverage restrictions, exclusions and denials that would protect company liquidity. Insurance discrimination based on genetics has already become the subject of an active debate in Great Britain. British insurers were widely employing predictive genetic testing by the late 1990s to underwrite life and medical insurance, and utilizing the results to increase premiums and deny coverage. The science of such testing is by no means authoritative or even reliable, but it allows insurers to justify higher prices and exclusions. Complaints of genetic discrimination have already become widespread. A third of those polled from genetic disorder support groups in Britain reported difficulties obtaining insurance, compared to just 5 percent from a general population survey. Similarly, a U.S. study cited by the American insurance publication Risk Management found that 22 percent of nearly one thousand individuals reported genetic discrimination.

A British Medical Journal study paper asserted, "our findings suggest that in less clear cut instances, where genes confer an increased susceptibility rather than 100% or zero probability, some people might be charged high premiums that cannot be justified on the actuarial risk they present". Nearly three-quarters of a group surveyed by Britain's Human Genetics Commission (HGC) objected to insurer access to genetic testing. One man who tested positive for Huntington's told of being denied insurance when his genetic profile became known; later, when he did obtain a policy, it was five times more expensive. One forty-one year-old London woman recalled that after her genetic report showed a gene associated with breast cancer, she was unable to buy life insurance. In consequence, when she attempted to purchase a home in 1995, it was more costly. Chairman of the HGC Helena Kennedy said: "most of us are nervous and confused about where technology might be leading, and the potential challenges to privacy and confidentiality.. we know from our survey that people are worried that these developments might lead to discrimination or exploitation, and are skeptical of the law's ability to keep up with human genetics".

A Code of Practice for genetic testing by British insurers was established in 1997, but in 2001, Norwich Union Insurance, among other firms, admitted it had been using unapproved genetic tests for breast and ovarian cancers, as well as Alzheimer's. British insurers began widely utilizing genetic tests after a leading geneticist consulting for the industry's trade association recommended the action, a Norwich Union executive explained. The widespread concern in England is generation-to-generation discrimination pivoting not on race, color or religion, but on genetic caste. "We are concerned, of course", warned Dr. Michael Wilks, of the British Medical Association's Medical Ethics Committee, "that the more we go down the road of precision testing for specific patients for specific insurance policies the more likely we are to create a group who simply will not be insurable". Wilks called such a group a genetic "underclass". A member of Parliament characterized Norwich Union's actions as an attempt to construct a "genetic ghetto".

Prominent voices in the genetic technology field believe that mankind is destined for a genetic divide that will yield a superior race or species to exercise dominion

(continued)

over an inferior subset of humanity. They speak of "self-directed evolution" in which genetic technology is harnessed to immeasurably correct humanity, and then immeasurably enhance it. Correction is already underway. So much is possible: genetic therapies, embryo screening in cases of inherited disease and even modification of the genes responsible for adverse behaviors, such as aggression and gambling addiction. Even more exotic technologies will permit healthier babies and stronger, more capable individuals in ways society never dreamed of before the Human Genome Project was completed. These improvements are coming this decade. Some are available now. But correction will not be cheap. Only the affluent who can today afford personalized elective health care will be able to afford expensive genetic correction. Hence, economic class is destined to be associated with genetic improvement. If the genetically "corrected" and endowed are favored for employment, insurance, credit and the other benefits of society, then that will only increase their advantages. But over whom will these advantages be gained? Those who worry about "gene lining", "genetic ghettos" and a "genetic underclass" see a sharp societal gulf looming ahead to rival the current inequities of the health care and judicial systems. The vogue term designer babies itself connotes wealth. The term designer babies is by and large just emblematic of the idea that genetic technology can do more than merely correct the frail aspects of human existence. It can redress nature's essential randomness.

Purely elective changes are in the offing. The industry argues over the details, but many assure that within our decade, depending upon the family and the circumstances, height, weight and even eye color will become elective. Gender selection has been a fact of birth for years with a success rate of up to 91 percent for those who use it. It goes further – much further. A deaf lesbian couple in the Washington, D.C., area sought sperm from a deaf man determined to produce a deaf baby because they felt better equipped to parent such a child. A child was indeed born and the couple rejoiced when an audiology test showed that the baby was deaf. A dwarf couple reportedly wants to design a dwarf child. A Texas couple reportedly wants to engineer a baby who will grow up to be a large football player. One West Coast sperm bank caters exclusively to Americans who desire Scandinavian sperm from select and screened Nordics. All of us want to improve the quality of our children's futures. But now the options for purely cosmetic improvements are endless. A commercialized, globalized genetic industry will find a way and a jurisdiction. It will be an international challenge to successfully regulate such genetic tampering and the permutations possible because few can keep up with the moment-to-moment technology. It goes much further than designer babies. Mass social engineering is still being advocated by eminent voices in the genetics community. Celebrated geneticist James Watson, co-discoverer of the double helix and president of Cold Spring Harbor Laboratories, told a British film crew in 2003, "if you are really stupid, I would call that a disease. The lower 10 per cent who really have difficulty, even in elementary school, what's the cause of it? A lot of people would like to say, 'Well, poverty, things like that'. It probably isn't. So I'd like to get rid of that, to help the lower 10 per cent". For the first half of the twentieth century, Cold Spring Harbor focused on the "submerged tenth"; apparently, the passion has not completely dissipated.

"Was Hitler a Darwinian?": A Biased Question About Scientific Biases

> *Struggle is always a means for improving a species' health and power of resistance and, therefore, a cause of its higher development.. those who want to live, let them fight, and those who do not want to fight in this world of eternal struggle do not deserve to live.* (Adolf Hitler, *Mein Kampf*)

After the above discussions about eugenics and how racism and ethnocentrism are still commonly seen among the prominent members of the scientific community, we are now better equipped to discuss Richards' 2013 book *Was Hitler a Darwinian? Disputed Questions in the History of Evolutionary Theory*, mentioned in Chap. 1. Interestingly, the line of defense used by Richards to contradict the historical links between some key Darwin ideas and certain central points of Nazi ideology is very different than the one he used in his 1987 book about Darwin. As we have seen, in that 1987 book, he used the more typical line of defense employed by most scholars to defend anything "negative" about Darwin: that was not Darwin's fault but the fault of his society. Although Richards' 2013 line of defense is different, it also involves the type of mental gymnastics commonly used in books idolizing Darwin. That is why Richards' 2013 book provides such a fascinating case study within the context of the present volume and its discussions on biases: Richards' knowledge of the history of science, and of Darwin's life and works, is exceptional, stressing that such type of mental gymnastics is not performed by only a few scholars or "straw-men" but instead by top scholars as well.

It is true that, due to his deep historical knowledge and scholarly skills, Richards does not fall into those mental gymnastics traps so often, or so deeply, as many other scholars discussed in this volume do. However, he does fall into them when it comes to discus Darwin's less "positive" aspects in his 1987 and 2013 books that are particularly focused on Darwin, contrary to what he does in his books focusing on other scientists, such as his 2008 book about Haeckel. In that excellent 2008 book, Richards used a very different and more sober tone when it discussed at length *both the strengths and the weaknesses*, at *both the personal and scientific levels*, of Haeckel. One thing is to talk about Haeckel without taboos, another very different one is to do the same with Darwin: a fact that, by itself, shows us the huge *power* that is still exerted by Darwin's name and symbol within academia, nowadays.

This of course does not mean that Richards went as far, in that 2008 book, to link Haeckel's evolutionary ideas to Nazi ideology, as many scholars do. After all, Richards is well aware that if he would do so, he would have to also recognize that there would be a link between Darwin's ideas and Nazi ideology. Richards knows too well that after reading Haeckel's ideas about biological evolution – including those on human evolution and "races" – Darwin explicitly wrote that he essentially agreed with them and felt that they were extremely similar to his own ideas. So, how did Richards, in his 1987 and 2013 books, addressed the very well-reported fact that the German Haeckel and the Darwinian Austrian Konrad Lorenz publicly supported, or provided "scientific" support to, Nazi ideology? He used the strategy that he used in his 1987 book about Darwin: "everybody did so back then" in Austria and Germany. Again, this is historically not true: not 100% of Austrian and German scientists back then publicly supported, or provided "scientific" support to, the ideologies defended by the Nazis.

If one would apply such a line of reasoning, then the Nuremberg trials – held for the purpose of bringing Nazi war criminals to justice after the Second World War – would be absurd, as many of those convicted in those trials were in a way much more "the product of their time." That is, contrary to people such as Lorenz or Haeckel, some of them were literally obliged to follow orders from higher officers to behave as they did. Someone obliged by his superior to kill, or plan how to kill, Jews or Gypsies clearly had less agency and choice than scientists such as Lorentz and Haeckel did when they first *chose* to write "scientific" texts or public statements supporting Nazi ideology. Fortunately, this "everybody was like that back then" line of defense was not applied during the Nuremberg trials, which did convict many criminals of war, giving a very important sign that such atrocities should *not* be permitted, defended, or repeated and that, if they were, they would *not* remain unpunished.

Coming back to Richards' 2013 book, *Was Hitler a Darwinian?*, its very title is an emblematic example of the type of mental gymnastics performed in that book because it was clearly chosen so that the answer was, as Richards enthusiastically states in the last sentence of the book, "a very loud and unequivocal no!" The way the question is framed, in particular the way Richards answers to it, shows that basically every politician or white supremacy leader would always qualify as an "unequivocal no." After all, "Darwinian," "Newtonian," or "Lamarckian" are names usually employed to designate scholars and thinkers, not commonly political leaders. Was Obama a Darwinian? Was Mandela a Newtonian? The answer would be a no. Even David Duke, who explicitly recognized the contributions of Darwin's books to his "awakening" as a white supremacist, as seen before, would not qualify as a "Darwinian" under the very strict definition that Richards constructed to define a "Darwinian." Under such a definition, even many of my biologist and anthropologist colleagues would probably not pass Richards' "Darwinian" test.

For instance, Richards criticizes scholars that argue that Nazi ideology was related to Darwin's non-teleological, materialist, and non-moralistic view of nature because, as he rightly points out, Darwin's view of human evolution was actually highly teleological. However, Richards also recognizes – rightly so – that many scholars that define themselves as Darwinian, idolize Darwin, and are influenced by his ideas claim that his writings are not teleological and even that Darwin "finished up with teleology" as we have seen. So, it is logically inconsistent to argue, as Richards did, that Hitler could not be influenced by the ideas of Darwin because some aspects of Nazi ideology do not match with some aspects of Darwin's ideas that are unknown even to countless scholars that define themselves as "Darwinian."

Richards admits this paradox, after referring to some very specific details about Darwin's ideas – that most people, and surely the vast majority of politicians, have no idea about – to argue that Hitler was not a "Darwinian;" he states, "it might be thought that I am simply quibbling about technicalities.. Hitler after all used a phrase of Darwinian provenance, which points to the ultimate source of his ideas." However, he then points out, "we are talking about ideas, not mere words; and the ideas that Hitler deploys are not Darwin's.. if words alone are to be the criterion, one might just as easily ascribe his enthusiasm for struggle to Christianity, the greatness of which he explicitly identified with its constant struggle against other religions and its efforts to extirpate them." Richards was referring to Hitler's obsessive use of the term "struggle" in his *Mein Kampf* – which literally means "my struggle" or "my battle" – including the fact that two times that Hitler used it, he used a phrase that was similar, or almost similar, to a central evolutionary concept in Darwin's writings: the struggle for existence. Actually, as recognized by Richards, the very phrase used in the German translation of Darwin's *Origin of Species* for the "struggle for existence" was "*Kampf um's Dasein*," so, literally, both books used the very same word in the title, *Kampf*:

> Most authors who try to connect Darwin with Hitler focus on Hitler's idea of "struggle", as if this implied Darwin's principle of "struggle for existence", that is, natural selection. The very title of Hitler's book, *My Battle* (or Struggle, War) hardly resonates of Darwinian usage – especially when one considers the title he originally planned: *A Four and a Half Year Battle [Kampf] Against Lies, Stupidity and Cowardice*. A simple word count indicates that Hitler had a mania for the notion of struggle that no simple acquaintance with the idea in a scientific work could possibly explain. The term appears in one form or another some 266 times in the first 300 pages of the 800 page book: from the simple Kampf (struggle) to "Bekämpfung" (a struggle), "ankämpfen" (to fight), "Kampffeld" (field of struggle), "Kampfeslust" (joy of struggle), etc.. The phrase used in the German translation of the *Origin of Species* for "struggle for existence" is "Kampf um's Dasein". Hitler uses that phrase, or one close to it, twice in *Mein Kampf*. Those two instances occur in an almost 800-page book in which some form of the word appears on almost every page; by sheer accident such a phrase might spill from the pen of an obsessed individual who seems to know hardly any other word.
>
> Those two instances yet do have a Darwinian ring. Both come in a context in which Hitler is worried about the apparent reduction in births in Germany due to lack of land. He deployed the terms in an effort to justify annexing "unused" land to the east (e.g., Poland, Ukraine). His convoluted argument runs like this: if Germans stay within their own borders, then restraint on propagation will be necessary, and compassion will require that even the weak will be preserved; moreover, barbarians lacking culture but strong in determination will take the unused land; hence Germans, the bearers of culture, ought to appropriate the area needed for living (Lebensraum). Hitler's argument makes little sense from a Darwinian perspective. If living conditions became restricted within closed borders, it would be the more fit who would survive; while if conditions became relaxed by moving into an unoccupied and fruitful land, then the fit and the less fit (by some measure) ought to have fairly equal chances. Hence, from a Darwinian point of view the conclusion ought to be just the opposite to that which Hitler drew. Be that as it may, Hitler did argue that maintaining current borders allowed the weaker to survive "in place of the natural struggle for existence, which lets live only the strongest and healthiest.

This example shows how Richards is indeed setting the bar so high that basically the vast majority of people that are not evolutionary biologists or historians of science – and even many of them – would not be called a "Darwinian." Obviously, what most people, including lay people, politicians, and white supremacist leaders, easily recognize about Darwin are mostly his catchy metaphors – which in great part were precisely used by Darwin so that they could be read and easily understood by the broader public. This includes the central metaphor of the *Origin*: the struggle for existence. Hitler used this metaphor – concerning, for instance, reproduction and in particular a reduction in births – in exactly the same type of context in which Darwin often used it. For instance, Hitler stated that "this earth is a trophy cup for the industrious man.. and this rightly so, in the service of *natural selection*.. he who does not possess the force to secure his Lebensraum in this world, and, if necessary, to enlarge it, does not deserve to possess the necessities of life.. he must step aside and allow stronger peoples to pass him by." Accordingly, Hitler stated that "our nation's *struggle for existence* forces us to utilize all means, even within Reich territory, to weaken the fighting power of our enemy and to prevent further advances." As another example, among many others, in *Mein Kampf* he wrote that "*Struggle* is always a means for improving a species' health and power of resistance and, therefore, a cause of its higher development.. those who want to live, let them fight, and those who do not want to fight in this world of eternal *struggle* do not deserve to live."

There is no doubt, at a factual level, that the notion of natural selection mainly driven by the struggle for existence that Hitler is referring to is the same notion that was defended in Darwin's books, which was central to his evolutionary ideas. As we have seen above, Richards himself had acknowledged in his 1987 book that Darwin conceived human evolution, including even the evolution of morals, as the product of a natural selection that "is daily and hourly scrutinizing, throughout the world, every variation, even the slightest; rejecting that which is bad, preserving and adding up all that is good." Indeed, Darwin referred to an *eternal* struggle for existence is *always* scrutinizing, every day, every hour, selecting what is "good," and condemning to extinction what is "bad," exactly as put by Hitler. It does seem rather odd that, despite these similarities, in particular despite the fact that Hitler *directly* used two of the most critical terms in Darwin's writings – "natural selection" and "struggle for existence" – to support his white supremacy and Nazi ideologies, one would argue that there is no link at all between Darwin's writings and Hitler's ideas.

Within his line of defense of Darwin in his 2013 book, Richards argued that Hitler's obsession with "struggle," and his use of the struggle for existence and natural selection notions that were central in Darwin's works, could be as easily ascribed to Darwin's ideas as to Christian ideas. Such a line of argumentation, by an otherwise brilliant historian of science, illustrates the power of biases: when you hear words such as "struggle," "struggle for existence," and "natural selection," would the first thing that comes to your mind be the *Bible*? Are these notions really as central to the *Bible* as they are to Darwin's books? As we have seen throughout this volume, the vast majority of authors, including both those that idealize and demonize Darwin, recognize that the concept of natural selection related to a struggle for existence is the most – or at least one of the most – central ideas in Darwin's evolutionary writings. As we have also seen above, in the *Descent*, Darwin explicitly stated the "fact" that in such a war-like struggle for existence, most non-"white" societies were naturally doomed to become extinct or to suffer the types of atrocities that the Nazis committed against the Jews, Gypsies, and other "others."

Struggle for existence was not merely a detail in Darwin's writings: it is instead a central metaphor that became – and still is – very powerful in popular culture. It is a factually inaccurate metaphor that influenced, and continues to influence, many people, including those eugenicists that directly quoted Darwin to support their ideas, which, in turn, clearly influenced the eugenicist ideas of many politicians, including those of the Third Reich. In fact, many of those eugenicists that quoted Darwin to support their ideas directly supported Hitler's eugenicist ideas, when he came to power, which is not surprising at all because Hitler's ideas were influenced by Darwin's ideas as well. This is the vicious cycle of indoctrination that results from the links between scholars, politicians, and the broader public. Together with the support of racism, colonialism, and slavery by countless biologists, the support provided by so many biologists to eugenicist ideas similar to those defended by Hitler, and in particular to Hitler himself, were among the darkest moments resulting from such a vicious cycle.

We tend to forget that there were only 7 years between Darwin's death (1882) and Hitler's birth (1889). The way that one learns world history in school often gives us the perception that those were two completely different worlds and eras. We associate Hitler with recent history, the middle of the twentieth century, and with people like Churchill, whereas we associate Darwin with very old events that occurred in the nineteenth century, the Industrial Revolution, and people like Queen Victoria. However, the reality is that if one had died or the other had been born less than a decade after or before, then the two would have coexisted. What happened in England in the nineteenth century, what happened in Germany in the twentieth century, what were the English scientists saying, what were the politicians doing, and so on are not part of completely different worlds: all these things are instead profoundly interconnected.

Indoctrination to old biases, prejudices, and factually wrong narratives is not necessarily, and even often, specific to a single country. Take, for example, the types of things people hear when they go to a protestant church on a Sunday in England, or several thousands of kilometers away from it, in the United States. Some of the things that will be read or sung in those two churches will likely be exactly the same, such as certain psalms, which were probably also read on Sundays at churches in the United States and England many decades ago. Now, do those people that went to churches on a Sunday and read those psalms, in various countries and at very different epochs, all know exactly the same thing about the Word of God, all the details, and so on? Of course not. However, still, one calls them – and they define themselves – as Christians because many of the things they do – such as going to church on Sundays, reading, and singing – are influenced by Christian ideas. So, it does seem a bit odd to argue that Hitler was not at all influenced by Darwinism, when he *did* cite ideas and defended – and literally used the same words – that Darwin did in contexts that were extremely similar to those in which Hitler defended them, particularly in Darwin's *Descent*. Often, indoctrination is mainly carried out through sound bites, metaphors, and the very basic ideas that stick in our heads, often from a young age. How did the priests that went to Africa or Latin America try to teach the locals about the Word of the Christian God? Did they explain to them all the details of the scriptures and require them to know all those countless details in order to call them Christians? Of course not, they often taught them a few key, simple Christian ideas and metaphors, many times even using no words at all but instead paintings and other forms of representation.

Hitler himself provides one of the most powerful, and tragic, examples of the importance of using such simple ideas, metaphors, and sound bites. What he wrote in *Mein Kampf* was not original and surely not brilliant. He merely combined some factually wrong and simplistic sound bites that were already immersed in popular culture – such as Darwin's evolutionary struggle for existence – with long-standing anti-Semitic ideas – if something goes wrong, always blame the Jews – and so on, in a way that was simplistic and catchy enough to make those ideas stick in the heads of millions of people, from many different backgrounds and social strata, who *believed* in them back then. Many millions of neo-Nazis still believe in them, today, across the globe: literally, every single Western country has at least a few neo-Nazi groups nowadays. Many people among those groups have never read *Mein Kampf*, and most of them do not know the details of Hitler's ideologies and the Third Reich but still define themselves as neo-Nazis. This is what makes indoctrination so dangerous – a tragic reminder of the power of catchy and simplistic sound bites and of how they are hugely efficient tools to help make people *believe* in factually wrong ideas, *idolize* those that proclaimed them, and, accordingly, support the atrocities that they defended or committed in the past. When a group of neo-Nazis kills an African American boy based on such Nazi sound bites, it would be rather odd – and would be completely missing the point – to argue that what they did has nothing to do with Hitler and his ideas, just because they did not read all the pages of *Mein Kampf* or know all the details and nuances of Nazi ideology. Instead, what should be done in such a case is trying to avoid that our youth accepts those ideas and sound bites as dogma by deconstructing the idealized way in which they see Hitler and his ideas, as well as the scientific "giants" that provided "scientific support" to those ideas, by showing them that such ideas are factually inaccurate.

Remarkably, only very few authors have attempted to do this in a proper, detailed way. One of the rare examples is Daniel Gasman, who published, in 1971, the book *The scientific origins of national socialism.* As recognized by scholars such as Richards, Gasman is not a creationist with an agenda to try to demonize Darwin and his followers and accuse them of being Antichrists that led to everything bad that happened on the planet in the last 150 years. Not at all. Gasman's book, which defends that social Darwinist ideas and Haeckel's Darwinian ideas in great part fostered Nazi biology and ideology, is based on a superb compilation of historical facts undertaken by someone that clearly has a deep knowledge about the history and ideology of Nazism. Of course, this does mean that Gasman is always right about these topics – nobody is. However, even Richards, in his 2013 book, in which he argues against the influence of Darwinism and social Darwinian ideas on Nazi ideology, admits that Gasman's monograph "has orchestrated a strong case."

Interestingly, after recognizing this point, Richards starts to use the type of vague, and sometimes off the point, argumentation that is so often used when scientists and historians of science react defensively to the very few scholars that dare to put in question Darwin's ideas and their societal implications. For instance, Richards stated that "the influence of Haeckel.. on Nazi ideology was hardly straightforward.. the Nazi elite resisted evolutionary theory, despite its scientific charms.. after all, could the Aryan race have descended from a tribe of baboons?" Such an argumentation is rather weak, particularly coming from someone as knowledgeable as Richards, who in his other books recognized that there is much more to Darwinism than the statement that humans came from monkeys. Many right-wing conservatives in the United States that do not accept that humans evolved from other primates label themselves as social Darwinists. As explained above, social Darwinism is much more related to the "facts" constructed by Darwin and Spencer about human humans to justify political conservatism, imperialism, racism, and societal hierarchies than it is to the evolution of nonhuman animals. As shown in Gasman's book, there are countless passages of texts written by the Nazi elite, including by Hitler himself, citing and defending Darwinian and social Darwinian ideas. In fact, as Richards' admits just a few sentences later, "some Nazi propagandists did eulogize Haeckel for having supplied scientific support for central ideas of the new regime.. Haeckel and some of the Monists undoubtedly created an intellectual environment congenial to the growth of Nazi pseudoscience."

After discussing Haeckel, Richards then examined the links between another prominent Darwinian biologist, Lorenz, and Nazi ideology:

> In 1938 Lorenz, like many other German scientists – and particularly physicians – joined the NSDAP [the Nazi party]. Prior to the war he had spoken at a respect scientific congress (a meeting of the German Psychological Association), which was, however, sponsored by the Nazis. He had also published two articles in a journal having explicit Nazi connections (*Der Biologe*). On these few occasions of public Nazi association, he had touched on a theme that Kalikow [an author that wrote about Lorenz] identifies as the principal evidence of endorsement of National Socialist ideology: that like domesticated animals, civilized, urban men and women were in peril of biological degeneration. Passages such as the following, she believes, reveal the Haeckel-Nazi source of the main elements of Lorenz's biology of behavior: "*Whether we share the fate of the dinosaurs or whether we raise ourselves to a higher level of development, scarcely imaginable by the current organization of our brains, is exclusively a question of biological survival power and the life-will of our people. Today especially the great difference depends very much on the question whether or not we can learn to combat the decay phenomena in our people and in humanity which arise from the lack of natural selection. In this very contest for survival or extinction, we Germans are far ahead of all other cultural peoples*". But this is a gossamer thread by which to tie Lorenzian biology to the Nazis. The vast bulk of Lorenz's work on instinct rests squarely in the Darwinian evolutionary tradition. Even his concern about behavioral and mental degeneration has deep roots in that tradition.
>
> The above passage, *mutatis mutandis*, could have been written by any number of British or American evolutionary biologists in the last [19th] century or the early part of this one [20th]. And again, recall that Darwin too warned of the dangers to human progress consequent on the disengagement of natural selection in civilized societies. Lorenz undoubtedly descended to accommodate some of his biological views to the ideology of his time and place. He may have reached his nadir in 1940, when in an article on domestication, he wrote: "*If it should turn out.. the mere removal of natural selection causes the increase in the number of existing mutants and the imbalance of the race, then race-care must consider an even more stringent elimination of the ethically less valuable than is done today, because it would, in this case, literally have to replace all selection factors that operate in the natu-*

> *ral environment*". In view of the history we have traversed [in Richards book], even this stain of ideology is hardly surprising. At this point in Lorenz's career, certain well-entrenched evolutionary ideas happened to intersect with despicable Nazi dogma. Certainly he fostered the union of biology and propaganda, but I doubt that his main concerns would have been markedly different had the Weimar Republic survived. Nonetheless, just this sort of public association of Nazism with human evolutionary biopsychology froze any enthusiasm for the discipline immediately after the war, and continues to chill its development within contemporary biology of behavior as well as within the social sciences.

Once again, Richards uses a defensive tone in his arguments against the "strong case" formulated by Gasman. In fact, a fascinating aspect of the case study provided by Richards' 2013 book is that in the above excerpt, he literally postulated the key answer to the main issue discussed in that book, but the combination of his idealization of Darwin and the defensive stance used by him seemingly did not allow him to realize that. That is, he provided this answer when he tried to defend Lorenz by stating that what Lorenz wrote could have been written by many British and American Darwinians in the nineteenth century or the early twentieth century. This is precisely the point. Many Darwinists would have said the same thing because, as we have seen above, the very first Darwinist – Charles Darwin – wrote exactly the same type of stuff, in the *Descent*, about the inferior "others," "savages," and so on. This was the very Word of Pope Darwin, so it is expected that it would become the "Darwinian evolutionary tradition" that was defended by so many other Darwinians subsequently, including those that directly defended the Nazi ideology. In fact, on some occasions, Richards seems to openly admit this crucial point, for instance, when he recognizes that "Darwin too warned of the dangers to human progress consequent on the disengagement of natural selection in civilized societies." Moreover, as noted above, in his 1987 book, Richards *did* recognize the ideological links between Darwinism, Lorenz's writings, and Nazi ideology, but, then, in his 2013 book, he argues that Darwin's ideas and writings have no links to, nor did they influence in any significant way, Nazi ideology, in opposition to the "strong" historical data compiled by scholars such as Gasman, which show otherwise.

Another major logical inconsistency between Richards' 1987 and 2013 books concerns a million dollar question that Richards fails to address in the latter book: where do the scientific ideas used by the Nazis to justify their ideology come from? From a *tabula rasa*? On the one hand, at the beginning of his 1987 book, Richards has stated, as we have seen, that Darwin was "triumphant" in the sense that his ideas became hugely influential across the globe, particularly within Western countries. However, on the other hand, he suddenly argues, in his 2013 book, that Darwin's racist, ethnocentric, and war-like struggle-for-existence evolutionary ideas did not have any influence at all on major events such as the subsequent widespread rise of white supremacy eugenic ideas within those countries, including Germany where it led to many aspects of Nazi ideology. Is this not a paradox? It is like saying that Marx's writings were "triumphant" in countries such as Russia, China, North Korea, Mozambique, Cuba, and so on but then arguing that they did not have any significant impact on the history of those countries. This is another example of the typical type of mental gymnastics employed when one is trying to have the cake and eating it too, when it comes to discussing Darwin's societal legacies. If Darwin's ideas were "triumphant" and became widely influential, particularly in Western countries, then one cannot say that they only influenced people that were "noble" or "good": this is just not possible, logically. His ideas obviously influenced people from all backgrounds, from "noble" professors to less "noble" authoritarian political leaders, exactly as Marx's ideas did. We know that this is factually true because numerous white supremacists such as David Duke explicitly quoted Darwin's racist, ethnocentric, and war-like struggle-for-existence ideas to "scientifically" justify the atrocious things that they defended or committed.

In summary, concerning Richards' 2013 book's title *Was Hitler a Darwinian?*, a historically and scientifically more relevant, and less biased, title would be *Were Hitler's Ideas Influenced by Darwinian Ideas?* After all, as recognized by Rochard, and by other scholars that tend to idealize Darwin, such as Bowler in his *Darwin Deleted*, Hitler's ideas about history, race, and struggle for existence are indeed very similar to some of those defended by Darwin in the *Descent*. Richards actually seems to recognize that "most scholars" would actually answer yes to the question "Were Hitler's ideas influenced by Darwinian ideas?" – for instance, when he declares that "most scholars of Hitler's reign don't argue for a strong link between Darwin's biology and Hitler's racism, but they will often deploy the vague concept of 'Social Darwinism' when characterizing Hitler's racial ideology." Considering that countless scholars still idealize Darwin, the fact that Richards recognizes that "most" of them would still recognize such a link between social Darwinism and Hitler's ideology truly means a lot. Richards also hints at this when he notes that many authors do *try* to dissociate Darwin's racist ideas from Haeckel's ones in order to "offer up Haeckel to save Darwin['s legacy] by claiming significant differences between their views." This is a claim that, as seen above and as also emphasized by him, "cannot be sustained":

> Scholars like [Stephen Jay] Gould, Bowler.. as a host of others.. attempt to distinguish Haeckel's views from Darwin's, so as to exonerate the latter while sacrificing the former to the presumption of a strong causal connection with Hitler's anti-Semitism. I don't believe this effort to disengage Darwin from Haeckel can be easily accomplished, since on central matters – descent of species, struggle for existence, natural selection, inheritance of acquired characters, recapitulation theory, progressivism, hierarchy of races – no essential differences between master and disciple exist. So if Hitler endorsed Haeckel's evolutionary ideas, he thereby also endorsed Darwin's.

This is indeed true. Moreover, following exactly the same line of reasoning used by Richards in this excerpt, one could say exactly the same about Hitler – one could say: "I don't believe this effort to disengage Darwin from Hitler can be easily accomplished, since on central matters – struggle for existence, natural selection, progressivism, hierarchy of races – essential similarities between the two exist." So, why did Richards not apply the same logic that he applied to the similarities between Darwin's and Haeckel's ideas, to the similarities between Darwin's ideas and Hitler's ideology? Instead of doing so, after he admitted that the ideas of Haeckel – who, one needs to emphasize, is considered by a huge number of historians of science and biologists to be someone with direct links to Nazi ideology – and Darwin were very similar, he started again to engage in rather vague and off-the-point divagations. For instance, he engaged in a discussion about how many degrees of separation there were between the ideas of Hitler and those of Darwin, that is, if Hitler directly read Darwin's books, or if he had instead two degrees of separation by reading a book written by someone that read Darwin, and so on. This is a somewhat odd discussion because even many scholars that declare themselves as Darwinians nowadays never read Darwin's books; instead, they choose to know about the sound bites and metaphors included in those books by reading the more recent books that talked about them or that referred to other books that talked about them.

So, in a nutshell, the historical truth is that if one was to raise the more fair, balanced, and relevant question "Were Hitler's ideas influenced by Darwinian ideas?", then the answer would be a "loud" yes, as pointed out in Daniel Gasman's well-informed 1971's book *The Scientific Origins of National Socialism* and in other publications cited above. As scientists, we can be ashamed to recognize this. I have heard myself many scholars arguing that, while they recognize this fact, one *should not* admit that this is true because this can play into the hands of creationists and therefore of the non-intelligent arguments made within the "field" of "intelligent design." However, to not recognize the truth would be to continue to engage in the also non-intelligent behavior of burying our heads in the sand. We need to stop being so defensive about these historical facts and start to actually actively engage in discussing them within the academia and the general public, without any taboos, once and for all. If we fail to do so, then we will not be able to end once and for all many of the "scientific" myths constructed by such "scientific" giants, which are still believed by a huge number of people concerning the erroneous notions of "human races," the biological inferiority of "savages" and other "others," and the suffocating, implacable, eternal, war-like, selfish "struggle for existence" between such "races", and between "us" and "others."

When I openly discuss these topics without taboos in my talks across the globe, as well as with some of my colleagues privately, a typical comment that I hear, particularly by those from Western countries, is: you are comparing Darwin with Hitler, that is just wrong, and what they *did* is completely different. Yes, that is true, what they *did* is completely different, and I am obviously not arguing that it is not so. This is a typical argument of those idolizing Darwin, or for that matter, anybody else: to not even engage in a constructive discussion by implying that saying anything that they perceive as "bad" about their idols equals to saying that they are "demons." This is how it goes: you are comparing Darwin with this horrible person, which is appalling, and, so, I will not even engage in this discussion anymore. However, as explained above, that is not at all what is being discussed in this chapter. What the discussion provided here shows is that some *ideas* that Hitler used, to justify the atrocious things that he did, were strikingly similar to, and seemingly even based on – either directly or indirectly, we cannot say for sure – *ideas* that Darwin defended as "scientific" in his books and that became part of Western popular culture and indoctrination. These include, among others, the so-called "evolutionary" war-like, selfish struggle for existence between different human "races," the notion that Europeans are mentally and "morally" superior, that "others" are "naturally" condemned to become extinct or massacred, and so on.

These are exactly the same ideas that persist in the minds of many white supremacists and neo-Nazis even today, as attested by the *fact* that many white supremacy websites continue to directly cite Darwin, as David Duke and the KKK did in the past, as we have seen. That is, white supremacists *do* think that Darwin's ideas and writings are ideal weapons to convince other people that their racist and ethnocentric – and often also sexist – ideas are correct and are backed by science: that this is just the way things naturally are, as Darwin argued. We cannot simply ignore this *fact*. However, having said that, the fact that some of the *key ideas* of Darwin and Hitler about human "races" and the struggle for existence are very similar obviously does not mean that I am comparing Darwin and Hitler at *an individual level*, nor am I comparing what Darwin *did* with what Hitler *did*. Darwin *did not* give orders to others to create concentration camps or to use gas to kill millions of Jews or Gypsies, including small kids. Darwin used *words* to express and justify racist and ethnocentric ideas in books, as Hitler did, but, in addition, Hitler, as a political leader, did atrocious things in *practice* as well – and this is a huge, abysmal difference. Darwin never was, or wanted to be, a political leader, and therefore never had the power to give orders to others to commit such atrocities. Moreover, as noted above, Darwin was seemingly mainly a pacifist, although he did openly defended British colonialism and imperialism, which were related to a huge number of atrocities around the globe.

Having said that, it is also important to emphasize that even if Hitler had not given orders to his subordinates to commit such atrocities to "others," his theoretical contributions to Nazi ideology, including the publication of his *Mein Kampf*, would still be highly criticizable. This is because they deeply influenced, and continue to influence, millions of people, with many of them doing horrible things based on what he said and wrote about "others" such as Jews. Atrocities are to be blamed not only on those that give orders or to the people executing those orders but also on people that defended or accepted them, publicly, as well as on those that constructed the biased theoretical framework that supported such ideologies. Actually, in a sense, on some occasions, ideas and ideologies can be much more dangerous than weapons because contrary to weapons, which can be used by elites to *force* some people to do what they do not want to do, ideologies can make people *want* to commit such atrocities by believing that doing them is the right thing to do. We have seen this in many of the genocides that have occurred in the last 150 years. For instance, in the Rwandan genocide, the radio played a critical role in dehumanizing the Tutsis by using hateful language and disseminating propaganda. Should we say that those that said those things on the radio did not play any significant role in those genocides? Of course we should not. This is precisely why many countries nowadays have laws against hate speech. As we will see in the section below, such type of dehumanization, or the ideas that some human groups are "superior" to others, or that "races" are engaged in a struggle for existence, have provided the theoretical framework to justify not only atrocities such as mass killings and genocides but also other horrific practices such as the use of "other" human groups for medical experimentation.

Medical Experimentation, Scientific Biases, and "Giants"

> *In (those) experiments with African Americans.. subjects were given experimental vaccines known to have unacceptably high lethality, were enrolled in experiments without their consent or knowledge, were subjected to surreptitious surgical and medical procedures while unconscious, injected with toxic substances, deliberately monitored rather than treated for deadly ailments, excluded from lifesaving treatments, or secretly farmed for sera or tissues that were used to perfect technologies such as infectious-disease tests.* (Harriet Washington)

The history of racism, scientific biases, and indoctrination is profoundly linked to the history of both forced medical experimentation and genocides. Although some of the case studies that I will mention below are not necessarily directly related to Darwin's ideas, they provide critical examples to better understand the fallacies used in current discussions about "canceling culture" or "rewriting history," which, in turn, are related to fallacies about the idolization of many scientific "giants." Many of these case studies are particularly atrocious, including some of the most horrible things ever done by humans, and raise very difficult philosophical and ethical questions. As pointed out by Harriet Washington in her excellent 2006 volume *Medical Apartheid*, there are several books that discuss such highly sensitive and disturbing issues but only a few actually discuss them in a broader way within a comprehensive historical and societal context.

Although Washington's book was published more than a decade ago, it directly relates to events that occurred in the last years, related to the rise of the Black Lives Matter movement, as discussed at the beginning of the present volume. Some of these case studies are directly related to Western "giants" that are often seen by many European descendants as "heroes" and are portrayed as such in public statues that some within the Black Lives Matter movement want to see removed. An emblematic example of such a "giant" is James Marion Sims (Fig. 3.13), as explained by Washington:

> On a sylvan stretch of New York's patrician upper Fifth Avenue, just across from the New York Academy of Medicine, a colossus in marble, august inscriptions, and a bas-relief caduceus grace a memorial bordering Central Park. These laurels venerate the surgeon James Marion Sims, M.D., as a selfless benefactor of women. Nor is this the only statuary erected in honor of Dr. Sims. Marble monuments to his skill, benevolence, and humanity guard his native South Carolina's statehouse, its medical school, the Alabama capitol grounds, and a French hospital. In the mid-nineteenth century, Dr. Sims dedicated his career to the care and cure of women's disorders and opened the nation's first hospital for women in New York City. He attended French royalty, his Grecian visage inspired oil portraits, and in 1875, he was elected president of the American Medical Association. Hospitals still bear his name, including a West African hospital that utilizes the eponymous gynecological instruments that he first invented for surgeries upon black female slaves in the 1840s. But this benevolent image vies with the detached Marion Sims portrayed in Robert Thom's J. Marion Sims: Gynecologic Surgeon, an oil representation of an experimental surgery upon his powerless slave Betsey [Fig. 3.13].
>
> Sims stands aloof, arms folded, one hand holding a metroscope (the forerunner of the speculum) as he regards the kneeling woman in a coolly evaluative medical gaze. His tie and morning coat contrast with her simple servants' dress, head rag, and bare feet. The painting, commissioned and distributed by the Parke-Davis pharmaceutical house more than a century after the surgeries as one of its *A History of Medicine in Pictures* series, takes telling liberties with the historical facts. Thom portrays Betsey as a fully clothed, calm slave woman who kneels complacently on a small table, hand modestly raised to her breast, before a trio of white male physicians. Two other slave women peer around a sheet, apparently hung for modesty's

sake, in a each woman's body was a bloodied battleground. Each naked, unanesthetized slave woman had to be forcibly restrained by the other physicians through her shrieks of agony as Sims determinedly sliced, then sutured her genitalia. The other doctors, who could, fled when they could bear the horrific scenes no longer. It then fell to the women to restrain one another.. Betsey's voice has been silenced by history, but as one reads Sims's biographers and his own memoirs, a haughty, self-absorbed researcher emerges, a man who bought black women slaves and addicted them to morphine in order to perform dozens of exquisitely painful, distressingly intimate vaginal surgeries. Not until he had experimented with his surgeries on Betsey and her fellow slaves for years did Sims essay to cure white women. Was Sims a savior or a sadist? *It depends, I suppose, on the color of the women you ask.* Marion Sims epitomizes the two faces – one benign, one malevolent – of American medical research.. a doctor could be open about buying slaves for experiments, or locating or moving hospitals to areas where blacks furnished bodies for experimentation and dissection.

Public Health Service physician Thomas Murrell could brashly insist in the 1940s, "The future of the Negro lies more in the research laboratory than in the schools.. When diseased, he should be registered and forced to take treatment before he offers his diseased mind and body on the altar of academic and professional education". Even more recently, the segregated nature of U.S. medical training emboldened some physicians to speak with candor of misusing black subjects. "[It was] cheaper to use Niggers than cats because they were everywhere and cheap experimental animals", neurosurgeon Harry Bailey, M.D., reminisced in a 1960s speech he delivered while at Tulane Medical School. But as societal attitudes changed, so did physician reticence, and most became more circumspect. However, as late as 1995, radiation scientist Clarence Lushbaugh, M.D., explained that he and his partner, Eugene Saenger, M.D., chose "slum" patients as radiation subjects because "these persons don't have any money and they're black and they're poorly washed".. [there were] numerous instances of such shocking frankness on the part of white researchers and physicians when they thought that nobody outside of their peer group was listening.

In this case, as in many others concerning Western "giants," Sims is not at all a "hero" when seen from the perspective of "others," particularly the slaves that he abused, or their descendants. The fact that he has been idealized by numerous European descendants for his "humanity" – using information gained from those atrocious experiments conducted on slaves to help the life of "white" women – can only make sense in a context in which the lives of "others" are deemed less important than those of the "ingroup" or simply as not important at all. In my opinion, in a world in which all lives matter, including the lives of "others," it clearly makes no sense to have such a statue displayed in public spaces, particularly in front or across public buildings such as the New York Academy of Medicine. These are public buildings where African-Americans that are descendants of those slaves that suffered immensely from such studies, pass by, on a daily basis – just to be reminded of the powerful message displayed in such statues. That is, that such statues were done by, and *for*, "whites," that "others" do not matter at all (see also Box 3.8).

Box 3.8 Racial Studies, Scientific Biases, Beliefs, and Medical Experimentation

Washington's 2006 book *Medical Apartheid* provides many disturbing examples that are crucial to highlight the direct links between the so-called 'racial studies', scientific biases, and medical *beliefs* about, and experimentation forced upon, African Americans, during and after slavery:

(continued)

In 1839, Morton published *Crania Americana*, a book written to demonstrate how human skull measurements indicated a hierarchy of racial types. Morton determined that Caucasians had the largest skulls, and therefore the largest brains, and blacks the smallest. His tests were the forerunner of phrenology, which sought to determine character and intelligence by interpreting the shape of the skull. By 1848, Louisiana's Samuel A. Cartwright, M.D., had gained renown by publishing a plethora of articles on Negro medicine in southern medical journals, leading the Medical Association of Louisiana to appoint him chair of its committee to investigate black health and physiology. That same year, Cartwright published his paper "*The Diseases and Physical Peculiarities of the Negro Race*".. By 1851, Cartwright had also discovered and described a host of imaginary "black" diseases, whose principal symptoms seemed to be a lack of enthusiasm for slavery. Escape might have seemed normal behavior for a slave in ancient Greece or Rome, but Cartwright medically condemned such behavior in American blacks, offering a diagnosis of drapetomania, from the Greek words for flight and insanity.

Hebetude was a singular laziness or shiftlessness that caused slaves to mishandle and abuse their owners' property. Dysthesia Aethiopica was another black behavioral malady, which was characterized by a desire to destroy the property of white slave owners. Cartwright claimed that it "differs from every other species of mental disease, as it is accompanied with physical signs or lesions of the body discoverable to the medical observer". Struma Africana was a form of tuberculosis that physicians misdiagnosed as a peculiarly African disease. Cachexia Africana referred to blacks' supposed propensity for eating nonfood substances such as clay, chalk, and dirt. Actually, this disorder, which is called pica today, is not racially specific and the cravings it inspires were probably related to the rampant malnutrition among slaves. Tellingly, Dr. Cartwright recommended that these ailments be treated with corporal punishment or with internment in "work camps": "put the patient to some hard kind of work in the open air and sunshine.. the compulsory power of the white man, by making the slothful negro take active exercise, puts into active play the lungs, through whose agency the vitalized blood is sent to the brain to give liberty to the mind".

Other medical disorders were thought to manifest differently, usually less severely, in blacks. Syphilis, for example, was held to be racially dimorphic. Physicians believed it worked its most feared damage within the neurological system of whites but that the less evolved nervous system of blacks was left relatively unimpaired. In blacks, syphilis was thought to attack the muscles, including the heart. This belief that syphilis in blacks differed dramatically from the disease in whites provided a rationale for the infamous U.S. Public Health Service's (PHS) *Tuskegee Study of Syphilis in the Untreated Negro Male*. Between 1932 and 1972, six hundred black men, their wives, and their children were deceived into participating in a research study that denied them treatment, so that PHS scientists could trace the progress of the disease in blacks. Allegedly inferior cognition was only the tip of the iceberg. In 1854, several years after Cartwright published '*Report on the Diseases and Physical Peculiarities of the Negro Race*,' and five years before Darwin published *On the Origin of Species*.. Alabama, physician Josiah Nott, M.D., and George R. Gliddon produced an equally popular screed entitled *Types of Mankind*. In it, they claimed that blacks' physical and mental differences signaled their polygenic origins and proved black inferiority. For example, Nott theorized that the distinctive knee joint and 'long heel' of the black man proved he had been created as a 'submissive knee-bender' – a servant to whites. Scientists adjudged the dark skin of Africans as a biblical curse that set them aside as eternal servants to other men.. As late as 1903, Dr. W. T. English observed, 'a careful inspection reveals the body of the negro a mass of imperfections from the crown of the head to the soles of the feet.' Even biological advantages were cast as racial flaws: in discussing the tendency of blacks to survive yellow fever epidemics that killed whites, one physician denounced the 'inferior susceptibility' of black slaves.

The last sentence of Box 3.8 is a powerful illustration of a key topic discussed in this book, which crucially affected Darwin's evolutionary ideas and writings, as seen above: the confirmation bias. In the specific case study to which that sentence refers, even when empirical data show that "others" tend to survive more a certain disease, that information is changed upside down in order to "confirm" the "inferiority" of those "others": in this case, by arguing that they have an "inferior susceptibility." Washington's book provides many examples in which Western academicians and physicians applied such type of mental gymnastics to discard relevant or potential contributions of "Africans" or African Americans to medicine and to other important aspects of daily life:

> Interestingly, the contradiction of the black slave as both "riddled with imperfections from head to toe" and as a hardy laborer who was impervious to most illness escaped the scientific racists. Scientists expressed whichever opinion fit their political needs at the moment, as abolitionist Frederick Douglass suggested when he observed that ninety-nine of one hundred polygenists were Anglo-Saxon slave owners. Scientists also claimed that the primitive nervous systems of blacks were "immune" to physical and emotional pain and to mental illness. This belief, which will be discussed at greater length in the next chapter, released physicians and owners from the responsibility of shielding black slaves from painful medical procedures and justified torture such as branding, whipping, hobbling, and maiming. All these precepts of scientific racism, although convenient for the slave owner and physician, were highly illogical articles of faith. So was the supposed inferior intelligence of blacks, because planters and doctors behaved in many contexts as though they held the abilities and judgment of blacks in high regard, employing slaves in responsible positions as nurses, cooks, herbalists, midwives, overseers, leaders of work gangs, accountants, and operators of farm and factory implements. Owners reaped profits from the many patents on slave inventions, and physicians used slaves as skilled apprentices, who often went on to practice independently. White households depended upon the specialized skills and discernment of slaves, not the other way around..
>
> [Yes], appeals to God, the importance of moral fitness, and enlisting the help of departed spirits, especially the intercession of ancestors, were all key to the African-based healing process. Ancestors who were angered by disrespect or neglect could cause illness, alienation, and other troubles for the living. This is one reason the respectful ritual treatment of the dead was so important to slaves and why they reviled Western medicine when they discovered that physicians appropriated the bodies of dead slaves for display and dissection. Western medicine was thought ineffective against spirit-caused illness, and slaves often lacked confidence in a Western doctor's ability to cure them: if a doctor did not believe that one could be cursed or "conjured", how could he remove the threat? This is a wide generalization, because some slaves mistrusted African practitioners, who sometimes used their skills to harm as well as to heal. But in planters' farm books and in medical journals, physicians and slave owners repeatedly berated the ignorance and superstition that led slaves to conceal illness and to shrink from "scientific" Western medicine in favor of conjure women and witch doctoring.
>
> However, whites had no monopoly on science. The African tradition involved physiological as well as spiritual approaches to healing, including an encyclopedic knowledge of herbs, roots, and other natural medicaments. This detailed knowledge was continually passed down along lines of apprenticeship from wise women and male herb doctors to gifted young members of the community. Despite their characterization as primitive, African healers first employed citrus juice for scurvy and inoculation for smallpox and other viral illnesses; midwives used African techniques, herbs, and medicines so successfully – without dangerous tools of the day, such as forceps – that many white women called them to attend births. Some whites were impressed by the success rate of Negro doctors and

"doctresses", consulted them, and placed their medication recipes in the family book, and Western doctors faced brisk competition from black herb doctors. In an 1855 journal article, Dr. R. H. Whitfield of Alabama railed against "unscientific" midwives: "*[There are no practices wherein which] the female practitioners are less educated, being chiefly negresses or mulatresses, or foreigners without anatomical, physiological and obstetrical education.. that such uneducated persons should be generally successful is owing to the fact that [in] a great majority of cases no scientific skill is required, and thus a lucky negress become[s] the rival of the most learned obstetrician*". For all their complaints, physicians in the early to midnineteenth century were happy to leave the business of birthing in the hands of black midwives. However, physicians wanted black healers under the scrutiny and supervision of white physicians. White doctors denigrated black midwives and healers, calling them "uneducated", but white physicians themselves usually had no academic preparation beyond a few months in proprietary medical school or a few years of apprenticeship, which many blacks also shared. So until the mid-1800s, such claims of superior education rang hollow.

What Do We Really Know About Human Evolution and "Races"?

When I'm born I'm black, when I grow up I'm black, when I'm in the sun I'm black, when I'm sick I'm black, when I die I'm black, and you.. when you're born you're pink, when you grow up you're white, when you're cold you're blue, when you're sick you're blue, when you die you're green and you dare call me colored. (Oglala Lakota)

So, after seeing how so many authors, including Darwin, have argued that there are major differences between "races," concerning not only anatomical but also mental, moral, and even medical traits, it is worthy to briefly discuss in this section what the factual data from fields such as biology and anthropology truly tell us about these topics. We have already seen that most of the scientific "facts" about human "races" described in Darwin's books such as the *Voyage*, the *Descent*, and the *Expression* are contradicted by a plethora of empirical data. Importantly, it needs to be stressed that many of these empirical data were already available *before* he wrote his books. In addition, many others facts were just before his eyes when he encountered "others" in his travels, but he opted to "see" instead what his preconceived biases and prejudices *expected* him to see. In this section, I will therefore not repeat those cases already discussed above. Instead, in this Section I will use as a basis an interesting, and revealing book that was published recently, in 2021, edited by Jeremy DeSilva, entitled *A Most Interesting Problem: What Darwin's Descent of Man Got Right and Wrong About Human Evolution*.

This book is revealing in the sense that it does show that something is changing, together with the recent rise of movements such as Black Lives Matter and Me Too and a call for action from many other sectors of the society as a whole. As it was not a coincidence that Darwin was influenced by Victorian biases, or that he wrote about biological evolution when there was already a lot of buzz about it in England back then, it is surely not a coincidence that *A Most Interesting Problem* was published in 2021, and not, let us say, in the 1930s. It is also not a coincidence that the ideas of the *Extended Evolutionary Synthesis* are starting to be accepted by more and more researchers in the last few decades. Hard-core blind idealization of Darwin, Darwinism, and Neo-Darwinism is still hugely common, but more and more people are starting to dare to question it, within natural sciences in particular. This is another reason why *A Most Interesting Problem* is revealing: while most of the authors of its chapters make a huge effort – and, often, perform some mental gymnastics – to provide a very benign way of portraying Darwin's *Descent* and its ideas, a minority of them have no problems to openly discuss, without taboos, what the available historical and scientific data actually show. An example of this is Agustin Fuentes, who in his chapter "On the races of man" did not refrain from explicitly stating that Darwin was a racist and that the racist "facts" that he defended or constructed were factually wrong. And, importantly – and in line to what is shown in this volume – that such factually inaccurate ideas *did* have huge scientific and societal repercussions that continue to be felt even today. For instance, Fuentes wrote:

> Over the past century and a half, many scholars have invoked the perspective of a scientist from another planet as narrator, assuming that such a "view from outside" keeps the science neutral and unbiased. However, in reality, this is a poor move. We know that humans are enmeshed, enculturated, and always shaped by their life experiences, language, and history. We are never fully objective when talking about humanity, even when we try to be, and a good scientist recognizes that. This was Darwin's first mistake.. [For instance] he reveals his bias, stating, "Even the most distinct races of man, with the exception of certain negro tribes, are much more like each other in form than would at first be supposed". Darwin demonstrates a consistent bias against people from sub-Saharan Africa and those of African descent.. To his credit, Darwin's review of the differences and similarities across human groups comes down strongly on the side of similarities, with a few exceptions. He spends a number of pages highlighting specific differences in bodies and behaviors and even claims that different species of lice infect different humans. This is incorrect. There are three kinds of human-specific lice, and they infect all humans.
>
> In his overview of differences (and similarities), Darwin draws from published books and studies but relies heavily on individual accounts of personal experience (including his own). This reliance on individual accounts poses a problem, given the substantive bias shown by the individuals on whom he relies (European colonialists, scientists, and travelers).. Given his overview of the "data", Darwin concludes it is most appropriate to see the "races" as subspecies. Subspecies are clusters of groups that, while in the same species (same common ancestor), have important and evolutionarily derived differences that set them apart from one another. This view, unfortunately is exactly the same argument coopted by racists and separatists today and remains incorrect. But.. [Darwin] goes beyond noting subspecific classifications. Darwin tells us "*Their [the "races"] mental characteristics are likewise very distinct; chiefly as it would appear in their emotional, but partly in their intellectual, faculties.. everyone who has had the opportunity of comparison, must have been struck with the contrast between the taciturn, even morose, aborigines of S. America and the lighthearted, talkative negroes*".
>
> So, while overall similarities dominate the initial discussion, Darwin asserts that mental abilities are key differences and the most different (and in his view the most deficient) are

people from sub-Saharan Africa and those of African descent. For example, he states, "It can hardly be considered as an anomaly that the Negro differs more", and asserts that "mulatto" women are characterized by their "profligacy", and that "hottentot" women "offer peculiarities, more strongly marked than those occurring in any other race". Darwin's ethnocentric, Eurocentric, and anti-African biases come through loud and clear throughout, despite his attempt at a neutral "science", looking at *Homo sapiens* in the same spirit as a naturalist would look at other organisms. And yet, when Darwin reflects on all he's covered – even the differences in intellect -he digs deep into his own experience and reveals that he is, maybe subconsciously, battling with the incongruence of what he presents as scientific "fact" and what he himself has experienced. He reflects: "The American aborigines, Negroes and Europeans differ as much from each other in mind as any three races that can be named; yet I was incessantly struck, whilst living with the Fuegians on board the "Beagle", with the many little traits of character, shewing how similar their minds were to ours; and so it was with a full-blooded negroe with whom I happened once to be intimate". He sees as a scientist, even if briefly, what he cannot see as a Briton immersed in structures and histories of European racism and bias: that the differences between people might not be what the "science" of the time states that they are.

Here, Fuentes makes a point that is central to the present volume: he does not argue that Darwin was always wrong, but instead that some of his ideas, particularly about human evolution, were wrong because of his biases and that this led, in turn, to a huge number of logical incongruities that plagued his works. To defend his theory, Darwin needed to emphasize that modern humans (*Homo sapiens*) are a single group, a relatively recent one with minor differences between its subgroups – real or imaginary. However, his Victorian biases, prejudices and racism, and lack of critical philosophical depth then led him to often go against all those scientific ideas in many excerpts of his "scientific" books. As pointed out by Fuentes, "Darwin remained committed to the premise that humans are divided into significantly distinct 'races' – even when his own scientific analyses suggest otherwise." As Fuentes noted, "human bodies are, in part, so variable because of our species' very wide distribution across the planet, with its diverse ecologies and landscapes.. however, this variation is not distributed in racial patterns." We know, he added, "that skin color, hair type, facial features, and body shape vary quite a bit across our species but not in any pattern that clusters into continental groups (e.g., African, European, Asian, etc.).. dark skin, for example, occurs in distinct populations in Africa, South Asia, Southeast Asia, Micronesia, and South America.. skin color is not a characteristic that pinpoints a person to a specific geographic place of origin." Indeed, saying that "blacks" form a true monophyletic group defined by an adaptation to a higher exposure to ultraviolet rays near the equator would be as absurd as to say that people adapted to live with less oxygen levels at high altitudes, such as in the Andes, Alps, and Himalayas, form a true, monophyletic "race" of "high-alts". The adaptations to living in high altitudes, as to living in places with a very high exposure to ultraviolet rays, are biological convergences that were acquired independently many times in different places of the globe and that have nothing to do with coming from a single group of ancestors that already had those adaptations in the first place, nor with forming a single "race."

As explained by Fuentes, other parts of the *Descent* were even more scientifically incorrect and unfortunate, such as those about the war-like struggle for existence between – and consequent unavoidable extinction of – certain human "races":

> Darwin opens up [his] key [the extinction of races] section [of the *Descent*] by telling us: "The partial and complete extinction of many races and sub-races of man are historically known events". Here he is referring to populations and communities of people who usually, on European contact, or shortly after, diminish radically in numbers or are completely wiped out. Darwin has already identified that humans (of all "races") are amazingly capable of living in the most challenging of environments, so from his perspective, as a naturalist, the fact that many groups have gone extinct offers a quandary. He suggests a solution to this quandary by asserting, as fact, "extinction follows chiefly from the competition of tribe with tribe, and race with race". He argues that this general pattern is very old and has been characteristic of the human lineage. However, his analysis is inspired by the specific case of European expansion, which he explains by telling the reader, "when civilised nations come into contact with barbarians the struggle is short, except where a deadly climate gives its aid to the native race". He adds: "the grade of civilisation seems a most important element in the success of nations which come in competition". He sees this outcome (group or "racial" extinction) as the result of direct competition: a group or "race" wins because it is more "civilized". He also assumes that when "civilized" (meaning European) groups change the landscape, it "will be fatal in many ways to savages, for they cannot, or will not, change their habits". He acknowledges that diseases can, and do, play a role and that "the evil effects from spirituous liquors, as well as with the unconquerably strong taste for them shewn by so many savages", also has deleterious impact. Here Darwin is asserting the specific (and still common) belief that the genocide of indigenous peoples at the hands of European colonizers is due to: a) the native peoples being naturally outcompeted by more "civilized" groups, and b) weakness on the part of the indigenous peoples – of mind, of constitution, of an inability to forgo spirituous liquors, of an inability to adapt to the "civilized" lifestyle.
>
> In the first edition of *Descent* (1871), Darwin spends little space on this section, offering a summary of his thoughts on the matter rather than a more comprehensive review of information. However, it is extremely relevant in this one instance to mention what Darwin added to this section of Chapter 7 in the second edition of *Descent*, published a few years later, in 1874, which is the version most commonly reprinted and read. He lengthened the section and offered multiple examples in support of his assertions, more expansively arguing his case. He also crystalized his erroneous, but powerful, "natural" argument for genocide and colonialism. In the second edition, a chunk of this section outlines "case studies" of the post-European contact devastation of the Tasmanians, the Maori of New Zealand, the New Hebrideans (today Vanuatuans), the Andaman Islanders, and others. In these pages, as Darwin describes the crashes and extinctions of populations, he also clearly documents the horrors and atrocities of European colonial contact. He describes massive stress leading to widespread infertility and infant mortality. He identifies the introduction and impact of infectious diseases, acknowledges displacement and forced movements, and suggests the inability of the indigenous populations to adapt.
>
> Darwin sets up these genocides to be seen as outcomes that ensue due to natural selection, the natural outcome of competition between "races". He argues that the entire process can be compared to the functioning of systems of the "lower" animals, maintaining that certain animals do better, are better able to adapt to challenging circumstances, and possess higher levels of health and vigor than others (citing his earlier work in that vein). He compares indigenous populations to certain lower animals, suggesting that "savages" are likely to respond poorly when challenged with a sudden change of lifeways. He claims, "civilised races can certainly resist changes of all kinds far better than savages". Here Darwin comes

> very close to asserting that the genocidal effects of expansion and colonialism are the logical and expected outcomes of natural laws.

This very last sentence is the type of sentence that, despite being factually correct, causes particular discomfort among Darwin's enthusiasts, and that will surely lead many of them to include Fuentes in the group of Darwin's "heretics." In fact, just a few months after the publication of Fuentes' book chapter, and of a similar editorial he wrote for one of the top scientific journals – *Science* – a huge number of people, including scientists – particularly Western ones – have already said the most atrocious things about him, publicly. The very fact that his last name is Fuentes has made such Western enthusiasts, particularly Anglo-Saxons, even more outraged, with some of them literally writing "who is this Fuentes to criticize *our* Darwin?" *Our* Darwin: this phrase alone summarizes in a nutshell what was emphasized in Chaps. 1 and 2 of this book, and by authors such as Browne in previous books, about some of the main reasons that led Westerners, in particular Anglo-Saxons, to idolize Darwin in the first place. That is, Darwin is often used as a symbol of Western, in particular Anglo-Saxon, supremacy, be it at a societal, moral, or scientific level. Another type of public attack against Fuentes concerns the building of factually inaccurate narratives such as that Fuentes might have a deep sympathy for creationism or creationists or that he is just a second-rate scientist, or an outcast, or something else. One of the merits of Fuentes is to go beyond a specific discussion about Darwin's racism and to explain how these issues have indeed a lot to do with enculturation. Enculturation was crucial to make Darwin *believe* in the tales created within the vicious cycle of systemic racism and subsequently to make others believe in the biased scientific "facts" constructed by him based on those tales, as well as idolize him, and use those "facts" to further reinforce such an enculturation process in popular culture, to the next generations. As stated by Fuentes:

> All humans are identical across more than 99 percent of their genome, and the <1 percent of genetic variation in our species is widely distributed. Thousands of populations of humans across the planet can be differentiated by clusters of patterns of genetic variation, but none of these variations are evolutionarily exclusive and none of the clusters define any populations as discrete enough to be considered a distinct biological lineage (a subspecies). The distribution of human DNA sequence variation does not map continental groupings, such as "African", "Asian", and "European", as distinct, relative to other possible groupings. Movement, migration, and population mixing are characteristic of much of human history, and our genome diversity and distribution of genetic variation demonstrates that.. [But] while "race" is not a biological category, race as a social reality – as a way of seeing people, structuring societies, and experiencing the world – is very real. Societies construct racial classifications not as units of biology but as ways to lump together groups of people with varying historical, linguistic, ethnic, religious, and other backgrounds. These categories are not static. They change over time as societies grow and diversify and alter their social, political, and historical makeups. The American and European histories of creating social races and of structuring their societies around racial inequalities are well documented. These processes and patterns are deeply rooted in the assumptions of "natural" differences between Europeans and those people from all other places on the globe.
>
> The differences evident in Darwin's biases (and in those of so many people today) are not present in the actual data of human variation. These systems of racial classification are tied to histories of expansion and contact, of colonialism, empire, and slavery. They are

rooted in classifications of different human beings as systems of justification for exploitation and oppression. However, while "race" is not biology, racism can certainly affect our biology, especially our health and well-being. Substantial research demonstrates that racialized social structures, from overt oppression and physical subjugation to access to health care to economic and educational discrimination to histories of segregation and material deprivation to one's own racialized self-image as a result of such systems, can impact the ways our bodies, immune systems, and even our cognitive processes react and develop. This means that "race", while not a biological division, can have important biological implications because of the effects of racism. The belief in "races" as natural divisions of human biology and the structures of inequality (racism) that emerge from such beliefs are among the most damaging elements of the human experience both today and 150 years ago.

At the end of his chapter, Fuentes does something that was unthinkable, coming from a top Western evolutionary scientist, just some years ago – he asks, plainly, whether Darwin was a racist or not and answers that Darwin was not only a racist but also an active participant in promoting systemic racism and that doing so subsequently has had important societal impacts, to this very day:

Darwin grew up with, and was educated into, the belief that humans are divided into biological "races" and that these "races" are ranked from the lower "savages" (most of the world) to the higher "civilized" Europeans. He believed that the "race" called "African" was at the bottom of the primitive-to-civilized hierarchy. He believed that there were evolutionary (biological) reasons for the existence of differences between the "races". He was wrong on all counts. Unfortunately, some of these beliefs are still present in society today, with only slight modifications. While Darwin's attempt to explain the origins of "races" was unsuccessful (as he himself noted), he did propose that sexual selection (differential mating pressures) had an influence in forming the races. This has been, and still is, used in some contemporary racist and nationalist (separatist) thought about miscegenation (race-mixing) and its threat to "racial purity". This interpretation is incorrect but remains a strong and lasting racist myth.. It is well documented that Darwin was an abolitionist and saw slavery and general race based cruelty as horrific and unjust. However, if "racism" is any prejudice against someone because of his or her race, when those views are reinforced by systems of power, then yes, Darwin was racist. His overt bias in regard to the mental, moral, and social capacities of humans from the continent of Africa, Afro-descendant populations, and indigenous peoples of the Americas was clear in *Descent of Man* and other writings.

Darwin's racism was neither intentional nor malicious, but it is an example of how racism is maintained – not by the vitriolic screaming and overt acts of violence by a minority but rather by passive acceptance of a particular "reality" and promulgation of the *status quo* by a majority. *Participation* in this pattern of racism was Darwin's greatest failure as a scientist and the singular missed chance for good with Chapter 7 [of his *Descent*] when it was originally published. It's clear that Darwin saw that the "race" ideology he accepted and endeavored to explain did not fit with the available data or his own life experiences. Yet he stuck to it. One does not have to harbor malice for one's racism to have truly malevolent and significantly damaging effects. If one has respect and prominence, then the damage is done. And Darwin had both. His words in Chapter 7 acted to bolster racist (and false) ideologies. To this day racists and nationalist/separatist ideologues use Darwin's words and general arguments as basis for their erroneous and intentionally hurtful and hateful positions and actions. Darwin was, like much of humanity, a biased human being who's at least a little bit racist. So many humans are that way because the societies that raised them are deeply structured with racist, classist, and gendered divisions central to their histories and contemporary functioning.

Within *A Most Interesting Problem*, no other chapter was as bold as Fuentes' one. As expected, many of them were actually very defensive concerning Darwin's legacy, despite the fact that in the preface of the book, DeSilva – the editor of the book – wrote: "our authors [of the book chapters] would not shy away from confronting him with data demonstrating how incorrect he was when he wrote that men were more intelligent than women.. he would learn how his words were used to justify the eugenics movement of the early twentieth century." In reality, only a minority of the book authors did not shy away. However, that is already a huge change: what DeSilva did as an editor, and the book as a whole, is a very important landmark that hopefully will contribute to further change the still prevailing status quo concerning these topics. One of those authors that did not shy away was Janet Browne who, as we have seen throughout this book, commonly provides very elegant and astute observations about Darwin's *Power of Place*. For instance, in the introduction of *A Most Interesting Problem*, she noted that:

> Today the fame of the Beagle voyage sometimes makes it hard to remember that its purpose was not to take Darwin around the world but to carry out British Admiralty instructions. The ship had been commissioned to extend an earlier hydrographic survey of South American waters that had taken place from 1825 to 1830. The area was significant to the British government for commercial, national, and naval reasons, buttressed by the Admiralty's preoccupation with providing accurate sea charts and safe harbors for its fleet in the world's oceans.. [About the *Descent*], it is clear that Darwin thought there had been a progressive advance of moral sentiment from the ancient 'barbaric' societies described in Victorian history books, such as those of ancient Greece or Rome, to the civilized world of nineteenth-century England that he inhabited. In this manner, he kept the English middling classes to the front of his readers' minds as representative of all that was best. The higher moral values were, for him, self-evidently the values of his own class and nation.. In this way Darwin made human society an extension of biology and saw in every human group a 'natural' basis for primacy of the male. After *Descent of Man*'s publication, early feminists and suffragettes bitterly attacked this doctrine, feeling that women were being 'naturalized' by biology into a secondary, submissive role. Indeed, many medical men asserted that women's brains were smaller than those of men, and they were eager enough to adopt Darwin's suggestion that women were altogether less evolutionarily developed and that the 'natural' function of women was to reproduce, not to think. For several decades, Anglo-American men in the medical profession thought that the female body was especially prone to medical disorders if the reproductive functions were denied. Something of this belief can be traced right through to the 1950s and beyond.
>
> In *Descent of Man*, Darwin also made concrete his thoughts on human cultural progress and civilization. The notion of a hierarchy of races informed his discussion and took added weight from being published at a time when the ideology of extending one nation's rule over other nations or peoples was unquestioned. Darwin stated that natural selection and sexual selection combined with cultural shifts in learned behavior to account for the differences that he saw between populations. The racial hierarchy, as Darwin called it, ran from the most primitive tribes of mankind to the most civilized and had emerged over the course of eons through competition, selection, and conquest. Those tribes with little or no culture (as determined by Europeans) were, he thought, likely to be overrun by bolder or more sophisticated populations. 'All that we know about savages, or may infer from their traditions and from old monuments,' he wrote, 'shew that from the remotest times successful tribes have supplanted other tribes' Darwin was certain that many of the currently existing peoples he called primitive would in time similarly be overrun and perhaps destroyed by more advanced races, such as Europeans; he had in mind particularly Tasmanian, Australian,

and New Zealand aboriginal peoples. This to him was the playing out of the great law of 'the preservation of favoured races in the struggle for life,' as expressed in the subtitle of his earlier book *On the Origin of Species*. Such an emphasis on the natural qualities underpinning social cultural development explicitly cast the notion of race into biologically determinist terms, reinforcing then contemporary ideas of a racial hierarchy.

It is notable that Browne reasserts, as Fuentes did, that Darwin's racism, sexism, and ethnocentrism, combined with his *power of place*, did indeed contribute to intensify and aggravate scientific, medical, and societal malpractices, and abuses and atrocities, including those historically related to the eugenics movement:

Partly because of Darwin's endorsement and partly because of the influential writings of others, these views intensified during the high imperialism of the early twentieth century. Herbert Spencer's doctrine of "survival of the fittest", as used by Darwin, Wallace, Spencer, and others, in *Descent of Man* and elsewhere, became a popular phrase in the development of social Darwinism. Embedded in powerful class, racial, and gender distinctions, social Darwinism used the prevailing ideas of competition and conquest to justify social and economic policies in which prosperity and success were the exclusive aim. "Survival of the fittest" was a phrase well suited to encourage hard-nosed economic expansion, rapid adaptation to circumstance, and colonization. Karl Pearson, a committed Darwinian biologist, expressed it starkly in Britain in 1900: no one, he said, should regret that "a capable and stalwart race of white men should replace a dark-skinned tribe which can neither utilise its land for the full benefit of mankind, nor contribute its quota to the common stock of human knowledge". Several of Darwin's remarks in *Descent of Man* captured anxieties that were soon to be made manifest in the eugenics movement. Darwin feared that what he called the "better" members of society were in danger of being numerically swamped by the "unfit". In this latter category Darwin included men and women of the streets, the ill, indigents, alcoholics, and those with physical disabilities or mental disturbances. He pointed out that medical aid and charity given to the sick and the poor ran against the fundamental principle of natural selection. Evidently torn between his social conscience and what he understood about evolutionary biology, he went on to declare that it was a characteristic of a truly civilized country to aid the sick and help the weak.

In these passages Darwin anticipated some of the problems that his cousin Francis Galton would try to alleviate through the eugenics movement. Galton was an enthusiastic convert to Darwin's theories and had little hesitation in applying the concept of selection to human populations. He aimed to improve human society though the principles of natural selection: in essence, by reducing the rate of reproduction among those he categorized as the poorer, unfit, profligate elements of society and promoting higher rates of reproduction among the middle classes. Galton hoped that the men he called highly gifted – the more successful men – should have children and pass their attributes on to the next generation. Galton did not promote policies of incarceration or sterilization ultimately adopted by the United States, nor did he conceive of the possibility of the whole-scale extermination of "undesirable" groups as played out during World War II. But he was a prominent advocate of taking human development into our own hands and the necessity of improving the human race. Darwin referred to Galton's point of view in *Descent*. While Darwin's *Descent of Man* can hardly account for all the racial stereotyping, nationalist fervor, and prejudice expressed in years to come, there can be no denying the impact of his work in providing a biological backing for notions of racial superiority, reproductive constraints, gendered typologies, and class distinctions.

Another very informative chapter of *A Most Interesting Problem* was Kristina Killgrove's take on Darwin and the concepts of cultural evolution, civilization, intelligence, and white nationalism. She did not shy away from making statements such as the "European thinkers of the nineteenth century used and misused Darwin's new ideas to discuss the 'savages' they had met and 'conquered,' whereas American scientists began to generate a race-based biological anthropology, in part to morally justify the enslavement of Africans." She also criticized Darwin's portrait of nomadism as something inferior and of sedentism, agriculture and property possession as something evolutionarily "good," and his "problematic assumptions about capitalism, religion, and marriage being natural and the result of cultural progress." She further reminds us that Darwin attributed "England's success as a colonial power to 'daring and persistent energy' while also suggesting that 'the wonderful progress of the United States, as well as the character of the people, are the results of natural selection; the more energetic, restless, and courageous men from all parts of Europe having emigrated during the last ten or twelve generations to that great country, and having there succeeded best.'" Such statements about Darwin further remind us of the logical incongruities and a certain lack of societal depth of his works about humans because at the time that he wrote such sentences, the "wonderful progress of the United States" was above all related to one of the biggest and longest atrocities committed in human history: the enslavement of countless Africans. Darwin knew very well the horrible details about enslavement and publicly opposed slavery, as we have seen. However, he still preferred to call the ancestors of those people that were enslaved – the Africans – as "savages" and barbaric, while saying that those that enslaved them – the Westerners – were "civilized" and the pinnacle of "civilization" and "human progress." As Killgrove notes, "Darwin's *Descent* will give any social scientist pause.. given his employment of patriarchal language, his conflation of religion with morality, and his uncritical naturalizing of the Western European and colonialist way of life, it is relatively easy to poke holes in Darwin's explanation of cultural evolution and civilization." Importantly, she reflects on how Darwin's legacy, as well as that of those scholars that he *actively chose* to cite in his works, such as the racist scholar Paul Broca, are felt even today, including the way in which we continue to engage in malpractices such as using "IQ" to measure intelligence or falling into the trap of prejudices in archeological works (see also Box 3.9):

> Darwin also cites the scientific understanding of his day, which held that, "there exists in man some close relation between the size of the brain and the development of the intellectual faculties.. supported by the comparison of the skulls of savage and civilised races, of ancient and modern people, and by the analogy of the whole vertebrate series". He cites as his sources of information people like Paul Broca and Johann Friedrich Blumenbach, the latter of whom is perhaps best known for creating and popularizing the five-fold grouping of human races, and who held that Adam and Eve were Caucasians and every other race was produced by degeneration. Both Broca and Blumenbach were pioneers in cranial anthropometry (skull measuring), which ultimately formed the basis for scientific racism in early biological anthropology, when physical traits were used to justify cultural and structural violence, including slavery, patriarchy, and colonialism. Three major issues with Darwin's use of intelligence in explaining culture and civilization are: 1) his assumption that brain size is a proxy for intelligence, 2) his assumption that intelligence can be quantified, and 3) his inference that quantifications of intelligence reflect heritable and immutable traits in humans.. Our contemporary understanding of intelligence is based on equally flawed but more recent tests devised in the first half of the twentieth century. Measuring what is widely known as IQ, or intelligence quotient, a term that dates back to 1912, the earliest modern

test was developed in France by Alfred Binet and later became the Stanford-Binet scale still in use today. Binet, attempting to identify children who might need scholastic intervention in order not to fall behind their peers in their age cohort, developed a test with questions about recognition of standard objects, verbal definitions, execution of simple commands, and working memory tests.

A child's IQ under this scheme was his or her ratio of tested mental age to expected chronological age. Around the same time, psychologist Henry Goddard began to use the IQ test to classify people with intellectual challenges; his terms "moron" (51–70 IQ), "imbecile" (26–50 IQ), and "idiot" (0–25 IQ) were used by psychologists and governments for decades. A true eugenicist, Goddard believed that low-IQ individuals should be removed from society by institutionalization or sterilization. Goddard backed up his eugenics with a healthy dose of racism and classism; in a study he ran on immigrants to New York's Ellis Island he found that roughly 80 percent of Island, he found that roughly 80 percent of those in steerage class were "feeble-minded". The most famous scientific criticism of IQ testing and the general quantification of human intelligence came in 1981, with the publication of Stephen Jay Gould's *The Mismeasure of Man*, in which Gould tackled the phenomenon of intelligence quantification and traced its ill effects on human society.

Killgrove further notes that:

While Darwin suggested that there were natural limits on the "downward tendency" of the human population, Gould a century later identified more specific political mechanisms that have lasted well into the twenty-first century.. Gould criticized men such as Paul Broca who were engaged in scientific racism and whose work was based on culturally biased, *a priori* expectations. At the heart of what Gould critiqued was not only biological determinism but also, particularly in his response to [book] *The Bell Curve*, the unstated assumptions held by many white the unstated assumptions held by many white Westerners that make IQ testing so insidious. Gould isn't alone. Data scientist Eric Siegel has summarized Herrnstein and Murray's book and its implicit racism by noting, "The Bell Curve endorses prejudice by virtue of what it does not say.. Nowhere does the book address why it investigates racial differences in IQ.. The net effect is to tacitly condone the prejudgment of individuals based on race". Psychologist Howard Gardner, famed for his work on multiple intelligences, further avers that the rhetoric in *The Bell Curve* encourages readers to align themselves with extreme positions – such as the abolition of affirmative action and the curbing of reproduction by people with low IQs.

Darwin clearly shares with Herrnstein and Murray these *a priori* assumptions about intelligence and similarly conflates intelligence with socioeconomic superiority, health, and well-being in the form of longevity, civilization, and progress. They're not the only men to assume racist perspectives on intelligence. In 2007, James Watson, the famous American molecular biologist who won the Nobel Prize for his work with DNA, told a reporter that "all our social policies are based on the fact that their ['Africans'] intelligence is the same as ours – whereas all the testing says not really". While Herrnstein and Murray, Galton, and others may have found data to support different IQs in different racial groups, those data are meaningless and cannot be interpreted without social and economic context. It's the context – the socioeconomic hardships, the discriminatory health-care system, the lack of access to education, the criminal justice system – that social scientists have shown to be impossible to separate from any measure of human intelligence, as inequality is a cultural construct, not a natural, biologically deterministic pattern. The idea of major differences in intelligence by race and/or social class, although thoroughly debunked, has reemerged in the United States, England, and elsewhere in the past few years. Although Gould passed away in 2002, his warnings about "political retrenchment and destruction of social generosity" are unfortunately once again relevant amid the resurgent popularity of biological determinism and white nationalism.

Box 3.9 Darwin, Cultural Evolution, Scientific Biases, Enculturation, and Politics

In her chapter of *A Most Interesting Problem*, Kristina Killgrove discusses some of the key topics of this book, particularly systemic racism and enculturation, which are often neglected in the relatively few discussions available on Darwin's racism, sexism, and ethnocentrism:

> Just as scientific racism surrounding intelligence belies its proponents' assumptions and prejudices about other groups, the contemporary discussion of civilization and progress has similar themes. Take former U.S. congressman Steve King, a man whose fifteen-year history of racist polemics, specifically surrounding civilization and birth rates, has warranted its own time line at the *New York Times*. In 2002, King tried to get a bill passed requiring public schools to teach that the United States is the greatest nation in the world, thanks to Christianity and Western civilization. In 2011, King spoke out against hormonal birth control, noting, "If we let our birthrate get down below the replacement rate, we're a dying civilization". And if there's any question that King was speaking specifically about the white birth rate, in 2017 he wrote on Twitter that "we can't restore our civilization with somebody else's babies". In early 2019, King told a reporter, "white nationalist, white supremacist, Western civilization – how did that language become offensive?".. The archaeology of complexity that contemporary researchers engage with is not as simple as what Darwin wrote of in the nineteenth century nor the same as what Mauss and Durkheim noted in the early twentieth century. Rather, culture change is addressed according to one or more archaeological theories, each offering a perspective that helps archaeologists identify and explain past behavior of individuals and cultures. These theories – neoevolutionism, historical particularism, and practice theory – help archaeologists address artifacts used in a complex society, the power and ideology behind the rise of the society, and how individual and collective action worked together to form it. Use of these theories requires archaeologists to be reflexive and to try to eliminate their own cultural biases when interpreting the remains of the past. But since we are all a product of our cultural upbringing, sometimes archaeologists slip up and reveal their biases. For example, Chapman notes that archaeologist Ian Hodder, regarding his wellknown research at Çatalhöyük in Turkey, writes about a "low degree of social complexity", and this, for Chapman, "raises the question as to how far such notions of social evolution are embedded in everyday thought and action in Western society"- today, as they were in Darwin's time as well..
>
> Given the Western cultural assumption that there is constant forward progress, and given the lack of collective memory of societal transformation, many people today tend to believe that their society is the norm and that it is not going anywhere. This widely held view also correlates with "civilization" and with cultural "progress" – and today these ideas have been used quite often by North American and European politicians to pit "us" (white upholders of the Western, Judeo-Christian tradition) against "them" (nonwhites, immigrants, nonChristians, and others). U.S. congressman Steve King may be a notorious and unrepentant racist, but his spoken and unspoken views on "civilization" and cultural "progress" are widely held among twenty-first century white Americans, just as these same ideas were widely held by Europeans in Darwin's time. The increase, decrease, stasis, and collapse of human cultural complexity is an ongoing research topic of interest within the field of anthropology. Whereas human biological evolution is far better understood today than it was in Darwin's time, thanks primarily to the discovery of genes and the

(continued)

burgeoning of that field of research, human societies are still messy. We can explain biological characteristics of humans through natural selection, mutation, genetic drift, and gene flow, but attempts to do this with culture have not borne the same fruit, as there is very little variation within cultures that can be seen as "competition" in the Darwinian sense, and therefore *there is no analogy for the "struggle for existence"*. And while a Darwinian approach has been attempted in measuring and explaining human intelligence, anthropology and other social scientific fields have similarly shown that the results of that approach are related in no small part to cultural correlates such as inequality rather than assumed biological correlates such as brain size. What anthropology has taught us, since the time of Darwin and Wallace, is that the Enlightenment ideal of human progress and the capitalism and inequality arising from the industrial revolution that underlie many early theories about human society are not "natural" nor are they necessarily something all humans on the planet should strive for. As the eminent author and lapsed anthropologist Kurt Vonnegut once commented: *I didn't learn until I was in college about all the other cultures, and I should have learned that in the first grade.. A first grader should understand that his or her culture isn't a rational invention; that there are thousands of other cultures and they all work pretty well; that all cultures function on faith rather than truth; that there are lots of alternatives to our own society.. cultural relativity is defensible and attractive.. it's also a source of hope. It means we don't have to continue this way if we don't like it"*. As more academics and scientists wade into social media, attempting to convince a growing number of doubters about the importance of vaccinations, climate catastrophe, historical facts, and biological discoveries, it becomes clear that many of us are trying to stem a rising tide of white nationalism barely disguised as "cultural heritage". If politicians and voters, however, fail to pay attention, the complex, global society we live in – our current civilization – may falter.

Some other chapters of *A most interesting problem* deal with far more neutral topics – in regard to their direct implications for broader societal issues – but are often also very informative, such as Yohannes Haile-Selassie's discussion on fossil evidence for human evolution. He reminds us that Darwin was right about important points such as that humans originated in Africa. However, he also reminds us that Darwin was wrong about various aspects of our evolutionary history. For example, he noted that "Darwin thought that the major reason the human lineage became bipedal was to free the hands for tool use, and that this was followed by reduction of the size of our canines and enlargement of our brain.. the fossil evidence that we currently have in hand indicates that in fact walking on two legs preceded tool use, canine reduction, and brain enlargement." He also points out that – contrary to Darwin's idea of a linear, gradual evolution – we now know that there were many different lineages within our human evolutionary history. The members of each of these lineages are now extinct, except our species, although we have some genes of other species – such as the Neanderthals – due to the intermix between different human species in the past. Within those lineages, there were cases of not only gradualism, as expected by Darwin, but also of punctuated equilibrium – a concept developed by Stephen Jay Gould and colleagues, as we have seen in Chap. 1. For instance, during the first three or four million years of our evolutionary history, cultural innovations happened at a much slower pace than they did in the last two or three million years, particularly in the last few hundreds of thousands of years. Apart from these points made by Haile-Selassie, we now also know that, contrary to what Darwin suggested, sexual selection was very likely not the main driver of most anatomical and cultural changes that happened in human evolutionary history, such as those concerning the evolution of bipedalism and skin color differences, as discussed above. Moreover, we now know that phenomena such as neutral drift, as well as other traits not directly related to specific adaptations, probably also played a major role in human evolution, particularly when the population sizes of certain groups were very small. We will further discuss the book *A Most Interesting Problem in the next chapter, which will focus on misogyny and Darwin's contribution to systemic sexism.*

Chapter 4
Misogyny and Its Damaging Legacy

> *From this [Darwin's] view, cutting edge science justified limiting the freedom of all but upper-class white men.. women evolved to be wives (and not scientists or scholars) and to carry out evolution's plan.. natural and sexual selection conveniently favored what society already did.. the scientific value of [Darwin's] Descent is impossible to untangle from the oppression that it inspired.*
>
> *(Holly Dunsworth)*

Ancient Greece, Christianity, and Darwin's Misogyny

> *As Darwin said, by keeping women at home their achievements were paltry compared to men's which proved women were biologically inferior.. and he should know because he was a Genius.. you probably learned about him at school.* (Jacky Fleming)

The *Cambridge Dictionary* definition of misogyny is: "feelings of hating women, or the belief that men are much better than women." As explained in my *Meaning of Life, Human Nature and Delusions* book, sedentism and in particular agriculture and the societal changes related to them, including the origin of major organized religions, strongly contributed to many of the misogynistic narratives that became so prevalent in popular culture of agricultural states. In this book, I will briefly refer to some of the discussions and data that I included in that book, which analyzed the history of misogyny, and the scientific biases related to it, in a much more detailed way.

In her outstanding book *A Brief History of Misogyny*, Holland shows that ancient agricultural societies such as ancient Greece and Judaea were already characterized by many of the sexist tales that subsequently became so widely accepted in many regions of the planet, particularly in the West (see Box 4.1).

R. Diogo, *Darwin's Racism, Sexism, and Idolization*,
https://doi.org/10.1007/978-3-031-49055-2_4

Box 4.1 Ancient Greece, Pandora, Aristotle, and Misogyny

Holland's *A Brief History of Misogyny* provides a very useful background to better understand the type of sexist tales that are so familiar to us nowadays, particularly in the West or in regions of the globe that were highly influenced by Western ideas. This does not mean, of course, that the West is unique about this. Basically, all sedentary agricultural societies are in general markedly misogynistic, but it is important to understand the origins of misogyny in the West because, as explained by Holland, such narratives deeply influenced authors such as Darwin:

> It is hard to be precise about the origins of a prejudice. But [in the West] if misogyny has a birthday, it falls sometime in the eighth century BC. If it has a cradle, it lies somewhere in the eastern Mediterranean. At around that time in both Greece and Judaea, creation stories that were to acquire the power of myth arose, describing the Fall of Man, and how woman's weakness is responsible for all subsequent human suffering, misery and death. Both myths have since flowed into the mainstream of Western civilization, carried along by two of its most powerful tributaries: in the Jewish tradition, as recounted in Genesis (which a majority of Americans still accept as true) the culprit is Eve; and in the Greek, Pandora. But in the history of misogyny, the Greeks also occupy a unique place as the intellectual pioneers of a pernicious view of women that has persisted down to modern times, confounding any notion we might still have that the rise of reason and science means the decline of prejudice and hatred. The myth of Pandora was first written down in the eighth century BC by Hesiod, a farmer turned poet, in two poems: 'Theogony' and 'Works and Days'. In spite of Hesiod's considerable experience as a farmer, his account of mankind's creation ignores some of the basic facts of life. The race of men exists before the arrival of woman, in blissful autonomy, as companions to the gods, 'apart from sorrow and from painful work/free from disease'. As in the Biblical account of the creation of man, woman is an afterthought. But in the Greek version, she is also a most malicious one. Zeus, the father of the gods, seeks to punish men by keeping from them the secret of fire, so that, like the beasts, they must eat their meat raw. Prometheus, a demi-god and the creator of the first men, steals fire from heaven and brings it to earth. Furious at being deceived, Zeus devises the supreme trick in the form of a 'gift' to men, 'an evil thing for their delight', Pandora, the 'all giver'. The Greek phrase used to describe her, 'kalon kakon', means 'the beautiful evil'. Since then, according to Greek mythology, mankind has been doomed to labour, grow old, get sick, and die in suffering..
>
> As well as burdening Pandora with responsibility for the mortal lot of man, the Greeks created a vision of woman as 'the Other', the antithesis to the male thesis, who needed boundaries to contain her. Most crucially, Greece laid the philosophical-scientific foundations for a dualistic view of reality in which women were forever doomed to embody this mutable, and essentially contemptible world. Any history of the attempt to dehumanize half the human race is confronted by this paradox, that some of the values we cherish most were forged in a society that devalued, denigrated and despised women. 'Sex roles that will be familiar to the modern reader were firmly established in the Dark Ages in Athens', wrote the historian Sarah Pomeroy. That is, along with Plato and the Parthenon, Greece gave us some of the cheapest sexual dichotomies of all, including that of 'good girl *versus* bad girl'. Having violent warrior divinities, however, is not necessarily an indication of a misogynistic culture. In the older civilizations the Greeks encountered, such as those

(continued)

of Egypt and Babylon, there was an abundance of war gods, but no equivalent of the Fall of Man myth. In Mesopotamia, the Sumerian poem 'The Epic of Gilgamesh', which dates back to the third millennium BC, has a hero who like Prometheus aspires to rival the gods. Gilgamesh does so by seeking to share in their immortality; but women are not made the instrument of revenge by some vindictive deity seeking to punish man for challenging his mortal lot. Nor does Gilgamesh castigate women for being to blame for 'the lot of man'; the gods are to blame for our mortality.

Homer based both *The iliad* and *The odyssey* (the latter recounting the long journey home of Odysseus, one of the Greek kings) on material which dates back to the earlier dynastic period. In these works, women are generally portrayed sympathetically; they are complex and powerful, and among the most memorable characters in all literature. The end of this era was accompanied by a move from a pastoral to a labour intensive agricultural economy, one concerned about the conservation of property. Laws regulating women's behaviour and opportunities give the most graphic and pertinent examples of how Hesiod's allegory of misogyny became a social fact. Legally speaking, Athenian women remained children, always under the guardianship of a male. A woman could not leave the house unless accompanied by a chaperone. She seldom was invited to dinner with her husband and lived in a segregated area of the house. She received no formal education: 'Let a woman not develop her reason, for that would be a terrible thing', said the philosopher Democritus. Women were married when they reached puberty, often to men twice their age. Such a difference in age and maturity, as well as in education, would have enhanced the notion of women's inferiority. The husband was warned: 'He who teaches letters to his wife is ill advised: he's giving additional poison to a snake'..

Aristotle has been described as one of the most ferocious misogynists of all time. His views on women take two forms: scientific and social. Although at times Aristotle was a precise observer of the natural world – his descriptions of various species impressed Charles Darwin – his observations of women were decidedly warped. As a sign of women's inferiority, he referenced the fact that they did not grow bald – proof of their more childlike nature. He also claimed that women had fewer teeth than men, about which Bertrand Russell is said to have commented: 'Aristotle would never have made this mistake if he had allowed his wife to open her mouth once in a while'. Aristotle introduced the concept of purpose as fundamental to science. The purpose of things, including all living things, is to become what they are. In the absence of any knowledge of genetics, or of evolution, Aristotle saw purpose as the realization of each thing's potential to be itself. In a sense, this is a materialistic version of Plato's *Theory of forms*: there is an Ideal Fish of which all the actual fishes are different realizations. The ideal is their purpose. When applied to human beings, notably to women, this has unfortunate but predictable results; it becomes a justification of inequality rather than an explanation for it. The most pernicious example is seen in Aristotle's theory of generation. This assumes different purposes for men and women: 'the male is by nature superior and the female inferior; and the one rules, and the other is ruled; the principle of necessity extends to all mankind'. Therefore, according to Aristotle, the male semen must carry the soul or spirit, and all the potential for the person to be fully human. The female, the recipient of the male seed, provides merely the matter, the nutritive environment. The male is the active principle, the mover, the female the passive, the moved. The full potential of the child is reached only if it is born male; if the 'cold constitution' of the female predominates, through an excess of menstrual fluid in the womb, then the child will fail to reach its full human potential and the result is female. 'For the female is, as it were, a mutilated male', Aristotle concludes.

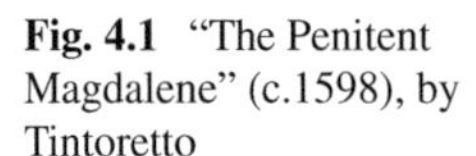

Fig. 4.1 "The Penitent Magdalene" (c.1598), by Tintoretto

Concerning Christianity in particular, an illustrative example of how manly made inaccurate tales can be constructed and subsequently changed to justify and promote misogynistic views concerns the case of Maria Magdalena (Fig. 4.1). Originally, according to the four canonical gospels, Magdalene traveled with Jesus and was a witness to his crucifixion, burial, and resurrection. That is, she clearly occupied a prominent role within the first followers of Jesus. However, later, the earlier Church Fathers did not mention her at all in their writings, or only briefly. In contrast, Ephrem the Syrian, who lived in the fourth century, makes one of the first identifications of Magdalena as a redeemed sinner. Later, in the Middle Ages, she is conflated with Mary of Bethany and thus with a "sinful woman." The tale that Mary Magdalene was as a prostitute held on for centuries after Pope Gregory the Great made it official in a sixth-century sermon, although such a view was not adopted by Orthodoxy nor by Protestantism when those faiths later split from the Catholic Church. Only many centuries later, about five decades ago, in 1969, the label "sinful woman" was removed in the General Roman Calendar, but the view that she was a prostitute, or at least not a "good", "coy" woman, is still common within popular culture in many countries. In a nutshell, within the writings or sermons done by men, Magdalena's story became more and more sexist, from her being one of the favorites – if not *the* favorite – follower of Jesus, to her being a repentant, penitent, prostitute.

Apart from contributing – together with the subjugation based on differences in physical force, as will be explained below – to the exclusion of women in public life or various types of jobs and tasks, historically, such manly made misogynistic narratives have moreover also constantly minimized the historical importance of women – real or imaginary – that were able to liberate themselves from the societal chains of male subjugation, oppression, and prejudice. As often occurs in cases of systemic sexism and racism, such a minimization was done by political and religious leaders, scientists, and the broader public, in a dynamic, interactive way. An emblematic example of this is one of the most renowned biologists and philosophers of all times, Aristotle, whose ideas hugely influenced Darwin's ones as noted above. Some of the scientific "facts" postulated by Aristotle were so misogynistic that he went all the way to even minimize the contribution of women to reproduction. Yes, you read correctly: to reproduction – a powerful example of the power of biases, stereotypes and prejudices in science. As put in the book *Biology and Feminism*, edited by Linn Nelson:

> Aristotle (384-322 BCE) was a logician, philosopher, physicist, and biologist.. But like many scientists, his views were in part informed by his historical and social context. For Aristotle, this context was ancient Greece, and fundamental differences between men and women, and males and females generally, were widely accepted and argued for. In brief, from at least Plato (428-348 BCE) forward, it was believed that men are superior to women in myriad ways. Like other philosophers of his time, Aristotle theorized about the differences between the sex/genders and sexes, largely in his biological research.. But he faced a conundrum putting together his theories concerning the "nature" of each sex/gender and his explanation of human reproduction. Women, Aristotle held, are "incomplete [or distorted] men", physically and intellectually inferior to them. And his explanation of sexual reproduction was androcentric (i.e., "male centered"), apparently reflecting the view that women's role, in all areas of human endeavor, was of far less consequence than that of men. Aristotle maintained that women provide the space and necessary physical matter for a human to grow. Sperm contain the "form" of a tiny human male. Given this one would expect all babies to be male. But this is obviously not the case. Aristotle hypothesized that, while the form a man contributes is always male, female offspring result when a woman's womb is insufficiently warm – resulting in an incomplete or distorted child – that is, a female.
>
> [More recently] in "*The energetic egg*" an article written for the lay public and published in 1983, developmental biologists Gerald Schatten and Heidi Schatten argued that there were striking parallels between the Grimm Brothers' fairytale "Sleeping Beauty" and the then accepted account of fertilization. They challenged the account's portrayal of the egg as passive and dormant until "penetrated and activated by a sperm" based on observations they understood to indicate that the egg's role in fertilization was as active as that of sperm.. Based on their observations using scanning electron microscopy, and on their reevaluation of reports of observations others had earlier made, Schatten and Schatten maintained that "it is becoming clear that the egg is not merely a large yolk-filled sphere into which the sperm burrows to endow new life". Rather, they contended, "recent research suggests the almost heretical view that sperm and egg are mutually active partners". What Schatten and Schatten observed is that when the egg and sperm interact, the sperm does not "burrow into the egg"; rather the egg "directs" the growth on its surface of small finger-like projections (called "microvilli") to clasp the sperm so that it can draw the sperm into itself. Interestingly, Schatten and Schatten also noted that as early as 1895 E. B. Wilson had published photographs of sea urchin fertilization in which the egg's extension of microvilli to

the sperm was visible. And there was more evidence to come that would support Schatten and Schatten's observations and interpretation of them.

In 1991, feminist anthropologist Emily Martin, who also compared the classic account of fertilization to "Sleeping Beauty" and maintained that scientists "had constructed a romance" between egg and sperm "based on stereotypical male-female differences", studied research into fertilization undertaken in the late 1980s and early 1990s in a lab at Johns Hopkins University. It had long been assumed that sperm used mechanical means to get through the zona (a thick membrane surrounding the ovum) and penetrate the egg. As Martin chronicles, earlier investigations had emphasized "the mechanical force of the sperm's tail" in enabling fertilization. To their great surprise, Martin states, investigators at Johns Hopkins Discovered.. that the forward thrust of sperm is extremely weak, which contradicts the assumption that sperm are forceful penetrators. Rather than thrusting forward, the sperm's head was now seen to move back and forth. The sideways motion of the sperm's tail makes the head move sideways with a force that is ten times stronger than its forward movement. In fact, its strongest tendency, by tenfold, is to escape by attempting to pry itself off the egg. Given these observations, Martin noted, the scientists concluded that the egg traps the sperm and adheres to it so tightly that the sperm's head is forced to lie flat against the surface of the zona.

The same book, *Biology and Feminism*, also provides examples of even more "modern" and so-called "scientific" narratives that are misogynistic and not based on empirical data at all (see also Box 4.2):

By 1978 when he published *On human nature* [a book intended for a general audience], [Edward] Wilson and other human sociobiologists had increasingly turned to explaining social behaviors they attributed to contemporary humans as adaptations selected for during the Pleistocene [often referred to as the 'Ice Age' including 'cave-men', this geological epoch lasted from about 2580000 to 11700 years ago]. In addition to proposing genetic bases for xenophobia and war, Wilson and other human sociobiologists devoted a good deal of attention to proposing adaptations of what they assumed or argued to be sex/gender differences in mating strategies and behavior.. of why men are "promiscuous" and "undiscriminating" and women are "coy" and "choosy"; of rape; and of societal practices and institutionalized norms that.. Wilson described as a universal "Double Standard" that allows men to engage in a range of activities frowned upon or outright denied to women..

[David] Barash offered the following argument in a book written for the lay public: "because men maximize their fitness differently from women, it is perfectly good biology that business and profession taste sweeter to him, while home and child care taste sweeter to women.. while it may be true that it's "not nice to fool Mother Nature", it can be done.. biology's whispers can be denied, but in most cases at a real cost.. although women who participate [in work outside the home] may be attracted by the promise of "liberation", they are in fact simply adopting a male strategy while denying their own.. Cavalier female parenting is maladaptive for all mammals; for humans, it may be a socially instituted trap that is harmful to everyone concerned".. He also suggested that rape may be a reproductive strategy: "rape in humans.. is by no means as simple [as the rape among mallard ducks I have observed], influenced as it is by an extremely complex overlay of cultural attitudes.. nevertheless, mallard rape and bluebird adultery may have a degree of relevance to human behavior.. perhaps human rapists, in their own criminally misguided way, are doing the best they can to maximize their fitness.. if so, they are not that different from the sexually excluded bachelor mallards", Barash notes.

Box 4.2 Sexism, Darwinism, Science, and Medicine

The book *Biology and Feminism*, edited by Lynn Nelson, emphasizes how many of the misogynistic narratives that became so predominant within Western science and medicine were constructed in nineteenth-century works, including those of Darwin:

> The changes the nineteenth century brought to medical views about women's biology and health were dramatic and multifaceted. First, differences between men's and women's organs and biological processes related to reproduction came to be viewed as significant and of consequence in relation to their "roles" and health. Second, women's organs and biological processes related to reproduction came to be understood as detrimental to women's health, if not pathological. Writing in 1900 about the "ravages" thought to be wrought by the "sexual storms" of female puberty, menstruation, pregnancy, and childbirth. The president of the American Gynecology Society used an analogy with shipwrecks to describe the dangers women encounter during each phase of their reproductive years. Third, medical experts came to view women's reproductive organs, her ovaries and uterus, as the source of disease and illness involving other organs, and as dictating that her role is that of wife and mother. Of course, this view of women's proper role was not new. What were new were the ideas that it is women's ovaries and uteruses that determine that role, and that their proper or improper functioning determines women's health or illness in every respect – even in terms of illnesses and diseases to which men are also subject. As one medical professor declared in 1870, it is "as if the Almighty in creating the female sex, had taken the uterus and built up a woman around it". Many physicians and psychologists emphasized women's ovaries as dictating all aspects of femininity, including women's psychology. One physician, speaking of women's ovaries and uteruses, declared "Women's reproductive organs are pre-eminent".
>
> From the perspective of nineteenth and early twentieth century medicine, menstruation was a unique kind of biological process, a process that was debilitating if not pathological – putting women at high risk from the onset of menses until menopause. Women (at least those belonging to the middle or upper class), medical experts maintained, require extensive rest before and during their monthly periods. In 1870, the zoologist, Walter Heape, expressed a common view when he described menstruation as a "severe, devastating, periodic action.. [that leaves behind] a ragged wreck of tissue, torn glands, ruptured vessels, jagged edges of stroma, and masses of blood corpuscles". Not surprisingly, Heape argued that menstruation requires medical treatment. "It would seem hardly possible", he maintained, that a woman could "heal satisfactorily without the aid of surgical treatment". Popular books advising women about menstruation sounded similar cautions.
>
> In the second half of the century, what Douglas-Wood describes as .. "fumbling experiments with the female interior," gave way to surgeries, many of which were undertaken to deal with diseases in other organs as well as "female personality disorders". For a brief period in the 1860s, some physicians treated "nymphomania" and "intractable masturbation" by removing the clitoris, although many physicians disapproved of the surgery. The most common surgery for women's diseases, including "personality disorders", was removal of the ovaries. Ehrenreich and English note that in 1906, a leading American gynecological surgeon estimated that 150,000 women in the country had had their ovaries removed. In the last third of the century, physicians' arguments.. came to reflect their understandings of the implications of Darwinism. Ehrenreich and English note that "civilization" was taken to explain

(continued)

why "the middle-class woman [was] sickly; her physical frailty went hand-in-white glove-hand with her superior modesty, refinement, and sensitivity". In contrast, working class women "were robust, just as they were supposedly 'coarse' and immodest". Assuming Darwin's general characterization of the significant sex/gender differences that evolution produced because of the greater selection pressures to which men were subject, it seemed, as feminists describe the issue, that in terms of the wealthy classes, that some came to believe that "men evolve and women devolve". Interests in maintaining the current "social order", as well as economics practices, also appear to have been factors in how poor and working class women were viewed. After all, someone had to do the work of scrubbing floors and other physical tasks; so, viewing poor and working class women as "robust", functioned to support the social order.

Feminists also cite growing interest on the part of middle class women in pursuing education and careers, and opposition against changes to sex/gender roles, as factors contributing to the perception of what might otherwise be viewed as normal and natural processes, such as puberty and menstruating, as requiring that women avoid intellectual activities. Arguments offered by medical experts against admitting women to college frequently cited the dangers to their reproductive organs. Among the most influential were arguments offered by Dr. Edward H. Clarke in his book, *Sex in education: or a fair chance for girls*. Clarke, a professor at Harvard, published the book in 1873 when pressure to admit women to that institution was at its height. As Ehrenreich and English chronicle, Clarke was opposed to women's admission. He appealed to what were common medical views, warning that women who engaged in strenuous mental activity – who studied in a "boy's way" – risked atrophy of their ovaries and uteruses, insanity, and sterility. Studying would cause their brains to drain energy and/or blood from their reproductive organs. Clarke's arguments persuaded many, even some successful women who came to believe education had harmed them; and some colleges cut back on the number of courses women could take in a year. A warning issued by R.R. Coleman, M.D. to women entering colleges or seeking to be admitted was representative. "Women beware"m he wrote, "you are on the brink of destruction.. science pronounces that the woman who studies is lost". Clearly, at least some were using medical theories to reinforce notions about women's proper role and to prevent them from entering spheres that men had dominated.

As it happened with Darwin's racist ideas, the factually inaccurate misogynistic evolutionary "facts" included in his books were used by many subsequent authors to "scientifically support" their sexist ideas as well as the continuation of women's discrimination, subjugation, and oppression, from the Victorian Era until the present time (see Box 4.2). Also, as it happened with his racist ideas, the most markedly sexist evolutionary statements made by Darwin were published in the *Descent*, in which he wrote things such as:

> Man is more courageous, pugnacious, and energetic than woman, and has a more inventive genius.. [there is a] difference in the mental powers of the two sexes.. woman seems to differ from man in mental disposition, chiefly in her greater tenderness and less selfishness.. The chief distinction in the intellectual powers of the two sexes is shewn by man attaining

> to a higher eminence, in whatever he takes up, than woman can attain – whether requiring deep thought, reason, or imagination, or merely the use of the senses and hands.. Man is more powerful in body and mind than woman, and in the savage state he keeps her in a far more abject state of bondage than does the male of any other animal; therefore it is not surprising that he should have gained the power of selection.. It is generally admitted that with woman the powers of intuition, of rapid perception, and perhaps of imitation, are more strongly marked than in man; but some, at least, of these faculties are characteristic of the lower races, and therefore of a past and lower state of civilisation. The chief distinction in the intellectual powers of the two sexes is shewn by man's attaining to a higher eminence, in whatever he takes up, than can woman--whether requiring deep thought, reason, or imagination, or merely the use of the senses and hands. If two lists were made of the most eminent men and women in poetry, painting, sculpture, music (inclusive both of composition and performance), history, science, and philosophy, with half-a-dozen names under each subject, the two lists would not bear comparison. We may also infer, from the law of the deviation from averages, so well illustrated by Mr. Galton, in his work on 'Hereditary Genius,' that if men are capable of a decided pre-eminence over women in many subjects, the average of mental power in man must be above that of woman. Amongst the half-human progenitors of man, and amongst savages, there have been struggles between the males during many generations for the possession of the females..
>
> That is, from success in the general struggle for life; and as in both cases the struggle will have been during maturity, the characters gained will have been transmitted more fully to the male than to the female offspring. It accords in a striking manner with this view of the modification and re-inforcement of many of our mental faculties by sexual selection, that, firstly, they notoriously undergo a considerable change at puberty, and, secondly, that eunuchs remain throughout life inferior in these same qualities. Thus, man has ultimately become superior to woman. It is, indeed, fortunate that the law of the equal transmission of characters to both sexes prevails with mammals; otherwise, it is probable that man would have become as superior in mental endowment to woman, as the peacock is in ornamental plumage to the peahen. It must be borne in mind that the tendency in characters acquired by either sex late in life, to be transmitted to the same sex at the same age, and of early acquired characters to be transmitted to both sexes, are rules which, though general, do not always hold. If they always held good, we might conclude (but I here exceed my proper bounds) that the inherited effects of the early education of boys and girls would be transmitted equally to both sexes; so that the present inequality in mental power between the sexes would not be effaced by a similar course of early training; nor can it have been caused by their dissimilar early training. In order that woman should reach the same standard as man, she ought, when nearly adult, to be trained to energy and perseverance, and to have her reason and imagination exercised to the highest point; and then she would probably transmit these qualities chiefly to her adult daughters. All women, however, could not be thus raised, unless during many generations those who excelled in the above robust virtues were married, and produced offspring in larger numbers than other women. As before remarked of bodily strength, although men do not now fight for their wives, and this form of selection has passed away, yet during manhood, they generally undergo a severe struggle in order to maintain themselves and their families; and this will tend to keep up or even increase their mental powers, and, as a consequence, the present inequality between the sexes.

Once again, none of the assertions included in the excerpts above had nothing to do with direct observations or statistical studies or experiments done by Darwin. In fact, most of them refer to an imaginary time in our human past, which he populated with rather vague just-so-stories, without specifying when or where they were supposed to have occurred. Take, for example, his assertion that "man has ultimately become superior to woman." When, specifically? Where? At all places around the globe at a single time? At all places at different times? At a single place at a single time? Darwin does not specify this because, obviously, he could not, as there was, and there is, no evidence at all that at a certain time, or times, or a certain place, or places, "men" became superior to "women." We can clearly see, in his writings, how his *a priori* ideas about male 'superiority' led him to create *a posteriori* such imaginary "evolutionary facts." As revealed by his private writings and notes, since he was young he had already internalized the biased idea that women are inferior to men. Such an idea was widespread within schools, universities, the media, and popular cultural in general in England back then, and he became indoctrinated with it. Things could have been different because, as noted above, others that lived before him, or at the same time, sometimes even in the same region – such as Alfred Russel Wallace, Mary Shelley, Mary Wollstonecraft, or William Godwin – did not accept so easily such sexist tales as if they were dogma. Darwin did accept them as dogma, and, much worse than that, and tragically, subsequently constructed factually wrong evolutionary "facts" in his scientific writings to support them, such as his statement that "man has ultimately become superior to woman." This "fact" is then "scientifically" supported by him by using another "fact" constructed by him: we know that "man has ultimately become superior to woman" because women nowadays have "lower mental" powers than men. Darwin did not cite any statistical comparative study, or any cognitive experiment, supporting such "facts": he merely stated them, case solved.

The sexist evolutionary "facts" included in the *Descent* are one of the most powerful examples of the dichotomy between Darwin the biased anthropologist and Darwin the sharp-eyed biologist. This is because in that same book Darwin included numerous strikingly detailed and factually accurate accounts about the sexual evolution of a wide range of nonhuman animals. But suddenly, when it came to the moment to talk about humans, he completely flipped upside down some of those accurate facts. For instance, as explained in Chap. 1, after correctly stating that in most mammals, including numerous primates, the females are often the ones that play a *most active role* in selecting their sexual partners, he then stated that this happened only at the earlier stages – another vague statement – of human evolution and that subsequently men began to be the active sexual selectors. He did not specify exactly when and where that happened, and did not provide any kind of factual evidence to support that "fact." Why did Darwin made such statements? The major reason clearly seem to have been that, within Victorian manly made sexist tales, women were supposed to be *passive* members of society, and accordingly to be sexually passive as well. That is, the true facts that Darwin the attentive biologist described for other primates and mammals about females being often the active sexual selectors did not conform with the *sexist beliefs* of Darwin the biased anthropologist about women. And, once again, as it happened for most of the things he wrote about women and about non-European groups in the *Descent*, the biases of Darwin the biased anthropologist prevailed over the observations of Darwin the biologist.

How could Darwin state as an evolutionary "fact" that one of the sexes of a certain species was "inferior" – including "mentally inferior" – to the other sex of the same sexual species, when both sexes are obviously necessary to procreate? It would be like saying that the brain is "inferior" to the heart, or the kidney is "superior" to the liver. But such awkward "facts" were indeed repeated over and over by Darwin in his writings about human evolution. And, due to the sexist biases that he and many – but, importantly, not all, as we have seen above – other male Victorian scientists had back then not only he affirmed such "facts" in the *Descent* but also such "facts" became uncritically accepted as dogmas by many of his scientific peers. This is another example illustrating how the evolutionary "facts" postulated in his books about human evolution were in great part accepted by his peers not because they were truly accurate, but instead because they were music to the hears of most of the male scientists, and most males in general, within the Victorian society. Those males were extremely happy to justify that the fact that most scientists, politicians, bakers, and so on within that society were men by arguing that this was just a *natural* fact of life. Women were the ones to blame, because they were biologically inferior, with inferior mental powers, and were naturally inclined to be passive, sexually coy, to not engage in scientific activities, and to be domestic, as the great Victorian scientist Darwin affirmed.

Darwin and the Feminist and Anti-Vivisection Movements

> *Even he (Charles Darwin), the father of evolutionary biology, was so affected by a culture of sexism that he believed women to be the intellectually inferior sex.* (Angela Saini)

While a huge number of Victorian scientists and lay people, in particular men, were happy with – and therefore blindly accepted – the type of sexist evolutionary inaccurate "facts" constructed by Darwin, some did not, as noted above, in particular women. Many of those women could not really do much about the way in which Darwin's sexist "facts" were so widely welcomed, accepted, and praised within the scientific community because obviously that community was basically dominated by men and *excluded* women. Outside the academy, there were however some feminist female writers that had made a way to have their voices heard within such a male-dominated society. In fact, if Darwin had not been so blinded by his misogynistic biases, he would have realized that what is remarkable in the history of agricultural states is that *despite* such biases, prejudices, and false narratives about women, and the historical subjugation of women to which they have contributed to, many women actually still found a way to do truly remarkable things. One beautifully illustrated book that I highly recommend about this topic because it is written for both adults and kids, is Rachel Ignotofsky's 2016 *Women in Science: 50 Fearless Pioneers Who Changed the World*. Obviously, apart from being able to do amazing science despite being mostly excluded from it by men, there is a wide range of other intellectual tasks that were done by women, even during the European Middle Ages, such as manuscript production, as pointed out in a 2019 article by Radini and colleagues. This also applies to many women activists. For instance, at Darwin's time, many of such activists openly criticized – correctly so – the sexist evolutionary ideas of Darwin the biased anthropologist, as we will see. Other activists opted instead to use some of the broader, in general less biased, ideas of Darwin the biologist about organismal evolution and natural selection as a way to defend their feminist thoughts. That is, such feminists, white not happy at all about Darwin's evolutionary ideas about the natural passivity, lack of imagination and overall lower mental powers of women, understood that Darwin's broader theories of evolution driven by natural selection could be a better tool for their arguments than religious ideas were. This subject was discussed in detail in books such as Samantha Evans's 2017 *Darwin and Women* and Hamlin's *From Eve to Evolution*.

As put by Hamlin, what "primarily motivated many women [such as feminists and reformers] to enlist Darwinian evolution" was not the specific evolutionary ideas of Darwin *per se* but instead "their deep frustration with Christian ideology based on Eve" (Fig. 4.2). That is, these women could have mentioned *Darwin*'s *Origin*, or Wallace's works, or Huxley's books: any book that would defend that organisms evolve, that their traits are not *fixed* as it is often defended in many religious texts, was a useful tool for their cause. As noted by Hamlin:

> Prominent women's suffragists did not invoke evolutionary theory very often, and Charles Darwin's ideas about women, like those of most of his fellow evolutionists, were largely

Fig. 4.2 Adam and Eve, by Richard Rothwell

> shaped by the ideology of "separate spheres" for men and women that dominated the Victorian era in which he lived. Visionary scientist, yes; feminist, no. Furthermore, since the 1970s, feminist historians have frequently argued that Darwinian evolutionary science, at least as it was articulated in the nineteenth century, should be considered, in the words of one scholar, "intrinsically anti-feminist".. This book suggests that Charles Darwin.. the most influential evolutionist of the nineteenth (or any) century, did not intentionally upend traditional ideas about gender and sex, but that is precisely what his writings helped to do, as many American women's rights activists immediately recognized.. *Reform Darwinists* defined themselves as progressive evolutionists, in favor of things such as worker's and women's rights and in opposition to social Darwinists, who tended to support Gilded Age industrial inequities and the *status quo*. These women, including Antoinette Brown Blackwell, Helen Hamilton Gardener, Eliza Burt Gamble, and Charlotte Perkins Gilman, forged an evolutionary feminism that grappled with questions of biological sex difference, the extent to which maternity did (and should) define women's lives, the equitable division of household labor, and female reproductive autonomy. The practical applications of this evolutionary feminism came to fruition in the early thinking and writing of the American birth control pioneer Margaret Sanger..
>
> Darwin's own views on gender, at least as expressed in his published writings, often rearticulated the dominant, patriarchal views of his era. In the nineteenth century, prescriptive literature and social customs dictated that men inhabit the worlds of commerce, labor, and politics, while women controlled the home, the family's spiritual life, and the children. Such a gendered division of labor was considered natural, civilized, and in accordance with God's will, and, at first glance, Darwin's writings about evolution did little to challenge these long-standing beliefs. Darwin's.. explained that, throughout the animal kingdom, the male "has been the more modified" due to the males' having "stronger passion than the females", which tend to retain "a closer resemblance" to the young. Among humans, Darwin believed that "owing to her maternal instincts" woman differs from man chiefly in her "greater tenderness and less selfishness" and lack of intellectual attainments. Overall, Darwin believed that female intellectual inferiority was natural and, most likely, immutable; he imposed Victorian gender roles and mating behavior on animals – combative male insects, strutting peacocks, and coy peahens; and he espoused patriarchal marriage as the epitome of civilization. Such descriptions inspired at least one generation of naturalists to conclude that women's inferiority was a permanent and necessary part of the evolutionary process and a later generation of feminists to reason that evolutionary science was inherently misogynistic. Yet, Darwin's writings, especially *The Descent of Man*, and *Selection in Relation to Sex* and its cornerstone theory of "sexual selection", were multivalent. Even though Darwin and most other nineteenth-century scientists believed that evolution, like Genesis, demanded women's subservience to men and total devotion to maternity, his theory of evolution contained the seeds of radical interpretations as well as conventional ones. Many feminists and other reformers were keen to these revolutionary insights and embraced evolutionary science as an ally.

In other worlds, some feminists at that time realized that evolutionary ideas *could be* a better ally for their cause than religious concepts because within biological evolution things can *change*, while the Word of God is supposedly eternal. As explained above, within most of the religious stories originally constructed or subsequently manipulated by men, women were seen as innately inferior, so it was indeed very difficult to use such stories to change the prevailing *status quo* of male domination and oppression. In contrast, misogynistic evolutionists such as Darwin argued that originally, in the first sexual species, Mother Nature did not "do"' females to be inferior to males – in the *Descent* the marked inferiority of women mainly arose within the earlier stages of human evolution. As discussed above, within the typical logical inconsistencies between Darwin the biologist and Darwin the anthropologist, Darwin's writings were confusing about this point. On the one hand, he stated that this marked inferiority of women, acquired in early human evolution, was now somewhat fixed, innate. On the other hand, in the same book, he also suggested that this could eventually change if something special would happen – when women will become the "breadwinners," as he stated.

Apart from such logical inconsistencies between his ideas of adaptations to local environments at specific times and of "races" and genders that are in "inferior", including their "mental" capacities, one can see here other of the main fallacies about Darwin's reasoning of what is supposed to be "superior" and "inferior" in human evolution: as a true Social Darwinian, he equated superiority to dominance, to being at the top of the ladder, to be well-off, to be a "breadwinner." Instead of engaging in a deep historical, anthropological, and sociological quest to try to understand the societal mechanisms that led to most "breadwinners" within the Victorian society being men, or being born in families that were already well-off as it was his case, he reversed this up-side-down: men were superior because they were the breadwinners, period. Exactly in the same way that, as stated by him as a "scientific fact", British were superior to Aboriginal Australians because the former were the ones that dominated and oppressed the latter, even if they did so by using brutal force and engaging in horrible massacres (Fig. 1.15). We can see here how Darwin did often use his combined concepts of natural selection and struggle-for-existence in a way that is very similar to the way in which sexists and white supremacists did in the past and still to do: if you dominate, oppress, and massacre others, that means you are *superior* to them, that you are the victorious "race," gender or nation within the eternal struggle-for-existence. Music to the hears of sexist and racist people such as James Watson or, tragically, such as David Duke and Hitler for that matter, as we have seen.

Even authors that tend to idealize Darwin, such as Gruber in *Darwin on Man*, have detected such major logical confusions and inconsistencies in Darwin's works, from his very first writings about the Beagle voyage to the last ones, including the *Descent*:

> Do Habits Become Hereditary? Several passages in the *Beagle Diary* tempt one to believe that Darwin was, during the Beagle voyage, already examining the hypothesis that habits may become hereditary.. Although Darwin speaks of the importance of "habit" at many different points in his notes, what he means by the term is shifting. In the Beagle Diary the term habit refers to a behavior pattern, without always distinguishing learned from hereditary patterns. This is not an unusual usage; even today naturalists informally describing the "habits" of an animal may mean simply its characteristic behavior patterns, without regard to their genesis.. Even in the *Origin* Darwin used habit in a number of different senses. In the chapter on "the laws of variation" there is a section entitled "*Effects of Use and Disuse*" in which, in a representative sentence, he writes, "We may imagine that the early progenitor of the ostrich had habits like those of a bustard". In the same chapter, in a section entitled "*Acclimatisation*", he begins, "Habit is hereditary with plants, as in the period of flowering, in the amount of rain requisite for seeds to germinate, in the time of sleep".. The significant point is that he uses habit in a way that makes it sometimes hard to tell whether he is speaking of a learned or an unlearned pattern of behavior.. An early expression in Darwin's notes of the hypothesis of habits becoming hereditary occurs in a quotation from F. Cuvier about "one of the most general laws of life – the transmission of a fortuitous modification into a durable form, of a fugitive want into a fundamental propensity, of an accidental habit into

> an instinct." Darwin adds, "I take higher grounds and say life is short for this object and others, viz. not too much change". He means to say that if such changes accumulate indefinitely in an individual organism, it will lose its identity as a member of a species, and so will its progeny if the changes are heritable. Consequently, the process of death serves the adaptive function of limiting change and maintaining the integrity of species. The changes in question are all the psychological developments that occur in the life of an individual.

As astutely noted by Dame Gillian Beer in the Foreword of Samantha Evans' 2017 *Darwin and Women*, apart from this logical confusion between "habits" and innate traits and Darwin's typical use of much more "positive" terms to refer to anything that was male within the natural world in natural, Darwin also clearly applied his typical circular reasoning – or "argumentative loop" paraphrasing Gillian Beer – to this topic of women's supposed inferiority:

> The sexual behaviour of different human groups is studied in the *Descent* alongside that of other kinds, as also are the physical differences between sexes in a range of creatures. And here we begin to see the problem that Darwin has not so much introduced as illuminated by setting the human among other kinds. In his descriptions of behaviour it is often difficult to discriminate human interpretation from physical structures. For example, discussing the secondary sexual characters of insects, he contrasts "the pectinated and beautifully plumose antennae of the males of many species" with the meagreness of the females: "the male has great pillared eyes, of which the female is entirely destitute". His children told him that his descriptions sounded like advertisements and here the males benefit from the enthusiasm of his language. And where he finds not physical difference but likeness between the sexes he comments, using the observations of colleagues, on contrasted behaviour: "In one of the sand-wasps (*Ammophila*) the jaws in the two sexes are closely alike, but are used for widely different purposes: the males, as Professor Westwood observes, 'are exceedingly ardent, seizing their partners round the neck with their sickle-shaped jaws'; whilst the females use these organs for burrowing in sand-banks and making their nests". Darwin's fundamental insistence in all his arguments on the similitudes between the human and other kinds inclines him to accept the fixed differences, for example in sand-wasps, as a model for human capabilities, rather than as the outcome of human behaviours in current social conditions..
>
> [Moreover] when Darwin in the *Descent* commented that women would never equal men until they became the breadwinners he seems not to have noticed that his wife's extraordinary time management smoothed his uninterrupted researches and writing and thus underpinned the 'breadwinning' of the household. He repeats this view in his late correspondence with Caroline Kennard: "women must become as regular 'bread-winners' as are men' to avoid 'the laws of inheritance'." But then he demurs further: "we may suspect that the early education of our children, not to mention the happiness of our homes, would in this case greatly suffer". Thus women are corralled within an argumentative loop: they cannot catch up with men until they are active in the wider world but if they are so active, they will lose the innate moral superiority with which he at present endows them, because they will be obliged to sacrifice their families to their ambitions.

Darwin's fascination for male traits indeed applied both to non-human organisms and to humans. This is further evidenced in countless passages of his books – not only the *Descent* but also books such as the *Voyage of the Beagle*, in which he stated that he "was much disappointed in the personal appearance of the women (in Tahiti): they are far inferior in every respect to the men". Strikingly – but not really surprising – even some authors that tend to be more nuanced in their books about Darwin, such as Desmond and Moore, have constructed idealized versions of Darwin's attitude toward women. For instance, in the preface of their 1994 book about *Darwin* they state that he "subordinated women but was totally dependent on his redoubtable wife." This is another powerful illustration of the type of mental gymnastics that are often used in order to idealize Darwin, because the latter statement – he depending on his wife – does not compensate the former – he subordinating women –, well on the contrary. One of the main reasons why so many men constructed just-so-stories about women's "inferiority" or "lower mental powers" similar to those constructed by Darwin in the *Descent* was precisely to be able to subjugate, or to justify the subjugation, of women so they cook for them, do domestic tasks for them, and so on (see Fig. 4.9). This is precisely one of the main goals and results of systemic sexism and racism, and of subjugation in general, as one can see in European colonialism, in which the European "masters" were "totally dependent" on the colonized populations to take care of their houses, to build their rail trails and roads, and so on. This is also evident when one sees what happens nowadays in countries such as Pakistan or Afghanistan, in which Desmond and Moore's rather odd defense of an idealized Darwin would apply to most men: they "subordinate women" *and accordingly* – instead of the "but" used by Desmond and Moore's – they are "totally dependent on" their "redoubtable wives."

Saini's 2017 book *Inferior* further highlights Darwin's sexism and, at the same time, emphasizes how some of the ideas that came from scholars such as Darwin are slowly starting to be more and more questioned, in certain regions of the globe, in particular due to the continuous and persistent effort of bright and brave women (see also Box 4.3):

> I'm holding three letters, all yellowing, the ink faded and the creases brown. Together they tell a story of how women were viewed at one of the most crucial moments of modern scientific history, when the foundations of biology were being mapped out. The first letter, addressed to Darwin, is written in an impeccably neat script on a small sheet of thick cream paper. It's dated December 1881 and it's from a Mrs Caroline Kennard, who lives in Brookline, Massachusetts, and a wealthy town outside Boston. Kennard was prominent in her local women's movement, pushing to raise the status of women (once making a case for police departments to hire female agents). She also had an interest in science. In her note to Darwin, she has one simple request. It is based on a shocking encounter she'd had at a meeting of women in Boston. Someone had taken the position, Kennard writes, that 'the inferiority of women; past, present and future' was 'based upon scientific principles'. The authority that encouraged this person to make such an outrageous statement was no less than one of Darwin's own books. By the time Kennard's letter arrived, Darwin was only a few months away from death. In her letter, Kennard naturally assumes that a genius like Darwin couldn't possibly believe that women are naturally inferior to men. Surely his work had been misinterpreted? 'If a mistake has been made, the great weight of your opinion and authority should be righted', she entreats. 'The question to which you refer is a very difficult one', Darwin replies the following month from his home at Downe in Kent. If polite Mrs Kennard was expecting the great scientist to reassure her that women aren't really inferior to men, she was about to be disappointed. 'I certainly think that women though generally superior to men [in] moral qualities are inferior intellectually', he tells her, 'and there seems to me to be a great difficulty from the laws of inheritance, (if I understand these laws rightly) in their becoming the intellectual equals of man'. It doesn't end there. For women to overcome this biological inequality, he adds, they would have to become breadwinners like men. And this wouldn't be a good idea, because it might damage young chil-

dren and the happiness of households. Darwin is telling Mrs Kennard that not only are women intellectually inferior to men, but they're better off not aspiring to a life beyond their homes. It's a rejection of everything Kennard and the women's movement at the time were fighting for. Darwin's personal correspondence echoes what's expressed quite plainly in his published work.

In *The Descent of Man* he argues that males gained the advantage over females across thousands of years of evolution because of the pressure they were under in order to win mates. Male peacocks, for instance, evolved bright, fancy plumage to attract soberlooking peahens. Similarly, male lions evolved their glorious manes. In evolutionary terms, he implies, females are able to reproduce no matter how dull their appearance. They have the luxury of sitting back and choosing a mate, while males have to work hard to impress them, and to compete with other males for their attention. For humans, the logic goes, this vigorous competition for women means that men have had to be warriors and thinkers. Over millennia this has honed them into finer physical specimens with sharper minds. Women are literally less evolved than men. 'The chief distinction in the intellectual powers of the two sexes is shown by man attaining to a higher eminence, in whatever he takes up, than woman can attain – whether requiring deep thought, reason, or imagination, or merely the use of the senses and hands', Darwin explains in *The Descent of Man*. The evidence appeared to be all around him. Leading writers, artists and scientists were almost all men. He assumed that this inequality reflected a biological fact. Thus, his argument goes, 'man has ultimately become superior to woman'. This makes for astonishing reading now. Darwin writes that if women have somehow managed to develop some of the same remarkable qualities as men, it may be because they were dragged along on men's coattails by the fact that children in the womb inherit attributes from both parents. Girls, by this process, manage to steal some of the superior qualities of their fathers. 'It is, indeed, fortunate that the law of the equal transmission of characters to both sexes has commonly prevailed throughout the whole class of mammals; otherwise it is probable that man would have become as superior in mental endowment to woman, as the peacock is in ornamental plumage to the peahen'. It's only a stroke of biological luck, he implies, that has stopped women from being even more inferior to men than they are. Trying to catch up is a losing game – nothing less than a fight against nature. To be fair to Darwin, he was a man of his time. His traditional views on a woman's place in society don't run through just his scientific works, but those of many other prominent biologists of the age. His ideas on evolution may have been revolutionary, but his attitudes to women were solidly Victorian.

Box 4.3 Misogyny and Darwinism

In her book *Inferior*, Saini explained that:

> Unconventional ideas can appear from anywhere, even the most conventional places. The township of Concord in Michigan is one of those places. In 1894.. a middleaged schoolteacher from right here in Concord published some of the most radical ideas of her age. Her name was Eliza Burt Gamble. Gamble believed there was more to the cause than securing legal equality. One of the biggest sticking points in the fight for women's rights, she recognized, was that society had come to believe that women were born to be lesser than men. Convinced that this was wrong, in 1885 she set out to find hard proof for herself. She spent a year studying the collections at the Library of Congress in the U.S. capital, scouring the books for evidence. She was driven, she wrote, 'with no special object in view other than a desire for information'. Although not a scientist herself, through Darwin's work Gamble realized just how devastating the scientific method could be. If humans were descended from lesser creatures, just

(continued)

like all other life on earth, then it made no sense for women to be confined to the home or subservient to men. These obviously weren't the rules in the rest of the animal kingdom.. But, for all the latent revolutionary power in his ideas, Darwin himself never believed that women were the intellectual equals of men. This wasn't just a disappointment to Gamble, but judging from her writing, a source of great anger. She believed that Darwin, though correct in concluding that humans evolved like every other living thing on earth, was clearly wrong when it came to the role that women had played in human evolution. Her criticisms were passionately laid out in a book she published in 1894, called *The evolution of woman: an inquiry into the dogma of her inferiority to man*. Marshalling history, statistics and science, this was Gamble's piercing counterargument to Darwin and other evolutionary biologists.

She angrily tweezed out their inconsistencies and double standards. The peacock might have had the bigger feathers, she argued, but the peahen still had to exercise her faculties in choosing the best mate. And while on the one hand Darwin suggested that gorillas were too big and strong to become higher social creatures like humans, at the same time he used the fact that men are on average physically bigger than women as evidence of their superiority. He had also failed to notice, Gamble wrote, that the human qualities more commonly associated with women – cooperation, nurture, protectiveness, egalitarianism and altruism – must have played a vital role in human progress. In evolutionary terms, drawing assumptions about women's abilities from the way they happened to be treated by society at that moment was narrow-minded and dangerous. Women had been systematically suppressed over the course of human history by men and their power structures, Gamble argued. They weren't naturally inferior; they just seemed that way because they hadn't been allowed the chance to develop their talents. Gamble also wrote that Darwin hadn't taken into account the existence of powerful women in some tribal societies, which might suggest that the present supremacy of men now was not how it had always been. The ancient Hindu text the Mahabharata, which she picked out as an example, speaks of women being unconfined and independent before marriage was invented. So she couldn't help but wonder, if 'the law of equal transmission' applied to men as well as women, might it not be possible that males had been dragged along by the superior females of the species? [In fact] parts of science remained doggedly slow to change. Evolutionary theory progressed pretty much as before, learning few lessons from critics like Albert Wolfe, Caroline Kennard and Eliza Burt Gamble. It's hard to picture the directions in which science might have gone if, in those important days when Charles Darwin was developing his theories of evolution, society hadn't been quite as sexist as it was. We can only imagine how different our understanding of women might be now if Gamble had been taken a little more seriously. Historians today have regretfully described her radical perspective as the road not taken. In the century after Gamble's death, researchers became only more obsessed by sex differences, and by how they might pick them out, measure and catalogue them, enforcing the dogma that men are somehow better than women.

Once again, we can see that the widespread acceptance and idolization of Darwin and his ideas within the scientific and societal "cream" of the Victorian society was in great part related to how well those ideas resonated to those "well-offs" – mostly "white" males – and were particularly useful for them to justify their subjugation of "others" – the underprivileged, the "other" races, and women. Once again, Darwin mostly sided with those well-offs, in great part because he was precisely one of them. As recognized by Samantha Evans in *Darwin and Women*: "although no campaigner, Emma [Darwin's wife] certainly didn't view the women's suffrage movement with the abhorrence that some more conservative [Victorian] women did.. Darwin's own views on such pragmatic support for women's suffrage would have been slightly different.. although both he and Emma were fundamentally opposed to cruelty to animals, Darwin was alarmed at the popular success of the anti-vivisection movement, led by campaigners such as Frances Power Cobbe." At that time, the "fight" against the vivisection of non-human animals – that is, against opening up non-human animals when they were still alive to study them, in particular their physiology (see Fig. 4.5) – was in great part promoted by women. Darwin, using his power of place, wrote a rather sexist, patronizing and paternalistic letter about this topic to one of the most influential English newspapers back then, the *Times*, which was published:

> TO THE EDITOR OF THE TIMES: Sir,—I do not wish to discuss the views expressed by Miss Cobbe in the letter which appeared in *The Times* of the 19th inst.; but as she asserts that I have "misinformed" my correspondent in Sweden in saying that "the investigation of the matter by a Royal Commission proved that the accusations made against our English physiologists were false," I will merely ask leave to refer to some other sentences from the Report of the Commission. (1) The sentence—"It is not to be doubted that inhumanity may be found in persons of very high position as physiologists," which Miss Cobbe quotes from page 17 of the report, and which, in her opinion, "can necessarily concern English physiologists alone and not foreigners," is immediately followed by the words "We have seen that it was so in Majendie." Majendie was a French physiologist who became notorious some half-century ago for his cruel experiments on living animals. (2) The Commissioners, after speaking of the "general sentiment of humanity" prevailing in this country, say (p. 10):— "This principle is accepted generally by the very highly educated men whose lives are devoted either to scientific investigation and education or to the mitigation or the removal of the sufferings of their fellow-creatures; though differences of degree in regard to its practical application will be easily discernible by those who study the evidence as it has been laid before us."
>
> Again, according to the Commissioners (p. 10):— "The secretary of the Royal Society for the Prevention of Cruelty to Animals, when asked whether the general tendency of the scientific world in this country is at variance with humanity, says he believes it to be very different, indeed, from that of foreign physiologists; and while giving it as the opinion of the society that experiments are performed which are in their nature beyond any legitimate province of science, and that the pain which they inflict is pain which it is not justifiable to inflict even for the scientific object in view, he readily acknowledges that he does not know a single case of wanton cruelty, and that in general the English physiologists have used anesthetics where they think they can do so with safety to the experiment." I am, Sir, your obedient servant, CHARLES DARWIN.

Darwin used a similar tone in another letter to *The Times*, about the same topic:

> To the Editor of the *Times*: Sir, as every one who is capable of forming a sound judgment on the subject is convinced that the relief of human suffering in future depends chiefly on the progress of physiology, I hope that you will find space in your columns for the enclosed article by Dr. Richardson, which has just appeared in "*Nature*". The article shews in a practical manner, and more conclusively than anything that has been published elsewhere, the necessity of experiments on living animals. Women, who from the tenderness of their hearts and from their profound ignorance are the most vehement opponents of all such experiments, will I hope pause when they learn that a few such experiments performed under the

> influence of anesthetics, have saved and will save through all future time, thousands of women from a dreadful and lingering death. It is humiliating to reflect that those to whom mankind owe the deepest debt of gratitude should now be overwhelmed by falsehood and calumny. I am, sir, | your obedient servant. Charles Darwin, Down, Beckenham, June 23rd. 1876.

"*Women, who from the tenderness of their hearts and from their profound ignorance are the most vehement opponents of all such experiments, will I hope pause when they learn that..*": Darwin's sexism, mixed with the use of his privileged position within Victorian society to recurrently defend the Victorian *status quo*, all reflected in a single sentence. In fact, it was about the same topic that Darwin famously asserted, in an 1877 letter, that "no woman has advanced the science." He wrote: "I make the distinction between a boy & a girl, because as yet no woman has advanced the science.. I believe much physiology cd. be learned without seeing any experiments performed or any organ in action; but I do not believe that a person could learn several parts of the subject with [the] vividness & clearness, which is necessary for well instructing others, unless he saw some of our organs in action."

It is important to put Darwin's sentences about women's ignorance, mental inferiority, and lack of scientific skills in historical context, to understand how both his public statements and scientific writings were not only biased but also neglected knowledge that was already known back then about human history, culture, arts, and society in general. Darwin wrote those letters to the *Times* a few years after he published the *Descent*, in 1871. By then, despite the profound sexism prevailing within the Victorian society, various English women had already made a way to reach prominent social positions, many of them by writting books that were amazingly complex intellectually. To give just an example, decades before Darwin wrote those letters, Mary Shelley (Fig. 4.3) had published *Frankenstein – or, the modern*

Fig. 4.3 Mary Shelley, by Richard Rothwell

Prometheus. She started writing that book when she was 18. That is, in contrast to Darwin, who according to his own autobiography at 18 often considered that many of the intellectually complex things he was forced to learn at university were "dull" and "was as ignorant as a pig about his subjects of history, politics, and moral philosophy," the 18-year-old Mary Shelley started to write such an intellectual masterpiece. A masterpiece that is consensually considered to be extraordinarily bright and rich because it precisely combines philosophy with elements of Gothic novels and the Romantic movements, as well as science – being in part inspired by experiments done by Darwin's grandfather, Erasmus Darwin –, science fiction, and even politics. The first edition of *Frankenstein* was published anonymously in 1818, when she was only 20, and her name then appeared in the second edition, published in 1821. That is, exactly 50 years before Darwin published the *Descent* and 55 years before he publicly stated that women are "ignorant."

Mary Shelley was the daughter of Mary Wollstonecraft (Fig. 4.4), who was a well-known writer, philosopher, and advocate of women's rights that is nowadays consensually considered to be one of the first modern Western feminist philosophers. This, itself, further shows the importance of the *power of place* and its social networks, within English society back then, because Mary Shelly had both a prominent father – William Godwin, a journalist, novelist, and political philosopher that is often considered to be one of the first advocates of utilitarianism and modern promoters of anarchism – and mother, who died less than a month after giving birth to her. As explained in the 2019 book *Making the Monster: The Science Behind Mary Shelley's Frankenstein* by Kathryn Harkup, Mary Shelley's masterpiece included references to many societal and scientific topics that she originally heard directly from her father or from the numerous prominent visitors that he received at his house during her youth. If she were born in let us say a very small village from an underprivileged family of

Fig. 4.4 Mary Wollstonecraft

farmers, it would be much less likely that she would have written such a book, or that she had the needed societal networks to publish and promote it.

In *Darwin and Women*, Samantha Evans emphasizes the point that Darwin's biases prevented him to recognize that the existence of women as intellectually remarkable as Mary Shelley – and even his "beloved" Jane Austen, whose books Darwin avidly consumed – showed that his public and published assertions about women being "ignorant" and mentally "inferior" were factually wrong. In fact, Darwin omitted from his scientific books not only the existence of such women and their excellent writings, but also, importantly, that many of them – such as Mary Wollstonecraft – had already explained very clearly in those writings why women often did not have the opportunity to reach societal prominence within the sexist and hierarchical English society. Evans wrote:

> Darwin was not overtly opposed to women's higher education. Towards the end of his life, events seemed to catch up with him when women's influence in the anti-vivisection movement became clear. He thought women underestimated the medical benefits of vivisection, and attributed this to their lack of education in physiology. He became a supporter of physiological education for women. In 1881, women were given the right to take examinations at Cambridge University. Darwin commented in a letter to his son George: "You will have heard of the triumph of the Ladies at Cambridge.. the majority was so enormous that many men on both sides did not think it worth voting.. the minority was received with jeers". Horace [Darwin] was sent to the Lady's College to communicate the success & was received with enthusiasm.. [Therefore] when Darwin wrote on women's capabilities in *Descent*, he couldn't have been unaware of the problems of bias and social disadvantage. However, he kept his argument strictly biological, and he relied on two factors he himself wasn't entirely sure about: the inheritance of acquired characteristics, and inheritance limited by sex.
>
> His theory was that men had long undergone more severe testing than women as adults in competition for wives, and that the qualities they acquired as a result – energy and perseverance – were passed on to their sons but not their daughters (the principle being that qualities that manifest later in life tend to be limited to one sex: as, for instance, colourful plumage in the male peacock). This accounts for his notion that for women to equal men in intelligence, they would not just have to be educated as young adults; they would have to pass on the effects of that education to their daughters and repeat the process for many generations.. [Thus Darwin wrote] "we may also infer.. that if men are capable of decided eminence over women in many subjects, the average standard of mental power in man must be above that of woman". It was a surprising thing to write at a time when there was already much discussion of the social disadvantages faced by women; their lack of education, their exclusion from the professions and politics, their legal disabilities. Darwin's own beloved Jane Austen had pointed out, through her heroine Anne Elliot in *Persuasion*, "Men have had every advantage of us in telling their own story.. education has been theirs in so much higher a degree; the pen has been in their hands.. I will not allow books to prove any thing". How could Darwin be unaware of the social bias that doomed most women to underachievement, and the bias of perception that caused even high achievers to be considered second rate compared with men?

The above discussion further illustrates how the typical line of defense used by Darwin's idolizers about Darwin's sexism and racism that "everybody was like that, back then" – that he was just a passive automaton replicating what he learned about those topics –, is flawed. This is because he was very *active*, to the point of sending two letters to a very influential newspaper that portrayed women in a way that clearly contrasted with the way in which they were portrayed by a *significant*

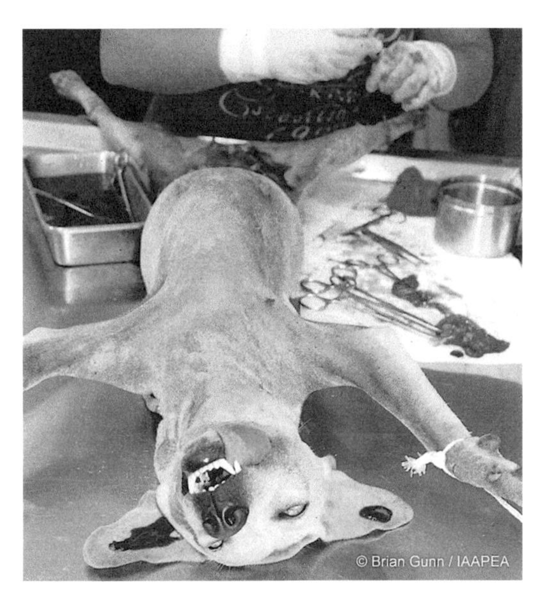

Fig. 4.5 A dog restrained in a brutal bloody experimental vivisection research testing lab: as noted in the caption provided by the IAA PEA – The International Association Against Painful Experiments on Animals – "atrocities are not less atrocities when they occur in laboratories and are called medical research"

number of people – women, but also many men – back then. That is, by those that contrary to Darwin were fighting against the Victorian *status quo* concerning the discrimination and subjugation of women. This was a widespread fight that comprised numerous crucial topics and struggles, including the women's suffrage movement. In England, the fight for women's right to vote became a national movement in 1872: just 1 year after Darwin published several sexist evolutionary "facts" in the *Descent* and a few years before he wrote those letters to the *Times* calling women naive and ignorant. The demands of the movement were finally put through laws in 1918 and 1928.

Apart from demands such as those of the anti-vivisection moment (Fig. 4.5), or to be able to vote, another type of demand that many women – as well as men – were doing at Darwin's time, not only in England but also in other countries such as the United States, was for women to be allowed to engage in science in the same way that men did. One the earliest and most prolific advocates for that cause in the United States was Antoinette Brown Blackwell, who was also the first woman to be ordained as a mainstream Protestant minister in that country (Fig. 4.6). Blackwell correctly pointed out that Darwin's writings were detrimental for that cause. As put by her, Darwin's sexist ideas were particularly problematic because Darwin was a famous scientist and used his power of place to promote such inaccurate ideas as if they were "scientific facts," and therefore he "more profoundly influenced the opinions of the civilized world than perhaps any other man [together with Spencer]." As explained by Hamlin in *From Eve to Evolution*:

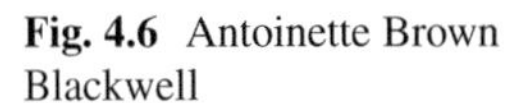

Fig. 4.6 Antoinette Brown Blackwell

Antoinette Brown Blackwell knew first hand what it felt like to be barred from intellectual and professional goals because of her sex, and, like her acquaintance Helen Gardener, she ultimately looked to science for recourse. When Blackwell entered Oberlin College in 1845, she intended to become the nation's first ordained female minister, despite the fact that her advisor.. and pretty much everyone else.. disapproved of women speaking in public, to say nothing of a woman leading a congregation. On September 15, 1853, she succeeded, ascending to the pulpit of her own Congregationalist parish in South Butler, New York. However, her hard-won and historic tenure lasted only a few months. After all the years of fighting the church and educational establishment for the right to preach, Blackwell began to lose her faith in Christian orthodoxy. While she retained her belief in an omnipotent higher power, she officially resigned from her pulpit on July 20, 1854, and turned her attention to science.. While Blackwell was enthusiastic about science, she had her doubts about the authority invested in the findings of male scientists. As Blackwell insisted, "[I]t is to the most rigid scientific methods of investigation that we must undoubtedly look for a final and authoritative decision as to woman's legitimate nature and functions". Whatever the results, she implored women to "most confidently appeal" to "Nature as umpire – to Nature interpreted by scientific methods."

The problem, according to Blackwell, was that scientific methods were often perverted by male scientists. In particular, Blackwell lamented that evolutionary theory had been misinterpreted by "the wisest, the highest, the most progressive and the most influential authorities in science to-day". Because they were "standing on a learned masculine eminence, looking from their isolated male standpoints through their men's spectacles and through the misty atmosphere of entailed hereditary glamour", these authorities could see only evidence of women's "natural" inferiority. Especially egregious in this regard were Charles Darwin and Herbert Spencer, "thinkers who have more profoundly influenced the opinions of the civilized world than perhaps any other two living men.. and [who] endorsed by other world-wide authorities, are joined in assigning the mete and boundary of womanly capacities". Unfortunately, Darwin and Spencer accepted the theory of "male superiority" as a "foregone conclusion" rather than establish it scientifically using "adequate tests" and "careful and exact calculation". Furthermore, Blackwell fumed, what exactly did men like Darwin and Spencer know about "the normal powers and functions of Woman"? The problem, then, was not science but the fact that science was being conducted mainly by men, for men, and did not include accurate studies of women.

Although too slowly, the narrative that Darwin's sexist and racist evolutionary ideas were just a mere product of his time, as if he had no agency at all when he published them as "facts" in his scientific books, is becoming to be more and more criticized in both the academia and the press. For instance, in a 2021 article entitled "*How Darwin got sex and race wrong*" published by Zulfikar Abbany in DW (Deutsche Welle), it is stated that "the problem [with Darwin's sexist evolutionary 'facts'] is that Darwin could have done better at the time.. he could have known better." Abbany cites DeSilva, the editor of the book *A Most Interesting Problem: What Darwin's Descent of Man Got Right and Wrong About Human Evolution*: "He [Darwin] had the data to do it, it's not like he couldn't go against the tide of the times" says DeSilva – "I mean, he wrote *Origin of Species*!" DeSilva is right to the point. Darwin's idolizers cannot praise, on the one hand, Darwin for being so courageous and visionary and going against the tide of the times when he wrote the *Origin* while arguing, on the other hand, that the sexist and racist ideas that he actively chose to write in the *Descent* as scientific "facts" were just a reflection of him passively going with the flow within the current created by that tide.

Inaccuracies of Darwin's Evolutionary "Facts" About Women

> *In the history of science, we have to hunt for the women – not because they weren't capable of doing research, but because for a large chunk of time they didn't have the chance.. We're still living with the legacy of an establishment that's just beginning to recover from centuries of entrenched exclusion and prejudice.* (Angela Saini)

As it usually happened in human history, the type of false narratives defended by misogynistic men – for instance that women are evolutionary passive, or sexually passive, as affirmed by Darwin – were widely disseminated to the broader public by the use of art. Science, art, the media, and education play a crucial role within the vicious cycle of both systemic sexism and racism. This was explained, and visually shown, in Stephanie Moser's wonderful 1998 book *Ancestral Images: The Iconography of Human Origins*. As she notes, most images about human evolution have historically focused – and continue to focus – on men. We all heard about our "cave-men" prehistory – as if women were completely inert, or even totally inexistent, for millions of years. This applies, for example, to three of the so-called most crucial moments of human evolution, as noted in Chap. 1: discovery of fire (Fig. 1.10), stone tool building and use (Fig. 1.11) and cave painting (Fig. 1.12). Of course, there is no empirical data showing that these things were originally, or exclusively, done by men, contrary to what is shown in such images.

This topic was recently discussed in Marylene Patou-Mathis' book *L'homme préhistorique est aussi une femme – Une histoire de l'invisibilité des femmes* [*Prehistoric man is also a woman – A history of the invisibility of women*]. The book presents numerous examples of sexist practices, biases, and narratives concerning scientific research about our evolutionary past. It considers the results of more recent and less biased scientific studies that are contradicting such narratives. For instance, recent studies have shown that many fossil skeletal remains found near sites of prehistorical stone tool production are from women. This would be expected from what we now know about the lifeways of nomadic hunter-gatherer groups, in which women typically tend to be more active in gathering plants, taking care of kids, and other activities done near the temporary settlements, while men tend to be more active in hunting more far away from those settlements. It is logical that stone tool production were often situated near such settlements, so it is likely that women played an important, or even predominant, role in stone tool manufacture. In fact, there is empirical data showing that that this is indeed what happens in some living human groups that produce stone tools, even in those cases in which they have already been influenced by some of the sexist agricultural narratives disseminated by other groups – such as the data provided in a 2003 work by Weedman and colleagues. As a further illustration of how sexist narratives are still so predominant in the media and popular culture within agricultural and industrialized states, that paper became globally widespread in the press because it was said to contain "shocking" news. Of course, they were only "shocking" for those that were indoctrinated to blindly *believe* the inaccurate misogynistic stories that were contradicted by that work, as explained in a press-release about it published by the American website *ScienceDaily*:

> Man the toolmaker: the idea of men as stone tool producers may need some rechiseling, say University of Florida (UF) scientists who found women sometimes are the masters. The research among an Ethiopian group indicates stone tool working is not just a male activity, but rather that women probably had an active part in creating stone tools, one of the most ubiquitous materials found on prehistoric sites. It really gives women a presence in the archaeological record and a chance for us to reflect upon a place in prehistory where women basically have been invisible", said Kathryn Weedman, a UF anthropology lecturer who led the National Science Foundation-funded research. There has always been this image of 'man the toolmaker' because it's generally perceived by the public, and many archaeologists, that males were the ones who made stone tools", said Steve Brandt, a UF anthropology professor and co-leader of the research team. "But we found that among one ethnic group, the Konso of Ethiopia, women dominate the activity". The Konso women create a stone tool called a scraper to clean animal hides to be made into bedding and clothing, he said. Stone tools are important because they were the first recognizable object people made, marking the beginning of the archaeological record dating back as early as 2.6 million years ago, Weedman said. Not until 5,000 to 10,000 years ago were pottery and metal tools introduced, she said. "Stone artifacts are critical for identifying a wide range of activities that will help us learn what life was like", she said. "Basically, they trace the evolution of human culture because, for better or worse, they are often the only things preserved."
>
> If people were found to be scraping antelope hides 100,000 years ago, for example, that might help tell us when they started making prepared clothing, Brandt said. The Konso project "is vitally important both in documenting how stone tools are made and used – most people who used stone tools have been dead for hundreds or thousands of years – and in the social context of their use", said Michael Shott, an anthropology professor at the University of Northern Iowa. The tradition of stone toolmaking continued into the 20th century in isolated parts of Africa, Australia and Siberia, but in the last couple of decades it has virtually disappeared as an everyday activity, except perhaps for the hide workers of Ethiopia, Brandt said. The project is unique because it provides evidence that women actively flake stone to produce tools, Brandt said. "This project changes our perspective dramatically because theoretically we can talk about gender issues – the role of men and women in ancient societies", he said. In the study described in the September/October issue of *Archaeology magazine*, the UF researchers identified 119 Konso hide workers who used flaked stone, glass or iron to scrape hides. Seventy-five percent of the hide workers were

women, and most – 73 percent – were 40 or older, Weedman said. Members of this group are born into the hide-working profession and remain locked into it, Weedman said. The products they make have not yet been completely replaced by Western industrial products, she said. "No one is living in the stone age", Brandt said of today's stone toolmakers. "They're wearing Western clothes, they have radios, they may even know about 9–11", he said. "They make these stone tools because the tools work. Stone is still the superior material – they prefer it over glass and iron".

More recently, a 2020 paper by Martínez-Sevilla and colleagues – entitled "*Who painted that? The authorship of schematic rock art at the Los Machos rockshelter in southern Iberia*" – pointed out that women likely – probably, but not surely, in the specific example examined in that paper – have also contributed to cave painting. As summarized in an article by Sam Jones in *The Guardian*, in an article in which he interviewed Martínez-Sevilla:

One day, perhaps a little over 7,000 years ago, a man in his 30s and a younger companion dipped their fingers in ochre pigment and set about daubing the walls of a shallow cave in southern Spain with anthropomorphic, circular and geometric designs. Today, thanks to the fingerprints they left behind in the natural shelter of Los Machos in the province of Granada, researchers have been able to determine their sexes and ages. The study, carried out by experts from the University of Granada, Durham University and the Autonomous University of Barcelona, has established that the pictures were painted by a 36-year-old man and, most probably, by a woman a little less than half that age. Their findings, published in the journal *Antiquity*, suggest for the first time that both men and women took part in making rock art, and that it was a social, rather than an individual, act. Francisco Martínez Sevilla, a researcher at the University of Granada, said that while experts had long known about the 32 neolithic images painted on the slopes of the Cerro de Jabalcón some time between 5500 BC and 2500 BC, the figures had only recently begun to yield some of their many secrets. "We looked at the number of fingerprint ridges and the distance between them and compared them with fingerprints from the present day", he told the *Guardian*. "Those ridges vary according to age and sex but settle by adulthood, and you can distinguish between those of men and women.. you can also tell the age of the person from the ridges". While some of the fingers that painted the pictures belonged to a man aged at least 36, Martínez Sevilla and his colleagues cannot say for certain who his fellow artist was. There is a high likelihood that they were a woman aged under 20, or perhaps a juvenile male.

"From our point of view, if there are two people taking part in the creation of this pictorial panel, it means it must have been a social, rather than an individual, act, as we'd thought until now.. It shows us that these manifestations of art were a social thing and not just done by one individual in the community, such as the shaman or whoever". Although Martínez Sevilla acknowledges that he and his colleagues will never know the relationship between the pair – nor the significance of the images they left behind – the enduring art speaks for itself. "The area where they are, and the fact that they haven't been changed or painted over, gives you the feeling that this was a very important place and must have had a really important symbolic value for this community", he said. As an archaeologist, added Martínez Sevilla, he tends to see the person behind the stone or the tool rather than just the object itself. "And when I look at these pictures, there's a bit of an emotional response because I see a person, many thousands of years ago, painting symbols or designs that would have meant something to them, or which would have been a way for them to express themselves, or identify the territory, or communicate socially".

An even more recent paper, published in the same scientific journal, also put in question the narrative that men are the "natural" leaders of groups – a narrative that, as seen above, was presented as a "fact" in Darwin's writings about human evolution. The paper showed that a certain woman occupied, very likely, a particularly prominent position in a major European society about four millennia ago. As summarized by the National Geographic in a March 2021 article entitled "*Ancient woman may have been powerful European leader, 4000-year-old treasure suggest*":

> A trove of ornate jewelry, including a silver diadem, suggest a woman buried nearly 4,000 years ago in what is modern-day Spain was a ruler of surrounding lands who may have commanded the might of a state, according to a study published today in the journal *Antiquity*. The discoveries raise new questions about the role of women in early Bronze Age Europe, and challenge the idea that state power is almost exclusively a product of male-dominated societies, say the researchers. The remains of the woman, alongside those of a man who may have been her consort, were originally unearthed in 2014 at La Almoloya, an archaeological site among forested hills about 35 miles northwest of Cartagena in south-eastern Spain. Radiocarbon dating suggests the burial happened about 1700 B.C., and its richness suggests to the researchers that she, rather than he, may have been at the top of the local chain of command. "We have two ways of interpreting this", says archaeologist Roberto Risch of the Autonomous University of Barcelona, a co-author of the study. "Either you say, it's just the wife of the king; or you say, no, she's a political personality by herself". The graves of some El Argar women were re-opened generations later to inter other men and women, an unusual practice that likely conferred a great honor. And research published by Risch and his colleagues in 2020 showed that elite women in Argarian graves ate more meat than other women, which suggested they had real political power. "What exactly their political power was, we don't know", he says. "But this burial at La Amoloya questions the role of women in [Bronze Age] politics.. it questions a lot of conventional wisdom". Dubbed the "Princess of La Almoloya", the woman belonged to the Argaric culture, which is named after the archaeological site of El Argar some 50 miles to the south. Argaric culture flourished in the southeastern Iberian Peninsula between 2200 and 1500 B.C. Its people used bronze long before neighboring tribes; many lived in large hilltop settlements, rather than on isolated small farms; and items found in their gravesites indicate they had stratified classes of wealth and social status – including a ruling class..
>
> The woman.. was buried a short time later with particular splendor, including bracelets, earlobe plugs, rings, spirals of silver wire, and the silver diadem, which still adorned her skull when the grave was unearthed. It matches six other diadems found on wealthy women in Argaric graves; all have a distinctive disc-shaped projection usually worn downwards to cover the brow and nose. Using the price of silver quoted in Mesopotamian records from the time, the archaeologists estimate the grave goods of the La Almoloya woman were worth the equivalent today of many tens of thousands of dollars. Other burials of high-status El Argar women also indicate great wealth, but men were never buried with such riches. It "suggests that when [women] were alive they played a very important role in the political management of the community", Risch says. The location of the burial also indicate the woman had a political role. Many of the dead in El Argar communities were buried beneath the floors of buildings, and her grave was found beneath a room set with benches for up to 50 people, nicknamed the "parliament" by the researchers. The room itself was part of an elaborate building that may be the earliest-known palace in continental western European, Risch says – a place where rulers both lived and carried out their duties.

It is surprising that, in the 2020s, the media continues to portray such stories as if they were "shocking" or "surprising" – as if they were merely fantastic anecdotes –, because it is well known that, despite the systemic sexism that has prevailed since the rise of the first civilizations, women have often – not rarely, or anecdotally – find a way to have their voices heard and become leaders. For instance, within one of the most prominent ancient civilizations, Cleopatra became the last active ruler of the Ptolemaic Kingdom of Egypt about 21 centuries ago, and Nefertiti, MerNeith, Hatshepsut, and Twosret also had leading, or at least very prominent, positions within Egyptian society. Such examples show, in turn, the fallacy of yet another "scientific fact" constructed by Darwin: the existence of a unidirectional "progress"' within human evolution, particularly leading to the type of Anglo-Saxon societies that he so revered as the "pinnacle" of evolution. This is because while various ancient societies – as well as many so-called "backward sexist" non-Western modern countries such as India, Bangladesh, Pakistan, Mali, Panama, or Kyrgyzstan – had female leaders, a woman has yet to be elected to the highest office in the so-called most powerful democracy on the planet, which was a former British colony, the United States.

Similarly, empirical data obtained from several studies undertaken into numerous different regions of the globe also directly contradict the evolutionary "man-the-provider" narratives defended by authors such as Darwin. Such data are discussed by Sarah Hrdy in her outstanding 2009 book *Mothers and Others*:

> From the outset, they [evolutionists] assumed.. that (the) provider must have been her [the wife's] mate, as Darwin himself opined in *The descent of man and selection in relation to sex*. Indeed, it was the hunter's need to finance slow-maturing children, Darwin thought, that provided the main catalyst for the evolution of our big brains: "the most able men succeeded best in defending and providing for themselves and their wives and offspring", he wrote.. it was the offspring of hunters with "greater intellectual vigor and power of invention" who "were most like to survive". According to this logic, males with bigger brains would have been more successful hunters, better providers, and more able to obtain mates and thereby pass their genes to children whose survival was underwritten by a better diet. Meat would subsidize the long childhoods needed to develop larger brains, leading eventually to the expansion of brains from the size of an australopithecine's to the size of Darwin's own. Thus did the 'hunting hypothesis' morph into one of the most long-standing and influential models in anthropology.. at the heart of the model lay a pact between a hunter who provided for his mate and a mate who repaid him with sexual fidelity so the provider could be certain that children he invested in carried at least half of his genes. This 'sex contract' assumed pride of place as the "prodigious adaptation central to the success of early hominids"..
>
> [However] as it became apparent that among foragers (like the !Kung) plant foods accounted for slightly more calories than meat, researchers started paying more attention to female contributions.. [also] when Frank Marlowe interviewed Hazda still living by hunting and gathering, he learned that only 36% of children had fathers living in the same group.. a hemisphere away, among Yanomano tribespeople in remote regions of Venezuela and Brasil, the chance of a 10-year-old child having both a father and a mother living in the same group was 1/3, while the chance that a Central African Aka youngster between the ages of 11 and 15 was living with both natural parents was closer to 58%.. pity the Ongee foragers living on the Andaman Islands: none of the 11- to 15-years-olds in that ethnographic sample still lived with either natural parent. When anthropologists reviewed a sample of 15 traditional societies, in 8 of them the presence or absence of the father had no apparent effect on the survival of children to age 5, provided other caregivers in addition to the mother were on hand in a position of help.

In reality, no living apes can be considered to be truly mainly "active hunters." Chimpanzees hunt animals such as monkeys sometimes, but this only contributes to a minor part of their diets. Gorillas and orangutans rarely hunt for meat in their natural environments. So, obviously, meat, or active hunting, is not so crucial for them:

females, and their kids, can live very well without that – that is, without being "saved" by the males and their hunted meat. Active hunting only became more common in the last one or two million years of the about 7 million years of human evolution, since we split from the ancestors of chimpanzees. Meat was very likely important for humans before that time, but back then humans were mainly scavengers, a bit like hyenas are today. That is, they ate meat more opportunistically, rather than by actively hunting and killing animals as human hunter-gatherer groups do. Moreover, it is important to note that while for more recent nomadic hunter-gatherer societies meat is often important – particularly for societies that depend hardly on eating animals like the Nunamiut, semi-nomadic people located in Inland Alaska – in *most* nomadic hunter-gatherer societies hunting actually contributes to less than 50% of the diet, as noted by Zoontz and Hrdy. This is shown in Fig. 4.7, adapted from Kelly's 2013 book *The Lifeways of Hunter-Gatherers*.

Furthermore, even within those *few* hunter-gatherer groups in which hunting does contribute to more than 50% of the diet, sharing meat is usually very common, as noted by Kelly. That is, even within such groups one cannot say that if a father dies or leaves his group, the mother and kids will necessarily die from starvation. In most cases, they will still obtain at least some – often enough, as noted by Hrdy – meat from other hunters. In other words, even in such groups one can say that the dependence is more on the *group as a whole*, rather than on a specific "father-the-hunter-savior." In addition, this also relates to the power networks and diet types that are prevalent within different groups of hunter-gatherers. As noted in Kelly's 2013 book, the type of diet will be a major determinant on "whether men's or women's foraging determines camp movement: Agta camp members, for example, discuss for hours or days whether to move, and foraging efforts of men *versus* women play a role in these debates." Namely, the "effective foraging distance for plants is shorter, in general, than it is for large game since many plant foods provide lower return rates than those of large game.. since large game is usually procured by men, women's foraging should normally determine when and where camp is moved." Therefore, "among the Agta, since hunting depends on mobile animals, it is not an important consideration [in determining moves].. men and women freely voice their opinion on residence change, but women, who must carry out the most gathering, have the final say." In fact, in numerous hunter-gatherer groups, such as within the so-called Central African forest pygmies, "in many cases, women are equal partners with men in.. collective hunts," as explained by Bahuchet in the first chapter of the 2014 book *Hunter-Gatherers of the Congo Basin*. Such types of power relationships between genders are, therefore, generally very different from what happens in more sedentary groups, and particularly in agricultural groups. As noted above, the rise of agriculture within a society has in general led to an increased subjugation of women within that society – women started to be forced to do more "work," both concerning agricultural tasks and domestic ones.

This is a crucial historical point neglected by those people – either men or women – that criticize the so-called new trend seen in numerous countries in which women go to "work" outside of home, instead of "staying at home as they should." Apart from a few women that were married with rich men and thus did not need to "work" – for instance in the Victorian era, as shown over and lover within the type of sexist story that Hollywood movies are still so often obsessed with –, since the rise of humans the vast majority of women never did 'just stay at home'. A 2018 article published by the World Economic Forum, entitled "*Women grow 70% of Africa's food – but have few rights over the land they tend,*" explains that, even today, "studies show that women account for nearly half of the world's smallholder farmers and produce 70% of Africa's food." However, in a further example of the results of systemic sexism, the article explains that "yet, less than 20% of land in the world is owned by women." The article further points out that "over 65% of land in Kenya is governed by customary laws that discriminate against women, limiting their land and property rights.. this means that women farmers have to access land through either their husbands or sons.. sometimes these male family members move to the cities leaving women behind to tend the land – land they have no right to own, use as collateral or sell the output without consent from the men."

Recently, a 2020 paper by Haas and colleagues entitled "*Female hunters of the early Americas*" stated that the "man-the-provider" narrative is wrong not only because by gathering wild plants women usually contribute the bulk of the diet eaten by the whole group, but also because some women are additionally involved in hunting, within certain hunter-gatherer groups:

> Scholars generally accept that projectile points associated with male burials are hunting tools, but have been less willing to concede that projectile points associated with female burials are hunting tools.. [our case study] presents an unusually robust empirical test case for evaluating competing models of gendered subsistence labor.. It is possible that the burial [studied by Haas and colleagues] represents a rare instance of a female hunter in a male-dominated subsistence field, but such an outlier explanation diminishes with the observation of 11 female burials in association with hunting tools from 10 Late Pleistocene or Early Holocene sites throughout the Americas, including Upward Sun River, Buhl, Gordon Creek, Ashworth Rockshelter, Sloan, Icehouse Bottom, Windover, Telarmachay, Wilamaya Patjxa, and Arroyo Seco 2. These results are consistent with a model of relatively undifferentiated subsistence labor among early populations in the Americas. Nonetheless, hunter-gatherer ethnography and contemporary hunting practices make clear that subsistence labor ultimately differentiated along sex lines, with females taking a role as gatherers or processors and males as hunters. Middle Holocene females and males at the Indian Knoll site in Kentucky were buried with atlatls in a respective ratio of 17:63, suggesting that big-game hunting was a male-biased activity at that time. Thirty percent of bifaces, including projectile points, are associated with females in a sample of 44 Late Holocene burials from seven sites in southern California. A similar trajectory may be observed in the European Paleolithic, where meat-heavy diets and absence of plant-processing or hide-working tools among Middle Paleolithic Neandertals would seem to minimize potential for sexually differentiated labor practices. Economies diversified in the Upper Paleolithic.. with increasing emphasis on plant processing and manufacturing of tailored clothing and hide tents creating new contexts for labor division.
>
> When and how such differentiated labor practices emerged from evidently undifferentiated ones require further exploration. Scholars have long grappled with understanding the extent to which contemporary gender behavior existed in our species' evolutionary past. A number of studies support the contention that modern gender constructs often do not reflect past ones. Dyble *et al.* show that both women and men in ethnographic hunter-gatherer societies govern residence decisions. The discovery of a Viking woman warrior further highlights uncritical assumptions about past gender roles. Theoretical insights suggest that

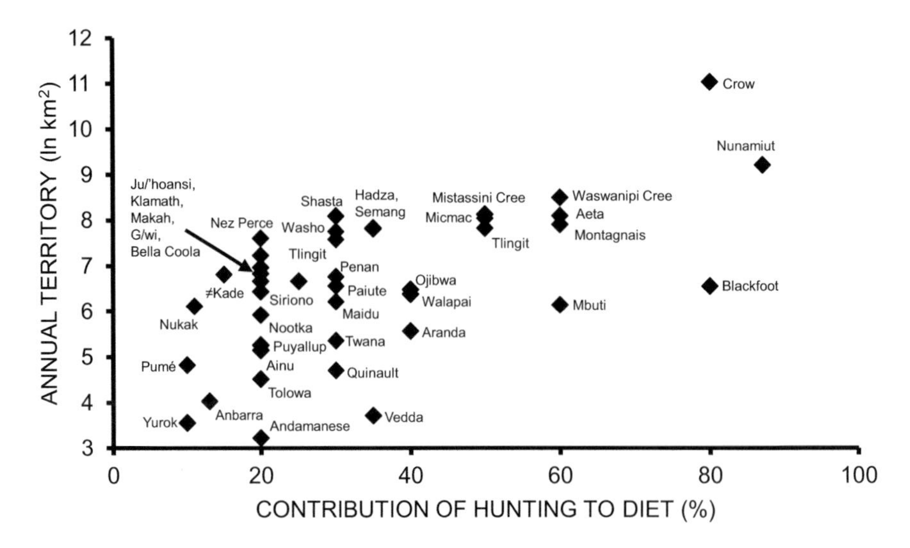

Fig. 4.7 The size of foragers' annual ranges plotted against the percent dependence on hunting – in general, as the dependence on hunting increases so does the size of the exploited territory. In the overall, it is, however, interesting to see how in the vast majority of groups hunting contributes less than 50% to their diet, contrary to the narrative of the "man-the-hunter-provider-savior"

> the ecological conditions experienced by early hunter-gatherer populations would have favored big-game hunting economies with broad participation from both females and males. Such models align with epistemological critiques that reduce seemingly paradoxical tool associations to cultural or ethnographic biases. [Our case study] and the sum of previous archaeological observations on early hunter-gatherer burials support this hypothesis, revealing that early females in the Americas were big-game hunters.

It is indeed very likely that at least some women have been hunters in various, if not most, hunter-gatherer groups in the past. However, one important distinction needs to be made. While there is no empirical evidence showing that women were less involved than men in the discovery of fire, stone tool production, and cave painting – they might have been actually *more* involved, in view of what we know of the lifeways of hunter-gatherers in general, as explained above – the scenario regarding hunting is very different. That is, there is solid evidence indicating that hunting by women was in general not as common as hunting by men. This concerns not only evidence available concerning human hunter-gatherer groups – including modern ones, or those described in historical records – but also concerning the few ape groups that often hunt for meat, such as chimpanzees, in which males tend to be the most active hunters. However, the reason for that does not seem to be simply because females merely "take care of their babies," as it is often affirmed in misogynistic tales that aim to reduce females to their maternal "obligations," as if a female cannot be, and therefore should not aspire to be, nothing more than a "caring mother." For example, in a 2017 paper by Gilby and colleagues entitled "*Predation by female chimpanzees,*" the authors explained that their results "suggest that before the emergence of social obligations regarding sharing and provisioning, constraints on hunting by females did not necessarily stem from maternal care." Instead, their results "suggest that a risk-averse foraging strategy and the potential for losing prey to males limited female predation [as is the often the case in chimpanzees].. sex differences in hunting behavior would likely have preceded the evolution of the sexual division of labor among modern humans."

The book *Evolution's Empress: Darwinian Perspectives on the Nature of Women* provides a fascinating overview of the fallacies of such sexist narratives about our evolutionary past, in particular those written by Darwin that are still so influential within various fields of science such as anthropology, biology, and psychology. As explained by Sarah Hrdy in the Foreword of that 2013 book edited by Maryanne Fisher, Justin Garcia, and Rosemarie Chang:

> For all its originality and power, Darwin's view of human nature was distorted by overly narrow, often misleading stereotypes about females. In writing his classic account of *The Descent of Man and Selection in Relation to Sex*, Darwin assumed that "the most able men will have succeeded best in defending and providing for themselves, their wives and off spring" and that off spring sired by hunters with "greater intellectual vigor and power of invention" would be most likely to survive.. Altogether, it was a very neat, internally consistent – if entirely androcentric – package, leaving out crucial female contributions to subsistence as well as all the strategizing females engage in to ensure their local clout and the survival of any young at all. Over time, this bias grew more pronounced among some of Darwin's disciples, persisting to the present day. No wonder many of those in the humanities and social sciences subscribed to Virginia Woolf's assessment: "Science it would seem, is not sexless; she is a man, a father, and infected too" (1938). Feminists were understandably put off by more than a century of male-centered constructs, including constructs whereby our ancestors evolved big brains so that males could outwit competitors or collaborate with one another the better to kill game or prevail over neighboring groups, or evolved to walk upright so males could carry meat back to females and off spring who waited back at camp (e.g., Lovejoy, 1981, among many other distinguished evolutionists). Feminists wrote off biology as a field unhelpful to women seeking either to improve their lot or to better understand themselves.. A few of these committed evolutionists are even wondering out loud how it was possible that sex differences apparent in some species could have been projected onto nature at large without taking into account just how flexible sex roles between and within species often are. How could widely accepted assumptions about universal sex differences have persisted and shaped evolutionary theorizing for so long after abundant evidence contradicting such presumptions had been reported?
>
> How could mainstream scientists ever have taken it as granted that females were too preoccupied with nurturing to compete in wider spheres, as in this statement: "primate females seem biologically unprogrammed to dominate political systems, and the whole weight of the relevant primates' breeding history militates against female participation in what we can call 'primate public life'" (Tiger, 1977)? How could the finest textbook in the field back in the 1970s have so casually pronounced that "most adult females.. are likely to be breeding at or close to the theoretical limit" while "among males by contrast there is the probability of doing better" (Daly & Wilson, 1978), with the obvious implication that somehow what matters most is competition between males for mates? (Hint: sometimes it does, except when it doesn't!). Given just how much evidence there was before our eyes, why did so many years elapse before stereotypes about sexually "ardent" males and universally discriminating "coy" females [as stated by Darwin] started to be challenged and before

long-standing stereotypes about evolved female nature were revised? .. How convenient to assume that if women behaved passively, opted out of "primate public life", or remained monandrous, it was because they were naturally inclined to do so! Such blinders would have eased the existence of a kindly Victorian gentleman, while also enhancing his professional as well as reproductive success. Whatever their sources, Darwin's blind spots constituted a highly adaptive obliviousness shared by a succession of brilliant researchers in the evolutionary sciences.

The fact that Darwin included in his book, as evolutionary "facts," inaccurate sexist tales such as the "men-the-provider" narrative cannot be blamed exclusively to the sexism of the Victorian society because in his travels he could easily have *seen* that such tales were factually wrong. He had the unique opportunity to directly observe, and interact with, nomadic hunter-gatherer groups, so he could have easily seen that in most of such groups men do *not* provide most of the calories consumed by the whole group (Fig. 4.7). In fact, Darwin himself recognized that in most of those groups there were less marked gender roles than in Victorian England, but because of his biases he saw that as necessarily "primitive" and "brute" and the Victorian sexist *status quo* as how gender relations should ideally be. As it often happened with the racist and ethnocentric "facts" described in his books, Darwin's sexist just-so-stories are the result of a combination of omissions, factual inaccuracies, and logical inconsistencies. One of the most illustrative, and influential, examples of this is the paragraph where Darwin referred, in the *Descent*, to the "natural" evolutionary "coyness" of women:

The female, on the other hand, with the rarest exception, is less eager than the male. As the illustrious Hunter long ago observed, she generally "requires to be courted"; she is coy, and may often be seen endeavouring for a long time to escape from the male. Every one who has attended to the habits of animals will be able to call to mind instances of this kind. Judging from various facts, hereafter to be given, and from the results which may fairly be attributed to sexual selection, the female, though comparatively passive, generally exerts some choice and accepts one male in preference to others. Or she may accept, as appearances would sometimes lead us to believe, not the male which is the most attractive to her, but the one which is the least distasteful. The exertion of some choice on the part of the female seems almost as general a law as the eagerness of the male. We are naturally led to enquire why the male in so many and such widely distinct classes has been rendered more eager than the female, so that he searches for her and plays the more *active part* in courtship.. It would be no advantage and some loss of power if both sexes were mutually to search for each other; but why should the male almost always be the seeker?

Unintentionally, in the last part of that paragraph, Darwin gives a clue to answer the question of why he *believed* that women are less eager to have sex than men – while logically it makes more evolutionary sense that both genders are eager to have sex, as sex is compulsory for the existence and evolution of sexual organisms. The clue is "the more *active part.*" Darwin did not *believe* that women can be as active as men, or even just slightly less active, because he blindly accepted the sexist narratives to which he was indoctrinated. If he had been able to question such narratives, as numerous women and many men were at that time, he could have understood that the examples provided by the societies that he observed across the globe provided very strong direct evidence to deconstruct the sexist narratives in which he *believed*. As noted by Hamlin in *From Eve to Evolution*:

> Women's intellectual capacities, according to Darwin and most other evolutionists, were permanently limited by their reproductive functions, which drew the lion's share of their energy and of evolutionary attention. Thus, over many thousands of years, Darwin concluded, "man has ultimately become superior to woman".. Darwin further believed that, technically speaking, women could possibly be educated to an equal intellectual standing with men over many generations but at too great a cost to the "easy education of our children" and the "happiness of our homes". Even though Darwin himself supported female education, opponents seized upon the peahen quote to argue that educating women went against nature's plan and was ultimately futile, if not injurious.. In *The Descent of Man*, Darwin asserted that sex differences promoted the evolutionary process by efficiently dividing labor and that the most advanced species were those in which the sexes were the most differentiated. As evolutionists saw it, animals progressed from asexual to sexual reproduction, developing increasingly complicated mating systems as they ascended the evolutionary ladder. At the very top of this ladder were those humans with the most strictly defined gender roles: married couples in which the husband worked outside the home and the wife tended to children and domestic tasks, couples that also tended to be middle or upper class and white. To those men steeped in evolutionary discourse and the attendant pride in being at the pinnacle of all living things, women going to college threatened to minimize sex differentiation, thwart evolutionary advancement, and diminish white racial superiority.

As noted by Hamlin, the logical inconsistencies of Darwin's sexist narratives about women's role in evolution were often so striking that even some scientist men openly criticized them, back then, including the sociologist Lester Frank Ward, who was:

> Disenchanted with the political focus of women's rights efforts and "saw greater opportunity for the emancipation of women through science education and outside work". In assessing recent scientific findings regarding sex difference, Ward was surprised that "those who start out avowedly from a Darwinian standpoint should so quickly abandon it and proceed to argue from pre-Darwinian premises". How could a theory based on reproduction deny the principal reproducers the lead role? And, if scientists had in fact established human-animal kinship, why should women's lives differ so drastically from men's when such gendered distinctions did not characterize other animal species? Ward argued that evolutionary science demonstrated beyond a doubt that "woman is the grandest fact in nature" and that "the elevation of woman is the only sure road to the evolution of man". A cornerstone of the future egalitarian society imagined by Ward was women working alongside men, participating in leisure activities along with men, and wearing simple, comfortable clothes like men.. Ward's influence on Darwinian feminists, especially Gilman, was foundational because he supplied them with a unified theory linking evolutionary progress with feminist goals, authored by the highly credible 'father of American sociology'.

Once again, the examples provided by such men contradict the typical line of defense used by Darwin's idolizers that "everybody back then was like that." This also applies to Alfred Russel Wallace, who apart from criticizing many Victorian racist, ethnocentric, and imperialist narratives also was in general more distrustful about many of the sexist narratives commonly accepted back then. As can be seen in Box 4.4, this does not mean that Wallace was flawless, obviously.

Box 4.4 Wallace, Socialism, and Feminism

Hamlin wrote, in *From Eve to Evolution*:

So strong was the appeal of female choice to socialists that by 1890 Alfred Russel Wallace, the leading scientific opponent of sexual selection, began advocating female choice among humans, the very concept he had spent years trying to discredit. After applauding the recent educational and professional advances made by women, Wallace suggested that the driving factor in any meaningful societal reform would be female choice of marriage partners: "I hope I make it clear that women must be free to marry or not to marry before there can be true natural selection in the most important relationship of life.. In order to cleanse society of the unfit [and allow natural selection to proceed]", he explained, "we must give to woman the power of selection in marriage, and the means by which this most important and desirable end can be attained will be brought about by giving her such training and education as shall render her economically independent".. he suggested that female choice could be a conscious, political attempt to thwart the counterevolutionary tendencies of capitalism. Wallace's position on female choice also depended on his acceptance of eugenic ideas about the "weeding out" of the "unfit", as evidenced in his statement above, but he rejected state-sponsored eugenics in favor of empowering individual women to make the best choices for themselves. Unlike organized eugenicists who campaigned for state, medical, and legal limits on marriage, Wallace and other feminist socialists believed future offspring would simply be better born when economically independent women could make reproductive decisions for themselves. In September of 1913, the *Masses*, a popular socialist magazine in the United States, cheered Wallace for uniting evolution with women's rights..

The *Masses*' editors reiterated Wallace's main point that it was unnatural and counterevolutionary for one sex to be dependent upon the other for survival, showing the appeal of female choice among feminist socialists in the United States. Moreover, the *Masses*' praise for Wallace "Female Choice" and the *Reproductive Autonomy of Women* demonstrated that left-leaning reformers felt affirmed by enlisting Wallace and evolutionary science more broadly to their cause. The working-class Wallace had long been drawn to socialist ideas but feared they were too authoritarian; likewise, he knew that human reproductive choices directed the future evolution of the species, but he rejected statesponsored controls on marriage as similarly authoritarian and also too elitist. In [Edward] Bellamy's socialist utopia, Wallace found precisely the balance of individual freedom and social equality for which he had long searched. As Fichman explains, Wallace thought Bellamy's brand of socialism would remove the "disparities of wealth and rank" and "eliminate the economic and political prejudices that.. dominated the selection of reproductive partners in Victorian society. In their place, mate choice would focus on those higher moral and intellectual traits often neglected (or rendered subservient) in competitive capitalist society". Wallace elaborated on his socialist ideas in several works published between 1890 and his death in 1913, beginning with his essay "*Human Selection*" in which he declared himself a socialist as a result of having read Looking Backward. Years later, as he surveyed his life's accomplishments, Wallace, the codiscoverer of natural selection, concluded that female choice by women in a socialist society was "by far the most important of the new ideas I have given to the world".

The last sentence of Box 4.4 – concerning the fact that when Wallace surveyed his life's accomplishments, he wrote that his support for female choice in a socialist society was "by far the most important of the new ideas I have given to the world" – does reiterate a huge difference with Darwin. As noted by Hamlin, that idea was not truly "new" – it was largely borrowed from Edward Bellamy, but the fact that Wallace refers to it as the most important idea defended by him, even more than his theory of evolution by natural selection, is something that one could hardly imagine hearing from Darwin. Not only because in his autobiography Darwin made clear that his biggest idea was his theory of natural selection as seen in Chap. 1 but also because Darwin did not loudly defend, at all, "female choice and the reproductive autonomy of women," as Wallace did. As shown above, Darwin often avoided to be publicly involved in societal matters, contrary to Wallace, but when he was, he tended to defend ideas that were more conservative than those defended by Wallace. This is evidenced not only in those cases in which Darwin defended the then prevailing misogynistic *status quo* in newspaper letters but also in the way in which he reacted to discussions on women's use of birth control, as noted by Hamlin:

> In pleading their case [in 1877], Charles Bradlaugh and Annie Besant wanted to establish that birth control was natural and that it accorded with evolutionary principles regarding population growth. To establish the movement's scientific legitimacy, Bradlaugh and Besant hoped especially to call one witness to the stand: Charles Darwin. The Neo-Malthusians believed that overpopulation presented a major problem for society and that it was far better to provide a check on birthrates rather than have death rates escalate. They hoped that Darwin would read excerpts from his books in the courtroom because they felt confident that his theories provided irrefutable evidence for the scientific basis of the birth control movement and that his scientific authority would lend credibility to their cause. Darwin politely asked to be excused from testifying on account of his ill health and the fact that he disagreed with artificial checks to fertilization. In his very cordial reply to Bradlaugh's invitation, Darwin explained that he opposed birth control because he feared "any such practices would in time spread to unmarried women & wd destroy chastity, on which the family bond depends; & the weakening of this bond would be the greatest of all possible evils to mankind". His response reveals that his objections to birth control were more cultural than scientific. He did not say that birth control failed to accord with evolutionary principles, simply that decoupling sex from reproduction would destroy the bonds of the family, a nuance that the next generation of evolutionists revisited.

"*Destroy chastity, on which the family bond depends.. would be the greatest of all possible evils to mankind.*" It is remarkable how Darwin evolutionary "facts" about human evolution so often involve a combination of moral constructs, sexist biases, ethnocentrism, and societal conservatism, as further evidenced in the following excerpt about this topic from the *Descent*:

> The other self-regarding virtues, which do not obviously, though they may really, affect the welfare of the tribe, have never been esteemed by savages, though now highly appreciated by civilised nations. The greatest intemperance with savages is no reproach. Their utter licentiousness, not to mention unnatural crimes, is something astounding. As soon, however, as marriage, whether polygamous or monogamous, becomes common, jealousy will lead to the inculcation of female virtue; and this being honoured will tend to spread to the unmarried females. How slowly it spreads to the male sex we see at the present day. Chastity

> eminently requires self-command; therefore it has been honoured from a very early period in the moral history of civilised man. As a consequence of this, the senseless practice of celibacy has been ranked from a remote period as a virtue. The hatred of indecency, which appears to us so natural as to be thought innate, and which is so valuable an aid to chastity, is a *modern virtue*, appertaining exclusively, as Sir G. Staunton remarks, to civilised life. This is shown by the ancient religious rites of various nations, by the drawings on the walls of Pompeii, and by the practices of many savages. We have now seen that actions are regarded by savages, and were probably so regarded by primeval man, as good or bad, solely as they affect in an obvious manner the welfare of the tribe,—not that of the species, nor that of man as an individual member of the tribe. This conclusion agrees well with the belief that the so-called moral sense is aboriginally derived from the social instincts, for both relate at first exclusively to the community.
>
> The chief causes of the low morality of savages, as judged by our standard, are, firstly, the confinement of sympathy to the same tribe. Secondly, insufficient powers of reasoning, so that the bearing of many virtues, especially of the self-regarding virtues, on the general welfare of the tribe is not recognised. Savages, for instance, fail to trace the multiplied evils consequent on a want of temperance, chastity, &c. And, thirdly, weak power of self-command; for this power has not been strengthened through long-continued, perhaps inherited, habit, instruction and religion. I have entered into the above details on the immorality of savages, because some authors have recently taken a high view of their moral nature, or have attributed most of their crimes to mistaken benevolence. These authors appear to rest their conclusion on savages possessing, as they undoubtedly do possess, and often in a high degree, those virtues which are serviceable, or even necessary, for the existence of a tribal community.. What ancient nation, as the same author asks, can be named that was originally monogamous? The primitive idea of justice, as shewn by the law of battle and other customs of which traces still remain, was likewise most rude. Many existing superstitions are the remnants of former false religious beliefs. The highest form of religion — the grand idea of God hating sin and loving righteousness—was unknown during primeval times.

In the last part of this excerpt, Darwin seems to unawarely contradict a key argument used by those that idolize him, in the sense that he points out that *not* everybody back then tended to defend the racist and sexist narratives defended by him. Moreover, he is informing us that "some authors have recently taken a high view of their moral nature" of "savages" and that, despite being exposed to the writings of such authors, he *continued* to actively go against their less ethnocentric and ideas. A similar point has been made by several female scholars, concerning Darwin's sexist ideas, including Lynn Nelson in the book *Biology and feminism*:

> Some feminists also argue that, contrary to the way he is generally portrayed by historians of science, there are several respects in which Darwin was decidedly not "swimming upstream" – that is, he was not critically taking on prevailing sociopolitical or scientific views. He was, as we have seen, assuming the gender stereotypes of his day. In addition, some feminists and others point out that Darwin's model of natural selection – which involves waves of competition for scarce resources, and "winners and losers" – paralleled then current arguments for capitalism. So, too, Darwin assumed then current sociopolitical beliefs about race differences. Although an abolitionist, he appealed to differences between "the races" in brain size and intelligence.. Many [if not most] Darwin scholars recognize that the claims Darwin makes about sex/gender, sex, and racial differences that feminists criticize are in fact unsupported assumptions, assumptions that were characteristic of Victorian England.

Darwin and the Supposed Sexual Passivity of Women

The emotional, sexual, and psychological stereotyping of females begins when the doctor says: it's a girl. (Shirley Chisholm)

One of the most enduring and damaging legacies of Darwin and the Victorian sexist ideas that he extrapolated to human evolution concerns the notion that women are basically evolutionary and sexually passive. This idea is still so prevalent in most countries that most people, including women themselves, commonly internalize and *believe* in it. That is the fascinating power of indoctrination.

Regarding the also factually wrong sexist idea that men have, in general, more sexual drive and are thus more prone to polygamy than women are, lay people, as well of journalists and even many scholars, often argue that this is a logical outcome of Darwin's evolutionary ideas. That is, that while men are able to "optimize"/"maximize" the number of their children by having many female partners, when women become pregnant they have to wait at least 9 months to have more children, so accordingly they tend to be "coyer," as described by Darwin. Such an assertion reflects two crucial problems regarding Darwin's writings. One concerns their sexist biases. The other concerns how they were influenced by both capitalist "more-is-better" and individualistic Victorian ideas, including the notion that evolution is mainly about individual survival and reproduction, about individual selfishness, and therefore about making more "copies" of oneself. Under such an erroneous evolutionary paradigm, for an individual human being it would indeed appear to be "better" to have as many children as possible, an idea that strikingly continues to be repeated over and over in popular culture, the media and even by numerous scientists, particularly within fields such as evolutionary psychology and evolutionary medicine.

In fact, one of the ways that were, and still are, often used by such scientists to support such misogynistic tales was to use polygynous – one male with various females – primate groups such as gorillas as *the* model for human societies, mixed with the erroneous extrapolation of capitalist "more is better" narratives to biological evolution. By combining these fallacious just-so-stories, it was assumed – not as as a testable idea, but mostly as an *a priori* dogma – that the best "adaptative optimization" for humans is to have a higher number of descendants. As put by Landau in *Narratives on Human Evolution*, Darwin's *Descent* basically argues that, ultimately, "success, in the long run, is measured by numbers of offspring." There are many flaws within such misogynistic narratives. First, gorillas are not our closest living relatives. Chimpanzees are – and they have a "multimale-multifemale" mating system, not a polygynous one. That is, in chimpanzees – as well as many other primate and non-primate groups – various individuals of each sex form large social groups in which both males and females are polygamous – they mate with multiple members of the opposite sex. As I explain in detail in *Meaning of Life, Human Nature, and Delusions*, a compilation of data from many fields of science strongly indicates that early humans had – and likely most of us, today, still have – the tendency to be polygamous, as chimpanzees are: this applies to both men *and* women.

Moreover, applying the capitalistic "more is better" notion to this topic is, in itself, paradoxical. This is because Malthus, who influenced Darwin so much, actually stated that a main *problem* that humanity would face would be overpopulation: it would lead to the "struggle-for-existence" that was so crucial, and exaggerated, in Darwin's writings, as we have seen above. In human history, there are in fact many cases of collapses of civilizations that were very likely mainly, or at least partially, related with overpopulation, as explained in detail in the 1999, 2005, and 2012 books of Jared Diamond. Unfortunately, nowadays we are remained almost daily about the fallacy of the "more children is better" notion, when we see the news about the huge problems related to human overpopulation: global warming, massive migrations, wars for territories or resources, alarming levels of pollution, and so on. Another paradox concerning this topic of reproduction and the ideas of Malthus is that historical records show that overpopulation has been mostly a problem characterizing sedentary, and in particular agricultural, "civilizations": not the small nomadic hunter-gatherer societies to which authors such as Malthus or Darwin called "brutish" or "wild" or "amoral primitive savages." So, the "more is better" type of "amoral" reproduction leading to overpopulation and therefore to the "struggle-for-existence" postulated by Malthus and extrapolated to human evolution and biological evolution in general by Darwin is clearly not the most typical "natural" pattern for humans, as it has not applied to 99.8% of our evolutionary (non-agricultural) history.

Indeed, such Darwinian sexist narratives about natural selection privileging a human polyginous model in which males would have sex with anything that moves while females would be "coy" are also contradicted by the fact that numerous human societies were, or are, polyandrous – one female with various males. Historical records show that many of such societies explicitly did, or do, so in order to precisely try to avoid environmental and societal collapse. For instance, the Guanches – indigenous people of the Canary Islands (Fig. 4.8) – took the *political decision* to change from an ancestral multimale-multifemale model to a polyandrous model with at least 5 husbands for a single wife to try to control population growth and avoid wars because those islands have scarce natural resources. By doing so, they would ensure that during the 9 months that the woman was pregnant, 6 people in total would have a single child: an efficient way to avoid overpopulation and societal collapse that is exactly the opposite of the dogmatic model that Darwinists often applied to humanity as a whole, everywhere, to reinforce manly made, biased sexist just-so-stories. Many other examples are given in Levine's 1998 book *The Dynamics of Polyandry*, the 2002 book *Cultures of Multiple Fathers* edited by Beckerman and Valentine, and Cerello and Kholoussy's edited book *Domestic Tensions, National Identities: Global Perspectives on Marriage, Crisis, and Nation*. About this topic, Coontz wrote, in *Marriage, a History*:

> Eskimo couples often had cospousal arrangements, in which each partner had sexual relations with the other's spouse. In Tibet and parts of India, Kahmir, and Nepal a woman may be married to two or more brothers [polyandry], all of whom share sexual access to her..

Fig. 4.8 The Guanches – indigenous people of the Canary Islands – are one of the many examples of cultures that adopted, for at least some time, a polyandrous model in which a female was with many males

The children of Eskimo cospouses felt that they shared a special bond, and society viewed them as siblings. Such different notions of marital rights and obligations made divorce and remarriage less emotionally volatile for the Eskimo than it is for most modern Americans. Among Tibetan brothers who share the same wife, sexual jealousy is rare. The expectation of mutual fidelity is a rather recent invention. Numerous cultures have allowed husbands to seek sexual gratification outside marriage. Less frequently, but often enough to challenge common preconceptions, wives have also been allowed to do this without threatening the marriage. In a study of 109 societies, anthropologists found that only 48 forbade extramarital sex to both husbands and wives.. in some societies the choice to switch partners rests with the woman. Among the Dogon of West Africa, young married women publicly pursued extramarital relationships with the encouragement of their mothers. Among the Rukuba of Nigeria, a wife can take a lover at the time of her first marriage..

Several small-scale societies in South America have sexual and marital norms that are especially startling for Europeans and North Americans. In these groups, people believe that any man who has sex with a woman during her pregnancy contributes part of his biological substance to the child. The husband is recognized as the primary father, but the woman's lover or lovers also have paternal responsibilities, including the obligation to share food with the woman and her child in the future. During the 1990s researchers taking life histories of elderly Bari women in Venezuela found that most had taken lovers during at least one of their pregnancies. Their husbands were usually aware and did not object. When a woman gave birth, she would name all the men she had slept with since learning she was pregnant, and a woman attending the birth would tell each of these men: you have a child.. When Jesuit missionaries from France first encountered the North American Montagnais-Naskapi Indians in the early 17th century, they were shocked by the native women's sexual freedom. One missionary warned a Naskapi man that if he did not impose tighter controls on his wife, he would never know for sure which of the children she bore belonged to him. The Indian was equally shocked that this mattered to the Europeans: "you French people.. love only your own children, but we love all the children of our tribe", he replied.

Coontz further explained that:

According to the protective or provider theory of marriage.. – still the most widespread *myth* about the origin of marriage – .. women and infants in early human societies could not survive without the men to bring them the meat of woolly mammoths and protect them, from marauding saber-toothed tigers and from other men seeking to abduct them. But males were willing to protect and provide only for their 'own' females and offspring they had a good reason to believe were theirs, so a woman needed to find and hold on to a strong, aggressive male. One way a woman could hold a mate was to offer him exclusive and frequent sex in return for food and protection. According to the theory, that is why women lost the estrus cycle that is common to other mammals, in which females come into heat only at periodic intervals. Human females became sexually available year-round, so they were able to draw men into long-term relationships. In anthropologist Robin Fox's telling of the story "the females could easily trade on the male's tendency to want to monopolize (or at least think he was monopolizing) the females for matting purposes, and say, in effect, 'okay, you get the monopoly.. and we get the meat'". The male willingness to trade meat for sex was, according to Fox, "the root of truly human society".

Proponents of this protective theory of marriage claim that the nuclear family, based on sexual division of labor between the male hunter and the female hearth keeper, was the most important unit of survival and protection in the Stone Age. People in the mid-20th century found this story persuasive because it closely resembled the male breadwinner/female homemaker family to which they were accustomed. The idea that in prehistoric times a man would spend his life hunting only for the benefit of his wife and children, who were dependent solely upon his hunting for survival, is simply a projection of 1950s marital norms onto the past. But since the 1970s other researchers have poked holes in the protective theory of marriage.. they argued that.. the origins of marriage lay not in the efforts of the women to attract protectors and providers but in the efforts of men to control the productive and reproductive powers of women for their own private benefit.. some (researchers) denied that male dominance and female dependence came from us from our primate ancestors. Studies of actual human hunting and gathering societies also threw doubt on the male provider theory – in such societies, women's foraging, not men's hunting, usually contributes the bulk of the group's food. Nor are women in foraging societies tied down by child rearing.

One anthropologist, working with an African hunter-gatherer society during the 1960s, calculated that an adult woman typically walked about twelve miles a day gathering food, and brought home anywhere from 15 to 33 pounds. A woman with a child under two covered the same amount of ground and brought back the same amount of food while she carried her child in a sling, allowing the child to nurse as the woman did her foraging. In many societies women also participate in hunting, whether as members of communal hunting parties, as individual hunters, or even in all-female hunting groups. Today most paleontologists reject the notion that early human societies were organized around dominant male hunters providing for their nuclear families.. [instead] there is strong evidence that in many societies.. [particularly] sedentary agriculturalists .. – *marriage was indeed a way that men put women's labor to their private use*. Women's bodies came to be regarded as the properties of their fathers and husbands.

Considering the empirical evidence mentioned above, we can now analyze in more detail the ideas that Darwin presented in the *Descent* as "evolutionary facts" about women sexuality and understand how biased and inaccurate they are, for instance when he wrote that:

With respect to the other form of sexual selection (which with the lower animals is much the more common), namely, when the females are the selectors, and accept only those males which excite or charm them most, we have reason to believe that it formerly acted on our progenitors. Man in all probability owes his beard, and perhaps some other characters, to inheritance from an ancient progenitor who thus gained his ornaments. But this form of selection may have occasionally acted during later times; for in utterly barbarous tribes the women have more power in choosing, rejecting, and tempting their lovers, or of afterwards changing their husbands, than might have been expected. As this is a point of some importance, I will give in detail such evidence as I have been able to collect. Hearne describes how a woman in one of the tribes of Arctic America repeatedly ran away from her husband and joined her lover; and with the Charruas of S. America, according to Azara, divorce is quite optional. Amongst the Abipones, a man on choosing a wife bargains with the parents about the price. But "it frequently happens that the girl rescinds what has been agreed upon between the parents and the bridegroom, obstinately rejecting the very mention of marriage". She often runs away, hides herself, and thus eludes the bridegroom. Captain Musters who lived with the Patagonians, says that their marriages are always settled by inclination; "if the parents make a match contrary to the daughter's will, she refuses and is never compelled to comply". In Tierra del Fuego a young man first obtains the consent of the parents by doing them some service, and then he attempts to carry off the girl; "but if she is unwilling, she hides herself in the woods until her admirer is heartily tired of looking for her, and gives up the pursuit; but this seldom happens". In the Fiji Islands the man seizes on the woman whom he wishes for his wife by actual or pretended force; but "on reaching the home of her abductor, should she not approve of the match, she runs to some one who can protect her; if, however, she is satisfied, the matter is settled forthwith". With the Kalmucks there is a regular race between the bride and bridegroom, the former having a fair start; and Clarke "was assured that no instance occurs of a girl being caught, unless she has a partiality to the pursuer". Amongst the wild tribes of the Malay Archipelago there is also a racing match; and it appears from M. Bourien's account, as Sir J. Lubbock remarks, that "the race, 'is not to the swift, nor the battle to the strong,' but to the young man who has the good fortune to please his intended bride". A similar custom, with the same result, prevails with the Koraks of north-eastern Asia.

Turning to Africa: the Kaffirs buy their wives, and girls are severely beaten by their fathers if they will not accept a chosen husband; but it is manifest from many facts given by the Rev. Mr. Shooter, that they have considerable power of choice. Thus very ugly, though rich men, have been known to fail in getting wives. The girls, before consenting to be betrothed, compel the men to shew themselves off first in front and then behind, and exhibit their paces. They have been known to propose to a man, and they not rarely run away with a favoured lover. So again, Mr. Leslie, who was intimately acquainted with the Kaffirs, says, "it is a mistake to imagine that a girl is sold by her father in the same manner, and with the same authority, with which he would dispose of a cow". Amongst the degraded bushmen of S. Africa, "when a girl has grown up to womanhood without having been betrothed, which, however, does not often happen, her lover must gain her approbation, as well as that of the parents." Mr. Winwood Reade made inquiries for me with respect to the negroes of western Africa, and he informs me that "the women, at least among the more intelligent pagan tribes, have no difficulty in getting the husbands whom they may desire, although it is considered unwomanly to ask a man to marry them. They are quite capable of falling in love, and of forming tender, passionate, and faithful attachments." Additional cases could be given. We thus see that with savages the women are not in quite so abject a state in relation to marriage as has often been supposed. They can tempt the men whom they prefer, and can sometimes reject those whom they dislike, either before or after marriage. Preference on the part of the women, steadily acting in any one direction, would ultimately affect the character of the tribe; for the women would generally choose not merely the handsomest men, according to their standard of taste, but those who were at the same time best able to

> defend and support them. Such well-endowed pairs would commonly rear a larger number of offspring than the less favoured. The same result would obviously follow in a still more marked manner if there was selection on both sides; that is, if the more attractive, and at the same time more powerful men were to prefer, and were preferred by, the more attractive women. And this double form of selection seems actually to have occurred, especially during the earlier periods of our long history.

That is, Darwin recognizes that in "early" human evolution women were likely *active* sexual selectors – as females commonly are in nonhuman mammal species, according to his own observations – but then argues that this pattern is "inferior" to the marked gender inequalities and sexually passive role attributed to women in Victorian society. Another fallacy of such narratives about the passivity and coyness of women – at least the "superior" ones of the Victorian society – and the huge sexual appetite of males is that if such narratives would be true this would mean that there would be a huge evolutionary unbalance of and mismatch between the sexual drive of males and females. As noted above, sex is not only mandatory but also a main motor of evolution within sexual animals, so it seems very unlikely that their evolution would repeatedly lead to a situation in which only one sex truly has the drive to do something that is required for reproduction. Imagine if this was truly the case: either one sex would always need to force the other to have sexual intercorse with it and every single individual would be born from a raped father or a raped mother, or the species would quickly become extinct. Amazingly, such notions that women "naturally" merely want to have a single male partner – and not so much because of the pleasure of having sex with that partner *per se*, but mainly as an "exchange" for food or any other kind of help they can have to raise the only thing they care in life, their kids – continues to be so in vogue today, even among scientists. That is, according to such a misogynistic tale, *the true desire of the father is to have sex*, while *the true desire of the mother is to take care of her kids*. Unfortunately, such erroneous sexist tales are not uncommonly used by men to justify the use of force towards, or even rape of, women: after all, women in general do not want to "do it" naturally, and even in the few cases they want they tend to be shy about it – when she says "no," she does not really mean "no" – so the man naturally has to take the "lead" on this matter, to take matters into his own hands.

It is important to stress that this is a relatively recent construction: Western misogynistic ideas about women being "less sexual" begun to be particularly prominent within the academia and popular culture only in the nineteenth century, in general, during the Victorian Era – Darwin's books being an illustrative example of this. As Coontz wrote:

> Throughout the Middle Ages women had been considered the lusty sex, more prey to their passions than men. Even when idealization of female chastity began to mount in the 18th century.. few of its popularizers assumed that women totally lacked sexual desire. Virtue was thought to be attained through self-control; it was not necessarily innate or biologically determined. The beginning of the 19th century, however, saw a new emphasis on women's innate sexual purity. The older view that women had to be controlled because they were inherently more passionate and prone to moral and sexual error was replaced by the idea that women were asexual beings, who would not respond to sexual overtures unless they had been drugged or depraved from an early age. This cult of female purity encouraged women to internalize limits on their sexual behavior that 16th and 17th century authorities

> had imposed by force. Its result was an extraordinary desexualization of women – or at least of good women, the kind of woman a man would want to marry and the kind of woman a good girl would wish to be. Given the deeply rooted Christian suspicion of sexuality, however, the new view of women as intrinsically asexual improved their reputation. Whereas women had once been considered snares of the devil, they were now viewed as sexual innocents whose purity should inspire all decent men to control their own sexual impulses and based appetites. The cult of female purity offered a temporary reconciliation between the egalitarian aspirations raised by the Enlightenment and the fears that equality would overturn the social order..
>
> Still, the new ideals of marriage and womanhood were more than simply a face-lift for patriarchy.. women gained the moral right to say no to sex even though husbands continued to have legal control over their bodies. Furthermore, the cult of female purity was not.. a one-way street.. men were called upon to emulate this purity themselves.. although they were thought to have strong sexual urges, these were seen as unfortunate impulses.. animal passions.. that had to be controlled and repressed. The sentimentalization of marriage made domestic violence much less acceptable as well. In addition, the unique moral influence accorded to mothers contributed to an expansion of educational opportunities for women. (However) for many women brought up with the idea that normal females should lack sexual passion, the wedding night was a source of anxiety or even disgust. In the 1920s, Katharine Davis interviewed 2200 American women, most of them born before 1890. Fully a quarter said they had initially been 'repelled' by the experience of sex. Even women who did enjoy sex with their husbands reported feeling guilt or shame about their pleasure, believing that 'immoderate' passion during the sex act was degrading. Of course many women *did* have sexual urges, and the struggle to repress them led to other problems.. physicians regularly massaged women's pelvic areas to alleviate 'hysteria'.. medical textbooks of the day make it clear that these doctors brought their patients to orgasm.. in fact the mechanical vibrator was invented at the end of the 19th century to relieve physicians of this tedious and time-consuming chore.

This Victorian way of seeing women as "asexual," mainly promoted by men, brought huge problems for women but also created some problems for men: if "decent wives" are chiefly asexual, with whom could their husbands fulfill their 'male-animal sexual instincts'? As noted in *Sex at Dawn*, they could "fool around," but if they would do so with "decent women.. [they would be] threatening familial and social stability" anyway, so "they were expected to purge their lust with prostitutes.. 19th century philosopher Arthur Schopenhauer observed that 'there are 80,000 prostitutes in London alone – and what are they if not sacrifices on the altar of monogamy'?"

A main example showing that such sexist narratives are still prevalent within academia is that, since Darwin's time until this very moment many scholars continue to try to answer to the supposedly "puzzling" question of why women have orgasms at all. After all, if one accepts such Darwinian sexist narratives, women would in theory not *need* to have orgasms, as their main goal is not having sexual pleasure but instead just having kids, which requires only that men have orgasms. For instance, at the beginning of the abstract of a 2016 paper published by two prominent biologists, Pavlicev and Wagner, they summarized the common views on this issue within a substantial part of the scientific community: "the evolutionary explanation of female orgasm has been difficult to come by.. The orgasm in women does not obviously contribute to the reproductive success, and surprisingly unreliably accompanies heterosexual intercourse.. Two types of explanations have been proposed: one insisting on extant adaptive roles in reproduction, another explaining female orgasm as a byproduct of selection on male orgasm, which is crucial for sperm transfer." Another recent example of such scientific sexist biases, and how they do influence the perception of the media and broader public about these topics, is the title of a recent paper by Brennan and colleagues, and in particular the way in which the paper was reported by the press. In the abstract of that 2022 paper, entitled "*Evidence of a functional clitoris in dolphins*," the authors explained that:

> In species that copulate during non-conceptive periods, such as humans and bonobos, sexual intercourse is known to be pleasurable for females. Dolphins also copulate throughout the year, largely to establish and maintain social bonds. In dolphins, the clitoris is positioned in the anterior aspect of the vaginal entrance, where physical contact and stimulation during copulation is likely. Clitoral stimulation seems to be important during female-female sexual interactions in common bottlenose dolphins (*Tursiops truncatus*), which rub each other's clitorises using snouts, flippers, or flukes. Determining a sexual pleasure response in animals not amenable to neurobehavioral examination is difficult, but investigation of the clitoris may elucidate evidence of functionality. In this study, we assessed macro- and micromorphological features of the clitoris in common bottlenose dolphins to examine functional features, including erectile bodies with lacunae, extensible collagen and/or elastin fibers, and the presence and location of sensory nerves. Our observations suggest the clitoris of dolphins has well-developed erectile spaces, is highly sensitive to tactile stimulation, and is likely functional.

As noted above, many Neo-Darwinists, particularly functional morphologists, tend to commit the fallacy of thinking that most if not all the anatomical traits present in a certain organism are both currently functional and optimal and at least suboptimal. Interestingly, in this case such functionalist biases seem to be weaker that the sexist narratives about females being sexually passive, in the sense that the authors of the paper seem to be surprised – or at least think that others will be surprised – by the fact that the clitoris of adult female dolphins is.. functional. This is the kind of issue in which the just-so-stories made by scientists based on their biases can become to sound as highly nonsensical. Ask a fisherman of a hunter-gatherer group, or a young kid that is not yet indoctrinated by such sexist biases, if they think the clitoris of a female dolphin is "functional," and most likely they would say something like: "of course, why would it not be, if those of the females of our species are functional?" But clearly this is not how many adult scientists, journalists and members of the broader public have been indoctrinated to think, as shown by the way in which the media disseminated that 2022 article. An illustrative example concerns a magazine that is highly respected among both academics and the broader public – the *Smithsonian Magazine* – which in a piece about that article published on January 13, 2022, stated enthusiastically, in its title, that "*Female Dolphins Have a Fully Functional Clitoris – A new study finds surprising similarities between human and cetacean sexual anatomy.*"

Before accepting such indoctrinated sexist Darwinist ideas *a priori* as scientific dogmas, researchers need to question them and test them empirically. In this sense this is the merit of that 2022 paper: it provided evidence that tested, and indirectly contradicted – based on anatomical analyses, rather than physiological studies – such sexist Darwinist just-so-stories and expectations. However, what is truly *surprising* in the title of the *Smithsonian Magazine* piece is that scientists and the media would think *a priori* that female adult dolphins having functional clitorises would be surprising at all, when there was no prior evidence at all indicating that this would be otherwise. This is a powerful reminder of the power of enculturation, and how it often leads to *inaccurate expectations* of what is supposed to be "normal" and "abnormal," anticipated and surprising. As noted in that piece:

> Evolutionarily, "it makes sense that intercourse is pleasurable for females, because that would lead to an increase in copulation and reproduction", Brennan says.. "The only thing that surprises me is how long it has taken us as scientists to look at the basic reproductive anatomy," Sarah Mesnick, an ecologist at NOAA Fisheries who was not involved with the research, tells the Times. She adds that studying social behavior in animals can help researchers better understand their evolution, which could help with management and conservation. Female sexual pleasure in nature has not been well-researched and scientists didn't even fully describe the human clitoris until the 1990s, says Brennan. Even in human medical research and in the medical curriculum, clitoral anatomy is largely missing, writes Calla Wahlquist for the *Guardian*. "This neglect in the study of female sexuality has left us with an incomplete picture of the true nature of sexual behaviors," says Brennan in a statement.

Another similar recent example, also from a 2022 paper that is related to sex and previous adaptationist just-so-stories constructed about it, was published by Roth and colleagues. Entitled "*Masturbation in the animal kingdom,*" its abstract directly referred to this kind of "one-sided theorizing that attributes a specific biological" to basically any trait, including the ones that give us pleasure, such as masturbation:

> Masturbation is one of the most common sexual behaviors in humans. It is also a phylogenetically widespread trait of various other mammalian and some non-mammalian species. Several hypotheses have been proposed aiming to explain the function of masturbation in primates and other species. These were mainly based on observations of nonhuman primates such as rhesus macaques or bonobos and rodents such as African ground squirrels. Based on these observations various scholars suggested that masturbation improves ejaculate quality, decreases the risk of contracting sexually transmitted infections or is merely a by-product of sexual arousal and thus an alternate outlet to copulation. While these theories may explain some facets of masturbation in some species, they do not explain why masturbation is so widespread and has developed in various species as well as our hominid ancestors. Moreover, the research on which these theories are based is scarce and heavily focused on male masturbation, while female masturbation remains largely unexplored. This sex difference may be responsible for the one-sided theorizing that attributes a specific biological benefit to masturbation. We propose that the widespread prevalence of masturbation in the animal kingdom may be better explained by viewing masturbation as a primarily self-reinforcing behavior that promotes pleasure both in human and in nonhuman species.

As it happens in various religions, under the "war"-like struggle-for-existence quasi-religious fundamentalist framework accepted by many adaptationist Neo-Darwinians there is no room at all for play, fun, or pleasure: everything is about surviving or becoming extinct, being optimal or being condemned to oblivion – as if there is nothing else in this amazingly diverse planet. Fortunately, more and more authors are starting to dare to criticize not only such a fundamentalist framework but also some of Darwin's just-so-stories about sex and gender, as exemplified by another 2022 paper, by Rosenthal and Ryan. Published in the prestigious scientific journal *Science*, the abstract of the paper – titled "*Sexual selection and the ascent of women: Mate choice research since Darwin*" – explicitly uses the words "Darwin's misogyny":

> Darwin's theory of sexual selection fundamentally changed how we think about sex and evolution. The struggle over mating and fertilization is a powerful driver of diversification within and among species. Contemporaries dismissed Darwin's conjecture of a "taste for the beautiful" as favoring particular mates over others, but there is now overwhelming evidence for a primary role of both male and female mate choice in sexual selection. *Darwin's misogyny* precluded much analysis of the "taste"; an increasing focus on mate choice mechanisms before, during, and after mating reveals that these often evolve in response to selection pressures that have little to do with sexual selection on chosen traits. Where traits and preferences do coevolve, they can do so whether fitness effects on choosers are positive, neutral, or negative. The spectrum of selection on traits and preferences, and how traits and preferences respond to social effects, determine how sexual selection and mate choice influence broader-scale processes like reproductive isolation and population responses to environmental change.

Regarding the way in which that paper was disseminated to the broader public, the popular science dissemination website *The Conversation* published a piece about it on January 20, 2022, that was also unambiguous. As explained by Matthew Wills in that piece, entitled "*Evolution – how Victorian sexism influenced Darwin's theories*":

> According to a new paper, published in *Science*, Charles Darwin's patriarchal world view led him to dismiss female agency and mate choice in humans. He also downplayed the role of female variation in other animal species, assuming they were rather uniform, and always made similar decisions. And he thought there was enormous variation among the males who battled for female attention by showing off stunning ranges of skills and beauty. This maintained the focus on the dynamics of male dominance hierarchies, sexual ornamentation and variation as drivers of sexual selection, even if females sometimes did the choosing.. Research since Darwin therefore reveals that mate choice is a far more complex process than he may have supposed, and is governed by variation in both sexes. So, is the accusation of sexism levelled at Darwin really valid, and did this cloud his science? There is certainly some evidence that Darwin underestimated the importance of variation, strategy and even promiscuity in most female animals. For example, Darwin – possibly as a result of a prevailing prudishness – placed little emphasis on mechanisms of sexual selection that operate after mating. Female birds and mammals may choose to mate with multiple males, and their sperm can compete to fertilise one or more eggs within the reproductive tract. Cats, dogs and other animals can have litters with multiple fathers (the gloriously named "heteropaternal superfecundation" – even though the sound of it is really quite atrocious!). There is even some suggestion that the human penis – being thicker than our nearest primate relatives – is an adaptation for physically displacing the sperm of competing males. Such earthy speculations were anathema to Darwin's sensibilities.
>
> Female blue tits often mate with multiple males in order to ensure their protection and support – a somewhat manipulative strategy when paternity for the prospective fathers is uncertain. All this challenges Darwin's assumption that females are relatively passive and non-strategic. Where males make a greater investment, they become more active in mate choice. Male (rather than female) poison dart frogs (*Dendrobates auratus*) protect the young, and therefore attract multiple females who compete to lay eggs for them to fertilise. Many bird species have biparental care, and therefore a richer diversity of mating systems. Inevitably, Darwin's world view was shaped by the culture of his time, and his personal writings make it difficult to mount a particularly robust defence. In a letter from 1882, he

> wrote "I certainly think that women, though generally superior to men to [sic] moral qualities are inferior intellectually; & there seems to me to be a great difficulty from the laws of inheritance ... in their becoming the intellectual equals of man". He also deliberated over the relative merits of marriage, famously noting: "Home, & someone to take care of house — Charms of music & female chit-chat. — These things good for one's health. — but terrible loss of time".

Another emblematic example of how things are slowly starting to change concerns a 2022 article by Emma Beddington in the popular newspaper *The Guardian* about Lucy Cooke's recent book *Bitch: A Revolutionary Guide to Sex, Evolution & The Female Animal.* As done in the book, that *Guardian* article does not shy away from directly criticizing Darwin's sexist evolutionary just-so-stories and pointing out how critical they have been in the perpetuation of systemic sexism to this day:

> For too long, Cooke argues, we have uncritically accepted a view of nature "through a Victorian pinhole camera" and worse, those misconceptions have been co-opted by ideologues "to claim that a host of grim male behaviours – from rape to compulsive skirt-chasing to male supremacy – were only natural for humans because Darwin said so." That is simply, demonstrably wrong, as she outlines, combining [in her book] colourful revelations with limpidly clear explanations of complex science. "Female animals," she writes, "are just as promiscuous, competitive, aggressive, dominant and dynamic as males".. "The Victorian era was all about imposing order on the natural world," she [Cooke] says, "but now the approach is to embrace the chaos and to realise the infinite possibilities of developmental plasticity".. Scientists not interested in the female experience failed to research it, creating a data gap that means we do not get the full picture. Cooke relates how often-female scientists have been fighting for decades to engineer the radical perspective shift that would allow us to see, and understand, the far more complex and less deterministic real story of sex and reproduction. In particular, Bitch is luminous with affection and admiration for a group of pioneering, now semi-retired professors – "The Broads", Sarah Blaffer Hrdy, Jeanne Altmann, Mary Jane West-Eberhard and Patricia Gowaty – whose formidable research challenged the Darwinian phallocracy.. For all their efforts, and those of the "brilliant young scientists" Cooke also met in her research, the battle against this male-oriented bias continues: "The struggle is real and it's happening now," she says.
>
> TV, too, Cooke thinks, still needs to do better. "I ended up screaming at the television," she says of *The Mating Game*, a recent BBC documentary on animal reproduction. "This idea that females are completely passive and it's the males that have all the strategy just gets trotted out again and again. God, can we change the record?" She's scornful of evolutionary psychologists making money trotting out the "Darwinian" theory that males are competitive and aggressive, females are coy and monogamous. "It's complete bullshit. Using animals as ideological weapons is a dangerous game". A sexist mythology has been baked into biology, and it distorts the way we perceive female animals. In the natural world female form and role varies wildly to encompass a fascinating spectrum of anatomies and behaviours. Yes, the doting mother is among them, but so is the jacana bird that abandons her eggs and leaves them to a harem of cuckolded males to raise. Females can be faithful, but only 7% of species are sexually monogamous, which leaves a lot of philandering females seeking sex with multiple partners. Not all animal societies are dominated by males by any means; alpha females have evolved across a variety of classes and their authority ranges from benevolent (bonobos) to brutal (bees). Females can compete with each other as viciously as males: topi antelope engage in fierce battles with huge horns for access to the best males, and meerkat matriarchs are the most murderous mammals on the planet, killing their competitors' babies and suppressing their reproduction.. Darwin's theory of sexual selection drove a wedge between the sexes by focusing on our differences; but these differences are greater culturally than they are biologically. Animal characteristics – be they

physical or behavioural – are both varied and plastic. They can bend according to a selection's whim, which makes sex traits fluid and malleable. Rather than predicting a female's qualities through the crystal ball of her sex, the environment, time and chance all play a significant role in shaping their form.

Cooke is right. Instead of often just parroting Darwin's inaccurate evolutionary ideas about women and their natural "coyness" or trying as hard as possible to find any kind of possible data to support them, scholars should rather focus on discussing the scientific data that truly refers to female sexuality, without *a priori* dogmatic sexist ideas, or taboos. For instance, they should instead discuss why are women the ones that actually have in general more frequently multiple orgasms, if according to such dogmatic narratives it is "puzzling" that women have orgasms at all because they are supposed to be "less sexual," "coy," and care less than men, or not care at all, about having sexual pleasure. Studies done in a wide range of backgrounds and different countries consistently find that multiple orgasms are indeed far more frequently reported in women than in men. For example, a study by Puts and colleagues in 2021 found that 43% of women reported usually experiencing multiple orgasms. Of course, the numbers in different studies change, depending on the geographical location and the cultural narratives commonly accepted in those locations, the context in which the questionnaires are done, and of course the type of questions. But what is clear, from the vast majority of such studies done so far, is that multiple orgasms are indeed much more frequent in women than in men, on average, for instance in their 2016 review titled "*Multiple orgasms in men – what we know so far,*" Wibowo and Wassersug wrote that "few men are multiorgasmic: <10% for those in their 20s and <7% after the age of 30."

Interestingly, these authors point out that a change of sexual partner is said to potentially increase how likely a male is to have multiple orgasms, or a similar sensation such as the so-called Coolidge effect: a re-arousal phenomenon where the refractory ("recovery") period from an orgasm is reduced in the presence of a *new sex partner*. As they further note, evidence of a similar phenomenon has been shown empirically in other animals: access to new or novel sexual partners may promote sexual desire and motivation to orgasm, truncating the refractory period. Such data therefore also put into question other types of narratives about our sexuality, for instance that our truly "human nature" is to be sexually, or even socially, monogamous, while empirical studies show, over and over, that risk and novelty – rather than having sex exclusively with the very same partner for various decades – are among the main drivers of sexual desire. As noted in *Sex at Dawn*:

Over fifty years ago, sex researchers Clellan Ford and Frank Beach declared, "in those [hunter-gatherer] societies which have no double standard in sexual matters and in which a variety of liaisons are permitted, the women avail themselves as eagerly of their opportunity as do the men". Nor do the females of our closest primate cousins offer much reason to believe the human female should be sexually reluctant due to purely biological concerns. Instead, primatologist Meredith Small has noted that female primates are highly attracted to novelty in mating. Unfamiliar males appear to attract females more than known males with any other characteristic a male might offer (high status, large size, coloration, frequent grooming, hairy chest, gold chains, pinky ring, whatever). Small writes, "the only consistent interest seen among the general primate population is an interest in novelty and variety.. In fact," she reports, "the search for the unfamiliar is documented as a female preference more often than is any other characteristic our human eyes can perceive".

Various studies indicate that women also experience, on average, more complex, elaborate, and intense orgasms than men, as reported in a 2002 paper by Mah and Binik. Ackerman, in her 1994 book *Natural History of Love*, noted that "men's oxytocin levels quintuple during orgasm.. but a Stanford University study showed that women have even higher levels of oxytocin than men do during sex, and that it takes more oxytocin for a woman to achieve orgasm.. drenched in this spa of the chemical, women are able to have more multiple orgasms than men, as well as full body orgasms." Furthermore, the clitoris normally has about 8000 sensory receptors – compared to half as many in an ordinary penis – and its sole function is to provide ecstatic pleasure, contrary to the penis, which is also related to urination, as emphasized in Browning's 2017a paper "*Survival secrets – what is about women that makes them more resilient than men.*" Based on these data, it seems that instead of asking "why women have orgasms," we should instead ask less biased questions such as: as woman have in average a higher orgasmic potential, why are there many studies showing that, in several countries and specific contexts, women orgasm on average less times than men, per male-female sexual act? I was actually asked this question, in a TV interview about this subject, by a sex therapist, some months ago. I answered that, indeed, as explained in a 2018 review by Blair and colleagues appropriately titled "*Not all orgasms are created equal,*" according to some of the most extensive studies done on the subject in Western countries, men report experiences of orgasm in 85% to 95% of their partnered sexual activities, while the percentage is only 40% to 65% in women. As those authors explain, this difference is *not* related to a lack of orgasmic potential by women – well on the contrary – being mainly due to other reasons, including precisely the prevailing manly made narratives about how "good" women should be more passive concerning their sexuality, or more ashamed of it. This women–men gap is further reinforced by the typical "women–men sexual scripts" that are also prevalent in our societies, which tend to favor sexual activities that are aimed to give pleasure to, and thus to more likely result in orgasms for, men, such as penile-vaginal intercourse.

In *Testosterone Rex*, Cordelia Fine discusses this topic, noting that "a large-scale study of thousands of female North American college students found that they had only 11% chance of experiencing an orgasm from a first casual 'hookup'.. follow-up interviews revealed why it was that women had such slim odds of reaching a climax.. students generally agreed that it was important for a man to be sexually satisfied in any context, and for women to be sexually satisfied in the context of a relationship." Indeed, "there was no perceived obligation to provide sexual satisfaction to a woman in hookup sex.. while many men felt that bringing their girlfriend to orgasm reflected well on their masculinity, they often didn't feel the same way about hookup partners." When interviewed for TVs and radio shows, I often use the following example: image that the opposite happens, that is, that a woman would say to a man "OK, we can have sex for hours and hours, but we can never touch, or use, your penis." Almost no men would be happy with this, as most of them would have a huge difficulty to have a single orgasm, even after several hours or sexual intercourses. Well, that is precisely what many women accept, when they have sex with men that penetrate them without ever stimulating their clitoris, which is the female region that embryologically corresponds to the penis. Many women, by trying to follow the script of being a "good, respectful woman," not only accept that man do that but also do not have the courage to even stimulate their clitoris themselves when that happens, particularly during the first time they have sex with a man. So, that U.S. female college students only have 11% chance of experiencing an orgasm from a first casual "hookup" with a man has mostly nothing to do with them, and their clitorises, having less orgasmic potential than men and their penises. Well on the contrary: it is instead related within the still prevailing manly made tales about how most women should follow a sexual script that cares mainly about the pleasure of men.

The fact that this topic is still plagued by erroneous tales in popular culture is illustrated by a sentence most of us have heard several times: "men have a penis, women have a vagina." This sentence emphasizes the passive role that is often assigned to women: that they basically have a "hole" to receive the active player, the man's penis. Although scientifically such tales are plainly wrong – biologically the penis corresponds to the clitoris, not to the vagina, as noted above – they have been and continue to be "scientifically supported," by several prominent researchers. A particularly conspicuous one was Freud, who defended the notion that contrary to "immature women," who mostly enjoy clitoral orgasms, "mature women" are able to fully enjoy vaginal orgasms. That is, only women that obey to men and the factually wrong tales created by them can achieve the recognition of men and be praised by them as "proper, mature, good women." For instance, Freud stated that "the woman who refuses to see her sexual organs as mere wood chips, designed to make the man's life more comfortable, is in danger of becoming a lesbian – an active, phallic woman, an intellectual virago with a fire of her own.. the lesbian body is a particularly pernicious and depraved version of the female body in general; it is susceptible to auto-eroticism, clitoral pleasure and self-actualization."

As put by Ackerman in *Natural History of Love*:

> It's worth noting that when we talk about gender we say that a man has a penis and a woman has a vagina. This distinction, which we take for granted, hides a prejudice about the baseness of women. A man's pleasure organ is the penis, and a woman's pleasure organ is her clitoris, not her vagina. Even if we're talking about procreation, it's not accurate: a man's penis delivers sperm and can impregnate, and a woman's womb contains eggs, which can become fertile. Equating the man's penis with the woman's vagina says, in effect, that the natural order of things is for a man to have pleasure during sex, and for a woman to have a sleeve for man's pleasure. It perpetuates the notion that women aren't supposed to enjoy sex, that they're bucking the natural and social order if they do. I don't think this will change very soon, but it reminds me how many of our mores travel almost invisibly in the plasma of language.

Regarding more recent scholarly works, one example evidencing that such sexist tales continue to permeate not only the media and popular culture – pornographic movies being an illustrative example of this – but also academia, is the 2016 paper by Prause and colleagues stated, entitled "*Clitorally stimulated orgasms are associated with better control of sexual desire, and not associated with depression or anxiety, compared with vaginally stimulated orgasms.*" As noted in the abstract of the paper, "recent claims that vaginal stimulation and vaginally generated orgasms are superior to clitoral stimulation and clitorally generated orgasms pathologize most women and maintain a clitoral vs vaginal dichotomy that might not accurately reflect the complexity of women's sexual experience." It is astonishing that scientific "claims" that clitorally stimulated orgasms are associated with depression or anxiety or are "inferior" to vaginal ones are still so widespread within academia and medicine in the twenty-first century. The empirical results of that paper obviously refuted such "claims": "most women (64%) reported that clitoral and vaginal stimulation contributed to their usual method of reaching orgasm.. women who reported that clitoral stimulation was primarily responsible for their orgasm reported a higher desire to self-stimulate and demonstrated greater control over their self-reported sexual arousal." As emphasized by the authors, "women experience orgasms in many varied patterns, a complexity that is often ignored by current methods of assessing orgasm source.. the reported source of orgasm was unrelated to orgasm intensity, overall sex-life satisfaction, sexual distress, depression, or anxiety". Importantly, "women who reported primarily stimulating their clitoris to reach orgasm reported higher trait sexual drive and higher sexual arousal to visual sexual stimulation and were better able to increase their sexual arousal to visual sexual stimulation when instructed than women who reported orgasms primarily from vaginal sources."

This point is further supported by empirical data showing that lesbians often have significantly higher rates of orgasm than self-identified heterosexual or bisexual women, a topic that we will discuss below. Numerous other lines of evidence contradict the fairytale that women are "naturally" less "sexual" than men. One of them, related to physiology, concerns the fact that several studies have indicated that women often outperform men in smell sensitivity tests – as noted in a 2014 paper by Oliveira-Pinto and colleagues – including those regarding scents associated to sexual arousal, as documented in Hirsch's 1998 book *Scentsational Sex*. One other line of evidence concerns studies on both sex fluidity and homosexuality. As pointed out in a 2018 review of many studies done on this topic, published by Jeffery and colleagues, "heterosexual women are more likely than heterosexual men to report same-sex sexual attractions." Individuals "with high sexual fluidity experience sexual responses toward a broad and shifting range of stimuli; male sexual desire is usually considered category-specific, as it strongly favors one sex; for men, there is a negative correlation between sexual attraction to one sex and sexual attraction to the other sex, and men of all sexual orientations with higher sex drives have more sex with a single preferred sex." As they note, "this is not true of women, who show a weak positive correlation between attraction to one sex and attraction to the other sex." Also, "women report more frequent shifts in sexual attractions than men across their lifetimes, particularly among non-heterosexuals." Importantly, such studies on homosexuality emphasize the complex links between "nature" and "nurture" that were often neglected by Darwin when he was discussing human evolution and sexuality. That is, such studies show that while women seemingly tend to be more sexually fluid by "nature" – biologically/evolutionarily – there is also an important social – environmental – component at play regarding sexual orientation. As the authors note, female prisons represent an extreme example of this, with "somewhere between 30% and 60% of female prison inmates engaging in same-sex sexual behaviors and relationships; most of these women did not identify as homosexual prior to incarceration and, the more time women spend in prison, the more accepting they become of sexual interactions between other inmates and of having a 'gay cellmate.'"

As also stressed by Jeffery and colleagues, various researchers argue that frustration with the typical model of "male–female intercourse" that is prevalent in most countries may contribute to homosexuality. For instance, "heterosexual men and women report that sexual relationships inflict greater costs than are reported by homosexual men and women; meanwhile, homosexual men enjoy more rapid and frequent sexual experiences than heterosexual men (on average) and homosexual women report greater sexual satisfaction, more sexual desire, more frequent orgasms, and greater satisfaction with their own bodies than heterosexual women." This point is further made in a detailed study about Brazilian female jails done by one of the most renowned medical researchers in Latin America, Drauzio Varella, during more than a decade. In his 2017 book *Prisioneiras*, he argues that one of the few places where women nowadays have "true sexual freedom" is in female jails, in the sense that they do not live on a daily basis with men so they are freer from manly made sexist scripts and gender violence, with sex being often more consensual. Accordingly, in such jails, he often observed a more profound, and complex, expression of feminine sexuality, with an incredibly wide range of identities and expressions. Varella noted that homosexuality is more frequent, and subtle, than in male prisons, and affirmed that although he had been a professional oncologist and gynecologist for many years, "in the female jails, I realized that I only truly knew about 10% of the variability of what women sexuality can be."

Another line of evidence comes from large anonymous online questionnaires, such as one made to 1300 persons, reported by the sexshop *Lovehoney*: 46% of the women that answered the questionnaire stated that they think about other people when they are having sex with their partners, while only 42% of men said the same. Another piece of evidence, among many others that could be cited here, comes from studies on phenomena such as the so-called hypersexual disorders – for instance "satyriasis" in men and "nymphomania" in women – which consistently show that, in different contexts and countries, women tend to score higher on "hypersexuality" than men and to engage more often in risky sexual behavior, contrary to what is often said in popular culture, as explained in a 2017 paper by Öberg and colleagues. Another example of women sexual fluidity is given in the book *Sex at Dawn*, which uses the term "erotic plasticity":

> This greater erotic plasticity appears to manifest in women's more holistic responses to sexual imagery and thoughts. In 2006, psychologist Meredith Chivers set up an experiment where she showed a variety of sexual videos to men and women, both straight and gay. The videos included a wide range of possible erotic configurations: man/woman, man/man, woman/woman, lone man masturbating, lone woman masturbating, a muscular guy walk-

ing naked on a beach, and a fit woman working out in the nude. To top it all off, she also included a short film clip of bonobos mating. While her subjects were being buffeted by this onslaught of varied eroticism, they had a keypad where they could indicate how turned on they felt. In addition, their genitals were wired up to plethysmographs (measures blood flow to the genitals.. think of it as an erotic lie detector). What did Chivers find? Gay or straight, the men were predictable. The things that turned them on were what you'd expect. The straight guys responded to anything involving naked women, but were left cold when only men were on display. The gay guys were similarly consistent, though at 180 degrees. And both straight and gay men indicated with the keypad what their genital blood flow was saying.

> As it turns out, men can think with both heads at once, as long as both are thinking the same thing. The female subjects, on the other hand, were the very picture of inscrutability. Regardless of sexual orientation, most of them had the plethysmograph's needle twitching over just about everything they saw. Whether they were watching men with men, women with women, the guy on the beach, the woman in the gym, or bonobos in the zoo, their genital blood was pumping. But unlike the men, many of the women reported (via the keypad) that they weren't turned on. As Daniel Bergner reported on the study in *The New York Times*, "With the women.. mind and genitals seemed scarcely to belong to the same person". Watching both the lesbians and the gay male couple, the straight women's vaginal blood flow indicated more arousal than they confessed on the keypad. Watching good old-fashioned vanilla heterosexual couplings, everything flipped and they claimed more arousal than their bodies indicated. Straight or gay, the women reported almost no response to the hot bonobo-on-bonobo action, though again, their bodily reactions suggested they kinda liked it. This disconnect between what these women experienced on a physical level and what they consciously registered is precisely what the theory of differential erotic plasticity predicts. It could well be that the price of women's greater erotic flexibility is more difficulty in knowing – and, depending on what cultural restrictions may be involved, in accepting – what they're feeling. This is worth keeping in mind when considering why so many women report lack of interest in sex or difficulties in reaching orgasm [particularly when having sex with men].

It is likely that many of the women that did not self-report to be aroused when there was a high blood flow to their genitals did that *consciously*, because a "good woman" is not supposed to be aroused, and even less to say so. This idea is further supported by many other studies. For instance, Browning's book *The Fate of Gender* stated that "in a series of controlled surveys published.. in 2011.. men recorded thinking about sex an average of 18 times a day while women recorded thinking about sex 10 times a day.. similarly, when men and women were asked orally how many partners they would like to have per year, young college-age men reported many more than young women did.. however, when both were attached to lie detector machines, the number of desired partners per year turned out to be about equal." However, coming back to the study of blood flow cited above, what is even more disturbing is that it is likely that at least some of the women that did not self-report to be aroused when there was a high blood flow to their genitals did *not even realize* that there was such a high blood flood to their genitals. That is, because women are bombarded since a young age with sexist narratives stating that they are sexually passive and "coy," as affirmed by Darwin, it is likely that some of them ultimately end of by somehow repressing their sexual feelings, somewhat disconnecting what they think they feel from the way their body is truly reacting physiologically. In this sense, such women have somewhat "reached" a state that is precisely the end goal

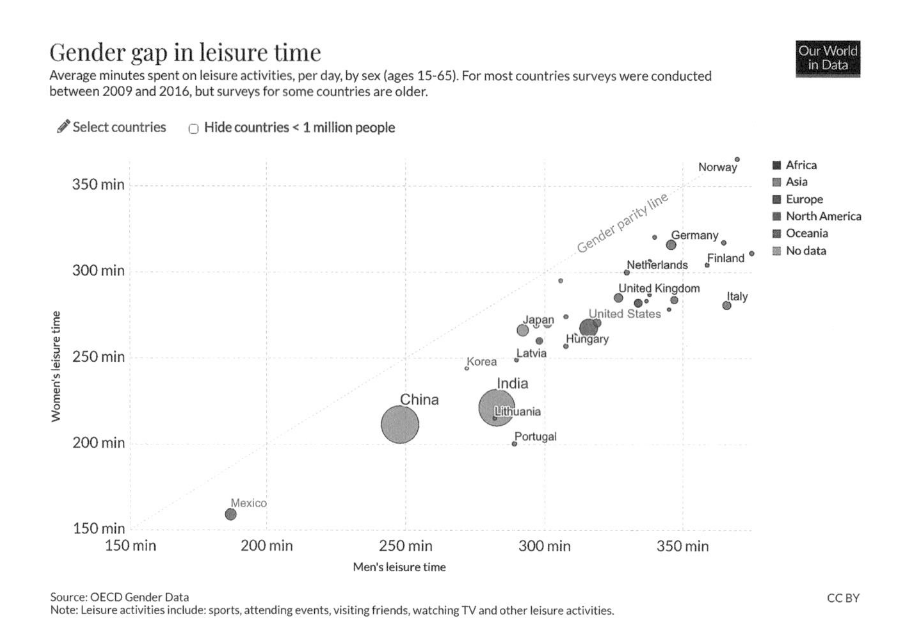

Fig. 4.9 Gender disparities in time spent within 33 countries – women spend nearly three times more in unpaid care work compared to men, a whopping total of 1.1 trillion hours each year, which means a lot less leisure time, except in Norway, which is a very rare exception

of many manly made misogynistic inaccurate narratives constructed by scientists such as Darwin and Freud, religious leaders, politicians, and so on.

The fact that men created narratives about women's natural "passivity," "coyness," and "tendency to do domestic tasks" in great part as a way to subjugate and domesticate them particularly since the rise of sedentarism and agriculture, and that they continue to promote such narratives to maintain that *status quo*, is illustrated in Fig. 4.9. As shown in that figure, even today, in the twenty-first century, in the vast majority of countries, including so-called "developed" countries such as the Netherlands, the United States, Germany, and so on, women indeed continue to have much less leisure time than men, on average, with a single exception, Norway. That figure, as well as an endless number of other figures about gender inequities – or for that matter about societal inequities in general – directly contradict those people that argue that we should no longer discuss the past – including authors such as Darwin and Freud – because the past is no longer relevant, everything is different now. That is, that we are already living in a "post-sexist" and "post-racist" epoch, at least in Western "developed" countries. Unfortunately, such figures and numerous other types of empirical data clearly indicate that this is not so. And a major reason for that is that many people, including scholars, continue to blindly accept the sexist and racist "facts" created, supported, or defended by scientists such as Aristoteles, Darwin, and Freud.

As noted in Saini's 2017 book *Inferior*, gender stereotypes, particularly those regarding "women's inferiority," are still so deeply embedded in our societies that they are often i*nternalized* not only by men but also by many women. She wrote: "in a study published in 2012, psychologist Corinne Moss-Racusin and a team of researchers at Yale University explored the problem of bias in science by conducting a study in which over a hundred scientists were asked to assess a résumé submitted by an applicant for a vacancy as a laboratory manager.. every résumé was identical, except that half were given under a female name and half under a male name." When "they were asked to comment on these supposed potential employees, scientists rated those with female names significantly lower in competence and hireability.. they were also less willing to mentor them, and offered far lower starting salaries." Strikingly, as recognized by the authors of the study, "the gender of the faculty participants did not affect responses, such that female and male faculty were equally likely to exhibit bias against the female student." That is, as put by Saini, "prejudice is so steeped in the culture of science, their results suggested, that women are themselves discriminating against other women."

What Do We Know About Human Evolution and Gender Stereotypes?

Girls hold their mother responsible for their lack of a penis and do not forgive her for their being thus put at a disadvantage. (Sigmund Freud)

A question that people – including scientists – often ask me during or after my talks on human evolution is: "so, if women have all the capacities and abilities you listed in your talk, why is it that men won more Nobel prizes and made more discoveries than women? Why were Da Vinci, Galileu, Newtown, Darwin, Einstein, and so on all men?" Sometimes I answer them by raising a new question: just 20 years ago, you could have just asked me a similar question about why all the "best" formula 1 drivers and golf players were all "white," right? But now we *know* that the "best" formula 1 drivers and golf players are not necessarily "white," because in recent years Tiger Woods and Lewis Hamilton have dominated those sports. So, what was truly happening for decades is that the vast majority of "blacks" *lacked the opportunities*, in Western countries, to show how good they *could be* at those sports, which were and continue to be mainly played by, and associated with, people that are "well off" within our Western societies. Some sports do have a lot to do with some specific physical attributes – Olympic weightlifting is obviously normally not won by very skinny people with scarce muscle mass. However, this does not apply to sports such as formula 1, golf, or chess, nor to things like winning a Nobel prize, to come back to the above question about women. Examples of female scientists such as Marie Curie shown, as do the case of "black" sportsmen such as Tiger Woods and Lewis Hamilton, that certain people of a so-called ethnic group "A" – remember that "whites" or "blacks" do not constitute real, natural biological taxa -, or of a socio-economic group A, or gender A, historically won more Nobel prizes or other scientific awards than people from a group or gender B mainly because of the narratives, social constructions and even laws that lead to and then maintained the segregation of, and therefore the inequities between, those groups.

Contrary to the distinction between groups such as "blacks" and "whites," the distinction between women and men as different biological sexes is real, although there are also many cases that lie in between, biologically, a point that continue to be too often neglected. That is, on average there are some significant differences between women and men not only concerning the shape and physiology of their genitals and ways of living their sexuality – for example, women tend to have a higher orgasmic potential as we have seen – but also regarding some specific mental capacities – with women often having higher performances, on average, than men in many types of mental tests, as I summarized in *Meaning of Life, Human Nature, and Delusions*. This makes sense. While the *Homo sapiens* ancestors of living human groups only started to diverge less than 250 thousand years ago, when our species arose, the evolution and biological distinction of female and male sexual beings has a much deeper evolutionary history. Within the fossil record of eukaryotes – organisms whose cells have a nucleus enclosed within a nuclear envelope, as is our case – sexual reproduction appeared earlier than 1 billion years ago, as noted above, and the Y chromosome that distinguishes males from females at the genetic level appeared at least 180 million years ago, that at least 2000 times the amount of time that has passed since *Homo sapiens* first left Africa about some 90 to 60 thousand years ago.

This is indeed a profound difference between the real biological differences between female and male, and many of the so-called gender differences that are usually blindly accepted as "real" by people of a certain society, which are mostly social constructions of that society. Women have been subjugated by men since many millennia ago, particularly since the rise of sedentism and agriculture. As noted above, until only some decades ago in many Western countries, women could not even vote, or have their own research laboratories. So obviously they could *not* receive the same amount of Nobel prizes that men received back then. When I make this point, people – mostly men, but also some women – often then ask me : "but why were women subjugated in the first place, if they are not mentally inferior?" This is a very common argument raised by Western people – particularly racist ones – about "others": for them, the fact that Europeans colonized such "others," and not the other way round, provides evidence that Europeans are "superior." Such arguments neglect the fact that many scholars have already provided very strong hypotheses – based on sound empirical data – about why were the Europeans and not the "others" that did so, which have nothing to do with so-called innate superiority. One example, among many others, is Diamond's superb1999 book *Guns, germs, and steel: the fates of human societies*. Moreover, such an argument implies that colonizing, using, abusing, and even exterminating "others" is something that shows that a society is "mentally superior" – and, strikingly, even "morally superior," as argued by Darwin in the *Descent* and the *Expression* – while one could argue exactly the opposite, as discussed above.

Furthermore, although the differences between human genders and so-called human "races" are both mostly socially constructed, the former are very different from the later because some "races" are mainly said to be from a certain region or continent – for instance, "white" Europeans *versus* "black" Africans – while women and men from a same society almost always co-existed physically. And this is precisely the point: contrary to the differences between, let us say, Europeans *versus* the Mayans, the subjugation of women by men is, in this sense, much easier to explain because it has been a *continuous* process that has often involved physical subjugation combined with sexist narratives that ultimately lead and justified a legal, social, and economic subjugation. That is, men used one of the very few features in which they do, *on average*, outcompete women: physical force. The very same feature that alpha males from certain mammalian groups, such as gorillas, tend to use at least in part to subjugate females as well as other males of their communities. Now, when there is a change of alpha-male in a gorilla group, for instance after a fight, does the winner of the fight suddenly become "more intelligent" than the previous alpha-male that lost the fight? Of course not, the previous alpha male just became weaker physically, or its adversary became stronger with time, or something else changed, showing that nothing is eternal, not even physical force: biological traits change with time, and depend on contingency, randomness, and on numerous other specific circumstances.

An example that is very disturbing, but in my opinion theoretically appropriate, to address such racist and sexist claims that men are "superior" to or "more intelligent" than women because they subjugate them, or that "whites" are superior to "others" because they colonized them, concerns what happened during the Second World War. When Jews were obliged, by physical force, to be confined to Nazi concentration camps, did they have the possibility to do any kind of scientific research that could later be awarded with a Nobel Prize? Of course not, because the Nazis *did not allow them* to use microscopes, laboratories, or any other type of scientific items in those camps. Does this mean that Nazis are "intellectually superior" to Jews? Apart from the fallacy of implying that a group that undertakes such an atrocious genocide is "superior" – similarly to the claims that Europeans are "superior" to others because they colonized and oppressed them - such an argument would also be fallacious because it would neglect the history of Nobel prizes, which sexist and racist people so often use in their biased arguments. This is because, on average, people that define themselves as Jews tend to have a very high ratio of Nobel prizes, per million people. But, obviously, those prizes do *not* refer to scientific work done by Jews in Nazi concentration camps. Similarly, the increased subjugation of women during the rise of sedentism and agriculture allowed men to create new forms of subjugation other than those directly related to the use of physical force, for instance via the creation of religious teleological narratives, social norms and stereotypes about women being "more passive," or "less intelligent," or "not naturally driven or able to govern." As an example, I will list here a few events that happened less than 8 decades ago, after the second world war: 1945, the United States – Harvard Medical School admitted women for the first time; 1955, Qatar – First public school for girls; 1966, Kuwait – University education open to women;

1983, the United States – Columbia College, Columbia University, allowed women to apply for admittance; 2016, Tibet – Women able to take the 'geshe' Tibetan Buddhist academic exams for the first time; 2023, nowadays, Afghanistan – under the *Taliban* women are not allowed to attend secondary school and higher education. Can those women that live currently in Afghanistan truly have a fair chance of winning Nobel Prizes?

An interesting point about this topic is that when I was doing an extensive literature review for my 2021 book *Meaning of Life, Human Nature, and Delusions*, I was particularly focused on the prevailing teleological narratives about love, marriage, and so-called gender differences, more than on potential biological differences between male and female sexual organisms. But within the numerous papers and books that I read for that book, and subsequently for this one, I begun to find a fascinating pattern within the literature: once and once again, empirical evidence was showing that women tend to outperform, on average, men in most of the tasks that were being compared. This was striking to me. Not because of the prevailing misogynistic narratives that are still so common in our societies, as there was a wide range of evidence showing that such narratives were mainly the result of inaccurate biased social constructions. But because one would think that the distribution of so-called biological strengths would be more equal between the two biological sexes, based on what we know about the evolutionary "trend-offs" that are so common in biological evolution. It is obvious that in order to discuss this subject in a more comprehensive way, one would need to compare *all* the biological items that have been studied in and compared between the two sexes. Clearly, this was not the main goal of my previous book, and some people would argue that it would better to not even try to do this anyway, as that could lead to the rise of discussions about the "superiority" of a certain sex, when one should instead focus on how men and women – as well as all groups of humans in general – are actually so strikingly similar. In this sense, the comparative information that I will briefly summarize in the next paragraphs is particularly relevant, precisely because that information is *not at all* the result of *a priori* planning or "cherry-picking." Having said this, it is important to emphasize again that in biological evolution there is truly no "better" or "worse," but mainly just adaptations to a certain place or specific time in history, so the type of differences that I am summarizing here have to be understood in that more holistic view of life, and not in the very simplistic way in which these topics are often discussed.

For instance, one example that seems to reflect a *current* difference between men and women in many countries and types of societies – although this difference is likely also related with local cultural factors – is that women tend, on average, to be more prone to have autoimmune diseases, at least those most commonly found in our species. A misogynistic man could argue that this shows that women are not only physically but also physiologically or medically, "weaker." However, this is precisely the type of cases in which there is a likely a myriad of complex evolutionary trade-offs contributing to this pattern. That is, within a "better" or "worse" fake dichotomy, this would be the "negative" – "too-much" – side of something that actually saves the lives of millions of women across the globe every year: concerning numerous types of dangerous infections that affect humans today, women tend to have a much more robust immune response to, and thus to die less as a result of, them (see also Box 4.5). In other words, if the immune system is "too weak" it might not be effective against dangerous germs and other organisms, but if it is "too strong" it might attack not only those organisms but also our own cells: a further example that in the natural world nothing is "perfect" or the "ideal solution," everything depends on very specific circumstances. Many biological reasons have been proposed by researchers to explain this current difference between women and men, and probably the reality is a combination of at least some of those reasons, so here I will just cite one of them, based on a study published in 2018 by Wilhelmson and colleagues. As summarized for the broader public on the website *ScienceDaily*, in a 2018 piece accordingly entitled "*New theory on why more women than men develop autoimmune diseases*":

> New findings are now being presented on possible mechanisms behind gender differences in the occurrence of rheumatism and other autoimmune diseases. The study, published in *Nature Communications*, can be of significance for the future treatment of diseases. "It's very important to understand what causes these diseases to be so much more common among women", says Asa Tivesten, professor of medicine at Sahlgrenska Academy, Sweden, a chief physician and one of the authors of the study. "In this way, we can eventually provide better treatment for the diseases". In autoimmune diseases, the immune system creates antibodies that attack the body's own tissue. Almost all autoimmune diseases affect women more often than men. The gender difference is especially great in the case of lupus, a serious disease also known as systemic lupus erythematosus or SLE. Nine out of ten of those afflicted are women. It has been known that there is a link between the male sex hormone testosterone and protection against autoimmune diseases. Men are generally more protected than women, who only have one tenth as much testosterone. Testosterone reduces the number of B cells, a type of lymphocyte that releases harmful antibodies.
>
> The researchers behind the study were trying to understand what the connection between testosterone and the production of B cells in the spleen actually looks like, mechanisms that have so far been unknown. After numerous experiments on mice and studies of blood samples from 128 men, the researchers were able to conclude that the critical connection is the protein BAFF, which makes the B cells more viable. "We have concluded that testosterone suppresses BAFF. If you eliminate testosterone, you get more BAFF and thereby more B cells in the spleen because they survive to a greater extent. Recognition of the link between testosterone and BAFF is completely new. No one has reported this in the past", says Asa Tivesten. The results correlate well with a previous study showing that genetic variations in BAFF can be linked to the risk of diseases such as lupus. That disease is treated with BAFF inhibitors, a medicine that has not, however, really lived up to expectations. "That's why this information about how the body regulates the levels of BAFF is extremely important, so that we can continue to put the pieces together and try to understand which patients should have BAFF inhibitors and which should not. Accordingly, our study serves as a basis for further research on how the medicine can be used in a better way."

These topics were the subject of a big fuzz, when I was writing this book during the COVID-19 pandemic, because within the same number of women and men infected with the virus, men were proportionally dying much more (Fig. 4.10). Not surprisingly, at the beginning, most journalists reported that such a pattern was "puzzling" because it contradicted the common *belief* that women are biologically "weaker" than men. By then, the media should have known better, because there was already a huge amount of scientific evidence showing that women are commonly far more resilient in terms of health than men are, particularly concerning infectious diseases precisely because they tend to have, in general, a stronger immune system. But most media sources neglected such evidence, instead preferring to create all types of unscientific narratives – many of them sexist ones – to try to "explain" such a "surprising, puzzling scenario."

For example, Sabra Klein, a Johns Hopkins University Professor that is specialist on gender differences and infectious diseases, was asked to explain this "puzzle," in an interview to the TV channel *France 24* aired the 13th of March 2020. She answered that in general women tend to be medically more resilient than men, so this was not surprising at all. Garima Sharma, also of Johns Hopkins University, who published a paper on sex differences in COVID-19 mortality with other colleagues, argued that one of the specific reasons that might contribute to the physiological toughness of women in general, and against COVID-19 in particular, is the fact that women have a "backup" X chromosome: "X chromosomes contain a high density of immune-related genes, so women generally mount stronger immune responses." But that was clearly not the type of discussion that most of the media platforms were *looking for*, or *wanted to engage in*. So, accordingly, they often ignored the explanations of such scholars and continued to instead discuss another wide range of "true reasons" leading to this "puzzling" pattern, including the "fact" that men are often more risk taking than women, and so on, ignoring the fact that not only most of those "facts" are innacurate but also that they would not explain at all why within a *same number of infected women and men*, men would tend to die more.

This pattern of reality "surprisingly" contradicting myths and stereotypes that have been also propagated by the media is very old, unfortunately. For instance, one well-know case concerning the supposed "natural passivity of women" dates back to the first time that women were allowed to vote in a parliamentary election in a currently existing independent country: namely, in New Zealand, the 28th of November 1893. As reported in the website *New Zealand History* (https://nzhistory.govt.nz/), this election occurred just 10 weeks after the governor, Lord Glasgow, signed the Electoral Act 1893 into law, after a longstanding, hard fight by women, including a third petition calling for them to be allowed to vote. Strikingly – but unfortunately not surprisingly, due to the reasons that have been explained so far – among the about 32,000 signatures that made part of this petition, just 21 are known to have been signed by men. When the time of the election was approaching, the newspaper *The Press* included an editorial that called into question the "interest of women in voting," stating that the vast majority of women would not want to vote as they would naturally prefer to stay at home and do "their domestic tasks." As it is usually the case, such a *belief* was completely deconstructed by empirical facts. Despite the short timeframe for voter registration and the warnings from opponents of women's suffrage that "lady voters" could be harassed at polling booths, 109,461 women – about 84% of the adult female population – enrolled to vote in the election and, on polling day, 90,290 of them casted their votes, a turnout of 82% that was far higher than the 70% turnout among registered male voters.

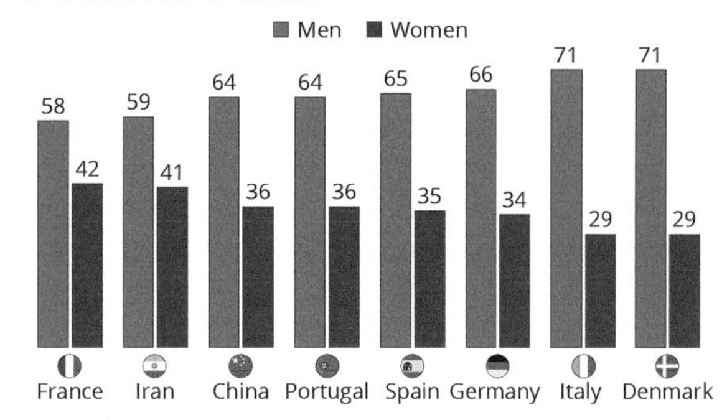

Fig. 4.10 Within the same number of women and men infected with COVID, many more men end up dying, than women

Coming back to the potential *current* biological differences between women and men, apart from – on average – having less physical force, one of the few other cases that seems to be potentially a valid case in which men *usually* outperform women concerns the capacity to handle certain specific spatial tasks, such as three-dimensional orientation. For instance, studies of brain development have shown that boys tend to develop, faster, areas of the brain involved in spatial memory and targeting and, concerning spatial-kinesthetic tasks such as building with blocks or hitting a target with a ball, they do tend to outperform girls, at an earlier age. This topic was discussed in some detail, for example, in a 2018 study by Boone and colleagues, entitled "*Sex differences in navigation strategy and efficiency.*" However, one should also emphasize that when one carefully reads the science behind some of these studies, it is clear that one needs to take them with a grain of salt, particularly the broader conclusions that they often do, which in at least some cases do not seem to be completely independent from prevailing misogynistic narratives and cultural stereotypes. That 2018 study done by Boone and colleagues included tests that mainly concerned navigation in virtual environments, using computers. The problem is that, particularly at younger ages, males tend to be more familiarized to using computers, in part because many parents tend to follow prevailing gender narratives and thus to give computers and videogames more often to boys than to girls. This possibility was addressed and recognized by Boone in an interview that accompanied the publication: "it is also possible that the sex difference in efficiency is due in part to facility with the interface or navigation in virtual environments, as men tend to spend more time playing video games." However, it should be said that, concerning the idea that men might have on average 'better 3D-navigation skills' than women, there are evolutionary reasons to think that this might be truly an accurate difference within at least our recent evolutionary history. As seen earlier, there is indeed empirical evidence that in ancient, as well as various current, non-agricultural societies men often were/are more involved than women in hunting – with exceptions, of course. Such an activity frequently requires the learning and use of complex 3D-navigational skills. Having said that, more studies – ideally aimed to minimize the possible influence of modern cultural stereotypes related to factors such as boys receiving more cars, using more computers and videogames, and so on – are needed to further analyze these topics in a more comprehensive, unbiased way, as acknowledged by Boone.

So, what about the opposite cases, in which women tend to outperform men, which seem to be more numerous according to the studies I have found so far? Before discussing them, one key point should be made: within the vicious cycle of systemic sexism, it is likely that the numerous studies showing cases in which women tend to outperform men were done and published not so much because of *a priori* biases but instead *despite* of them – that is, it is likely that such cases might be even more frequent than suggested by those studies. An illustrative example is that, as noted above, empirical studies consistently show that women tend to experience more complex, elaborate, and intense orgasms than men and to have multiple orgasms more frequently, *despite* the fact that for several decades many prominent evolutionary biologists have been arguing that it is a 'mystery' that women have orgasms at all. As also noted earlier, scientific studies have also shown that women consistently outperform men in smell sensitivity tests, including those regarding scents related to sexual arousal, *despite* the fact that prominent biologists such as stated that women are naturally 'less sexual' and 'coy' than men.

In a 2014 study by Oliveira-Pinto and colleagues about olfactory capacities, they show that there is seemingly a true sexual dimorphism in humans in the sense that females have, on average, more neurons and glial cells in the olfactory bulb than men do. Within the many other empirical case studies in which women tend to outperform men one concerns the ability to read facial expressions, as explained in Hrdy's book *Mothers and Others*. As noted in Shermer's 2004 *The Science of Good and Evil*, "to the extent that lie detection through the observation of body language and facial expressions is accurate (overall not very), women are better at it than men because they are more intuitively sensitive to subtle cues.. in experiments in which subjects observe someone either truth telling or lying, although no one is consistently correct in identifying the liar, women are correct significantly more often than men." This is an example of a point that is not often taken into consideration within the recruitment of intelligence and police officers, which continue to be mostly men. Shermer further added that "women are also superior in discerning which of two people in a photo was the other's supervisor, whether a male-female couple is a genuine romantic relationship or a posed phony one, and when shown a two-second silent video clip of an upset woman's face, women guess more accurately than men whether she is criticizing someone or discussing her divorce." Other examples concern the development of complex cognitive language skills and fine motor skills as well as optimization of brain connections: on average, girls tend to develop all these items faster than boys – see Boyatzis et al. 1993's, Labarthe's 1997, Hanlon et al.'s 1999, and Lenroot et al.'s 2007 works and also Box 4.5 for further examples.

Box 4.5 Gender Stereotypes, Health, and Health Research Biases

Because these topics are so important within the context of the present book, it is worthy to see in some detail what Saini wrote about them in her book *Inferior*:

> We often think of males as being the tougher and more powerful sex. It's true that men are on average six inches taller and have around double the upper body strength of women. But then, strength can be defined in different ways. When it comes to the most basic instinct of all – survival – women's bodies tend to be better equipped than men's. The difference is there from the very moment a child is born. 'When we were there on the neonatal unit and a boy came out, you were taught that, statistically, the boy is more likely to die', explains Joy Lawn. Besides her academic research into child health, she has worked in neonatal medicine in the United Kingdom and as a paediatrician in Ghana. The first month following birth is the time at which humans are at their greatest risk of death. Worldwide, a million babies die on the day of their birth every year. But if they receive exactly the same level of care, females are statistically less likely to die than males. Lawn's research encompasses data from across the globe, giving the broadest picture possible of infant mortality. And having researched the issue in such depth, she concludes that boys are at around a 10 per cent greater risk than girls in that first month – and this is at least partly, if not wholly, for biological reasons. Thus, in South Asia, as elsewhere in the world, the mortality figures should be in favour of girls. The fact that they're not even equal, but are skewed in favour of boys, means that girls' natural power to survive is being forcibly degraded by the societies they are born into. 'If you have parity in your survival rates, it means you aren't looking after girls', says Lawn. 'The biological risk is against the boy, but the social risk is against the girl'. Elsewhere, child mortality statistics bear this out. For every thousand live births in subSaharan Africa, ninety eight boys compared with eighty six girls die by the age of five. Research Lawn and her colleagues published in the journal *Pediatric Research* in 2013 confirmed that a boy is 14 per cent more likely to be born prematurely than a girl, and is more likely to suffer disabilities ranging from blindness and deafness to cerebral palsy when he's at the same stage of prematurity as a girl. In the same journal in 2012 a team from King's College London reported that male babies born very prematurely are more likely to stay longer in hospital, to die, or to suffer brain and breathing problems.
>
> 'I always thought that it was physically mediated, because boys are slightly bigger, but I think it's also biological susceptibility to injury', says Lawn. One explanation for more boys being born preterm is that mothers expecting boys are, for reasons unknown, more likely to have placental problems and high blood pressure. Research published by scientists from the University of Adelaide in the journal *Molecular Human Reproduction* in 2014 showed that newborn girls may be healthier on average because a mother's placenta behaves differently depending on the sex of the baby. With female foetuses, the placenta does more to maintain the pregnancy and increase immunity against infections. Why this is, nobody understands. It could be because, before birth, the normal human sex ratio is slightly skewed towards boys. The difference after birth might simply be nature's way of correcting the balance. But the reasons could also be more complicated. After all, a baby girl's natural survival edge stays with her throughout her entire life. Girls aren't just born survivors, they grow up to be better survivors too. 'Pretty much at every age, women seem to survive better than men', confirms Steven Austad, chair of the biology department at

(continued)

the University of Alabama at Birmingham, who is an international expert on ageing. He describes women as being more 'robust'. It's a phenomenon so clear and undeniable that some scientists believe understanding it may hold the key to human longevity. At the turn of the millennium, Austad began to investigate exactly what it is that helps women outlive men at all stages of life. 'I wondered if this is a recent phenomenon.. is this something that's only true in industrialised countries in the twentieth century and twenty-first century?' Digging through the Human Mortality Database, a collection of longevity records from around the world founded by German and American researchers in 2000, he was surprised to discover that the phenomenon really does transcend time and place. The database now covers thirty-eight countries and regions. Austad's favourite example is Sweden, which has kept some of the most thorough and reliable demographic data of any country. In 1800 life expectancy at birth in Sweden stood at thirty three years for women and thirty one for men. In 2015 it was around eighty three for women and around seventy nine for men. 'Women are more robust than men.. I think there's little doubt about that', Austad says. 'It was true in the eighteenth century in Sweden, and it's true in the twenty-first century in Bangladesh, and in Europe, and in the US'.

I ask Austad whether women might be naturally outliving men for social reasons. It's reasonable to think, for instance, that boys are generally handled more roughly than girls are. Or that more men than women take on risky jobs, such as construction and mining, which also expose them to toxic environments. And we know that in total across the world, far more men than women smoke, which dramatically pushes up mortality rates. But Austad is convinced that the difference is so pronounced, ubiquitous and timeless that it must mean there are features in a woman's body that underlie the difference. 'It's hard for me to imagine that it is environmental, to tell you the truth', he says. The picture of this survival advantage is starkest at the end of life. The Gerontology Research Group in the United States keeps a list online of all the people in the world that it has confirmed are living past the age of 110. I last checked the site in July 2016. *Of all these 'supercentenarians' in their catalogue, just two were men.* Forty six were women. 'It's a basic fact of biology', observes Kathryn Sandberg, director of the Center for the Study of Sex Differences in Health, Aging and Disease at Georgetown University in Washington, DC, who has explored how much of a role disease has to play in why women survive. 'Women live about five or six years longer than men across almost every society, and that's been true for centuries.. first of all, you have differences in the age of onset of disease.. so, for example, cardiovascular disease occurs much earlier in men than women.. the age of onset of hypertension, which is high blood pressure, also occurs much earlier in men than women.. there's also a sex difference in the rate of progression of disease.. if you take chronic kidney disease, the rate of progression is more rapid in men than in women'. Even in laboratory studies on animals, including mice and dogs, females have done better than males, she adds. By picking through the data, researchers like her, Joy Lawn and Steven Austad have come to understand just how widespread these gaps are. 'I assumed that these sex differences were just a product of modern Westernised society, or largely driven by the differences in cardiovascular diseases', says Austad. 'Once I started investigating, I found that women had resistance to almost all the major causes of death'. One of his papers shows that in the United States in 2010, women died at lower rates than men from twelve of the fifteen most common causes of death, including cancer and heart disease, when adjusted for age. Of the three exceptions, their likelihood of dying from Parkinson's or stroke was about the same. And they were more likely than men to die of Alzheimer's Disease.

(continued)

When it comes to fighting off infections from viruses and bacteria, women also seem to be tougher. 'If there's a really bad infection, they survive better.. If it's about the duration of the infection, women will respond faster, and the infection will be over faster in women than in men', says Kathryn Sandberg. 'If you look across all the different types of infections, women have a more robust immune response'. It isn't that women don't get sick. They do. They just don't die from these sicknesses as easily or as quickly as men do. One explanation for this gap is that higher levels of oestrogen and progesterone in women might be protecting them in some way. These hormones don't just make the immune system stronger, but also more flexible, according to Sabine Oertelt-Prigione, a researcher at the Institute of Gender in Medicine at the Charité University Hospital in Berlin. 'This is related to the fact that women can bear children', she explains. A pregnancy is the same as foreign tissue growing inside a woman's body that, if her immune system was in the wrong gear, would be rejected. 'You need an immune system that's able to switch from pro-inflammatory reactions to anti-inflammatory reactions in order to avoid having an abortion pretty much every time you get pregnant.. the immune system needs to have mechanisms that can, on one side, trigger all these cells to come together in one spot and attack whatever agent is making you sick.. but then you also need to be able to stop this response when the agent is not there any more, in order to prevent tissues and organs from being harmed'. The hormonal changes that affect a woman's immune system during pregnancy also take place on a smaller scale during her menstrual cycle, and for the same reasons. 'Women have more plastic immune systems.. they adapt in different ways', says Oertelt-Prigione.

In 2011 health researcher Annaliese Beery at the University of California, San Francisco, and biologist Irving Zucker at the University of California, Berkeley, published a study looking into sex biases in animal research in one sample year: 2009. Of the ten scientific fields they investigated, eight showed a male bias. In pharmacology, the study of medical drugs, the articles reporting only on males outnumbered those reporting only on females by five to one. In physiology, which explores how our bodies work, it was almost four to one. It's an issue that runs through other corners of science too. In research on the evolution of genitals (parts of the body we know for certain are different between the sexes), scientists have also leaned towards males. In 2014 biologists at Humboldt University in Berlin and Macquarie University in Sydney analysed more than three hundred papers published between 1989 and 2013 that covered the evolution of genitalia. They found that almost half looked only at the males of the species, while just 8 per cent looked only at females. One reporter, Elizabeth Gibney, described it as 'the case of the missing vaginas'. When it comes to health research, the issue is more complicated than simple bias. Until around 1990, it was common for medical trials to be carried out almost exclusively on men. There were some good reasons for this. 'You don't want to give the experimental drug to a pregnant woman, and you don't want to give the experimental drug to a woman who doesn't know she's pregnant but actually is', says Arthur Arnold. The terrible legacy of women being given thalidomide for morning sickness in the 1950s proved to scientists how careful they need to be before giving drugs to expectant mothers. Thousands of children were born with disabilities before thalidomide was taken off the market. 'You take women of reproductive age off the table for the experiment, which takes out a huge chunk of them', continues Arnold. A woman's fluctuating hormone levels might also affect how she responds to a drug. Men's hormone levels are more consistent. 'It's much cheaper to study one sex.. so if you're going to choose one sex, most people avoid females because they

(continued)

have these messy hormones.. so people migrate to the study of males.. in some disciplines it really is an embarrassing male bias'. This tendency to focus on males, researchers now realise, may have harmed women's health. 'Although there were some reasons to avoid doing experiments on women, it had the unwanted effect of producing much more information about how to treat men than women', Arnold explains. A 2010 book on the progress in tackling women's health problems, cowritten by the Committee on Women's Health Research, which advises the National Institutes of Health (NIH) in the USA, notes that autoimmune diseases – which affect far more women than men – remain less well understood than some other conditions: 'despite their prevalence and morbidity, little progress has been made toward a better understanding of those conditions, identifying risk factors, or developing a cure'.

In the *Fate of Gender*, Browning notes that "not only do little girls excel in preschools more quickly than little boys, but realms of data show that women's mental agility persists seriously longer than men's as everyone's life expectancy grows longer." An interesting point made by Browning, which should be studied in further detail, is that the better grades of women in school, including at a young age, are not necessarily related to an inferior cognitive capacity of men. He notes: "researchers found after examining 5800 students from kindergarten through 5th grade.. no particular evidence of inferior cognitive capacity among the boys.. the boys were simply more fidgety, less attentive, and less 'eager to learn', and their teachers, stressed by larger class sizes, graded them down for their comportment – a track that followed them on through high school." On the other hand, he explains that "MRI imaging.. [shows] that the ventral prefrontal cortex.. well known to relate to social awareness and interpersonal response, was in fact about 10% larger in women than in men." However, the same researchers found that "little boys actually had (a) larger (ventral prefrontal cortex) than the little girls but that this *smaller* (size) correlated with greater interpersonal activity – exactly the opposite of what they found in the sixty adults." That is, "their investigations showed a far muddier portrait of what makes men masculine and women feminine, and as they age not only do their interests, behavior, and mannerisms change, but so too can their neural biology change." This illustrates the very complex links between "nature" and "nurture." As put by Browning, "as neurologist Lise Eliot, author of *Pink brain, blue brain*, commented in a review of [the above] work, 'individuals gender traits – their preferences for masculine or feminine clothes, careers, hobbies and interpersonal styles – are inevitably shaped more by rearing and experience than by their biological sex." Likewise, "their brains, which are ultimately producing all this masculine or feminine behavior, must be molded – at least to some degree – by the sum of their experiences as a boy or a girl'.. we are all of us both male and female, and the way we express our 'masculinity' and 'femininity' depends on the *circumstances* in which we find ourselves living – and moreover those experiences can alter our neurobiology and physiology."

The power of plasticity and epigenetics, and particularly hormones, and how they change during lifetime, was also noted in Ryan's *The Virility Paradox*: "does it mean that larger brains make boys smarter than girls? Nope. In fact, having more testosterone and more active androgen receptors during adolescence may actually have a negative impact on brain function.. tests on teenage male subjects.. found .. a significant higher rate of depression and suicidal thinking in boys with a combination of higher testosterone." As noted above, women tend to be, on average, more resilient than men in terms of health in general, a difference that starts at a very early age, as discussed in Box 4.5. Strikingly, just a few weeks after gestation, when they are still in the wombs of their mothers, females already tend to get less sick and to have a quicker/stronger homeostatic response when exposed to conditions similar to those to which males are exposed. Women tend to have a higher life expectancy in almost every country, including those where they are highly oppressed, subjugated, and neglected (Fig. 4.11). Similarly, even in countries where education possibilities are commonly less available for women, they tend to be better students than men, on average, in most areas of knowledge, including mathematics. They also tend to outperform men in multitasking tests: something that would be expected even under the prevalent gender stereotypes and narratives, because according to them women are *expected to* be better at taking care of kids and at performing multiple types of domestic tasks. As noted in Browning's *The Fate of Gender*, "most male brains tend to have as much as 7 times more 'gray matter' associated with concentration

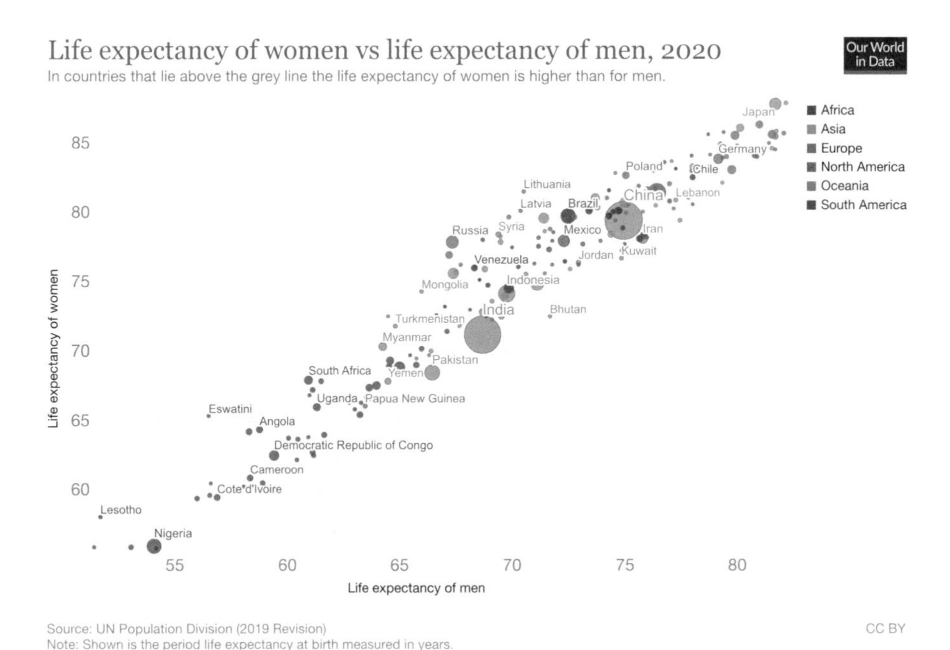

Fig. 4.11 A major fallacy of the "weaker sex" narrative: women have, in almost all countries, an average higher life expectancy than men

attention on a specific action or task while female brains can have ten times more white matter – associated with coordination between gray matter centers and, hypothetically, with women's alleged capacity to switch between tasks and to multitask, advantages that can be beneficial in infant nurturing."

More surprising, for those that *believe in* sexist narratives and stereotypes about "hysterical women," is that in *reality* women also tend to be less emotional and stressed than men in the sense that, when exposed to similar types of external stress, they often react in a less stressful and more pragmatic way than men tend to do. Such an "hysterical women" narrative is, together with the "coy asexual women" tale, among the sexist tales created by men that have been most interiorized by women, probably because both tales have been supported by scientific "giants" such as Darwin and Freud and continue to be so commonly repeated in movies, TV series, novels, and by the press in general. As explained above, one of the most tragic repercussions of systemic sexism and racism is precisely that those oppressed and subjugated often start internalizing the tales created by their oppressors. A powerful and profoundly sad illustration of this is the fact that, when African American girls and boys are asked to choose between a "white" doll and a "black" doll, or when asked about which of the two dolls is "better," they tend to choose the "white" one (see Fig. 4.12).

In Browning's book *The Fate of Gender*, as well as in a 2015 article that he wrote in the *California Magazine* of the *Cal Alumni Association of UC Berkeley*, he refers to the work of numerous researchers that study gender differences, particularly

Fig. 4.12 What systemic racism looks like

concerning resilience. It is fascinating to see the reasons provided by some of those researchers – from very different backgrounds and fields of science – to explain gender dissimilarities and how epigenetic factors are extremely important within discussions on the links between the so-called nature and nurture, as well as between the so-called mind and body. The 2015 paper states:

> Andy Scharlach, a UC Berkeley professor of aging and director of its Center for the Advanced Study of Aging Services, says.. [that] men, whatever age.. tend to blather about safe topics circling over and around their tender anxieties, while women are ever more pragmatic and direct about the foibles and frustrations brought on by the passage of time. Over and over again, Scharlach's research has shown that women generally retain far more resilience as they age than men. Biological difference and genetic inheritance clearly play important roles in our health as we age, but resilience, an admittedly fuzzy concept, can also affect our biological response to stress, and therefore to both cardio illness and cognitive failure. Where does resilience come from? It begins very early, he says – often even before boys and girls learn to read – "you become resilient by dealing with small-scale stressors that you're able to learn from.. women have many more opportunities to do that in their lives than men do, in part because they have more exposure to the stresses that come from being excluded from the privileges that come automatically to little boys; and that continues throughout women's lives as they carry different burdens and expectations from men.. women still carry more child rearing responsibilities.. they carry more of the emotional load in families.. the gender biases that exist either beat you down, or you develop a sense of yourself and others as being OK". A second source of female resilience concerns what many sociologists have noticed in gender relations across the lifespan. Says Scharlach: "women develop richer social networks than men that are not as work bound, and not as sports bound, or activity bound".
>
> Susan Folkman.. spent the last decade of her professional life as a distinguished professor of medicine at the University of California, San Francisco. Her first major research concerned how men learned to take care of each other during the worst years of the AIDS epidemic. From there she looked at the kinds of people who learn and succeed in caregiving – especially as they age. "From a very early age", she notes, "boys are indoctrinated with the athletic metaphor: you don't give up.. you keep going after that success.. you fight for it.. you don't take a second.. you just fight harder.. I don't think women are brought up with that metaphor". Nearly all the women understood threat and challenge as physical danger, something that had to be mastered, but more importantly, something that had to be understood. "I think all that's very embedded very early in males and females" she says. And there is something more profound at work: a marked difference in spousal mortality rates following the death of a partner. A man is much more likely to die in the first six months following the death of his wife than is a woman following the death of her husband. While there are slight differences concerning race and ethnicity in what's called the "widowhood effect," the greater death toll among men appears to be a worldwide phenomenon. Janice Schwartz, another gerontologist at UCSF, has also focused on gender gaps in health, longevity and caregiving. She is even more convinced of the negative consequences that result from conventional men's inability to form close friendships and strong social networks following retirement. Schwartz's research extended well beyond the upper-middle-class-territory of doctors, businessmen and technologists. She spent several years interviewing and following aging residents of a residential retirement trailer park north of San Francisco in Sonoma County, where she found the same behavior.

Browning's book *The Fate of Gender* provides further fascinating details about these topics:

> Boys born prematurely are 1.7 times less likely to survive than premature girls.. A Canadian study of mothers living in a chemically polluted area found that male embryos were far less likely to survive through gestation than female embryos.. [Susan] Pinker wrote in her 2009 book *The sexual paradox* [that] cognitive and attention deficit disorders.. were from 4 to 10 times more common in boys than in girls. Men develop cardiovascular disease on average 7 to 10 years earlier than women do. Men have strokes much more often than women – even though women strokes seem to be more severe. Cancer numbers are still worse: 4.5 to 5.5 times more throat and mouth cancers in men than in women and 3.3 to 1 for bladder cancer; 2.3 to 1 for lung cancer deaths and about the same for liver cancer. The notion that males as the tougher and sturdier to the two human animals is highly suspect.

In a nutshell, it can be said that the available scientific data do show that "gender roles" in particular, and to a certain extent also the so-called gender differences in general, are mainly socially constructed – a clear example is the stereotype that "kids like blue, girls like pink" (see Box 4.6). In contrast, biological differences between males and females concerning for instance sexual and physiological aspects often have a very deep evolutionary history. This concerns not only anatomical traits such as the shape of their genitals but also *certain* responses to infections or diseases or even *certain* mental abilities or at least propensities, as we have seen. For example, the fact that men seemingly tend to outperform women in 3-D navigation tasks could be partially the result of "nurture" but might well also be an evolutionary biological result of the fact that men most likely were traditionally more – but not exclusively – involved in hunting, for a long time during our evolutionary history. Similarly, the fact that women, *on average*, tend to be more resilient in terms of health, less emotional, and outperform men at multitasking, might be directly related to childcare and mother–child interactions. Within our closest living relatives, the great apes, males do tend to contribute much less than females to childcare. Archeological, ethnographic, and paleontological data about our human lineage indicate that this has happened for a long time and still does happen within most hunter-gatherer groups, as it does generally within agricultural societies. This means that if an adult mother died, or got very sick, or was exceedingly stressed, this could put in direct danger the life, or at least the well-being, of her children. This would happen even if other women would help taking care of those children, as they probably did, following Hrdy's concept of "mother and others" – that is, that babies and infants were mostly taken care of by the mother and by other women around them. Based on empirical data collected from studies of nomadic hunter-gathered societies, we know that the same does not often apply, at least so directly, to the father: in many of these societies, a diseased or even completely absent father does not often affect in a major way the well-being of his children, as noted above. In other words, contrary to a sick or absent father, a mother being very sick, or dying, would very likely lead to her children also becoming weak or sick, or even dying, thus directly affecting their subsequent reproduction and/or survival. Therefore, in the long run, natural selection tends to favor those cases in which the females are more resilient as well as less stressed in general.

This is a crucial point to better understand human evolution, and biological evolution in general: in a way, women were mainly the subject of a kind of "double selection," because if something serious happened to them it would often also strongly affect their children, while men were mostly the subject of a "single selection." This very likely does not apply only to humans but also to other primate and many other mammalian taxa in which the father does not contribute so much to raising babies and infants. Indeed, there are data showing that at least within various other species of mammals, females tend to also be more resilient than males, in terms of health. This idea is also supported, in a way, by several examples given in Klarsfeld and Revah's 2003 book *The Biology of Death*, in which males of several species tend to die earlier than females, in many cases literally after reproducing with them – within some of them, the males being literally killed, and even eaten, by the females. This makes sense biologically and evolutionarily in the sense that in general the females obviously need to live for at least some time after reproduction in order to lay their eggs or deliver their progeny and, in many taxa such as humans, to then take care of that progeny. For example, some "queen bees" use the sperm of the dead males to produce, during *various years* after their death, the "bee workers" that are needed to keep the beehives. Another well-known example is the southern black widow, *Latrodectus mactans*: the females of this venomous spider species occasionally eat their mate after reproduction. As noted by Klarsfeld and Revah's, contrary to popular culture, these females are actually *not* the champions of erotic cannibalism, but they do tend to live *on average* 10 times longer than the males. When we see Fig. 4.13 and know that "black widow" female spiders not only are much larger than males but also are the ones that are poisonous and that bite and in addition sometimes kill and eat the males, and that tend to live much longer than them, one realizes again how misogynistic tales about the "natural" "superiority" of males and of men in particular are actually not based at all in what often truly happens in the natural world.

Fig. 4.13 "Black Widow"

Box 4.6 Self Illusion, Gender Stereotypes, Blue and Pink, and Child Development

In his book *The Self Illusion*, Hood explains:

> Although not cast in stone, gender stereotypes do tend to be perpetuated across generations. Many parents are eager to know the sex of their children before they are born, which sets up gender expectations such as painting the nursery in either pink or blue. When they eventually arrive, newborn baby girls are described mainly in terms of beauty, whereas boys are described in terms of strength. In one study, adults attributed more anger to a boy than to a girl reacting to a jack-in-the-box toy even though it was always the same infant. Parents also tend to buy gender-appropriate toys with dolls for girls and guns for boys. In another study, different adults were introduced to the same child wearing either blue or pink clothes and told that it was either Sarah or Nathan. If adults thought it was a baby girl, they praised her beauty. If they thought it was a boy, they never commented on beauty but rather talked about what occupation he would eventually have. When it came to play, they were boisterous with the boy baby, throwing him into the air, but cuddled the baby when they thought it was a girl. In fact, the adults seemed to need to know which sex the baby was in order to play with them appropriately. Of course, it was the same baby, so the only difference was whether it was wearing either blue or pink. It is worth bearing in mind the association of the colour blue is only recent – a hundred years ago it would have been the boys wearing pink and the girls wearing blue. With all this encouragement from adults during the early months, is it any surprise that, by two years of age, most children easily identify with their own gender and the roles and appearances that they believe are appropriate? However, this understanding is still very superficial. For example, up until four years of age, children think that long hair and dresses determine whether you are a boy or girl. We know this because if you show four-year-olds a Ken Barbie Doll and then put a dress on the male doll, they think that he is now a girl. By six years, children's gender understanding is more sophisticated and goes over and beyond outward appearances. They know that changing clothes and hair does not change boys into girls or vice versa. They are already demonstrating an understanding of what it means to be essentially a boy or a girl. When they identify gender as a core component of the self, they will tend to see this as unchanging and foundational to who they and others are.
>
> As children develop, they become more fixed in their outlook about what properties are acquired and what seem to be built in. For example, by six years, children think that men make better mechanics and women are better secretaries. Even the way parents talk to their children reinforces this generalized view of what is essential to gender. For example, parents tend to make statements such as 'boys play soccer' and 'girls take ballet' rather than qualifying the statements with 'some boys play soccer' or 'some girls take ballet'. We can't help but fall into the gender trap. Our interaction with children reinforces these gender divisions. Mothers tend to discuss emotional problems with their daughters more than with their sons. On a visit to a science museum, parents were three times more likely to explain the exhibits to the boys than to the girls. And it's not just the parents. Teachers perpetuate gender stereotypes. In mixed classes, boys are more likely to volunteer answers, receive more attention from teachers and earn more praise. By the time they are eight to ten years old, girls report lower self-esteem than boys, but it's not because they are less able. According to 2007 UK National Office of Statistics data, girls outperform boys at all levels of education from preschool right through to university. There may be some

(continued)

often-reported superior abilities in boys when it comes to mathematics but that difference does not appear until adolescence, by which time there has been ample opportunity to strengthen stereotypes. Male brains are different to female brains in many ways that we don't yet understand (for example, the shape of the bundle fibres connecting the two hemispheres known as the corpus callosum is different), but commentators may have overstated the case for biology when it comes to some gender stereotypes about the way children should think and behave that are perpetuated by society.

Stereotypes both support and undermine the self illusion. On the one hand most of us conform to stereotypes because that is what is expected from those in the categories to which we belong and not many of us want to be isolated. On the other hand, we may acknowledge the existence of stereotypes but maintain that as individuals we are not the same as everyone else. Our self illusion assumes that we could act differently if we wished. Then there are those who maintain that they do not conform to any stereotypes because they are individuals. But who is really individual in a species that requires the presence of others upon which to make a relative judgment of whether they are the same or different? By definition, you need others to conform with, or rebel against. Consider another universal self stereotype – that of male aggression. Why do men fight so much? Is it simply in their nature? It's an area of psychology that has generated a multitude of explanations. Typical accounts are that males need physically to compete for dominance so that they attract the best females with whom to mate, or that males lack the same negotiation skills as women and have to resolve conflicts through action. These notions have been popularized by the 'women are from Venus, men are from Mars' mentality. It is true that men have higher levels of testosterone and this can facilitate aggressive behaviour because this hormone makes you stronger. But these may be predispositions that cultures shape. When we consider the nature of our self from the gender perspective, we are invariably viewing this through a lens, shaped by society, of what males and females should be. Males may end up more aggressive but surprisingly they may not start out like that. Studies have shown equal levels of physical aggression in one-year-old males and females, but by the time they are two years of age, boys are more physically aggressive than girls and this difference generally continues throughout development. In contrast, girls increasingly rely less on physical violence during conflicts but are more inclined to taunting and excluding individuals as a way of exerting their influence during bullying. Males and females may simply differ in the ways in which they express their aggression.

As we did for Chap. 3, in this section about what we now know concerning the type of biological and evolutionary issues discussed in Darwin's writings about women, I will briefly refer to the recent – 2021 – book *A Most Interesting Problem: What Darwin's Descent of Man Got Right and Wrong About Human Evolution* edited by DeSilva. Only one chapter of that book, authored by Holly Dunsworth, focuses particularly on Darwin's sexism, but that chapter is actually one of the few that has no problem in pointing out the inaccuracies of Darwin's writings. The chapter is particularly powerful not only because of this, but because Dunsworth links this topic with broader aspects of systemic sexism, scientific biases, and popular culture:

At a glance, Chapters 19 and 20 of *Descent of Man* look like Victorian Age–appropriate evolutionary explanations for skin and hair differences between the sexes and among peoples across the globe, and for how our naked, colorful skin developed out of our ape ancestry. But underneath Darwin's ambitious scientific contribution lies much more than surface beauty. This is Darwin begetting every caveman-inspired nugget of dating advice, every best-selling author's stance on innate gender roles, and every entertainer's sexist appeal to science. This is where Darwin first turned his concept of sexual selection loose on humans, launching the evolutionary narrative starring "the strongest and boldest men.. in contests for wives" that dominates popular culture. In *Descent*, Darwin parlayed visible anatomical differences between sexes – like those in skin color, hair patterns, and height – into the evolutionary "logic" behind why Man and Wife perform differently, in matters of love, sex, parenting, cognitive feats, and seemingly everything else, according to the contemporary world views he continues to shape. But pop culture has been slow to adopt new knowledge that has complicated and overturned old facts. Many of the novel insights on human evolution have come from women, despite persistent beliefs like Darwin's that, "with woman the powers of intuition, of rapid perception, and perhaps of imitation, are more strongly marked than in man; but some, at least, of these faculties are characteristic of the lower races, and therefore of a past and lower state of civilisation". In *Descent*, Darwin even concludes, as if by scientific logic, that "man has ultimately become superior to woman". For Darwin, women were wives, but men were so much more than husbands, an assumption seeding his science of sex differences. From his view of life, Darwin penned nature's seal of approval.. In the Darwinian context [regarding sexual selection], Western notions of masculinity and patriarchy are justified because gendered behavior is deemed the driving force of evolution – a process that Darwin valued as progress and improvement, and our culture still does..

Science is the human creation of knowledge. Prizing the completely objective scientific mind is foolish because such a mind does not exist. This myth is dangerous because it creates opportunity for scientists to exploit humanity.. As biological anthropologist Robin Nelson writes, "the idealization of an objective and apolitical science built on rational thought and deliberation has a face, and that face is white and male". An uncritical belief in scientific objectivity does not just continue to dominate the profession, it influences the public perception of what counts as science and what is valued. Too often when these issues are raised, critics, who may even be scientists themselves, are labeled "anti-science" or "science deniers" or considered to be too politically motivated to be taken seriously as scientists. Just going in nontraditional scientific directions in a nontraditional body can result in a person's rejection from science. On these issues, indigenous-studies scholar Kim Tallbear writes, "Being differently situated is advantageous for producing different insights but has its risks. when one fails to exemplify a white Western often straight and masculinist gaze that is ironically seen to embody 'objectivity', or if one researches too close to home, one gets accused of bias". The persistent myth of an unblemished science is what tricks us into believing that Darwin's ideas, like his harmful beliefs about women, can be challenged only with science and, ironically, by those scientists deemed to be objective about such issues: men.

Dunsworth then provides an interesting brief summary of some specific case studies that contradict misogynistic ideas postulated in Darwin's *Descent*, as well as some of Darwin's broader theories about biological evolution in general, such as his depiction of the natural world being mostly characterized by an incessant fierce war-like struggle-for-existence and by competition. It is worthwhile to finish this Chap. 4 with some of Dunsworth's excerpts about this topic and about the need to change the *status quo* concerning the subjugation, oppression, or exclusion of women based on such inaccurate tales:

For both natural and sexual selection, Darwin's view of life was couched in competition, especially as he saw humans.. "Darwinian" has become a synonym for competition. While it may be fitting, it is misleading about evolutionary theory both in Darwin's time and now.. cooperation [is] fundamental to evolution.. Competition has dominated the scientific narrative.. Complicating or potentially overturning Darwinian views of competition and dominance can take thousands of hours of careful primatological observations. Rebecca Lewis has demonstrated just how complex power dynamics can be in primates, especially with regard to females. Her work has also revealed just how difficult competition and dominance are to study, because of issues of quantification and human bias. From her fresh perspective, Lewis has observed how dominance can be expressed differently between the sexes. She noticed that "refusing to act is a means of power that is often overlooked in primate studies and may be a more common means of intersexual power used by females than is aggression". This is the case in primate species with considerable sex differences in body size, in which males are often assumed to have disproportionate power by observers who have ignored all but aggressive displays.. As one of the earliest scientists to overturn Victorian assumptions about female primates, Sarah Hrdy warned us all: "Chauvinists of both sexes have dipped into the primate literature to document their positions". That we can string the same data together with completely different narratives is demonstration not of nature's duplicity but of its complexity, as is increasingly understood by scientific progress. But this progress was delayed.. In the late nineteenth and early twentieth centuries, American opposition to female higher education prevented women from entering the same knowledge-creation business that built, claimed, and perpetuated Darwin's work – which conveniently suggested that women weren't suited for higher education.. Just after its publication, readers interpreted *Descent of Man* as a call to action to contribute to the evolutionary process by aiding selection's improvement of our species. Where there was once God's plan that we must carry out, now there was selection's. Talk about stringing the same data together with two different narratives. From this view, cutting edge science justified limiting the freedom of all but upper-class white men. Women evolved to be wives (and not scientists or scholars) and to carry out evolution's plan. Natural and sexual selection conveniently favored what society already did. *The scientific value of Descent is impossible to untangle from the oppression that it inspired..*

Anthropologist Venla Oikkonen describes the power of narrative as "one of the primary mechanisms through which assumptions about biological gender differences are produced as scientific fact in popular discourses.. if narrative naturalizes ideas and experiences, then narratives of gender turn gendered assumptions into gendered facts". Gender is culture, and culture has a talent for writing itself out of the story, for naturalizing itself, in deference to its preference for nature.. It is largely thanks to candid and diligent scientific study of animals like bonobos that we can bust Darwinian myths that women (but not men) are naturally passive, coy, and sexless, while men (but not women) are naturally big, strong, arrogant, sex machines. In one of the most recent endeavors to expose just how tenuous and otherwise flat out wrong many of science's and society's assumptions about human sexuality have been, [writter Wednesday] Martin jokes about women's libidos to make a serious point: "What seems beyond dispute is that women are.. super freaks.. we are sexual anarchists.. we might quite fairly be described as the largest group of perverts in America". This new evolutionary script inspires a question: If people of genders, sexualities, races, ethnicities, classes, cultures, and life experiences different from Darwin's, if women like those highlighted in this chapter, had been included in the scientific pursuit of knowledge since the beginning, would there have been a *Descent of Man*? .. The unscientific bias in *Descent of Man* that Darwin legitimized as science, and that plagues us all, may be unavoidable, but it is surmountable.. For far too long, science has excluded the lives that could make this view of life even dare to approximate the truth. What's more, we owe it to our species to break the link between Darwin's inchoate offerings and their perceived social implications. Outdated science-inspired narratives alienate difficult for many people to claim it for themselves. Inclusivity, all of humanity, is the only path toward a less false view of life, with everyone and for everyone. That's just one wife's view. What's yours?

Chapter 5
Bringing Reality to Society and Science

> *The saddest aspect of life right now is that science gathers knowledge faster than society gathers wisdom.*
>
> *(Isaac Asimov)*

One of the most accurate, balanced, non-idealized summaries about Darwin, and about the confusion that often plagues both the scientific literature and the popular culture ideas about him and his writings, was given in Quammen's *The reluctant Mr. Darwin*:

> Charles Darwin holds a peculiar position in the history of science and society. His name is a household word but his ideas - with a single exception - aren't household ideas. He's central, he's iconic, but that's not to say that he's widely and well understood. If the scientific community issued bank notes, true enough, the face on the dollar bill would be Darwin's. It's a good face, an amiably stolid face.. yet it conceals.. deep veins of complexity and tension. Everyone knows something about who Darwin was, what he did, what he said, and the thing that most people think they know is: he concocted "the theory of evolution". This isn't quite wrong, just confused and imprecise, but it misses those points about Darwin's work that are most profoundly original, and dangerous, and thrilling. Both as hero and as bugaboo, Darwin is taken for granted in a way that Copernicus, Kepler, Newton, Linnaeus, Charles Lyell, Gregor Mendel, Albert Einstein, Marie Curie, Niels Bohr, Werner Heisenberg, Alfred Wegener, Frederick Hubble, James Watson, and Francis Crick are not.
>
> One measure of his supposed familiarity is the careless use, within common discourse, of the terms "Darwinism" and "Darwinian", which presume at reducing to trademark simplicity a diverse body of work that can't be so easily reduced. Forget about Darwinism, it doesn't exist. Not unless you define it by arbitrary stipulation - these concepts included, those concepts not - in a way that Darwin himself never did. And what is Darwinian? Well, a fascination with fancy pigeons is Darwinian, in the sense that our man, during one period, became entranced by his aviary full of pouters and fantails and runts. A fondness for long solitary strolls, not far from home, is Darwinian.. My point is this: Charles Darwin didn't found a movement or a religion. He never assembled a creed of scientific axioms and chiseled them onto a stone tablet beneath his own name. He was a reclusive biologist who wrote books. Sometimes he made mistakes. Sometimes he changed his mind. Sometimes he worked on little subjects and sometimes on big ones. True, most of his published writings

R. Diogo, *Darwin's Racism, Sexism, and Idolization*,
https://doi.org/10.1007/978-3-031-49055-2_5

share a single underlying theme - the unity of all life, reflecting the processes of evolution. But he particularized that theme in a variety of concepts, some of which interlock nicely and remain valuable to biology, some of which don't. It's better to examine his ideas individually than to try to bundle them as a *brand*.

Still, many historians and scientists continue to precisely do this: to refer to Darwin's ideas as if they were part of a "brand," or a "movement," with Darwinism often seen as infallible or irrefutable. Darwin never explicitly tried to create such a movement and much less a Darwinian religion. As noted in Chap. 1, in this sense what happens today with Darwin and his ideas is somewhat similar to the idolization of Aristotle – by both Christian and Muslim scholars – that occurred for many centuries after Aristotle's death, before the Renaissance. In this sense, we badly need a new "Renaissance." That is, we need to be able to openly put in question the idolization of Darwin, and the blind acceptance of his ideas as if they were the Word of a deity, without being promptly accused of being "heretics" or traitors that are taking the side of creationists. As it occurred with Darwin, Aristotle's idolization was not merely about his scientific ideas per se, but about the fact that his *beliefs and biased views about the cosmos, about women, or about slaves*, were music to the ears of many of those that were in power during those centuries before the Renaissance. For instance, Aristotle's *belief* that the Earth was the center of the World and that the sun moved around it, as well as some of his writings suggesting that there was a single God – an "unmoved mover" – pleased Christian and Muslim leaders and theologians. As put by Thomas Aquinas: "*Aristotle's conclusion is that there is one ruler of the whole universe, the first mover, and one first intelligible object, and one first good, whom above he called God, who is blessed for ever and ever.*" This is one of the reasons why for centuries Aristotle was cited as an infallible authority and why scientists that would empirically show that Aristotle was wrong – for instance, that it was the Earth that moved around the sun – would often be criticized and even punished.

This topic is discussed in Mario Livio's book *Galileo's science deniers*. As explained by him, two major arguments used to criticize Galileo's empirical observations that the Earth moved around the sun were based on two of the main issues discussed in the present book about Darwin: *beliefs* and *idolization*. One of the major arguments was that Galileo's empirical observations *had to be wrong* because they contradicted the *Holy Scripture*. In reality, the idea that it was the Earth that moved around the sun had already been postulated in the Copernicus' model, which was conceptually extended by Giordano Bruno – who was burned to death by the Roman Inquisition for his "heretical" ideas. The other, related, major argument against Galileo's empirical observations was that they contradicted what the idolized Aristotle had said. Importantly, for the context of the present book, this latter argument was not only used by theologians but also by scientists. For instance, as noted by Livio, Cesare Cremonini, who was a colleague of Galileo at Padua's University, was an emblematic example of an atheist who desperately defended Aristotle's ideas about the Earth and the sun, against the empirical evidence obtained by Galileo. Cremonini apparently even refused to look at Galileo's telescope. Other scholars and philosophers reacted in a similar way, back them – for instance, Giuliu Libri. In this sense, another fascinating – but also distressing – similarity between Darwin's and Aristotle's idolization is that most of their idolizers were/are more papist than the Pope: from what we know about Aristotle, it seems rather unlikely that he would refuse to look at Galileo's telescope if he had the opportunity to do so. Both Aristotle and Darwin had a genuine desire to seek the truth and were not afraid of it. They did not always truly see what was before their eyes because of their biases, but they mostly tried to do so, to see what is out there. In this regard, another disconcerting similarity between Darwin's and Aristotle's idolization is that a huge number of those that were more Papist than Pope Aristotle and that were and still are more Papist than Pope Darwin never truly fully read the original works of Aristotle about the cosmos, nor Darwin's works about human evolution, "races," and gender differences.

A major difference between the narratives constructed about Aristotle and Darwin is that the ideas of the former were music to the hears to Christians and Muslims, while the latter is often said to have "changed the world" because his ideas were crucial for the demise of religious narratives about our place in nature. However, as we have seen, this assertion is mainly an ethnocentric and somewhat condescending narrative built by Western scholars, because the vast majority of people living on this planet nowadays are religious. Contrary to such narratives, the religiously unaffiliated population is actually expected to *decline*: from 16% of the global population in 2015 to 13% in 2060 (Fig. 5.1). This is particularly so because the higher rates of population increase are expected to occur in regions where most people are religious, such as sub-Saharan Africa. For instance, within polls made during the last decades, the results have indicated that only about 34% of Armenians, and about 26% of Afghans, accept that humans evolved from other animals. Most living people *do not* accept Darwin's evolutionary ideas in general, and in particular that humans evolved from other primates. Even within the economically most powerful Western country, the U.S., *most* people do not accept that humans derived from other primates. A 2009 Harris Poll of 2303 adults living in the U.S. in which people were asked to "indicate for each category below if you believe in it, or not," lead to these results: *God: 82%; Miracles: 76%; Heaven: 75%; Angels: 72%; Hell: 61%; The virgin birth of Jesus: 61%; The devil: 60%; Darwin' theory of evolution: 45%; Ghosts: 42%*. The U.S. is of course a main exception within Western nations in this regard, but the same pattern is seen in most non-Western countries. Ironically, if one would apply the idea still defended by many Neo-Darwinians that what truly matters in biological evolution is to have "bigger numbers," the fact that most humans are religious and that both the number and proportion of people in the globe that is religious will seemingly be increasing in the next decades, would mean that religious people are 'more evolutionarily successful' than people such as Darwin and many of his idolizers.

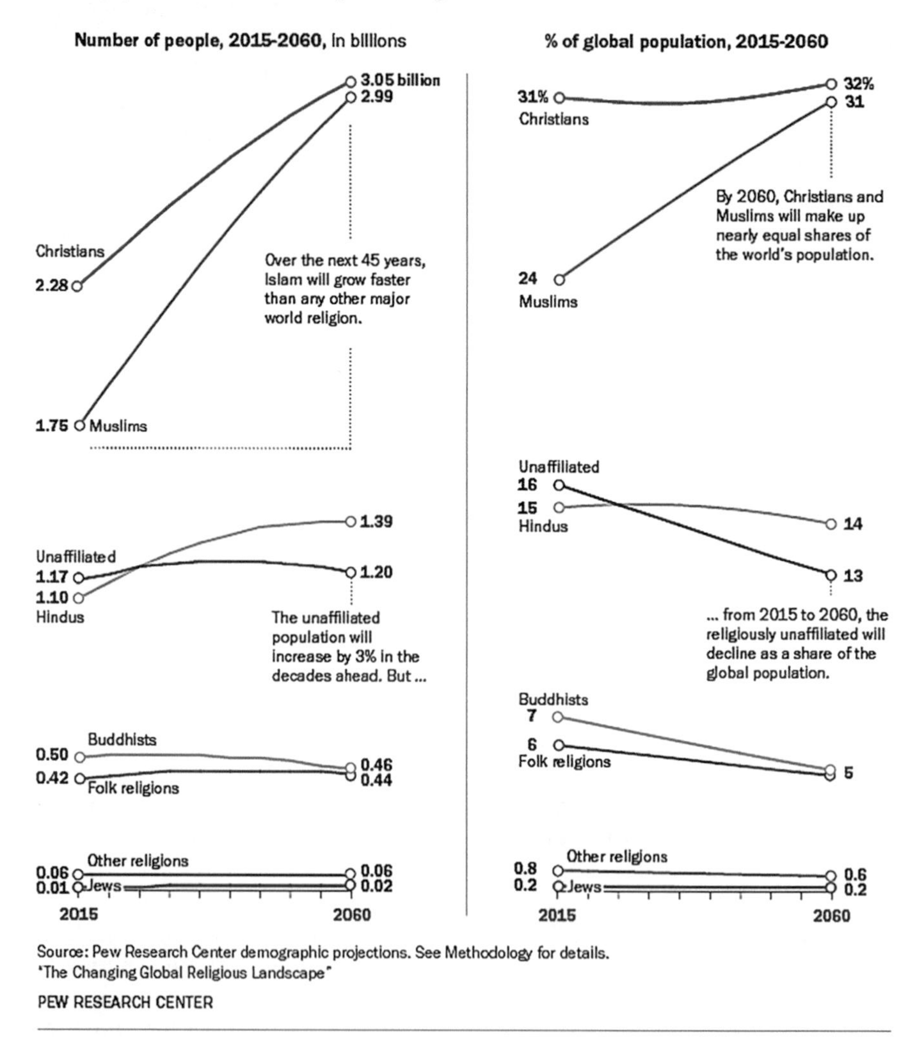

Fig. 5.1 Projected changes in global population and religiosity, 2015–2060

Darwin was obviously very important and influential, but as we have seen throughout the present volume he influenced in particular Western scientists, politicians, and popular culture, although of course that influence was and continues to be also felt in other countries because of Western colonialism and globalization.

Darwin's influential ideas had both positive and negative repercussions. Many of his ideas, or of the works of his followers, paved the way for huge, positive changes in fields such as biology and geology, and even medicine. This cannot be denied and should be emphasized. But it cannot also be denied – and should be underscored – that there is a huge amount of historical data that show that some of Darwin's ethnocentric, racist, sexist, hierarchical, and elitist ideas about human evolution had and continue to have tragic societal repercussions. This is an historical fact. As it is also an historical fact that, for instance, Wallace's writings about non-Europeans or about women tended to be less ethnocentric, racist, and sexist than those of Darwin. That is, considering such historical evidence, one could argue that, at least in theory, if Darwin had never existed and let's say Wallace would have been the main promoter and disseminator of the evolutionary theory of natural selection, it would have likely been more difficult for people like David Duke to directly cite Wallace to justify their ideas and the atrocities done against "others." This does not mean that many, or even most, such atrocities would have not occurred if Darwin was never born. There were numerous atrocities in human history much before Darwin was born, so surely Darwin cannot be blamed for all the atrocities that were done by Westerners since he wrote his books.

Demonizing Darwin is as absurd, and factually inaccurate, as to idealize him. This is the main point of the present book, to bring reality to these topics and deconstruct such inaccurate extreme and quasi-religious narratives. Reality tends to lie between extremes. Scientifically, one can easily falsify the assertion that Darwin was a God or a Demon. Similarly, it is absurd to simply state that Darwin was the "bad" guy and Wallace the "good" one, or vice-versa. Wallace was quicker to finish the first complete manuscript about evolution driven by natural selection, and his evolutionary ideas were in general less racist and ethnocentric than those of Darwin, but Darwin was far more articulate and consistent in what regards the explanation and dissemination of his evolutionary ideas – for better or worse – and was born in the right place, time, and family to assure a higher visibility of those ideas within the Victorian society and therefore within Western countries as a whole.

Discussing whether things would have or not been different if Darwin did not exist, or whether there would have been less or more atrocities or racism, sexism, and ethnocentrism in science, or whether "Social Darwinism" would have been as prevalent in Western society as it still is, is much less important than discussing what Darwin truly did and wrote and if his writings were accurate or not, precisely because we cannot change what happened in the past. In this sense, the present book is really not so much about the past, or about how things could have been, but instead focuses on the future: it aims to contribute to a future that will be less plagued by idolization, racism, sexism, ethnocentrism, and "Social Darwinism." That is why it is so important to emphasize both the positive and accurate aspects *and* the negative and erroneous ones of Darwin's writings, and in particular to specify which of his ideas were factually wrong, baseless, and/or biased, so the general public, media, historians, scientists, and next generations can learn how to not repeat the mistakes of the past.

For instance, while I was writing part of this book, in 2021, two independent studies have shown how baseless and biased are some of the key depictions of our past that many have taken for granted for a long time due to the repetition of such inaccurate evolutionary "facts." Importantly, these papers have also shown how such erroneous "facts" can be deconstructed, empirically. What is particularly disturbing with such erroneous depictions of our evolutionary past is that they are not only shown in scientific papers and textbooks, but literally everywhere, from scientific museums visited by millions of kids every year to educational materials, documentaries and TV shows seen and internalized by hundreds of millions of people worldwide. One of those papers, entitled "Visual Depictions of Our Evolutionary Past," was published by some colleagues and me, at the end of February 2021. It empirically shows that the reconstructions of human fossils as famous as Lucy and the Taung Child (see Fig. 5.2) are often not truly based on a deep knowledge of the soft tissues of our living relatives – that is, apes such as the chimpanzees and gorillas – and the application of robust technical methods. Instead, such depictions often continue to be influenced by factually inaccurate ethnocentric, racist, or sexist narratives such as those published in books such as the *Descent*. Because our species is about 250 to 300 thousand years old, while Lucy is about 3 million years older than that, a substantial portion of what the artists do is done "by eye" or is mostly mainly based on what they know about the muscles of living humans. Such depictions

Fig. 5.2 Casts of facial reconstructions of Lucy (left) and the Taung child (right): to highlight how the use of different models is crucial for the final image, the authors reconstructed Lucy's tone to appear more similar to that of bonobos, whereas the Taung child's tone is more similar to that of anatomically modern humans native to South Africa

therefore tend to be more subjective and more prone to be affected by their unconscious – or eventually conscious, in some cases – biases. Even more problematic is the fact that, when the general public sees such depictions in museums or documentaries, it tends to perceive them as if they were *reality*. This is exactly what frequently happens when lay people read or hear, in textbooks, documentaries, or TV shows, that Darwin's views about human evolution were "strikingly" correct or were not racist or sexist: they often tend to accept such inaccurate statements as a fact, as *reality*. That is exactly how biases and prejudices are often propagated, from one generation to the next.

A particularly powerful example provided in the press release of that 2021 paper published by my colleagues and me concerns one of the most renowned natural history museums, which receives about 4 to 5 million visitors every year at the very capital of the U.S., including a huge number of kids: the Smithsonian National Museum of Natural History. For instance, some depictions that have been displayed in that museum somewhat replicate the idea that human evolution was mainly a linear process leading to a progressive "whitening of the skin" (Fig. 5.3). However, scientific evidence clearly shows that human evolution, as biological evolution in

Fig. 5.3 Figure previously shown by the Smithsonian National Museum of Natural History, by John Gurche

Fig. 5.4 Marching toward evolutionary "progress," leading to the "pinnacle of evolution": "white" Western males

general, is not linear and there is no data supporting the idea that a progressive "whitening of the skin" occurred before humans left Africa, because a lighter skin pigmentation was chiefly only selected within people that migrated to non-African regions with high or low latitudes, as explained above. Similar depictions can basically still be found everywhere, from other Western museums to textbooks, media images, and a plethora of popular websites (see Fig. 5.4). In a way, such inaccurate depictions of our past are just one more "modern" version of the longstanding concept of a ladder of life (Fig. 1.7): a linear progressive process in which humans and specifically Westerners – in particular those that do not have a dark hair, as shown in Fig. 5.4 – are at the pinnacle of that ladder. The subliminal message is as powerful today as it was since these narratives were created: if you are not "white," then you are far down within the ladder of life, at a "lower," "inferior" position – *you should be aware of that and behave accordingly: know thy place*.

The second paper, published by Garcia-Campos and colleagues in 2021, was entitled "Indicators of Sexual Dimorphism in Homo Antecessor Permanent Canines." It revealed that one of the most famous fossils in Europe, found in the Sierra de Atapuerca, in Spain, and belonging to *Homo antecessor* (one of the several human species that is now extinct) is from a girl between 9 and 11 years old (Fig. 5.5), and not from a boy, as had been *believed* by most scientists until now. The new evidence came from an analysis of the canine teeth of the previously called "boy from the Gran Dolina," in reference to the title of an emblematic popularization book published by José María Bermúdez de Castro. What is particularly telling is that the authors of this 2021 paper, which include Bermúdez de Castro, recognized that there was no scientific reason for him to have previously asserted that the remains of the fossil were of a boy. "It arose randomly.. when José María [Bermúdez

Fig. 5.5 Analysis of the teeth of the "boy from the Gran Dolina" shows that the fossil was actually very likely.. a girl

de Castro] decided to make the book, he chose this masculine name, but for no specific reason.. it has been necessary to wait for these new techniques to be able to know the sex with certainty," explained García-Campos, one of the co-authors. The problem is that such decisions are too frequent to be truly "random." Almost all depictions of our past show men – "cavemen" – and, in the very few cases in which women are depicted, they are almost exclusively shown as passive mothers, not as active inventors or innovators or artists, as we have discussed above (Figs. 1.11, 1.12, and 1.13). That is, within such evolutionary depictions of our past women are portrayed exactly in the inaccurate sexist way in which Darwin described them in his writings about human evolution: as "naturally" passive, mentally inferior, lacking innovatory skills, being coy, and so on.

As explained above, actual scientific data from a plethora of fields such as archeology, psychology, neurobiology, and anthropology shows that such assertions were wrong, and this was also shown in that March 2021 paper, which showed that this girl was likely *actively* contributing to the interaction between different groups. As recognized by its authors, the findings of the paper put in question "the traditional gender roles that are still preserved in which the woman is at home and the man at work" and contributed "to change the collective imagination of the female in the cave with two young or tanning skins" and to "show us that women participated in hunting work and in disputes over territory."

Apart from the inaccurate sexist and ethnocentric inaccurate statements including in Darwin's writings about human evolution, we have also seen in this book that many other ideas defended by him about other topics, including some about the evolution of non-human organisms, have been also empirically shown to be wrong, partially inaccurate, or at least largely exaggerated. This includes his fixation on the Malthusian concept of a war-like struggle-for-existence, many of the teleological assertions he made about Mother Nature, and even some rather bizarre and completely unfounded theories such as his "*pangenesis theory.*" As explained by Slotten in *The heretic in Darwin's court*, even Wallace first embraced this bizarre theory, but ultimately abandoned it in favor of a "far superior one":

> Most naturalists believed in 'blending inheritance', but the mechanism remained a mystery. Darwin thought that he had hit on a solution and called his theory '*pangenesis*', which he published in 1868 as *The Variation of Animals and Plants Under Domestication*. At the end of the first volume of this book, he struck a distinct Lamarckian note: "*It is often sufficient for the inheritance of some peculiar character, that one parent alone should possess it.. but the power of transmission is extremely variable. In a number of individuals descended from the same parents, and treated in the same manner, some display this power in a perfect manner, and in some it is quite deficient.. the effects of injuries or mutilations are occasionally inherited; and.. the long continued use and disuse of parts produces an inherited effect*". Darwin postulated the existence of "gemmules," or microscopic germs, "cast off" into the circulation by the almost infinite number of cells that compose an organism and then concentrated in the generative cells, primarily the ova and sperm. If properly nourished, he argued, any gemmule could reproduce a portion of the organism or even the whole organism itself. These gemmules were inconceivably small - perhaps the size of atoms - since all could be contained in an ovum or a spermatazoon.
>
> Sometimes the gemmules remained dormant or undeveloped in an individual, only to be activated somehow in future generations. Gemmules also could be modified by external conditions. If multiplying sufficiently, they were capable of replacing the older gemmules and developing new structures. For example, a mutilation in an adult theoretically could be passed to its offspring. Pangenesis also explained the persistence of rudimentary and useless organs. Wallace initially embraced pangenesis as the best available theory to explain heredity. But the confusion the theory wrought among scientists caused Darwin to pronounce it "stillborn" and to predict, like a dejected Cassandra, that it eventually would reappear under some other name. He was wrong. One of the first experimentalists to disprove the theory of pangenesis was his cousin Francis Galton, who pioneered the study of heredity and performed blood transfusions on rabbits, replacing the entire blood volume of one animal with the blood of another and then breeding the transfused rabbits. The offspring were not in the least bit altered and resembled their biological parents, not the donor rabbits. Thus the blood did not carry the hereditary material from the body to the reproductive organs, as Darwin had speculated. When Weismann published his views on heredity nearly twenty years later, Wallace abandoned Darwin's "ponderous" theory for one that he deemed far superior.

Having said this, four crucial points need to be made, in this last chapter. First, I am not arguing, in the present book, that Darwin's theories and ideas, including those about human evolution, were all wrong. Well on the contrary. Most of the things he said and described about specific traits of plants, non-human animals, and other non-human organisms tended to be right, and obviously his grand idea that evolution is in great part driven by natural selection was right as well. As were some key aspects of his view of human evolution, for instance that humans very likely had originated from Africa. Second, it is important to point out that not all theories of Darwin that were factually inaccurate were wrong in a same way, or for the same reasons, or had the same type of societal repercussions. The pangenesis theory was wrong, but nobody really knew at that time how were the mechanisms of heredity, so Darwin basically made a wrong guess, which moreover did not lead to any major tragic societal and scientific repercussions. In contrast, many of his racist, sexist, and ethnocentric factually inaccurate ideas had already been criticized and deconstructed by others. Tragically, those inaccurate ideas are the ones that did have, and continue to have, tragic scientific and societal consequences. Third, it is important to stress that such racist, sexist, and ethnocentric inaccurate ideas had nothing to do with the "objective" observations of a "noble" scientist, because they could have been easily disproven if Darwin had considered what the historical records showed or what others had written about those topics before he did, and particularly what he truly saw before his eyes during his travels. That is, those inaccuracies have nothing to do with a "wrong guess," highlighting instead the powerful role played by biases and blind beliefs in science and, in this sense, the remarkable parallel between teleological religious narratives and many of the so-called "scientific" ideas that have been used or constructed by scholars such as Darwin or by those that have idolized him afterward and continue to idealize him today.

In this sense, as I anticipate that many creationists and so called "intelligent designers" will be tempted to use this book to criticize Darwin and evolutionary biology in general, and also that many Darwin's idolizers will criticize it for "taking the side" of creationists, I want to be clear about the following crucial point. This book criticizes the type of teleological narratives, biases, and beliefs used in Darwin's work, as well as the quasi-religious reasoning done by many of his idolizers. Therefore, it would be a huge paradox – and plainly wrong – to use such a criticism to justify creationism or intelligent design or to argue that the book is taking the side of creationists or intelligent designers. Furthermore, the fact that I – an evolutionist – am providing *empirical scientific data* to test some of Darwin's ideas that are often venerated by many other evolutionists, and showing that some of these ideas are refuted by such empirical data, shows a crucial difference between what science is and what religion is. Namely, science, when done properly and in an unbiased way, *involves the testing and eventual refutation of ideas based on empirical data and new knowledge, including ideas defended by prominent scientists such as Darwin.* That is, many scientists can – and many do, as seen in this book – fall into the trap of biases, systemic racism and sexism, and a quasi-religious type of circular reasoning, but *at least some scientists* are able to escape from them precisely because the scientific method gives them tools to test and eventually contradict their falsifiable theories. The proof is that scientific ideas have changed dramatically over the last millennia, in stark contrast with the "immutable" "truth," or "Word of God." Many religious ideas are unfalsifiable – for instance, the existence of an immaterial God – and even those that are falsifiable – for example, that the Earth is the center of the Cosmos – have historically been accepted even if there was a huge amount of evidence contradicting them. This is still the case today, as highlighted by the fact that most living people do not accept the huge amount of scientific data showing that humans evolved from apes, apes from monkeys, monkeys from other mammals, mammals from other tetrapods, and tetrapods from fish, and that none of these animals, or any other organism, was spontaneously created by a God.

The fourth point that I want to stress is that we need to recognize, once for all, that one of the major lines of defense commonly used in the publications, newspaper articles, documentaries, or educational materials that idolize Darwin is flawed. That is, as we have seen throughout this book, the argument that criticizing the racist, sexist, and ethnocentric ideas defended as facts in Darwin's books is not "historically fair" because at that time "everybody" was racist and sexist is literally wrong, empty, *and* moreover involves a critical logical fallacy. Empty because if we used the same logic, we could use the same argument to defend anybody that used racist ideas, such as David Duke did to justify and promote the horrible atrocities done by the KKK (Fig. 5.6), or Hitler, because we could argue that they also lived at a "different time." The argument is also logically fallacious because there are historical records showing that, at Darwin's epoch, many scholars and writers, including several very prominent ones such as Wallace, were not defending the sexist and racist ideas postulated in Darwin' books about human evolution. As explained in the present volume, that is precisely why societal changes occur. This is a typical

Fig. 5.6 Lynching of Elias Clayton, Elmer Jackson, and Isaac McGhie in Duluth: on June 15, 1920 – these African American circus workers, suspects in an assault case, were taken from jail and lynched by a "white" mob of thousands in Duluth, Minnesota, because there were rumors that six African Americans had raped and robbed a nineteen-year-old woman, although a physician who examined her found no physical evidence of rape

logical fallacy done in such idealizations of the past, which are the ones that truly tend to "cancel history," because they are not only omitting the flaws and inaccuracies written by "scientific giants" but also the fact that such "giants" were not obliged to commit such flaws and to write inaccuracies, as if they were passive automata, because others living at the same time did not do so. That is, by blindly excusing those that defended racist and sexist ideas, or colonialism, and so on, using such "back then that was normal," "everyone was doing the same" type of erroneous arguments, such idolizations are truly "canceling" history by omitting the obvious fact that – paraphrasing again Marguerite Yourcenar – *"at all times there are people who do not think like others.. that is, who do not think like those who do not think."*

This topic is directly related to another typical line of defense used by Darwin's idolizers discussed above: that before Darwin there were already wars, discrimination, white supremacy, and even genocides, so clearly one cannot "blame" Darwin for providing "scientific" weapons to those that justified or did such horrible things after he wrote his books. This is completely nonsensical. One could say exactly the same about Hitler, or Mao, or David Duke: before them, there were already atrocities, even genocides, so they really did nothing 'new' or 'particularly bad', right? No: it is not right. The fact that there were huge atrocities before them does not make them less responsible, theoretically and in practice, for the ideas that they defended and the horrible things they did. I would argue that it is actually the opposite: contrary to the first people that defended such ideas or justified similar atrocities in human history, scholars, politicians, and writers of the nineteenth and twentieth century had the privilege to *know about those historical ideas and events* before they wrote their biased and factually wrong ideas. As the saying goes, worse than doing something wrong, is to not even learn from that, and do it again.

For instance, Darwin not only knew about, but *saw* with his own eyes, the atrocities that Europeans were doing to "others" in so many regions of the planet, justified by factually inaccurate and biased racist and ethnocentric narratives about how Europeans were superior biologically and "morally," how they had the 'moral' duty to civilize "others," and so on. And what did Darwin write about such atrocities after seeing them with his own eyes? He *decided to actively write* scientific books that not only repeated, but actually "scientifically" supported, those very same biased ethnocentric and racist ideas. He did so and moreover often included catchy metaphors in his writings that could be easily absorbed by, and used to indoctrinate, the broader public, and that could therefore be easily used by politicians, white supremacists or misogynists such as David Duke or Hitler to further support their ideas and the atrocities they promoted or did. In other words, the fact that there were atrocities, racism, and ethnocentrism before Darwin does not mean that there is nothing wrong with the racist, ethnocentric, and sexist things that he wrote and disseminated as "scientific facts". On the contrary, what he wrote, particularly in the *Descent*, is highly problematic and criticizable, precisely because he wrote those false narratives as if they were "facts." That is the main point that those that idolize Darwin seem to forget: the crucial problem is not that he was sexist and racist and that he made sexist and racists comments at home or at a bar, but instead that he used his *power of place* to write them as evolutionary "facts" and to widely disseminate those "facts" to the broader public in the form of scientific books.

It is therefore urgent that *all of us*, together, start to "un-cancel" history. This should start by addressing such idealizations, longstanding denials, historically inaccurate narratives, and misconceptions, at a scientific and broader societal level, including changing educational materials and museum displays and producing new documentaries and textbooks about these topics. They should reflect Darwin's complexity, comprising his astonishingly accurate writings about some key aspects of the natural world *as well as* his inaccurate and biased ideas; his quest for reality *as well as* for scientific "priority" and immortality; his humble *as well as* patronizing attitudes; and his racist, sexist and ethnocentric *as well as* anti-slavery ideas. Systemic racism, sexism, and ethnocentrism can only stop when we are free to discuss the *whole story* without taboos: not just a part, or half, or even two-thirds of it.

To break this cycle of using old, biased ideas to justify discrimination, societal hierarchies, racism, sexist, and ethnocentrism, and eventually atrocities similar to those that have been done in the past, let us finally openly discuss our scientific legacies, in order to learn from them. "Those who cannot remember the past are condemned to repeat it" – a quote often credited to Spanish philosopher George Santayana. The main aim of the present book is therefore to contribute to a better understanding of our scientific past and to change the vicious cycle of idolization, systemic racism and sexism, and of indoctrination involving factually inaccurate narratives that are read, heard, or seen by our kids in school, textbooks, and TV since they are young. If we fail to do so, and do not emphasize how such narratives continue to plague science, education and the media and popular culture in general, and deconstruct such narratives, our kids will be condemned to internalize and propagate such ideas and therefore to impose or suffer the same type of discrimination, oppression, and atrocities that have been previously justified by them.

Figure Credits

Fig. 1.1 (https://upload.wikimedia.org/wikipedia/commons/f/f8/Portrait_of_Charles_Darwin_Wellcome_M0011462.jpg)
Fig. 1.2 (https://www.slideshare.net/jmlynch/11-huxley)
Fig. 1.3 (https://en.wikipedia.org/wiki/The_Creation_of_Adam)
Fig. 1.4 (https://www.pewresearch.org/fact-tank/2017/09/06/more-americans-now-say-theyre-spiritual-but-not-religious/)
Fig. 1.5 (Friedrich Georg Weitsch, Staatliche Museen zu Berlin, https://www.nybooks.com/articles/2019/12/05/alexander-von-humboldt-magnetic-polymath/)
Fig. 1.6 (https://allthatsinteresting.com/hypatia-of-alexandria)
Fig. 1.7 (https://www.dreamstime.com/ascent-life-vintage-engraving-engraved-illustration-earth-man-image162956266)
Fig. 1.8 (Joseph Wright of Derby, https://upload.wikimedia.org/wikipedia/commons/0/0a/Erasmus_Darwin%2C_after_Joseph_Wright.jpg)
Fig. 1.9 (https://upload.wikimedia.org/wikipedia/commons/9/95/Portrait_of_A.R._Wallace._Wellcome_L0000531.jpg)
Fig. 1.10 (https://historymadeeveryday.files.wordpress.com/2013/09/homo-erectus.gif)
Fig. 1.11 (https://fineartamerica.com/featured/prehistoric-man-tools-granger.html)
Fig. 1.12 (https://upload.wikimedia.org/wikipedia/commons/2/27/Font-de-Gaume.jpg)
Fig. 1.13 (https://upload.wikimedia.org/wikipedia/commons/5/56/Ushuaia-yamana7.jpg)
Fig. 1.14 (http://www.chileparaninos.gob.cl/639/w3-article-321019.html)
Fig. 1.15 (http://nationalunitygovernment.org/content/aborigines-question-1905-report-charges-against-police)
Fig. 1.16 (https://upload.wikimedia.org/wikipedia/commons/e/e7/Didgeridoo_%28Imagicity_1070%29.jpg)
Fig. 1.17 (https://www.flickr.com/photos/rnw/5883542473)

R. Diogo, *Darwin's Racism, Sexism, and Idolization*,
https://doi.org/10.1007/978-3-031-49055-2

Fig. 1.18 (https://api.ndla.no/image-api/raw/Eskimo-familie..jpg?width=10720)
Fig. 1.19 (Adapted from https://www.flickr.com/photos/tonythemisfit/3296436058)

Fig. 2.1 (https://upload.wikimedia.org/wikipedia/commons/9/91/Charles_Robert_Darwin%2C_as_an_ape%2C_holds_a_mirror_up_to_another._Wellcome_V0001472_%28cropped%29.jpg)
Fig. 2.2 (https://upload.wikimedia.org/wikipedia/commons/3/34/The_funeral_ceremony_of_the_Charles_Darwin_at_Westminster_Ab_Wellcome_V0018693.jpg)
Fig. 2.3 (https://upload.wikimedia.org/wikipedia/commons/2/20/Colonne_vendome.jpg)
Fig. 2.4 Painting by d'Alfred Darjou for L'Illustration du 10 juin 1871, displayed at the Bibliothèque historique de la Ville de Paris. (https://upload.wikimedia.org/wikipedia/commons/5/57/Commune_de_Paris_enl%C3%A8vement_des_cadavres_par_les_parisiens.jpg)
Fig. 2.5 (https://commons.wikimedia.org/wiki/File:A._R._Wallace,_%22The_Malay_archipelago%22_Wellcome_L0028206.jpg)
Fig. 2.6 (http://humanfoodproject.com/wp-content/uploads/2012/06/DSC01238.jpg)
Fig. 2.7 (https://commons.wikimedia.org/wiki/File:Jean_Baptiste_Pierre_Antoine_de_Monet_Lamarck._Stipple_engra_Wellcome_L0014871.jpg)
Fig. 2.8 (https://commons.wikimedia.org/wiki/File:Flap-necked_chameleon_(Chamaeleo_dilepis)_female.jpg)
Fig. 2.9 (https://en.wikipedia.org/wiki/Tulsa_race_massacre#/media/File:Tulsaraceriot1921-wounded-pickedup-fullpicture.jpg)

Fig. 3.1 (https://upload.wikimedia.org/wikipedia/commons/6/6c/Einsatzgruppen_murder_Jews_in_Ivanhorod%2C_Ukraine%2C_1942.jpg)
Fig. 3.2 (https://commons.wikimedia.org/wiki/File:Nicolaes_Tulp_1641_3de_capvt_lvi_satyr.JPG)
Fig. 3.3 (http://flashbak.com/wp-content/uploads/2016/08/Tyson-chimpanzee-cowper.jpg)
Fig. 3.4 (https://upload.wikimedia.org/wikipedia/commons/b/b1/A_Pair_of_Broad_Bottoms.jpg)
Fig. 3.5 (https://upload.wikimedia.org/wikipedia/commons/8/8a/Hoppius_Anthropomorpha.png)
Fig. 3.6 (https://knoji.com/images/user/kingkong austrianposter.jpg)
Fig. 3.7 (https://upload.wikimedia.org/wikipedia/commons/f/fe/St_louis_1904_mucha_poster.jpg)
Fig. 3.8 (https://www.theguardian.com/world/2015/jun/03/the-man-who-was-caged-in-a-zoo)
Fig. 3.9 (https://en.wikipedia.org/wiki/File:Shark_Island_-_Death_Camp_-_BBC.jpg)
Fig. 3.10 (https://www.brookings.edu/blog/up-front/2020/02/27/examining-the-black-white-wealth-gap/)
Fig. 3.11 (https://snl.no/James_Watson)

Fig. 3.12 (https://en.wikipedia.org/wiki/American_Indian_boarding_schools#/media/File:Carlisle_pupils.jpg)
Fig. 3.13 (https://imgur.com/a/qXSrn?grid)

Fig. 4.1 (https://upload.wikimedia.org/wikipedia/commons/4/40/Tintoretto_-_Penitent_Magdalene_-_Google_Art_Project.jpg)
Fig. 4.2 (By Richard Rothwell https://commons.wikimedia.org/wiki/File:Adam_et_Eve_Guido_Reni.jpg)
Fig. 4.3 (By Richard Rothwell https://commons.wikimedia.org/wiki/File:Rothwell_-_Mary_Shelley_(Enanced_Crop).jpg)
Fig. 4.4 (https://snl.no/Mary_Wollstonecraft)
Fig. 4.5 (Photo by Brian Gunn /IAAPEA, freely made available by IAAPEA at https://www.animalexperimentspictures.com/photo/797)
Fig. 4.6 (https://upload.wikimedia.org/wikipedia/commons/a/af/Antoinette_Louisa_Brown_Blackwell.jpg)
Fig. 4.7 (Modified from Kelly (2013))
Fig. 4.8 (https://commons.wikimedia.org/wiki/File:Guanche_-_Puerto_de_la_Cruz.jpg)
Fig. 4.9 (https://www.visualcapitalist.com/how-people-spend-their-time-globally/)
Fig. 4.10 (https://www.statista.com/chart/21345/coronavirus-deaths-by-gender/)
Fig. 4.11 (https://ourworldindata.org/life-expectancy)
Fig. 4.12 (https://www.mobilemuseumofart.com/wp-content/uploads/2020/06/37.007_2016-scaled.jpg)
Fig. 4.13 (http://entomology.wsu.edu/outreach/bug-info/black-widow/)

Fig. 5.1 (Freely available from Pew Research Center)
Fig. 5.2 (Image modified from our Campbell et al.'s 2021 paper)
Fig. 5.3 (https://www.reddit.com/r/Naturewasmetal/comments/kihhwi/the_evolutionary_history_of_the_human_face/)
Fig. 5.4 (https://medium.com/paperkin/where-is-evolution-taking-the-human-race-6ddaf7eaddba)
Fig. 5.5 (https://twitter.com/archaeologyEAA/status/1372105138236706818/photo/2)
Fig. 5.6 (https://en.wikipedia.org/wiki/Duluth_lynchings)

References and Suggested Further Reading

Achtner W. 2009. The evolution of evolutionary theories of religion. In *The biological evolution of religious mind and behavior* (Voland E, Schiefennhovel W, eds). Springer, New York. p. 257–273.

Ackerman D. 1994. *Natural history of love*. Random House, New York.

Alghamdi M, Ziermann JM, Diogo R. 2017. An untold story: the important contributions of Muslim scholars for the understanding of human anatomy. *Anat Rec* 300:986–1008.

Al-Gharbi M. 2021. People of the book: empire and social science in the Islamic Commonwealth period. *Socius* 7:https://doi.org/10.1177/23780231211021200

Al-Krenawi A. 2013. Mental health and polygamy: the Syrian case. *World J Psychiatr* 3:1–7.

Andreassen R. 2014. Danish perceptions of race and anthropological science at the turn of the twentieth century. In *The invention of race – scientific and popular representations* (Bancel N, David T, Thomas D, eds.). Taylor & Francis, London, p. 117–129

Asma ST. 2009. *On monsters – an unnatural history of our worst fears*. Oxford University Press, Oxford.

Aubin HJ, Berlin I, Kornreich C. 2013. The evolutionary puzzle of suicide. *Int J Environm Res Pub Health* 10:6873–6886.

Bahuchet S. 2014. Cultural diversity of African pygmies. In *Hunter-gatherers of the congo basin: cultures, histories, and biology of African pygmies* (Hewlett BS, ed.). Transaction Publishers, London, p. 1–29.

Baldwin JM. 1895. *Mental development in the child and race: methods and processes*. MacMillan, New York.

Baldwin JM. 1896a. A new factor in evolution. *Am Naturalist* 30:441–451.

Baldwin JM. 1896b. A new factor in evolution (continued). *Am Naturalist* 30:536–553.

Baldwin JM. 1896c. On criticisms of organic selection. *Science* 4:724–727.

Bancel N, David T, Thomas D, eds. 2014. *The invention of race – scientific and popular representations*. Taylor & Francis, London.

Barash D. 2012. Sex at Dusk. *The Chronicle of Higher Education*, 21 July 2012.

Barsanti G. 2009. *L'uomo dei boschi. piccola storia delle grandi scimmie da Aristotele a Darwin*. Università La Sapienza, Roma.

Becker E. 1973. *The denial of death*. The Free Press, New York.

Beckerman S, Valentine P, eds. 2002. *Cultures of multiple fathers*. University Press of Florida, Gainesville.

Bergreen L. 2003. *Over the edge of the world – Magellan's terrifying cirumnavigation of the globe*. Harper Perennial, New York.

R. Diogo, *Darwin's Racism, Sexism, and Idolization*,
https://doi.org/10.1007/978-3-031-49055-2

Bering J. 2011. *The belief instinct – the psychology of souls, destiny, and the meaning of life*. W. W. Norton & Company, New York.

Bering J. 2018. *Suicidal – why we kill ourselves*. The University of Chicago Press, Chicago.

Bernard J. 1982. *The future of marriage*. Yale University Press, New Haven.

Bethencourt F. 2013. *Racisms – from the crusades to the twentieth century*. Princeton University Press, Princeton.

Bill J. 2019. *Evolution before Darwin: theories of the transmutation of species in Edinburgh, 1804–1834*. Edinburgh University Press, Edinburgh.

Black E. 2003. *War against the weak – eugenics and America's campaign to create a master race*. Four Walls Eight Wondows, New York.

Blair KL, Cappell J, Pukall CF. 2018. Not all orgasms were created equal: differences in frequency and satisfaction of orgasm experiences by sexual activity in same-sex *versus* mixed-sex relationships. *J Sex Res* 55:719–733.

Blanco-Wells G, Libuy M, Harambour A, *et al.* 2021. Plagues, past, and futures for the Yagan canoe people of Cape Horn, southern Chile. *Maritime Studies* 20:101–113.

Blumenbach F. 1804. *De l'unite du genre humain*. Allut, Paris.

Blumberg MS. 2009. *Freaks of nature: what anomalies tell us about development and evolution*. Oxford University Press, New York.

Blume M. 2009. The reproductive benefits of religious affiliation. In *The biological evolution of religious mind and behavior* (Voland E, Schiefennhovel W, eds). Springer, New York. p. 117–149.

Boesch, C. 1992. New elements of a theory of mind in wild chimpanzees. *Brain Behav Sci* 15:149–150.

Boetsch G, Blanchard P. 2014. From cabinets of curiosity to the "Hottentot Venus": a long history of human zoos. In *The invention of race – scientific and popular representations* (Bancel N, David T, Thomas D, eds). Taylor & Francis, London, p. 185–194.

Bonner JT. 2013. *Randomness in evolution*. Princeton University Press, Princeton.

Boone AP, Hegarty M, Gong X. 2018. Sex differences in navigation strategy and efficiency. *Memory & Cognit* 46:909–922.

Bondeson J. 1997. *A cabinet of medical curiosities*. W. W. Norton & Company, New York.

Bondeson J. 2000. *The two-headed boy, and other medical curiosities*. Cornell University Press, Ithaca.

Boyatzis C, Chazan E, Ting CZ. 1993. Preschool children's decoding of facial emotions. *J Gen Psychol* 154:375–382.

Boyle PA, Buchman AS, Wilson RS, *et al.* 2012. Effect of purpose in life on the relation between Alzheimer disease pathologic changes on cognitive function in advanced age. *Arch Gen Psychiatry* 69:499–505.

Bowler PJ. 1987. *Theories of human evolution – a century of debate, 1844–1944*. John Hopkins University Press, Oxford.

Bowler PJ. 2013. *Darwin deleted*. University of Chicago Press, Chicago.

Bowler PJ. 2017. Alternatives to Darwinism in the early twentieth century. In *The darwinian tradition in context: research programs in evolutionary biology* (Delisle RG, ed.). Springer, New York, p. 195–218.

Brooks RC, Griffith SC. 2010. Mate choice. In *Evolutionary behavioral ecology* (Westneat DF, Fox CW, eds.). Oxford University Press, New York, p. 416–433.

Browne J. 1995. *Charles Darwin voyaging*. Princeton University Press, Princeton.

Browne J. 2003. *Charles Darwin – the power of place*. Princeton University Press, Princeton.

Browning F. 2017a. Survival secrets: what is about women that makes them more resilient than men. *Calif Mag Cal Alumni Assoc UC Berkeley*, 29 April 2018.

Browning F. 2017b. *The fate of gender: nature, nurture, and the human future*. Bloomsbury, New York.

Brady SP, Bolnick DI, Angert AL, *et al.* 2019. Causes of maladaptation. *Evol Appl* 12:1229–1242.

Brantlinger P. 2003. *Dark vanishings: discourse on the extinction of primitive races, 1800–1930.* Cornell University Press, Ithaca.
Brattain M. 2007. Race, racism, and anti-racism: UNESCO and the politics of presenting science to the postwar public. *Amer Hist Rev* 112:1386–1413.
Brennan PLR, Cowart JR, Orbach DN. 2022. Evidence of a functional clitoris in dolphins. *Curr Biol* 2022 32:R24–R26.
Brune M. 2009. On shared psychological mechanisms of religiousness and delusional beliefs. In *The biological evolution of religious mind and behavior* (Voland E, Schiefennhovel W, eds). Springer, New York, p. 217–228.
Buettner D. 2012. *The blue zones, second edition – 9 lessons for living longer.* National Geographic, Washington DC.
Buklijas T, Gluckman PD. 2013. From Evolution and Medicine to Evolutionary Medicine. In *The Cambridge Encyclopedia of Darwin and Evolutionary Thought* (Ruse M, ed). Cambridge University Press, Cambridge, p. 505–514.
Butler FP. 2012. *Evolution without darwinism – the legacy of Stephen Jay Gould.* CreateSpace, New York.
Burton GJ, Moffett A, Keverne B. 2015. Human evolution: brain, birthweight and the immune system. *Philos Trans R Soc Lond B Biol Sci* 370:20140061.
Edgar Cabanas and Eva Illouz's 2019 book *Manufacturing happy citizens: how the science and industry of happiness control our lives.* Polity, Cambridge.
Call V, Susan S, Pepper S. 1995. The incidence and frequency of marital sex in a national sample. *J Marriage Family* 57:639–52.
Campbell R, Vinas G, Henneberg M, Diogo R. 2021. Visual depictions of our evolutionary past: a broad case study concerning the need for quantitative methods of soft tissue reconstruction and art-science collaborations. *Frontiers Ecol Evol* 26:https://doi.org/10.3389/fevo.2021.639048.
Camper P. 1791 *Dissertation sur les variétés naturelles qui caractérisent la physionomie des hommes des divers climats et des différens ages: suivie de réflexions sur la beauté, particulièrement sur celle de la tête: avec une manière nouvelle de dessiner toute sorte de têtes avec la plus grande exactitude.* Chez H.J. Jansen, Paris.
Canizares-Esguerra J. 2006. *Nature, empire and nation – explorations of the history of science in the Iberian world.* Stanford University Press, Stanford.
Canudas-Romo V. 2010. Three measures of longevity: time trends and record values. *Demography* 47:299–312.
Carroll S. 2016a. *The big picture – on the origins of life, meaning, and the universe itself.* Dutton, New York.
Carroll SB. 2016b. *The Serengeti rules: the quest to discover how life works and why it matters.* Princeton University Press, Princeton.
Carroll SB. 2020. *A series of fortunate events: chance and the making of the planet, life, and you.* Princeton University Press, Princeton.
Carter CS, Perkeybile AM. 2018. The monogamy paradox: what do love and sex have to do with it. *Frontiers Ecol Evol* 6:202.
Casserius I. 1600–1601. *De Vocis Auditus Que Organis Historia Anatomica.* Ferrariae, Venice.
Catholic Church. 1983. *Code of canon law, Latin-English edition.* Catholic Church, Vatican.
Cavanaugh, J. 2018. *The prosocial paradox: Unraveling oxytocin's role in monogamous relationships.* PhD dissertation, University of Nebraska at Omaha.
Cerello K, Kholoussy H, eds. 2016. *Domestic tensions, national identities – global perspectives on marriage, crisis, and nation.* Oxford University Press, New York.
Church of England. 1844. *The Book of common prayer: printed by Whitchurch, March 1549; commonly called The first book of Edward VI.* William Pickering, London.
Clark AE, Georgellis Y. 2013. Back to baseline in Britain: adaptation in the British household panel survey. *Economica* 80:496–512.
Cloquet J. 1821–1831. *Anatomie de 'homme.* Charles-Philibert, Paris.

Cole FJ. 1975. *A history of comparative anatomy – from Aristotle to the eighteenth century.* Dowe Publications, New York.

Conley TD, Matsick J, Valentine B, *et al.* 2017. Investigation of consensually non-monogamous relationships: theories, methods and new directions. *Persp Psycholog Sci* 12:205–232.

Coontz S. 2005. *Marriage, a history: how love conquered marriage*. Penguin Books, New York.

Corbey RHA. 2005. *The metaphysics of apes: negotiating the animal-human boundary.* Cambridge University Press, Cambridge.

Corbey RHA, Theunissen B, eds. 1995. *Ape, man, apeman: changing views since 1600*. Leiden University, Leiden.

Cottonham DP, Madson MB, Nicholson BC, *et al.* 2018. Harmful alcohol use and alcohol-related sex expectancies as predictors of risky sex among african american female college drinkers. *J Ethn Subst Abuse* 17:389–400.

Coyne J. 2016. *Why do some scientists always claim that evolutionary biology needs urgent and serious reform?* Blogpost: https://whyevolutionistrue.wordpress.com/2016/12/26/why-are-scientists-always-saying-that-evolutionary-biology-needs-urgent-and-serious-reform/.

Crews F. 2017. *Freud: the making of an illusion*. Metropolitan Books, New York.

Croutier AL. 1991. Harem: the world behind the veil. Abbeville Press, New Work.

Csoka AC. 2016. Innovation in medicine: Ignaz the reviled and Egas the regaled. *Med Health Care Philos* 19:163–168.

Cunningham A. 1997. *The anatomical Renaissance: the resurrection of the anatomical projects of the ancients*. Scolar Press, Aldershot.

Cuvier G. 1797. *Tableau elementaire de 'histoire naturelle des animaux*. Bandouin, Paris.

Damasio A. 1994. *Descartes' error – emotion, reason and the human brain*. Putnam Publishing, New York.

Damasio A. 2021. *Feeling & knowing – making mind conscious*. Pantheon, New York.

Darling J, De Pijpekamp MV. 1994. Rousseau on the education, domination and violation of women. *Brit J Educational Stud* 42:115–132.

Darwin C. 1839. *Voyage of the Beagle-Charles Darwin's Journal of Researches*. Penguin, London.

Darwin C. 1842. *The structure and distribution of coral reefs*. Smith Elder and Co., London.

Darwin C. 1844. *Geological observations on the volcanic islands, visited during the voyage of HMS Beagle*. Smith Elder and Co., London.

Darwin C. 1859. *On the origin of species by means of natural selection, or, the* preservation of favored races in the struggle for life. J. Murray, London.

Darwin C. 1871. *The descent of man, and selection in relation to sex*. J. Murray, London.

Darwin C. 1872. *The expression of the emotions in man and animals*. J. Murray, London.

Darwin C. 1881. *The formation of vegetable mould, through the action of worms, with observations on their habits*. J. Murray, London.

Darwin C. (Darwin F., ed.) 1887. *The complete life and letters of Charles Darwin, including an autobiographical chapter*. J. Murray, London.

Darwin C. 1987. *Charles Darwin's Notebooks, 1836–1844: geology, transmutation of species, metaphysical enquiries* (Barrett, P.H., *et al.*, eds). Cornell University Press, Ithaca.

Darwin, F., ed., 1898, "The Life and Letters of Charles Darwin," Basic Books: New York, 2 Vols, Reprinted, 1959.

Darwin, F., ed., 1902, "The Life of Charles Darwin," Senate: London, Reprinted, 1995.

Darwin, F., ed., 1909, "The Foundations of the Origin of Species: Two Essays Written in 1842 and 1844 by Charles Darwin," Kraus Reprint Co: New York NY, Reprinted, 1969.

Darwin, F. & Seward, A.C., eds, 1903, "More Letters of Charles Darwin," John Murray: London, 2 Vols.

Daston L, Park K. 1998. *Wonders and the order of nature, 1150–1750*. Zone Books, New York.

Davis DM. 2018. *The beautiful cure – the revolution in immunology and what it means for your health.* The University of Chicago Press, Chicago.

Dawkins R. 1976. *The selfish gene*. Oxford University Press, Oxford.

Daynes K. 2019. *The dark side of the mind: true stories from my life as a forensic psychologist.* Octopus, London.

Delisle RG. 2007. *Debating humankind's place in nature, 1860–2000: the nature of paleoanthropology.* Pearson Prentice Hall, Upper Saddle River NJ.

Delisle RG, ed. 2017a. *The darwinian tradition in context – research programs in evolutionary biology.* Springer, New York.

Delisle RG. 2017b. From Charles Darwin to the evolutionary synthesis: weak and diffused connections only. In *The Darwinian tradition in context: research programs in evolutionary biology* (Delisle RG, ed.). Springer, New York, p. 133–167.

Delisle RG. 2019. *Charles Darwin's incomplete revolution – the origin of species and the static worldview.* Springer, New York.

De Queiroz A. 2014. *The monkey's voyage – how improbable journeys shaped the history of life.* Basic Books, New York.

Desilva JM, ed. 2021. *A most interesting problem: what Darwin's Descent of Man got right and wrong about human evolution.* Princeton University Press, Princeton.

Desmond A, Moore J. 1994. *Darwin: the life of a tormented evolutionist.* WWW Norton & Company, New York.

Desmond A, Moore J. 2009. *Darwin's sacred cause: race, slavery and the quest for Human Origins.* Houghton Mifflin Harcourt, London.

De Wall F. 2016. *Are we smart enough to know how smart animal are?* WWW Norton & Company, New York.

De Wall F. 2019. *Mama's last hug: animal emotions and what they tell us about ourselves.* WWW Norton & Company Inc, New York.

Depew DJ. 2017. Darwinism in the twentieth century: productive encounters with saltation, acquired characteristics, and development. In *The Darwinian tradition in context: research programs in evolutionary biology* (Delisle RG, ed.). Springer, New York, p. 61–68.

DeSesso JM. 2019. The arrogance of teratology: a brief chronology of attitudes throughout history. *Birth Defects Res* 111:123–141.

Diamond J. 1991. *The rise and fall of the third chimpanzee.* Hutchinson Radius, London.

Diamond J. 1999. *Guns, germs, and steel: the fates of human societies.* W. W. Norton & Company, New York.

Diamond, J. 2005. *Collapse: how societies choose to fail or succeed.* Viking Press, New York.

Diamond J. 2012. *The world until yesterday: what we can learn from traditional societies?* Penguin Books, New York.

Dibble HL, Aldeias V, Goldberg P, *et al.* 2015. A critical look at evidence from La Chapelle-aux-Saints supporting an intentional Neandertal burial. *J Archaeol Sci* 53:649–657.

Diogo R. 2010. Comparative anatomy, anthropology and archaeology as case studies on the influence of human biases in natural sciences: the origin of 'humans', of 'behaviorally modern humans' and of 'fully civilized humans'. *Open Anat J* 2:86–97.

Diogo R. 2017a. *Evolution driven by organismal behavior – a unifying view of life, function, form, trends and mismatches.* Springer, New York.

Diogo, R. 2017b. Etho-eco-morphological mismatches, an overlooked phenomenon in ecology, evolution and Evo-Devo that supports ONCE (Organic Nonoptimal Constrained Evolution) and the key evolutionary role of organismal behavior. *Front Ecol Evol – EvoDevo*:10.3389.

Diogo R. 2018a. Links between the discovery of primates and anatomical comparisons with humans, the chain of being, our place in nature, and racism. *J Morphol* 279:472–493.

Diogo R. 2018b. Where is, in 2017, the Evo in Evo-Devo (Evolutionary Developmental Biology)? *J Exp Zool B* 330:15–22.

Diogo R. 2019. Sex at Dusk, Sex at Dawn, selfish genes: how old-dated evolutionary ideas are used to defend fallacious misogynistic views on sex evolution. *J Soc Sci Humanit* 5:350–367.

Diogo R. 2021. *Meaning of life, human nature, and delusions – how tales about love, sex, races, Gods and progress affect our lives and Earth's splendor.* Springer, New York.

Diogo R, Abdala V. 2010. *Muscles of Vertebrates – comparative anatomy, evolution, homologies and development.* Taylor & Francis, Oxford.

Diogo R, Wood B. 2011. Soft-tissue anatomy of the primates: phylogenetic analyses based on the muscles of the head, neck, pectoral region and upper limb, with notes on the evolution of these muscles. *J Anat* 219:273–359.

Diogo R, Wood B. 2012. *Comparative anatomy and phylogeny of primate muscles and human evolution.* Taylor and Francis, Oxford.

Diogo R, Wood B. 2013. The broader evolutionary lessons to be learned from a comparative and phylogenetic analysis of primate muscle morphology. *Biol Rev* 88:988–1001.

Diogo R, Potau JM, Pastor JF, *et al.* 2010. *Photographic and descriptive musculoskeletal atlas of Gorilla – with notes on the attachments, variations, innervation, synonymy and weight of the muscles.* Taylor & Francis, Oxford.

Diogo R, Potau JM, Pastor JF, *et al.* 2012. *Photographic and descriptive musculoskeletal atlas of gibbons and siamangs (Hylobates) – with notes on the attachments, variations, innervation, synonymy and weight of the muscles.* Taylor & Francis, Oxford.

Diogo R, Potau JM, Pastor JF, *et al.* 2013a. *Photographic and descriptive musculoskeletal atlas of chimpanzees (Pan) – with notes on the attachments, variations, innervation, synonymy and weight of the muscles.* Taylor & Francis, Oxford.

Diogo R, Potau JM, Pastor JF, *et al.* 2013b. *Photographic and descriptive musculoskeletal atlas of orangutans (Pongo) – with notes on the attachments, variations, innervation, synonymy and weight of the muscles.* Taylor & Francis, Oxford.

Diogo R, Pastor JF, Hartstone-Rose A, *et al.* 2014. *Baby Gorilla: photographic and descriptive musculoskeletal atlas of the skeleton, muscles and internal organs – including CT scans and comparisons to other gorillas and primates.* Taylor & Francis, Oxford.

Diogo R, Ziermann JM, Linde-Medina M. 2015. Is evolutionary biology becoming too politically correct? A reflection on the scala naturae, phylogenetically basal clades, anatomically plesiomorphic taxa, and "lower" animals. *Biol Rev* 90:502–521.

Diogo R, Bello-Hellegouarch G, Kohlsdorf T, *et al.* 2016a. Comparative myology and evolution of marsupials and other vertebrates, with notes on complexity, Bauplan, and "Scala Naturae". *Anat Rec* 299:1224–1255.

Diogo R, Noden D, Smith CM, *et al.* 2016b. *Learning and understanding human anatomy and pathology: an evolutionary and developmental guide for medical students.* Taylor & Francis, Oxford.

Diogo R., Molnar JL, Wood B. 2017a. Bonobo anatomy reveals stasis and mosaicism in chimpanzee evolution, and supports bonobos as the most appropriate extant model for the common ancestor of chimpanzees and humans. *Sci Rep* 7:608.

Diogo R, Guinard G, Diaz R. 2017b. Dinosaurs, chameleons, humans and Evo-Devo-Path: linking Étienne Geoffroy's teratology, Waddington's homeorhesis, Alberch's logic of 'monsters', and Goldschmidt hopeful 'monsters'. *J Exp Zool B* 328:207–229.

Diogo R, Shearer B, Potau JM, *et al.* 2017c. *Photographic and descriptive musculoskeletal atlas of bonobos, with notes on the attachments, variations, innervation, synonymy and weight of the muscles.* Springer, New York.

Dittrich-Reed DR, Fitzpatrick BM. 2013. Transgressive hybrids as hopeful monsters. *Evol Biol* 40:310–315.

Driscoll CA, Thompson JC. 2018. The origins and early elaboration of projectile technology. Evol Anthropol 27:30–45.

Duke D. 1998. *My awakening: a path to racial understanding.* Free Speech Press, Covington.

Dyble M, Thorley J, Page AE, *et al.* 2019. Engagement in agricultural work is associated with reduced leisure time among Agta hunter-gatherers. *Nature Hum Behav*:s41562-019-0614-6.

Edwardes MPJ. 2019. *The origins of self – an anthropological perspective.* UCL Press, London.

Erb CM. 1998. *Tracking King Kong – a Holliwood icon of world culture.* Wayne State University Press, Detroit.

Eldredge N. 2014. *Extinction and evolution: what fossils reveal about the history of life*. Firefly Books, Toronto
Eldredge N, Gould SJ. 1972. Punctuated equilibrium: an alternative to phyletic gradualism. In *Models in paleobiology* (Schopf TJM, ed.). Freeman, Cooper and Co., San Francisco, p. 82–115.
Engelmeier H. 2016. *Der Mensch. Der Affe*. Böhlau Verlag, Köln.
Epstein GM. 2010. *Good without God – what a billion nonreligious people do believe*. Harper, New York.
Evans S. 2017. *Darwin and women – a selection of letters*. Cambridge Press, Cambridge.
Fábrega H. 1997. *Evolution of sickness and healing*. University of California Press, Berkeley.
Fabrici G. 1600. *De Formato Foetu*. Embryo Project Encyclopedia (2008-08-27), ISSN, 1940-5030.
Feiler B. 2017. *The first love story – Adam, Even and US*. Penguin Press, New York.
Figes O. 2007. *The whisperers – a private life in Stalin's Russia*. Picador, New York.
Fine C. 2017. *Testosterone Rex – myths of sex, science and society*. W. W. Norton & Company, New York.
Finkel EJ. 2017. *The all-or-nothing marriage – how the best marriages work*. Penguin Press, New York.
Fisher M, Garcia J, Rosemarie Chang, eds. 2013. *Evolution's empress – Darwinian perspectives on the nature of women*. Oxford University Press, Oxford.
Fleckenstein JR, Cox DW. 2015. The association of an open relationship orientation with health and happiness in a sample of older U.S. adults. *Sex Relation Ther* 30:94–116.
Fox CW, Westneat DF. 2010. Adaptation. In *Evolutionary behavioral ecology* (Westneat DF, Fox CW, eds.). Oxford University, New York, p. 16–32.
Frey U. 2009. Cognitive foundations of religiosity. In *The biological evolution of religious mind and behavior* (Voland E, Schiefennhovel W, eds). Springer, New York, p. 229–241.
Fuentes A. 2017. Human niche, human behaviour, human nature. *Interface Focus* 7:20160136.
Fuentes A. 2021. The Descent of Man, 150 years on. *Science* 372:769.
Futuyma DJ. 2017. Evolutionary biology today and the call for an extended synthesis. *Interface Focus* 7: 20160145.
García-Campos C, Martinén-Torres M, Modesto-Mata M, Martín-Francés L, Martínez de Pinillos M, Bermúdez de Castro JM. 2021. Indicators of sexual dimorphism in *Homo antecessor* permanent canines. *J Anthropol Sci*: https://doi.org/10.4436/JASS.99001.
Gardiner G, Lee D, Baranski E, *et al.* 2020. Happiness around the world: a combined etic-emic approach across 63 countries. *PLoS ONE* 15:e0242718.
Gasman D. 1971. *The scientific origins of national socialism*. Science History Publications, New york.
Gawande A. 2014. *Being mortal – medicine and what matters in the end*. Metropolitan Books, New York.
Gilby IC, Machanda ZP, O'Malley RC, *et al.* 2017. Predation by female chimpanzees: toward an understanding of sex differences in meat acquisition in the last common ancestor of *Pan* and *Homo*. *J Hum Evol* 110:82–94.
Gissis SB, Jablonka E, Eds. 2011. *Transformations of Lamarckism: from subtle fluids to molecular biology*. MIT Press, Cambridge.
Gladwell M. 2013. *David and Goliath – underdogs, misfits, and the art of battling giants*. Little, Brown, and Company, New York
Goodall J. 1988. *In the shadow of man*. Houghton Mifflin Company, Boston.
Goncalves A, Susana C. 2019. Death among primates: a critical review of non-human primate interactions towards their dead and dying. *Biol Rev* 94:1502–1529.
Gottschall J. 2012. *The storytelling animal – how stories make us human*. Houghton Mifflin Hartcourt, New York.
Gould SJ. 1981. *The mismeasure of man*. W. W. Norton & Company, New York.
Gould SJ. 1996. *Full House: the spread of excellence from Plato to Darwin*. Belknap Press, Cambridge.

Gould SJ. 2002. *The structure of evolutionary theory*. Belknap, Harvard.
Gray J. 2018. *Seven types of atheism*. Penguin Books, London.
Groves C. 2008. *Extended family: long lost cousins. A personal look at the history of primatology*. Conservation International, Arlington.
Guerrini A. 2003. *Experimenting with humans and animals – from Galen to animal rights*. The Johns Hopkins University Press, Baltimore.
Green T. 2019. *A fistful of shells – West Africa from the rise of the slave trade to the age of revolution*. The University of Chicago Press, Chicago.
Greenblatt S. 2011. *The swerve – how the world became modern*. W. W. Norton & Company, New York.
Greenblastt S. 2017. *The rise and fall of Adam and Eve*. W. W. Norton & Company, New York.
Gray J. 2013. *The silence of animals: on progress and other modern myths*. Farrar, Straus & Giroux, New York.
Gruber HE. 1981. *Darwin on man – a psychological study of scientific creativity*. The University of Chicago Press, Chicago.
Guedron M. 2014. Panel and sequence: classification and associations in scientific illustrations of the human races (1770–1830). In *The invention of race – scientific and popular representations* (Bancel N, David T, Thomas D, eds). Taylor & Francis, London, p. 60–67.
Guerrini A. 2015. *The Courtiers' Anatomists – Animals and Humans in Loius XIV's Paris*. The University of Chicago Press, Chicago.
Gupta M, Prasad NG, Dey S, *et al.* 2017. Niche construction in evolutionary theory: the construction of an academic niche? *J Genet* 96:491–504
Gurven M, Kaplan H. 2007. Longevity among hunter-gatherers: a cross-cultural examination. *Population & Devel Rev* 33:321–365.
Haas R, Watson J, Buonasera T, *et al.* 2020. *Female hunters of the early Americas. Science Adv* 6: eabd0310.
Haeckel E. 1868. *Die Natürliche Schöpfungsgeschichte*. Georg Reimer, Berlin.
Haeckel E. 1870. *Die Natürliche Schöpfungsgeschichte, 2nd edition*. Georg Reimer, Berlin.
Haeckel E. 1874. *Anthropogenie oder Entwickelungsgeschichte des Menschen. Keimes- und Stammesgeschichte*. Engelmann, Leipzig.
Haeckel E. 1887. *The history of creation, or the development of the earth and its inhabitants by the action of natural causes*. Appleton and Company, New York.
Halcrow S, Warren R, Kushnick G, *et al.* 2020. Care of infants in the past: bridging evolutionary anthropological and bioarchaeological approaches. *Evol Hum Sci*:https://doi.org/10.1017/ehs.2020.46.
Hamilton LD, Meston CM. 2013. Chronic stress and sexual function in women. *J Sexual Med* 10:2443–2454.
Hamlin KA. 2014. *From Eve to evolution – Darwin, science, and women's rights in gilded age America*. The University of Chicago Press, Chicago.
Hanlon H, Thatcher R, Cline M. 1999. Gender differences in the development of EEG coherence in normal children. *Dev Neuropsychol* 16:479–506.
Hannon E, Lewens T, eds. 2018. *Why we disagree about human nature*. Oxford University Press, Oxford.
Hanson R. 2012. Sex at Dusk v. Dawn. *Overcoming Bias (Blog),* 30 August 2012 (http://www.overcomingbias.com/2012/08/sex-at-dusk-v-sex-at-dawn.html).
Harari YN. 2017. *Homo Deus – a brief history of tomorrow*. Harper, New York.
Hare B, Woods V. 2020. *The survival of the friendliest – understanding our origins and rediscovering our common humanity*. Random House, New York.
Harkup K. 2019. *Making the monster: the science behind Mary Shelley's Frankenstein*. Bloomsbury Sigma, London.
Haught JE. 2000. *Science and religion in search for of cosmic purpose*. Georgetown University Press, Washington DC.
Hazlewood N. 2001. *Savage: the life and times of Jemmy Button*. Thomas Dunne Books, New York.

Henning BG, Scarfe AC, eds. 2013. *Beyond mechanism: putting life back into biology*. Lexington Books, Lexington.

Hewlett BS, ed. 2014a. *Hunter-Gatherers of the Congo basin: cultures, histories, and biology of African pygmies*. Transaction Publishers, London.

Hewlett BS. 2014b. Hunter-gatherer childhoods in the Congo basin. In *Hunter-Gatherers of the Congo Basin: Cultures, Histories, and Biology of African Pygmies* (Hewlett BS, ed.). Transaction Publishers, London, p. 245–275.

Hirsch AR. 1998. *Scentsational sex: the secret to using aroma for arousal*. Element Books, New York.

Hoßfeld U. 2010. Ernst Haeckel. Biographienreihe absolute. Orange press, Freiburg.

Hoßfeld U. 2016. *Geschichte der biologischen Anthropologie in Deutschland – Von den Anfängen bis in die Nachkriegszeit. 2nd ed.* Franz Steiner Verlag, Stuttgart.

Hochschild AR. 2016. *Stranger in their own land – anger and morning on the American right*. The New Press, New York.

Hoff EV. 2004. A friend living inside me – the forms and functions of imaginary companions. *Imagination, Cognit & Personal* 24:151–189.

Hoffman D. 2019. *The case against reality: why evolution hid the truth from our eyes*. W.W. Norton & Company Inc., New York.

Hoffmeyer J. 2013. Why do we need a semiotic understanting of life. In *Beyond mechanism: putting life back into biology* (Henning BG, Scarfe AC, eds). Lexington Books, Lexington, p. 147–168.

Holland J. 2012. *A brief history of mysogyny – the world's oldest prejudice*. Constable & Robinson Ltd., London.

Hood B. 2013. *The self illusion – how the social brain creates identity*. Oxford University Press, Oxford.

Hood RW, Hill PC, Spilka B. 2009. *The psychology of religion – an empirical approach*. The Guilford Press, London.

Hoquet T. 2014. Biologization of race and racialization of the human: Bernier, Buffon, Linnaeus. In *The invention of race – scientific and popular representations* (Bancel N, David T, Thomas D, eds). Taylor & Francis, London, p. 17–32.

Hrdy SB. 2009. *Mothers and others – the evolutionary origins of mutual understanding*. Belknap Press, Cambridge.

Huber E. 1931. *Evolution of facial musculature and expression*. The Johns Hopkins University Press, Baltimore.

Humboldt A. 1914. Views of the cordilleras and monuments of the indigenous peoples of the Americas. In *Views of the Cordilleras and monuments of the indigenous peoples of the Americas: a critical edition, 2012* (Kutzinski VM, Ette O, eds). University of Chicago Press, Chicago, p. 1–370.

Huneman P, Walsh DM, eds. 2017. *Challenging the modern synthesis – adaptation, development, and inheritance*. Oxford University Press, Oxford.

Hutton R. 2017. *The witch – a history of fear from ancient times to the present*. Yale University Press, New Haven.

Huxley TH. 1863. *Evidence as to man's place in nature*. Williams and Norgate, London.

Jablonka E, Lamb MJ. 2005. *Evolution in four dimensions – genetic, epigenetic, behavioral, and symbolic variation in the history of life*. MIT Press, Cambridge.

Janson HW, ed. 1952. *Apes and Ape lore in the Middle Ages and the Renaissance*. Warburg Institute University of London, London.

Jeffery AJ, Shackelford TK, Zeigler-Hill V, *et al.* 2018. The evolution of human female sexual orientation. *Evol Psychol Sci*:40806-018-0168-2.

Jensen JD, Payseur BA, Stephan W, *et al.* 2019. The importance of the neutral theory in 1968 and 50 years on: a response to Kern and Hahn 2018. *Evolution*. 73:111–114.

Jiang XQ, Mei XD, Feng D. 2016. Air pollution and chronic airway diseases: what should people know and do? *J Thorac Dis* 8:E31–E40.

Johnson MR. 2005. *Aristotle on teleology*. Clarendon Press, Oxford.

Johnson NA, Lahti DC, Blumstein DT. 2012. Combating the assumption of evolutionary progress: lessons from the decay and loss of traits. *Evol Educat & Outreach* 5:128–138.

Kasperbauer TJ. 2018. *Subhuman – the moral psychology of human attitudes to animals*. Oxford University Press, Oxford.

Kauffman SA. 2010. *Reinventing the sacred: a new view of science, reason, and religion*. Basic Books, New York.

Kelly J. 2006. *The great mortality: an intimate history of the Black Death, the most devastating plague of all time*. Harper Collins Publishers Inc., New York.

Kelly RL. 1995. *The foraging spectrum: diversity in hunter-gatherer lifeways*. Smithsonian Institution Press, Washington.

Kelly RL. 2013. *The lifeways of hunter-gatherers – the foraging spectrum*. Cambridge Press, Cambridge.

Kendy IX. 2016. *Stamped from the beginning – the definitive history of racist ideas in America*. Nation Books, New York.

Kevles DJ. 1995. *In the name of eugenics – genetics and the uses of human heredity*. Harvard University Press, Cambridge.

Kingsberg SA, Janata JW. 2007. Female sexual disorders: assessment, diagnosis, and treatment. *Urol Clin N Am* 34:497–506.

Kirkegaard EOW. 2019. Race differences: a very brief review. *Mankind Quart* 60.2:142–173.

Kirkham D. 2019. *Our shadowed world – reflections on civilization, conflict, and belief*. Cascade Books, Eugene.

Klarsfeld A, Revah F. 2003. *The biology of death: origins of mortality*. Cornell University Press, New York.

Klein N. 1999. *No logo – no space, no jobs, no choice*. Picador, Toronto.

Klein N. 2019. *On fire: the (Burning) case for a Green New Deal. Simon & Schuster, New York.*

Kposowa AJ. 2000. Marital status and suicide in the National Longitudinal Mortality Study. *J Epidemiol Community Health* 54:254–261.

Kteily N, Bruneau E, Waytz A, *et al.* 2015. The ascent of man: theoretical and empirical evidence for blatant dehumanization. *J Pers Social Psychol* 109:901–931.

Kühl HS, Kalan AK, Arandjelovic M, *et al.* 2016. Chimpanzee accumulative stone throwing. *Sci Rep* 6:22219.

Kuklick H, ed. 2008. *A New History of Anthropology*. Blackwell, Oxford.

Kull K. 2014. Adaptive evolution without natural selection. *Biol J Linn Soc* 112:287–294.

Kunz J. 2009. Is there a particular role for ideational aspects of religions in human behavioral ecology? In *The biological evolution of religious mind and behavior* (Voland E, Schiefennhovel W, eds). Springer, New York, p. 89–104.

Kvarnemo C. 2018. Why do some animals mate with one partner rather than many? A review of causes and consequences of monogamy. *Biol Rev* 93:1795–1812.

Labarthe JC. 1997. Are boys better than girls at building a tower or a bridge at 2 years of age? *Arch Diseases Childhood* 77:140–144.

Lagerkvist U. 2005. *The enigma of ferment – from the philosopher's stone to the first biochemical Nobel prize*. World Scientific, Hackensack.

Lahti DC. 2009. The correlated history of social organization, morality, and religion. In *The biological evolution of religious mind and behavior* (Voland E, Schiefennhovel W, eds). Springer, New York, p. 67–88.

Laland KN, Odling-Smee J, Turner S. 2014. The role of internal and external constructive processes in evolution. *J Physiol* 592:2413–2422.

Laland KN, Uller T, Feldman MW, *et al.* 2015. The extended evolutionary synthesis: its structure, assumptions and predictions. *Proc R Soc Lon B* 282:https://doi.org/10.1098/rspb.2015.1019.

Laland K, Matthews B, Feldman MW. 2016. An introduction to niche construction theory. *Evol Ecol* 30:191–202.

Landau, M. 1991. *Narratives on human evolution*. Yale University Press, New Haven.

Lee RB. 2018. Hunter-Gatherers and human evolution: new light on old debates. *Annu Rev Anthropol* 47:513–531.
Leeman RW. 2012. *The teleological discourse of Barack Obama*. Lexington Books, New York.
Lenoir T. 1982. *The strategy of life – teleology and mechanics in nineteenth-century german biology*. The University of Chicago Press, Chicago.
Lenroot RK, Gogtay N, Greenstein DK, *et al.* 2007. Sexual dimorphism of brain developmental trajectories during childhood and adolescence. Neuroimage 36:1065–1073.
Leroi AM. 2003. *Mutants: on the form, varieties and errors of the human body.* Harper Collins, London.
Leroi AM. 2014. *The lagoon: how Aristotle invented science.* Bloomsbury, London.
Levin J. 2020. *Religion and medicine – a history of the encounter between humanity's two greatest institutions*. Oxford University Press, Oxford.
Levi-Strauss C. 2011. *Tristes tropiques.* Penguin Books, New New York.
Levine, N. 1998. *The dynamics of polyandry: kinship, domesticity, and population on the Tibetan border*. University of Chicago Press, Chicago.
Levy JS, Thompson WR. 2010. *Causes of war*. Wiley-Blackwell, New York.
Lewis HS. 2001. Boas, Darwin, science, and anthropology. *Curr Anthropol* 42:381–406.
Lewis J. 2014. Egalitarian social organization: the case of the mbendjele BaYaka. In *Hunter-gatherers of the Congo basin: cultures, histories, and biology of African pygmies* (Hewlett BS, ed.). Transaction Publishers, London, p. 219–243.
Lewis J. 2018. *The science of sin: why we do the things we know we shouldn't.* Bloomsbury Sigma, New York.
Lewontin RC. 1992. *Biology as ideology – the doctrine of DNA*. HarperPerennial, New York.
Lieberman P. 1991. *Uniquely human: the evolution of speech, thought and selfless behavior.* Harvard University Press, Cambridge.
Lindholm M. 2015. DNA dispose, but subjects decide – learning and the extended synthesis. *Biosemiotics* 8:4431–4461.
Linnaeus C. 1735. *Systema naturae*. Laurentius Salvius, Stockholm.
Livio M. 2020. *Galileu's science deniers*. Simon & Schuster, New York.
Long M, Long E, Waters T. 2013. Suicide among the Mla Bri Hunter-Gatherers of Northern Thailand. *J Siam Soc* 101:156–176.
Lovejoy AO. 1936. *The great chain of being: a study of the history of an idea.* Harvard University Press, Cambridge.
Luhmann M, Hofmann W, Eid M, *et al.* 2012. Subjective well-being and adaptation to life events: a meta-analysis. *J Personal Social Psychol* 102:592–615.
Lupo K, Ndanga A J-P, Kiahtipes C. 2014. On Late Holocene population interactions in the Northwestern Congo Basin: when, how and why does the ethnographic pattern begin. In *Hunter-gatherers of the Congo Basin: cultures, histories, and biology of African Pygmies* (Hewlett BS, ed.). Transaction Publishers, London, p. 59–83.
Mah K, Binik YM. 2001. The nature of human orgasm: a critical review of major trends. *Clin Psychol Rev* 21:823–856.
Mah K, Binik YM. 2002. Do all orgasms feel alike? Evaluating a two-dimensional model of the orgasm experience across gender and sexual context. *J Sex Res* 39:104–113.
Malesevic S. 2011. *The rise of organised brutality – a historical sociology of violence*. Cambridge University Press, Cambridge.
Malik A, Ziermann JM, Diogo R. 2017. An untold story in biology: the historical continuity of evolutionary ideas of Muslim scholars from the 8th century to Darwin's time. *J Biol Educ*:https://doi.org/10.1080/00219266.2016.1268190.
Malik K. 2014. *The quest for a moral compass – a global history of ethics.* Melville House, Brooklyn.
Mancuso S, Viola A. 2015. *Brilliant green: the surprising history and science of plant intelligence*. Island Press, Washington DC.

Marshall JM. 2007. *The day the world ended at Little Bighorn – a Lakota history*. Penguin Books, London.

Martin D. 1984. *Primate Origins and Evolution*. Chapman and Hall, London.

Martínez-Sevilla F, Arqués M, Jordana X, *et al.* 2020. Who painted that? The authorship of Schematic rock art at the Los Machos rockshelter in southern Iberia. *Antiquity* 94:1133–1151.

Mayr E. 1976. *Evolution and the diversity of life: selected essays.* Harvard University Press, Cambridge.

McBrearty S, Brooks AS. 2000. The revolution that wasn't: a new interpretation of the origin of modern human behavior. *J Hum Evol* 39:453–563.

McCalman I. 2009. *Darwin's armada: four voyages and the battle for the theory of evolution.* W. W. Norton & Company, New York.

McCool-Myers M, Theurich M, Zuelke A, *et al.* 2018. Predictors of female sexual dysfunction: a systematic review and qualitative analysis through gender inequality paradigms. *BMC Womens Health* 18:108.

McMahon DM. 2006. *Happiness – a history*. Grove Press, New york.

McShea DW. 2012. Upper-directed systems: a new approach to teleology in biology. *Biol & Philos* 27:663–684.

Meijer MC. 2014. Cranial varieties in the human and orangutan species. In *The invention of race – scientific and popular representations* (Bancel N, David T, Thomas D, eds). Taylor & Francis, London, p. 33–47.

Meral Z. 2018. *How violence shapes religion – belief and conflict in the Middle East and Africa.* Cambridge University Press, Cambridge.

Minelli A. 2009. *Forms of becoming – the evolutionary biology of development.* Princeton University Press, Princeton.

Montagu MFA. 1943. Edward Tyson, M.D., F.R.S., 1650–1708. *Mem Am Philos Soc* 20:1–488.

Moore J., Desmond, A., eds. 2004. *The descent of man, and selection in relation to sex by Charles Darwin.* Penguin Classics, London.

Moravec H. 1990. *Mind children*. Harvard University Press, Harvard.

Morris D. 2013. *Monkey*. Reaktion Books Ltd, London.

Moser S. 1998. *Ancestral images – the iconography of human origins*. Cornell University Press, Ithaca.

Moss S. 2023. *Ten birds that changed the world.* Basic Books, New York.

Nanjundiah V. 2019. Many roads lead to Rome: Neutral phenotypes in microorganisms. *J Exp Zool (Mol Dev Evol)* 332:339–348.

Nee S. 2005. The great chain of being. *Nature* 435:429–429.

Nelson LH, ed. 2017. *Biology and feminism – a philosophical introduction.* Cambridge University Press, Cambridge.

Nesse RM. 2019. *Good reasons for bad feelings – insights from the frontier of evolutionary psychiatry*. Dutton, New York.

Nisbet R. 1980. *History of the idea of progress*. Basic Books, New York.

Nixey C. 2017. *The Darkening age – the Christian destruction of the classic world.* Macmillan, London.

Noble D. 2006. *The music of life: biology beyond the genome.* OUP, Oxford.

Noble D. 2017. *Dance to the tune of life – biological relativity.* Cambridge University Press, Cambridge.

Nordenskiold E. 1936. *The history of biology, 2nd ed.* Tudor, New York.

Nour NM. 2008. Female genital cutting: a persisting practice. *Rev Obstet Gynecol* 1:135–139.

Öberg KG, Hallberg J, Kaldo V, *et al.* 2017. Hypersexual disorder according to the hypersexual disorder screening inventory in help-seeking Swedish men and women with self-identified hypersexual behavior. *Sex Med* 5:e229–e236.

Odling-Smee FJ, Laland KN, Feldman MW. 2003. *Niche construction – the neglected process in evolution (Monographs in population biology 37).* Princeton University Press, Princeton.

Oliveira-Pinto AV, Santos RM, Coutinho RA, *et al.* 2014. Sexual dimorphism in the human olfactory bulb: females have more neurons and glial cells than males. *PLoS ONE* 9:e111733.
Omland KE, Cook LG, Crisp MD. 2008. Tree thinking for all biology: the problem with reading phylogenies as ladders of progress. *BioEssays* 30:854–867.
Oota H, Pakendorf B, Weiss G, *et al.* 2005. Recent origin and cultural reversion of a hunter-gatherer group. *PLoS Biol* 3: e71.
Panese F. 2014. The creation of the 'negro' at the turn of the nineteenth century: Petrus Camper, Johan Friedrich Blumenbach, and Julien-Joseph Virey. In *The invention of race – scientific and popular representations* (Bancel N, David T, Thomas D, eds). Taylor & Francis, London, p. 48–59.
Patou-Mathis M. 2020. *L'homme préhistorique est aussi une femme – Une histoire de l'invisibilité des femmes*. Allary, Paris.
Pavlicev M, Wagner G. 2016. The evolutionary origin of female orgasm. *J Exp Zool B* 326:326–337.
Peterson A. 2001. *Being human – ethics, environment, and our place in the world.* University of California Press, Berkeley.
Perrault C. 1676. *Memoires pour servir a l'histoire naturelle des animaux.* De l'Imprimerie Royale, Paris.
Persaud TVN. 1984. *Early history of human anatomy: from antiquity to the beginning of the modern era.* Charles C Thomas, Springfield.
Persaud TVN, Loukas M, Tubbs RS. 2014. *A history of human anatomy, 2nd ed.* Charles C Thomas, Springfield.
Piazza PV. 2019. *Homo Biologicus: Comment la biologie explique la nature humaine.* Albin Michel, Paris.
Pigliucci M. 2017. Darwinism after the modern synthesis. In *The Darwinian tradition in context: research programs in evolutionary biology* (Delisle RG, ed.). Springer, New York, p. 94–104.
Pigliucci M, Müller GB, eds. 2010. *Evolution – the extended synthesis.* MIT Press, Cambridge.
Pinker S. 2011. *The better angels of our nature: why violence has declined*'. Penguin Books, New York.
Potarca G. 2020. The demography of swiping right – an overview of couples who met through dating apps in Switzerland. *PLoS ONE* 15:e0243733.
Pratarelli ME, Chiarelli B. 2007. Extinction and overspecialization: the dark side of human innovation. *Mankind Quart* 48:83–98.
Prause N, Kuang L, Lee P, Miller G. 2016. Clitorally Stimulated Orgasms Are Associated With Better Control of Sexual Desire, and Not Associated With Depression or Anxiety, Compared With Vaginally Stimulated Orgasms. *J Sex Med* 13:1676–1685.
Pringle P. 2008. *The murder of Nikolai Vavilov – the story of Stalin's persecution of one of the great scientists of the twentieth century.* Simon & Schuster, New York.
Pruetz JD, LaDuke TC. 2010. Reaction to fire by savanna chimpanzees (*Pan troglodytes verus*) at Fongoli, Senegal: conceptualization of "fire behavior" and the case for a chimpanzee model. *Am J Phys Anthropol* 141:646–650.
Prum RO. 2017. *The evolution of beauty: how Darwin's forgotten theory of mate choice shapes the animal world – and us.* Anchor Books, New York.
Prushinskaya A. 2017. *A Woman is a woman until she is a mother: essays.* MG Press, Des Plaines.
Puchner M. 2017. *The written world: the power of stories to shape people, history, civilization.* Penguin Random House LLC, New York.
Puts DA, Dawood K, Welling LLM. 2012. Why women have orgasms: an evolutionary analysis. *Arch Sexual Behav* 41:1127–1143.
Quammen D. 2006. *The reluctant Mr. Darwin – an intimate portrait of Charles Darwin and the making of his theory of evolution.* Atlas Books, New York.
Radini A, Tromp M, Beach A, *et al.* 2019. Medieval women's early involvement in manuscript production suggested by lapis lazuli identification in dental calculus. *Science Adv* 5:eaau7126.
Ramsey G, Pence CH, eds. 2016. *Chance in evolution.* The University of Chicago Press, Chicago.
Reiss JO. 2009. *Not by design: retiring Darwin's watchmaker.* University of California Press, Berkeley.

Rendu W, Beauval C, Crevecoeur I, *et al.* 2014. Evidence supporting an intentional Neandertal burial at La Chapelle-aux-Saints. *PNAS* 111:81–86.
Reyes-García V. 2019. Did foragers enjoy more free time? *Nature Hum Behav*:s41562-019-0610-x.
Reynaud-Paligot C. 2014. Construction and circulation of the notion of 'race' in the nineteenth century. In *The invention of race – scientific and popular representations* (Bancel N, David T, Thomas D, eds). Taylor & Francis, London, p. 87–99.
Richards RJ. 1987 *Darwin and the emergence of evolutionary theories of mind and behavior.* University of Chicago Press, Chicago.
Richards RJ. 2008. *The tragic sense of life: Ernst Haeckel and the struggle over evolutionary thought.* University of Chicago Press, Chicago.
Richards RJ. 2013. *Was Hitler a Darwinian? Disputed questions in the history of evolutionary theory.* University of Chicago Press, Chicago.
Richert RA, Smith EI. 2009. Cognitive foundations in the development of a religious mind. In *The biological evolution of religious mind and behavior* (Voland E, Schiefennhovel W, eds). Springer, New York, p. 181–204.
Rigato E, Minelli A. 2013. The great chain of being is still here. *Evo Educ & Outreach* 6:18.
Richtel M. 2019. *An elegant defense – the extraordinary new science of the immune system.* William Morrow, New York.
Riva A, Orrù B, Pirino A, *et al.* 2001. Iulius Casserius (1552–1616): the self-made anatomist of Padua's golden age. *Anat Rec* 265:168–175.
Ritvo L. 1990. *Darwin's influence on Freud – a tale of two sciences.* Yale University Press, New Haven.
Rosenthal GG, Ryan MJ. 2022. Sexual selection and the ascent of women: Mate choice research since Darwin. *Science* 375:eabi6308.
Roth L, Briken P, Fuss J. 2022. Masturbation in the Animal Kingdom. *J Sex Res*https://doi.org/10.1080/00224499.2022.2044446.
Rovelli C. 2017. *Reality is not what it seems – the journey to quantum gravity.* Riverhead books, New York.
Ruse M. 1993. Will the real Charles Darwin please stand up? *Quart Rev Biol* 68:225–231.
Ruse M. 1996. *Monad to man: the concept of progress in evolutionary biology.* Harvard University Press, Cambridge.
Ruse M. 2003. *Darwin and design – does evolution have a purpose?* Harvard University Press, Cambridge.
Ruse M. 2013. From organisms to mechanisms – and halfway back? In *Beyond mechanism: putting life back into biology* (Henning BG, Scarfe AC, eds). Lexington Books, Lexington, p. 409–430.
Ruse M. 2018. *On purpose.* Princeton University Press, Princeton.
Ruse M. 2017. *Darwinism as religion – what literature tells us about evolution.* Oxford University Press, Cambridge.
Ruse M. 2019. *A meaning to life.* Oxford University Press, Cambridge.
Rutherford A. 2021. Race, eugenics, and the canceling of great scientists. *Am J Phys Anthropol* 175:448–452.
Ryan C. 2018. *The virility paradox: the vast influence of testosterone on our bodies, our minds, and the world we live in.* BenBella Books, Dallas.
Ryan C, Jetha C. 2010. *Sex at dawn: how we mate, why we stray, and what it means for modern relationships.* Harber Collins Publishers, New York.
Sagan C. 1997. *The demon-haunted world – science as a candle in the dark.* Ballantine Books, New York.
Sahlins M. 1976. *Use and abuse of biology.* University of Michigan Press, Ann Arbor.
Saini A. 2017. *Inferior – how science got women wrong, and the new research that's rewriting the story.* Beacon Press, Boston.
Sánchez-Villagra M, van Schaik CP. 2019. Evaluating the self-domestication hypothesis of human evolution. *Evol Anthropol* 28:133–143.
Sapolsky RM. 2017. *Behave – the biology of humans at our best and worst.* Penguin Press, New York.
Saxon L. 2012. *Sex at Dusk: lifting the shiny wrapping from Sex at Dawn.* CreateSpace, New York.

Walter Scheidel. 2017. *The great leveler: violence and the history of inequality from the stone age to the twenty-first century*. Princeton University Press, Princeton.
Schmitt S. 2004. *Histoire d'une question anatomique: la repetition des parties*. Museum National d'Histoire Naturelle, Paris.
Schueler GF. 2005. *Reasons & purposes – human rationality and the teleological explanation of actions*. Clarendon Press, Oxford.
Scott JC. 2017. *Against the grain – a deep history of the earliest states*. Yale University Press, New Haven.
Sechiyama K. 2014. Japan, the sexless nation. *Tokyo Business Today* 19 December 2014.
Shanahan T. 2017. Selfish genes and lucky breaks: Richard Dawkins' and Stephen Jay Gould's divergent Darwinian agendas. In *The Darwinian tradition in context: research programs in evolutionary biology* (Delisle RG, ed.). Springer, New York, p. 31–36.
Shermer M. 2004. *The science of good and evil – why people cheat, gossip, care, share, and follow the golden rule*. Times Books, New York.
Shermer M. 2011. *The believing brain – from ghosts and gods to politics and conspiracies, how we construct beliefs and reinforce them as truths*. Times Books, New York.
Shermer M. 2015. *The moral arc – how science makes us better people*. St. Martin's Griffin, New York.
Skinner BF. 1948. 'Superstition' in the pigeon. *J Exp Psychol* 38:168–172.
Singer C. 1959. *A history of Biology to about the year 1900*. Abelard-Schuman, London.
Singh U. 2008. A history of ancient and early medieval India: from the Stone Age to the 12th century. Pearson Education, New York.
Slotten RA. 2004. The heretic in Darwin's court – the life of Alfred Russel Wallace. Columbia University Press, New York.
Smith EA, Hill K, Marlowe F, *et al.* 2010. Wealth transmission and inequality among hunter-gatherers. *Curr Anthropol* 51:19–34.
Smith RJ. 2016. Freud and evolutionary anthropology's first just-so story. *Evol Anthropol* 25:50–53.
Sommer M. 2015. *Evolutionäre Anthropologie zur Einführung*. Junius, Hamburg.
Sommer V, Vasey PL, eds. 2006. *Homosexual behavior in animals – an evolutionary perspective*. Cambridge University Press, Cambridge.
Sorenson J. 2009. *Ape*. Reaktion Books Ltd, London.
Sproul B. 1991. *Primal myths – creation myths around the world*. HarperCollins, New York.
Schacht R & Kramer KL. 2019. Are we monogamous? A review of the evolution of pair-bonding in humans and its contemporary variation cross-culturally. *Front Ecol Evol*:https://doi.org/10.3389/fevo.2019.00230.
Stoltzfus A. 2017. Why we don't want another "synthesis". *Biology Direct* 12:23.
Strager H. 2016. *A modest genius – the story of Darwin's life and how his ideas changed everything*. Createspace Independent Publishing Platform.
Sultan SE. 2016. *Organisms & environment – ecological development, niche construction, and adaptation*. Oxford University Press, Oxford.
Sugiyama Y. 2017. Sex-biased dispersal of human ancestors. *Evol Anthropol* 26:172–180.
Suzman J. 2017. *Affluence without abundance – the disappearing world of the Bushmen*. Bloomsbury, New York.
Suzman J. 2021. *Work – a deep history from the Stone Age to the age of robots*. Bloomsbury, New York.
Swan L, Gordon R, Seckbach J, eds. 2012. *Origin(s) of design in nature – a fresh, interdisciplinary look at how design emerges in complex systems, especially life*. Springer, New York.
Taleb NN. 2010. *The black swan – the impact of the highly improbable, 2nd Edition*. Trader House Trade Paperback, New York.
Tallis F. 2018a. Women are more prepared for love. In *O Publico* 28 May 2018:14–15.
Tallis F. 2018b. *The incurable romantic: and other tales of madness and desire*. Basic Books, New York.
Thagard P. 2010. *The brain and the meaning of life*. Princeton University Press, Princeton.

Theiss JA. 2016. Frequency of sexual relations in marriage. In *Encyclopedia of family studies* (Shehan C, ed.). Wiley-Blackwell, Hoboken, p. 1–5.
Thomas K. 1983. *Man and the natural world – changing attitudes in England 1500–1800*. Oxford University Press, Oxford.
Tilley L, Oxenham M. 2011. Survival against the odds: modeling the social implications of care provision to seriously disabled individuals. *Internat J Paleopathol* 1:35–42.
Todes DP. 1989. *Darwin without Malthus*. Oxford University Press, Oxford.
Trüper H, Chakrabarty D, Subrahmanyam S, eds. 2015. *Historical teleologies in the modern world*. Bloomsbury, New York.
Tulp NP. 1641. *Observationes Medicae*. Vivie, Leiden.
Turner DD. 2017. Paleobiology's uneasy relationship with the Darwinian tradition: stasis as data. In *The Darwinian tradition in context: research programs in evolutionary biology* (Delisle RG, ed.). Springer, New York, p. 333–352.
Turner JS. 2000. *The extended organism – the physiology of animal-built structures*. Harvard University Press, Cambridge.
Turner JS. 2007. *The tinkerer's accomplice: how design emerges from life itself*. Harvard University Press, Cambridge.
Turner JS. 2013. Biology's second law: homeostasis, purpose and desire. In *Beyond Mechanism: Putting Life Back into Biology* (Henning BG, Scarfe AC, eds). Lexington Books, Lexington, p. 183–204.
Turner JS. 2016. Homeostasis and the physiological dimension of niche construction theory in ecology and evolution. *Evol Ecol* 30:203–219.
Tuttle RH, ed. 1975. *Primate Functional Morphology and Evolution*. Aldine, Chicago.
Tyson E. 1699. *Orang-Outang sive Homo sylvestris, or the anatomy of a pygmie compared to that of a monkey, an ape and a man*. T. Bennet, London.
UNESCO. 1950. U.N.E.S.C.O. on Race. *Man* 50:138–139.
UNESCO. 1951. U.N.E.S.C.O.'s New Statement on Race. *Man* 51:154–155.
UNESCO. 1952. U.N.E.S.C.O.'s New Statement on Race. *Man* 52:9.
Vaas R. 2009. Gods, gains and genes – on the natural origin of religiosity by means of bio-cultural selection. In *The biological evolution of religious mind and behavior* (Voland E, Schiefennhovel W, eds). Springer, New York, p. 25–49.
Valentine P, Beckerman S, Ales C. 2017. *The anthropology of marriage in lowland South-America: bending and breaking the rules*. University Press of Florida, Gainesville.
Van Arsdale A. 2017. Human evolution as a theoretical model for an extended evolutionary synthesis. In *The Darwinian tradition in context: research programs in evolutionary biology* (Delisle RG, ed.). Springer, New York, p. 105–130.
Van Schaik C, Michel K. 2016. *The good book of human nature – an evolutionary reading of the bible*. Basic Books, New York City.
Van Wyhe J, Kjaergaard PC. 2015. Going the whole orang: Darwin, Wallace and the natural history of orangutans. Studies Hist Philos Sci C 51:53–63.
Varella D. 2017. *Prisioneiras*. Companhia das Letras, Sao Paulo.
Varki A. 2009. Human uniqueness and the denial of death. *Nature* 460: 684.
Veracini C, Teixeira DM. 2016. Perception and description of New World non-human primates in the travel literature of the fifteenth and sixteenth centuries: a critical review. *Annals Sci* 74:25–65.
Vesalius A. 1543. *De humani corporis fabrica libri septem*. Ex officina Joannis Oporini, Basel.
Vinicius M. 2012. *Modular evolution: how natural selection produces biological complexity*. Cambridge University Press, Cambridge.
Voland E. 2009. Evaluating the evolutionary status of religiosity and religiousness. In *The biological evolution of religious mind and behavior* (Voland E, Schiefennhovel W, eds). Springer, New York, p. 9–24.
Voland E, Schiefennhovel W, eds. 2009. *The biological evolution of religious mind and behavior*. Springer, New York.
Von Hornbostel EM. 1948. The music of the Fuegians. *Ethnos* 13: 61–102.

Wagner A. 2014. *Arrival of the fittest: solving evolution's greatest puzzle*. Oneworld publications, London.
Walker W. 2017. *Why we sleep – unlocking the power of sleep and dreams*. Simon & Schuster Inc, New York.
Wallace AR. 1853. *Palm trees of the Amazon and their uses*. Biodiversity Heritage Library, London.
Wallace AR. 1869. *The Malay Archipelago*. Harper, London.
Wallace AR. 1870. *Contributions to the theory of natural selection, 2nd ed*. Macmillan and Company, London.
Wallace AR. 1876. *The geographical distribution of Animals*. Harper and brothers, London.
Wallace AR. 1878. *Tropical nature, and other essays*. Macmillan and Company, London.
Wallace AR. 1881. *Island life*. Harper and brothers, London.
Wallace AR. 1889a. *Darwinism: an exposition of the theory of natural selection, with some of its applications*. Macmillan and Company, London.
Wallace AR. 1889b. *Travels on the Amazon and Rio Negro*. Ward, Lock, Bowden & Co, London.
Wallace AR. 1903. *Man's place in the universe* (Gutenberg). Chapman & Hall, London.
Wallace AR. 1905. My life. Chapman & Hall, London.
Wang Y, Liu H, Sun Z. 2017. *Lamarck rises from his grave: parental environment-induced epigenetic inheritance in model organisms and humans. Biol Rev* 92:2084–2111.
Washington HA. 2006. *Medical apartheid: the dark history of medical experimentation on black Americans from colonial times to the present*. Anchor Books, New York.
Weber BH, Depew DJ, eds. 2003. *Evolution and learning: the Baldwin effect reconsidered*. MIT Press, Cambridge.
Weisbecker V, Nilsson M. 2008. Integration, heterochrony, and adaptation in pedal digits of syndactylous marsupials. *BMC Evol Biol* 8:160.
West-Eberhard MJ. 2003. *Developmental plasticity and evolution*. Oxford University Press, Oxford.
West-Eberhard MJ. 2004. Ryuichi Matsuda: a tribute and a perspective on pan-environmentalism and genetic assimilation. In *Environment, development and evolution: toward a synthesis (The Vienna Series in Theoretical Biology)* (Hall BK, Pearson RD, Müller GB, eds). A Bradford Book, Massachusetts, p. 109–116.
West-Eberhard MJ. 2007. Dancing with DNA and flirting with the ghost of Lamarck. *Biol & Philos* 22:439–451.
West-Eberhard MJ. 2014. *Darwin's forgotten idea: the social essence of sexual selection. Neurosci Biobehav Rev* 46:501–508.
Westneat DF, Fox CW, eds. 2010. *Evolutionary behavioral ecology*. Oxford University Press, New York.
Wetherington RK. 2011. *Readings in the history of evolutionary theory*. Oxford University Press, Oxford.
A. Whiten, W. Bodmer, B. Charlesworth, et al. Response to Fuentes' "The Descent of Man, 150 years on". *Science* E-letter, 6 June 2021.
Wilhelmson AS, Lantero Rodriguez M, *et al.* 2018. Testosterone is an endogenous regulator of BAFF and splenic B cell number. *Nat Commun* 9:2067.
Wilkins AS. 2009. Charles Darwin: genius or plodder? *Genetics* 183:773–777.
White C. 1799. *An account of the regular gradation in Man*. C. Dilly, London.
Wibowo E, Wassersug RJ. 2016. Multiple orgasms in men – what we know so far. *Sex Med Rev* 4:136–48.
Wilson EO. 2014 *The meaning of human existence*. W. W. Norton & Company, New York.
Wolfe SE, Tubi A. 2018. Terror Management Theory and mortality awareness: A missing link in climate response studies? *WIREs Clim Change* 2018:e566.
Wulf A. 2015. *The invention of nature – the adventures of Alexander von Humboldt, the lost hero of science*. Alfred A. Knopf, New York.
Wrangham RW. 2018. Two types of aggression in human evolution. *PNAS* 115:245–253.
Wray GA, Hoekstra HE, Futuyma DJ, *et al.* 2014. Does evolutionary theory need a rethink? No, all is well. *Nature* 514:161–164.
Wylie K. 2009. A global survey of sexual behaviours. *J Family Reprod Health* 3:39–49.

Index

R. Diogo, *Darwin's Racism, Sexism, and Idolization*,
https://doi.org/10.1007/978-3-031-49055-2

Printed in the United States
by Baker & Taylor Publisher Services